T0181662

Basic Classes of Linear Operators

Israel Gohberg
Seymour Goldberg
Marinus A. Kaashoek

Birkhäuser

Israel Gohberg
School of Mathematical Sciences
Raymond and Beverly Sackler
Faculty of Exact Sciences
Tel Aviv University
Ramat Aviv 69978
Israel
e-mail: gohberg@post.tau.ac.il

Marinus A. Kaashoek
Department of Mathematics and Computer Science
Vrije Universiteit Amsterdam
De Boelelaan 1081a
NL-1081 HV Amsterdam
The Netherlands

e-mail: m.a.kaashoek@cs.vu.nl

Seymour Goldberg
Department of Mathematics
University of Maryland
College Park, MD 20742-4015
USA
e-mail: sgoldber@umd.edu

2000 Mathematics Subject Classification 47-01

A CIP catalogue record for this book is available from the
Library of Congress, Washington D.C., USA

Bibliographic information published by Die Deutsche Bibliothek
Die Deutsche Bibliothek lists this publication in the Deutsche Nationalbibliografie;
detailed bibliographic data is available in the Internet at <http://dnb.ddb.de>.

ISBN 3-7643-6930-2 Birkhäuser Verlag, Basel – Boston – Berlin

This work is subject to copyright. All rights are reserved, whether the whole or part of the material
is concerned, specifically the rights of translation, reprinting, re-use of illustrations, recitation,
broadcasting, reproduction on microfilms or in other ways, and storage in data banks. For any
kind of use permission of the copyright owner must be obtained.

© 2003 Birkhäuser Verlag, P.O. Box 133, CH-4010 Basel, Switzerland
Printed on acid-free paper produced from chlorine-free pulp. TCF ∞
Cover design: Micha Lotrovsky, CH-4106 Therwil, Switzerland

ISBN 3-7643-6930-2

9 8 7 6 5 4 3 2 1   www.birkhauser.ch

Dedicated to our grandchildren

Table of Contents

Preface... xiii
Introduction xv

Chapter I   Hilbert Spaces... 1

1.1   Complex n-Space 1
1.2   The Hilbert Space ℓ_2 3
1.3   Definition of Hilbert Space and its Elementary Properties 5
1.4   Distance from a Point to a Finite Dimensional Space 8
1.5   The Gram Determinant 10
1.6   Incompatible Systems of Equations 13
1.7   Least Square Fit 15
1.8   Distance to a Convex Set and Projections onto Subspaces 16
1.9   Orthonormal Systems 18
1.10 Szegö Polynomials 19
1.11 Legendre Polynomials 24
1.12 Orthonormal Bases 26
1.13 Fourier Series 29
1.14 Completeness of the Legendre Polynomials 31
1.15 Bases for the Hilbert Space of Functions on a Square 32
1.16 Stability of Orthonormal Bases 34
1.17 Separable Spaces 35
1.18 Isometry of Hilbert Spaces 36
1.19 Example of a Non Separable Space 38
      Exercises 38

Chapter II   Bounded Linear Operators on Hilbert Spaces............. 51

2.1   Properties of Bounded Linear Operators 51
2.2   Examples of Bounded Linear Operators with Estimates of Norms ... 52
2.3   Continuity of a Linear Operator 56
2.4   Matrix Representations of Bounded Linear Operators 57
2.5   Bounded Linear Functionals 60
2.6   Operators of Finite Rank 63
2.7   Invertible Operators 64
2.8   Inversion of Operators by the Iterative Method 69
2.9   Infinite Systems of Linear Equations 71
2.10 Integral Equations of the Second Kind 73

2.11   Adjoint Operators 76
2.12   Self Adjoint Operators 80
2.13   Orthogonal Projections 81
2.14   Two Fundamental Theorems 82
2.15   Projections and One-Sided Invertibility of Operators 84
2.16   Compact Operators 91
2.17   The Projection Method for Inversion of Linear Operators 96
2.18   The Modified Projection Method 105
2.19   Invariant Subspaces 108
2.20   The Spectrum of an Operator 109
       Exercises 118

Chapter III   Laurent and Toeplitz Operators on Hilbert Spaces 135

3.1   Laurent Operators 135
3.2   Toeplitz Operators 141
3.3   Band Toeplitz operators 143
3.4   Toeplitz Operators with Continuous Symbols 152
3.5   Finite Section Method 159
3.6   The Finite Section Method for Laurent Operators 163
       Exercises 166

Chapter IV   Spectral Theory of Compact Self Adjoint Operators 171

4.1   Example of an Infinite Dimensional Generalization 171
4.2   The Problem of Existence of Eigenvalues and Eigenvectors 172
4.3   Eigenvalues and Eigenvectors of Operators of Finite Rank 174
4.4   Existence of Eigenvalues 175
4.5   Spectral Theorem 178
4.6   Basic Systems of Eigenvalues and Eigenvectors 180
4.7   Second Form of the Spectral Theorem 182
4.8   Formula for the Inverse Operator 183
4.9   Minimum-Maximum Properties of Eigenvalues 185
       Exercises 188

Chapter V   Spectral Theory of Integral Operators 193

5.1   Hilbert-Schmidt Theorem 193
5.2   Preliminaries for Mercer's Theorem 196
5.3   Mercer's Theorem 197
5.4   Trace Formula for Integral Operators 200
       Exercises 200

Chapter VI   Unbounded Operators on Hilbert Space 203

6.1   Closed Operators and First Examples 203
6.2   The Second Derivative as an Operator 204

6.3   The Graph Norm 206
6.4   Adjoint Operators 208
6.5   Sturm-Liouville Operators 211
6.6   Self Adjoint Operators with Compact Inverse 214
      Exercises 215

Chapter VII   Oscillations of an Elastic String...................... 219

7.1   The Displacement Function 219
7.2   Basic Harmonic Oscillations 220
7.3   Harmonic Oscillations with an External Force 222

Chapter VIII   Operational Calculus with Applications 225

8.1   Functions of a Compact Self Adjoint Operator 225
8.2   Differential Equations in Hilbert Space 230
8.3   Infinite Systems of Differential Equations 232
8.4   Integro-Differential Equations 233
      Exercises 234

Chapter IX   Solving Linear Equations by Iterative Methods.......... 237

9.1   The Main Theorem 237
9.2   Preliminaries for the Proof 238
9.3   Proof of the Main Theorem 240
9.4   Application to Integral Equations 242

Chapter X   Further Developments of the Spectral Theorem 243

10.1   Simultaneous Diagonalization 243
10.2   Compact Normal Operators 244
10.3   Unitary Operators 246
10.4   Singular Values 248
10.5   Trace Class and Hilbert Schmidt Operators 253
       Exercises 254

Chapter XI   Banach Spaces.. 259

11.1   Definitions and Examples 259
11.2   Finite Dimensional Normed Linear Spaces 262
11.3   Separable Banach Spaces and Schauder Bases 264
11.4   Conjugate Spaces 265
11.5   Hahn-Banach Theorem 267
       Exercises 272

Chapter XII   Linear Operators on a Banach Space 277

12.1   Description of Bounded Operators 277
12.2   Closed Linear Operators 279
12.3   Closed Graph Theorem 281
12.4   Applications of the Closed Graph Theorem 283
12.5   Complemented Subspaces and Projections 286
12.6   One-Sided Invertibility Revisited 288
12.7   The Projection Method Revisited 289
12.8   The Spectrum of an Operator 290
12.9   Volterra Integral Operator 293
12.10   Analytic Operator Valued Functions 295
        Exercises 296

Chapter XIII   Compact Operators on a Banach Space 299

13.1   Examples of Compact Operators 299
13.2   Decomposition of Operators of Finite Rank 302
13.3   Approximation by Operators of Finite Rank 303
13.4   First Results in Fredholm Theory 305
13.5   Conjugate Operators on a Banach Space 306
13.6   Spectrum of a Compact Operator 310
13.7   Applications 313
        Exercises 314

Chapter XIV   Poincaré Operators: Determinant and Trace 317

14.1   Determinant and Trace 317
14.2   Finite Rank Operators, Determinants and Traces 321
14.3   Theorems about the Poincaré Determinant 327
14.4   Determinants and Inversion of Operators 330
14.5   Trace and Determinant Formulas for Poincaré Operators 336
        Exercises 340

Chapter XV   Fredholm Operators 347

15.1   Definition and Examples 347
15.2   First Properties 347
15.3   Perturbations Small in Norm 352
15.4   Compact Perturbations 355
15.5   Unbounded Fredholm Operators 356
        Exercises 358

Chapter XVI   Toeplitz and Singular Integral Operators 361

16.1   Laurent Operators on $\ell_p(\mathbb{Z})$ 361
16.2   Toeplitz Operators on ℓ_p 364
16.3   An Illustrative Example 372
16.4   Applications to Pair Operators 377
16.5   The Finite Section Method Revisited 384
16.6   Singular Integral Operators on the Unit Circle 390
         Exercises 395

Chapter XVII   Non Linear Operators 401

17.1   Fixed Point Theorems 401
17.2   Applications of the Contraction Mapping Theorem 402
17.3   Generalizations 405

Appendix 1: Countable sets and Separable Hilbert Spaces 409

Appendix 2: The Lebesgue integral and L_p Spaces 411

Suggested Reading 415

References 417

List of Symbols 419

Index 421

Preface

The present book is an expanded and enriched version of the text *Basic Operator Theory*, written by the first two authors more than twenty years ago. Since then the three of us have used the basic operator theory text in various courses. This experience motivated us to update and improve the old text by including a wider variety of basic classes of operators and their applications. The present book has also been written in such a way that it can serve as an introduction to our previous books *Classes of Linear Operators*, Volumes I and II. We view the three books as a unit.

We gratefully acknowledge the support of the mathematical departments of Tel-Aviv University, the University of Maryland at College Park, and the Vrije Universiteit at Amsterdam. The generous support of the Silver Family Foundation is highly appreciated.

Amsterdam, November 2002   The authors

Introduction

This elementary text is an introduction to functional analysis, with a strong emphasis on operator theory and its applications. It is designed for graduate and senior undergraduate students in mathematics, science, engineering, and other fields.

The discussion below describes the main aims of the book. This is the same path we followed in our book *Basic Operator Theory* which was published more than twenty years ago. The present book differs substantially from the above mentioned one in a number of ways. First, the examples of concrete classes of linear operators which we study in this book are greatly enriched. For example, it contains the theory of Laurent, Toeplitz and singular integral operators. This book also presents the theory of traces and determinants in an infinite dimensional setting. Fredholm theory of operators is studied here and the theory of unbounded operators is expanded. The number of exercises and examples is also increased.

From the beginning of this book to its end, a great deal of attention is paid to the interplay between abstract methods and concrete problems. Motivation of the theory and its systematic applications are characteristic of all chapters.

Our aim has been to present the material in a form which the reader can understand and appreciate without too much difficulty. To accomplish this end, we have not stated the principal results in their most general form. Instead, statements were chosen which make more transparent the main ideas and the further generalizations as well.

The book covers only a limited number of topics, but they are sufficient to lay a firm foundation in operator theory and to demonstrate the power of the theory in applications. Much of the material in this volume is an essential portion of training in mathematics. It is presented as a natural continuation of the undergraduate courses in linear algebra and analysis.

In the past, graduate courses in integral equations were usually offered as part of the mathematics curriculum. Later, some instructors thought to develop the courses as special cases of results in functional analysis. However, this goal was not realized and courses in integral equations almost disappeared. One of our aims is to reestablish a theory of integral equations as a significant part of operator theory. A unified approach to some phases of differential equations, approximation theory and numerical analysis is also provided.

One of the main problems in operator theory is the problem of inverting linear operators on infinite dimensional spaces. This is achieved by means of projection methods and the theory of determinants. The projection methods appear in many chapters of the book.

A theme which also appears throughout this book is the use of one sided invertible operators. For example they appear in the theory of Fredholm, Toeplitz and singular integral operators.

This book consists, basically, of two unequal parts. The major portion of the text is devoted to the theory and applications of linear operators on a Hilbert space. We begin with a chapter on the geometry of Hilbert space and then proceed to the theory of linear operators on these spaces. The theory is richly illustrated by numerous examples.

In the next chapter we study Laurent and Toeplitz operators which provide further illustrations and applications of the general theory. Here the projection methods and one sided invertibility play an important role.

Spectral theory of compact self adjoint operators appears in the next chapter followed by a spectral theory of integral operators. Operational calculus is then presented as a natural outgrowth of spectral theory. The reader is also shown the basic theory of unbounded operators on Hilbert space.

The second part of the text concentrates on Banach spaces and linear operators, both bounded and unbounded, acting on these spaces. It includes, for example, the three basic principles of linear analysis and the Riesz-Fredholm theory of compact operators which is based on the theory of one sided invertible operators.

In later chapters we deal with a class of operators, in an infinite dimensional setting, on which a trace and determinant are defined as extensions of the trace and determinant of finite matrices. Further developments of Toeplitz operators and singular integral operators are presented. Projection methods and one sided invertible operators continue to play an important role. Both parts of the book contain plenty of applications and examples. All chapters deal exclusively with linear problems except for the last chapter which is an introduction to the theory of non linear operators.

In general, in writing this book, the authors were strongly influenced by recent developments in operator theory which affected the choice of topics, proofs and exercises.

One of the main features of this book is the large number of new exercises chosen to expand the reader's comprehension of the material, and to train him or her in the use of it. In the beginning portion of the book we offer a large selection of computational exercises; later, the proportion of exercises dealing with theoretical questions increases. We have, however, omitted exercises after Chapters VII, IX and XVII due to the specialized nature of the subject matter.

To reach as large an audience as possible, the material is self-contained. Beginning with finite dimensional vector spaces and matrices, the theory is gradually developed.

Since the book contains more material than is needed for a one semester course, we leave the choice of subject matter to the instructor and the interests of the students.

It is assumed that the reader is familiar with linear algebra and advanced calculus; some exposure to Lebesgue integration is also helpful. However, any lack of knowledge of this last topic can be compensated for by referring to Appendix 2 and the references therein.

It has been our experience that exposure to this material stimulated our students to expand their knowledge of operator theory. We hope that this will continue to be the case. Included in the book is a list of suggested reading material which should prove helpful.

Chapter I
Hilbert Spaces

In this chapter we review the main properties of the complex n-dimensional space $\mathbb{C}^n$ and then we study the Hilbert space which is its most natural infinite dimensional generalization. Many applications to classical problems are included (Least squares, Fourier series and others).

1.1 Complex n-Space

Let $\mathbb{C}$ denote the set of complex numbers and let $\mathcal{H} = \mathbb{C}^n$ be the set of all n-tuples $(\xi_1, \ldots, \xi_n)$ of complex numbers with addition and scalar multiplication defined as follows.

For $x = (\xi_1, \ldots, \xi_n)$ and $y = (\eta_1, \ldots, \eta_n)$,

$$x + y = (\xi_1 + \eta_1, \ldots, \xi_n + \eta_n) \text{ and } \alpha x = (\alpha \xi_1, \ldots, \alpha \xi_n), \alpha \in \mathbb{C}.$$

The complex valued function $<,>$, defined on the Cartesian product $\mathcal{H} \times \mathcal{H}$ by

$$\langle x, y \rangle = \sum_{i=1}^{n} \xi_i \bar{\eta}_i$$

is called an *inner product* on $\mathcal{H}$.

It is clear from the definitions that if x, y and z are in $\mathcal{H}$, then

(i) $\langle x, x \rangle > 0$ if and only if $x \neq 0$;

(ii) $\overline{\langle x, y \rangle} = \langle y, x \rangle$;

(iii) $\langle \alpha x, y \rangle = \alpha \langle x, y \rangle$;

(iv) $\langle x + y, z \rangle = \langle x, z \rangle + \langle y, z \rangle$.

It follows from (ii), (iii) and (iv) that

$$\langle x, \alpha y + \beta z \rangle = \bar{\alpha} \langle x, y \rangle + \bar{\beta} \langle x, z \rangle; \quad \langle x, 0 \rangle = \langle 0, x \rangle = 0.$$

Each n-tuple $x = (x_1, \ldots, x_n)$ is called a vector. The *length* or *norm*, $\|x\|$, of the vector x is defined by

$$\|x\| = \langle x, x \rangle^{1/2} = \left(\sum_{i=1}^{n} |x_i|^2 \right)^{1/2}.$$

If $x = (x_1, x_2)$ and $y = (y_1, y_2)$ are vectors in the plane, then $\langle x, y \rangle = \|x\| \|y\|$ $\cos \theta$, where θ is the angle between x and y, $0 \leq \theta \leq \pi$. Thus $|\langle x, y \rangle| \leq \|x\| \|y\|$.

Based only on the properties of the inner product, we shall prove this inequality in $\mathbb{C}^n$.

Cauchy-Schwarz Inequality: *If x and y are in $\mathbb{C}^n$, then*

$$|\langle x, y \rangle| \leq \|x\| \|y\|. \tag{1.1}$$

When $y \neq 0$, equality holds if and only if $x = \lambda y$ for some $\lambda \in \mathbb{C}$.

Proof: For any $\lambda \in \mathbb{C}$, it follows from (i)–(iv) that

$$\begin{aligned} 0 \leq \langle x - \lambda y, x - \lambda y \rangle &= \|x\|^2 - \bar{\lambda}\langle x, y \rangle - \lambda\langle y, x \rangle + \lambda\bar{\lambda}\|y\|^2 \\ &= \|x\|^2 - 2\operatorname{Re}\lambda\langle y, x \rangle + |\lambda|^2\|y\|^2. \end{aligned} \tag{1.2}$$

If $\langle x, y \rangle = 0$, the inequality is trivial. Suppose $\langle x, y \rangle \neq 0$. Substituting $\lambda = \frac{\|x\|^2}{\langle y, x \rangle}$ in (1.2) gives

$$0 \leq -\|x\|^2 + \frac{\|x\|^4 \|y\|^2}{|\langle x, y \rangle|^2},$$

from which (1.1) follows.

Suppose $y \neq 0$. If $\langle y, x \rangle = 0$, then equality holds in (1.1) if and only if $x = 0$. If $\langle y, x \rangle \neq 0$, then equality in (1.1) holds if and only if equality in (1.2) holds for $\lambda = \frac{\|x\|^2}{\langle y, x \rangle}$, in which case $x = \lambda y$.

As a consequence of the Cauchy-Schwarz inequality applied to $x = (|x_1|, \ldots, |x_n|)$ and $y = (|y_1|, \ldots, |y_n|)$, we have

$$\sum_{i=1}^n |x_i y_i| \leq \left(\sum_{i=1}^n |x_i|^2 \right)^{1/2} \left(\sum_{i=1}^n |y_i|^2 \right)^{1/2}. \tag{1.3}$$

The following inequality, when confined to vectors in the plane, states that the sum of the lengths of two sides of a triangle exceeds the length of the third side. $\qquad\square$

Triangle inequality: *For x and y in $\mathbb{C}^n$,*

$$\|x + y\| \leq \|x\| + \|y\|. \tag{1.4}$$

Proof: By the Cauchy-Schwarz inequality,

$$\begin{aligned} \|x + y\|^2 = \langle x + y, x + y \rangle &= \|x\|^2 + 2\operatorname{Re}\langle x, y \rangle + \|y\|^2 \\ &\leq \|x\|^2 + 2\|x\|\|y\| + \|y\|^2 = (\|x\| + \|y\|)^2. \end{aligned}$$

$\qquad\square$

1.2 The Hilbert Space ℓ_2

The passage from $\mathbb{C}^n$, the space of n-tuples $(\xi_1, \ldots, \xi_n)$ of complex numbers, to the space consisting of certain infinite sequences $(\xi_1, \xi_2, \ldots)$ of complex numbers is a very natural one.

If $x = (\xi_1, \xi_2, \ldots)$ and $y = (\eta_1, \eta_2, \ldots)$, define $x + y = (\xi_1 + \eta_1, \xi_2 + \eta_2, \ldots)$ and $\alpha x = (\alpha \xi_1, \alpha \xi_2, \ldots)$ for $\alpha \in \mathbb{C}$. The inner product $\langle x, y \rangle = \sum_{i=1}^{\infty} \xi_i \bar{\eta}_i$ and the corresponding norm $\|x\| = \langle x, x \rangle^{1/2} = (\sum_{i=1}^{\infty} |\xi_i|^2)^{1/2}$ make sense provided $\sum_{i=1}^{\infty} |\xi_i|^2 < \infty$ and $\sum_{i=1}^{\infty} |\eta_i|^2 < \infty$. Indeed, by (1.3) in Section 1,

$$\sum_{i=1}^{n} |\xi_i \eta_i| \le \left(\sum_{i=1}^{\infty} |\xi_i|^2 \right)^{1/2} \left(\sum_{i=1}^{\infty} |\eta_i|^2 \right)^{1/2},$$

and by the triangle inequality (1.4) in Section 1,

$$\left(\sum_{i=1}^{\infty} |\xi_i + \eta_i|^2 \right)^{1/2} \le \left(\sum_{i=1}^{\infty} |\xi_i|^2 \right)^{1/2} + \left(\sum_{i=1}^{\infty} |\eta_i|^2 \right)^{1/2}.$$

Thus

$$|\langle x, y \rangle| \le \|x\| \|y\| \quad \text{and} \quad \|x + y\| \le \|x\| + \|y\|.$$

Definition of ℓ_2: The vector space ℓ_2 consists of the set of those sequences $(\xi_1, \xi_2, \ldots)$ of complex numbers for which $\sum_{i=1}^{\infty} |\xi_i|^2 < \infty$, together with the operations of addition and scalar multiplication defined above.

The complex valued function $\langle \, , \rangle$ defined on $\ell_2 \times \ell_2$ by $\langle x, y \rangle = \sum_{i=1}^{\infty} \xi_i \bar{\eta}_i$, where $x = (\xi_i)$, $y = (\eta_i)$, is called the *inner product* on ℓ_2. Clearly, the inner product satisfies (i)–(iv) in Section 1.

The *norm*, $\|x\|$, on ℓ_2 is defined by

$$\|x\| = \langle x, x \rangle^{1/2} = \left(\sum_{i=1}^{\infty} |\xi_i|^2 \right)^{1/2}.$$

The *distance* between the vectors $x = (\xi_1, \xi_2, \ldots)$ and $y = (\eta_1, \eta_2, \ldots)$ is $\|x - y\| = (\sum_{i=1}^{\infty} |\xi_i - \eta_i|^2)^{1/2}$.

The extension of $\mathbb{C}^n$ to ℓ_2 is not just an interesting exercise. As we shall see, ℓ_2 arises from the study of Fourier series, integral equations, infinite systems of linear equations, and many applied problems.

In a treatment of advanced calculus it is shown that every Cauchy sequence in $\mathbb{C}^n$ converges, i.e., if $\{x_k\}$ is a sequence of vectors in $\mathbb{C}^n$ such that $\|x_m - x_n\| \to 0$ as $m, n \to \infty$, then there exists an $x \in \mathbb{C}^n$ such that $\|x_n - x\| \to 0$. This enables one to prove, for example, the convergence of certain series.

The space ℓ_2 has the same important property which is called the *completeness* property of ℓ_2.

Theorem 2.1 (Completeness of ℓ_2). *If $\{x_n\}$ is a sequence of vectors in ℓ_2 such that $\|x_n - x_m\| \to 0$ as $n, m \to \infty$, then there exists an $x \in \ell_2$ such that $\|x_n - x\| \to 0$.*

Proof: Suppose $x_n = (\xi_1^{(n)}, \xi_2^{(n)}, \ldots) \in \ell_2$. The idea of the proof is to show that for each fixed k, the sequence of components $\xi_k^{(1)}, \xi_k^{(2)}, \ldots$ converges to some number ξ_k and that $x = (\xi_1, \xi_2, \ldots)$ is the desired limit of $\{x_n\}$.

For k fixed,

$$|\xi_k^{(n)} - \xi_k^{(m)}| \le \|x_n - x_m\| \to 0, \quad \text{as } n, m \to \infty. \tag{2.1}$$

Thus for each k, $\{\xi_k^{(n)}\}_{n=1}^\infty$ is a Cauchy sequence of complex numbers which therefore converges. Take $\xi_k = \lim_{n \to \infty} \xi_k^{(n)}$ and $x = (\xi_1, \xi_2, \ldots)$. First we show that x is in ℓ_2.

For any positive integer j,

$$\sum_{k=1}^{j} |\xi_k|^2 = \lim_{n \to \infty} \sum_{k=1}^{j} |\xi_k^{(n)}|^2 \tag{2.2}$$

and

$$\sum_{k=1}^{j} |\xi_k^{(n)}|^2 \le \|x_n\|^2. \tag{2.3}$$

Now $\sup_n \|x_n\| = M < \infty$ since $|\|x_n\| - \|x_m\|| \le \|x_n - x_m\| \to 0$ as m, $n \to \infty$. It therefore follows from (2.2) and (2.3) that $\sum_{k=1}^\infty |\xi_k|^2 \le M^2$, i.e., x is in ℓ_2.

Finally, we prove that $\|x_n - x\| \to 0$. Let $\varepsilon > 0$ be given. There exists an integer N such that if $m, n \ge N$, and p is any positive integer,

$$\sum_{k=1}^{p} |\xi_k^{(n)} - \xi_k^{(m)}|^2 \le \|x_n - x_m\|^2 \le \varepsilon. \tag{2.4}$$

Fix $n \ge N$. From (2.1) and (2.4) we have that for all p,

$$\sum_{k=1}^{p} |\xi_k^{(n)} - \xi_k|^2 = \lim_{m \to \infty} \sum_{k=1}^{p} |\xi_k^{(n)} - \xi_k^{(m)}|^2 \le \varepsilon.$$

Thus for $n \ge N$,

$$\|x_n - x\|^2 = \sum_{k=1}^{\infty} |\xi_k^{(n)} - \xi_k|^2 \le \varepsilon.$$

Sometimes we shall also deal with $\ell_2(\mathbb{Z})$, the two-sided version of ℓ_2. The vector space $\ell_2(\mathbb{Z})$ consists of all doubly infinite sequences $(\ldots, \xi_{-1}, \xi_0, \xi_1, \ldots)$ of complex numbers that are square summable, that is, $\sum_{j=-\infty}^{\infty} |\xi_j|^2 < \infty$. As its one-sided version ℓ_2, the space $\ell_2(\mathbb{Z})$ is a vector space with addition and scalar multiplication being defined entry wise. Also, $\ell_2(\mathbb{Z})$ has a natural inner product, namely $\langle x, y \rangle = \sum_{j=-\infty}^{\infty} \xi_j \bar{\eta}_j$, where $x = (\xi_j)_{j \in \mathbb{Z}}$ and $y = (\eta_j)_{j \in \mathbb{Z}}$. The corresponding norm is given by

$$\|x\| = \langle x, x \rangle^{1/2} = \left(\sum_{j=\infty}^{\infty} |\xi_j|^2 \right)^{1/2},$$

and Theorem I.2.1 above remains valid with $\ell_2(\mathbb{Z})$ in place of ℓ_2. In other words, the space $\ell_2(\mathbb{Z})$ has the completeness property. $\qquad\square$

1.3 Definition of Hilbert Space and its Elementary Properties

The vector spaces $\mathbb{C}^n$ and ℓ_2 each have an inner product which satisfy (i)–(iv) in Section 1. In applications, other vector spaces which possess an inner product also need to be considered. For this reason we introduce the definition of an inner product space.

It is obvious from the definitions of addition and scalar multiplication on $\mathbb{C}^n$ and ℓ_2 that if E is either of these spaces, then for x, y and z in E, $x + y$ and αx are also in E and

(a) $x + y = y + x$;

(b) $(x + y) + z = x + (y + z)$;

(c) There exists a vector $0 \in E$ such that $x + 0 = x$ for all $x \in E$.

(d) For each $x \in E$ there exists a vector $-x \in E$ such that $x + (-x) = 0$.

For every α and β in $\mathbb{C}$,

(e) $\alpha(x + y) = \alpha x + \alpha y$;

(f) $(\alpha + \beta)x = \alpha x + \beta y$;

(g) $(\alpha \beta)x = \alpha(\beta x)$;

(h) $1x = x$.

These properties are used to define an abstract vector space.

Definition: A *vector space* over $\mathbb{C}$ is a set E, whose elements are called *vectors*, together with two rules called addition and scalar multiplication. These rules associate with any pair of vectors x and y in E and any $\alpha \in \mathbb{C}$, unique vectors in E, denoted by $x + y$ and αx, such that (a)–(h) hold.

A subset M of E is called a *subspace* of E provided the following two conditions hold.

(i) If u and v are in M, then $u + v$ is in M.

(ii) If u is in M and α is in $\mathbb{C}$, then αu is in M.

Given a set $S \subset E$, the *span* of S, written sp S, is the subspace of E consisting of all linear combinations of vectors in S, i.e., all vectors of the form $\sum_{i=1}^{n} \alpha_i s_i$, $\alpha_i \in \mathbb{C}$, $s_i \in S$.

Vectors $v_1, \ldots, v_n$ are called *linearly independent* if $\sum_{i=1}^{n} \alpha_i v_i = 0$ holds only if each $\alpha_i = 0$. Otherwise, the vectors are called *linearly dependent*. Infinitely many vectors $v_1, v_2, \ldots$ are linearly independent if for each k, $v_1, \ldots, v_k$ are linearly independent.

A subset $\{v_1, \ldots, v_n\}$ of a vector space V is a *basis* for V if $V = \text{sp}\{v_1, \ldots, v_n\}$ and $v_1, \ldots, v_n$ are linearly independent. In this case we say that V has dimension n, written dim $V = n$. V is *infinite dimensional* if it contains an infinite linearly independent set of vectors. Thus ℓ_2 is infinite dimensional since $e_1 = (1, 0, 0, \ldots)$, $e_2 = (0, 1, 0, 0, \ldots), \ldots$ are linearly independent.

Definition: Let E be a vector space over $\mathbb{C}$. An *inner product* on E is a complex valued function, $\langle \, , \, \rangle$, defined on $E \times E$ with the following properties.

(a) $\langle x + y, z \rangle = \langle x, z \rangle + \langle y, z \rangle$;

(b) $\langle \alpha x, y \rangle = \alpha \langle x, y \rangle$;

(c) $\overline{\langle x, y \rangle} = \langle y, x \rangle$;

(d) $\langle x, x \rangle \geq 0$ and $\langle x, x \rangle > 0$ if $x \neq 0$.

E, together with an inner product, is called an *inner product space*.

The *norm*, $\|x\|$, on E is given by $\|x\| = \langle x, x \rangle^{1/2}$, and the *distance* between vectors x and y is $\|x - y\|$.

It follows from (a)–(c) that

$$\langle x, \alpha y + \beta z \rangle = \bar{\alpha}\langle x, y \rangle + \bar{\beta}\langle x, z \rangle \quad \text{and} \quad \|\alpha x\| = |\alpha|\,\|x\|.$$

For $x_0 \in E$ and $r > 0$, the sets $\{x : \|x - x_0\| \leq r\}, \{x : \|x - x_0\| = r\}$ are called the *r-ball* and *r-sphere* of E, respectively, with center x_0.

Examples: 1. $\mathbb{C}^n$ and ℓ_2 are inner product spaces. Given $x_0 = (\eta_1, \eta_2, \ldots) \in \ell_2$, the set of all sequences $\{\xi_k\} \in \ell_2$ such that $(\sum_{k=1}^{\infty} |\xi_k - \eta_k|^2)^{1/2} \leq r$ is the r-ball in ℓ_2 with center x_0.

2. Let $L_2([a, b])$ be the vector space of all complex valued Lebesgue measurable functions f defined on the interval $a \leq x \leq b$ with the property that $|f|^2$ is Lebesgue integrable. By identifying functions which are equal almost everywhere,

$$\langle f, g \rangle = \int_a^b f(x)\bar{g}(x) \, dx$$

defines as inner product on $L_2([a, b])$. A discussion of this space appears in Appendix 2. Given $f_0 \in L_2([a, b])$, the set of all vectors $f \in L_2([a, b])$ for which $(\int_a^b |f(x) - f_0(x)|^2 dx)^{1/2} = r$ is the r-sphere of $L_2([a, b])$ with center f_0.

3. Let ℓ_2^D be the subspace of ℓ_2 consisting of all sequences (ξ_k), where $\xi_k = 0$ for all but at most a finite number of k. With the inner product inherited from ℓ_2, ℓ_2^D is an inner product space.

4. The vector space $C_2([a, b])$ of continuous complex valued functions with inner product defined in Example 2 is an inner product space.

A number of the geometric properties of the plane carry over to inner product spaces. For example, we have the following basic inequalities.

Theorem 3.1 *Let E be an inner product space. For x and y in E,*

(i) $|\langle x, y \rangle| \leq \|x\|\|y\|$ *(Cauchy-Schwarz inequality)*

(ii) $\|x + y\| \leq \|x\| + \|y\|$ *(Triangle inequality)*

(iii) $\|x + y\|^2 + \|x - y\|^2 = 2(\|x\|^2 + \|y\|^2)$ *(Parallelogram law).*

(iv) $\|x - y\| \geq |\|x\| - \|y\||$

Proof: The proofs of (i) and (ii) are exactly the same as the proofs of the inequalities (1.1) and (1.4) in Section 1. Also,

$$\|x + y\|^2 + \|x - y\|^2$$
$$= \langle x + y, x + y \rangle + \langle x - y, x - y \rangle$$
$$= \|x\|^2 + 2 \operatorname{Re} \langle x, y \rangle + \|y\|^2 + \|x\|^2 - 2 \operatorname{Re} \langle x, y \rangle + \|y\|^2$$
$$= 2(\|x\|^2 + \|y\|^2).$$

The proof of (iv) follows from (ii).

Most of the fundamental theorems which are concerned with operators defined on an inner product space $\mathcal{H}$ require that $\mathcal{H}$ be complete. In fact, we rarely consider inner product spaces which do not have this property.   □

Definition: A sequence $\{x_n\}$ in an inner product space $\mathcal{H}$ is said to *converge* to $x \in \mathcal{H}$, written $x_n \to x$, if $\|x_n - x\| \to 0$.

A sequence $\{x_n\}$ in $\mathcal{H}$ is called a *Cauchy sequence* if $\|x_n - x_m\| \to 0$ as $n, m \to \infty$.

If every Cauchy sequence in $\mathcal{H}$ converges to a vector in $\mathcal{H}$, then $\mathcal{H}$ is called *complete*. A complete inner product is called a *Hilbert space*. It is the most natural generalization of $\mathbb{C}^n$.

Examples: 1. $\mathbb{C}^n$ is an n-dimensional Hilbert space.

2. ℓ_2 is an infinite dimensional Hilbert space.

3. $L_2([a, b])$ is a Hilbert space (cf. Appendix 2). This space is infinite dimensional since the functions $1, x, x^2, \ldots$ are linearly independent. For if $\sum_{k=0}^{n} a_k x^k$ is the zero function, then $a_k = 0,\ 0 \le k \le n$, since any polynomial of degree n has at most n zeros.

4. ℓ_2^D is not complete. Indeed, let $x_n = (\frac{1}{2}, \frac{1}{2^2}, \ldots, \frac{1}{2^n}, 0, 0, \ldots) \in \ell_2^D$. Then for $n > m$,

$$\| x_n - x_m \|^2 < \sum_{k=m+1}^{\infty} \frac{1}{2^k} = \frac{1}{2^m} \to 0.$$

Thus $\{x_n\}$ is a Cauchy sequence which converges in ℓ_2 to $x = (\frac{1}{2}, \frac{1}{2^2}, \ldots) \notin \ell_2^D$. Consequently, $\{x_n\}$ cannot converge to a vector in ℓ_2^D since limits of sequences in ℓ_2 are unique.

5. The inner product space $\mathcal{P}$ of all polynomials with $\langle f, g \rangle = \int_0^1 f(x)\bar{g}(x)\, dx$ is not complete. To see this, let

$$P_n(x) = \sum_{j=0}^{n} \frac{1}{2^j} x^j.$$

Then for $g(x) = \frac{1}{1-\frac{1}{2}x},\ 0 \le x \le 1, P_n \to g$ in $L_2([0, 1])$. Thus, $\{P_n\}$ is a Cauchy sequence in $\mathcal{P}$ which does not converge to a vector in $\mathcal{P}$ since $g \notin \mathcal{P} \subset L_2([0, 1])$.

We shall see later that every finite dimensional inner product space is complete.

1.4 Distance from a Point to a Finite Dimensional Subspace

Throughout this section, E denotes an inner product space

Definition: The *distance* $d(v, S)$ from a point $v \in E$ to a set $S \subset E$ is defined by

$$d(v, S) = \inf\{\|v - s\| : s \in S\}.$$

We shall show that if M is a finite dimensional subspace of E, then for each $v \in E$ there exists a unique $w \in M$ such that $d(v, M) = \|v - w\|$. Hence $\|v - w\| < \|v - z\|$ for all $z \in M,\ z \ne w$.
The following preliminary results are used.

Definition: Vectors u and v in E are called *orthogonal*, written $u \perp v$, if $\langle u, v \rangle = 0$.
The vector v is said to be orthogonal to a set $M \subset E$ if $v \perp m$ for all $m \in M$. We write $v \perp M$.

A system of vectors $\{\varphi_1, \varphi_2, \ldots\}$ is called *orthogonal* if $\varphi_i \perp \varphi_j$, $i \neq j$. If, in addition, $\|\varphi_i\| = 1$, $1 \leq i$, the system is called *orthonormal*.

An orthonormal system $\{\varphi_i\}$ is linearly independent since

$$0 = \sum_{j=1}^{n} a_j \varphi_j \quad \text{implies} \quad 0 = \left\langle \sum_{j=1}^{n} a_j \varphi_j, \varphi_k \right\rangle = a_k.$$

Every finite dimensional inner product space has an orthonormal basis. This result is a special case of Theorem 9.1.

The following simple theorem is useful for calculations.

Pythagorean Theorem 4.1 *If $u \perp v$, then*

$$\|u + v\|^2 = \|u\|^2 + \|v\|^2.$$

Proof:

$$\|u + v\|^2 = \langle u + v, u + v \rangle = \|u\|^2 + \langle u, v \rangle + \langle v, u \rangle + \|v\|^2$$
$$= \|u\|^2 + \|v\|^2.$$

To return to our problem, we are given a finite dimensional subspace M of E and a vector $v \in E$. We wish to find a $W \in M$ such that $d(v, M) = \|v - w\|$. If there exists a $w \in M$ such that $v - w \perp M$, then w is the unique vector in M such that $d(v, M) = \|v - w\|$. Indeed, if $z \in M$ and $z \neq w$, then $w - z$ is in M and by the Pythagorean theorem applied to $v - w$ and $w - z$,

$$\|v - z\|^2 = \|v - w + w - z\|^2 = \|v - w\|^2 + \|w - z\|^2 > \|v - w\|^2. \quad (4.1)$$

To find w so that $v - w \perp M$, let $\varphi_1, \ldots, \varphi_n$ be an orthonormal basis for M. Then $w = \sum_{j=1}^{n} \alpha_j \varphi_j$ and for each k,

$$0 = \langle v - w, \varphi_k \rangle = \langle v, \varphi_k \rangle - \left\langle \sum_{j=1}^{n} \alpha_j \varphi_j, \varphi_k \right\rangle = \langle v, \varphi_k \rangle - \alpha_k.$$

Thus

$$w = \sum_{j=1}^{n} \langle v, \varphi_j \rangle \varphi_j. \quad (4.2)$$

Conversely, if w is given by (4.2), then $v - w \perp M$; for if $z \in M$, say $z = \sum_{k=1}^{n} \beta_k \varphi_k$, then $\langle v - w, z \rangle = \sum_{k=1}^{n} \bar{\beta}_k \langle v - w, \varphi_k \rangle = 0$.

Thus we have the following result.   □

Theorem 4.2 *Let M be a finite dimensional subspace of E and let $\{\varphi_1, \ldots, \varphi_n\}$ be an orthonormal basis for M. For each $v \in E$, the vector $w = \sum_{j=1}^{n} \langle v, \varphi_j \rangle \varphi_j$ is the unique vector in M with the property that $\|v - w\| = d(v, M)$.*

The equivalence of a closest vector and orthogonality is given in the next theorem.

Theorem 4.3 *Let M be a subspace of E. Suppose $v \in E$ and $W \in M$. Then $v - w \perp M$ if and only if $\|v - w\| = d(v, M)$.*

Proof: If $v - w \perp M$, then it follows from (4.1) that $d(v, M) = \|v - w\|$. Suppose $\|v - w\| \leq \|v - z\|$ for all $z \in M$. Since M is a subspace, $w + \lambda z$ is in M for all $z \in M$ and $\lambda \in \mathbb{C}$. Therefore,

$$\|v - w\|^2 \leq \|v - (w + \lambda z)\|^2 = \langle v - w - \lambda z, v - w - \lambda z \rangle$$
$$= \|v - w\|^2 - 2 \operatorname{Re} \lambda \langle z, v - w \rangle + |\lambda|^2 \|z\|^2.$$

Hence

$$2 \operatorname{Re} \lambda \langle z, v - w \rangle \leq |\lambda|^2 \|z\|^2.$$

Taking $\lambda = \overline{r \langle z, v - w \rangle}$, where r is a real number, we get

$$2r|\langle z, v - w \rangle|^2 \leq r^2 |\langle z, v - w \rangle|^2 \|z\|^2.$$

Since r is arbitrary, it follows that $\langle z, v - w \rangle = 0$. $\square$

1.5 The Gram Determinant

Throughout this section, E denotes an inner product space. Let $y_1, \ldots, y_n$ be a basis (not necessarily orthogonal) for a subspace M of E. We know from Theorem 4.2 that there exists a unique $w = \sum_{i=1}^{n} \alpha_i y_i$ in M such that $\|y - w\| = d(y, M)$ or, equivalently, $y - w \perp M$. Our aim in this section is to find each α_i and $d(y, M)$ in terms of the basis.

Now

$$0 = \langle y - w, y_j \rangle = \langle y, y_j \rangle - \sum_{i=1}^{n} \alpha_i \langle y_i, y_j \rangle, \quad 1 \leq j \leq n,$$

or

$$\alpha_1 \langle y_1, y_1 \rangle + \alpha_2 \langle y_2, y_1 \rangle + \cdots + \alpha_n \langle y_n, y_1 \rangle = \langle y, y_1 \rangle$$
$$\vdots \qquad\qquad \vdots \qquad\qquad \vdots \qquad\qquad \vdots \tag{5.1}$$
$$\alpha_1 \langle y_1, y_n \rangle + \alpha_2 \langle y_2, y_n \rangle + \cdots + \alpha_n \langle y_n, y_n \rangle = \langle y, y_n \rangle.$$

Since the set of α_i is unique, we have by Cramer's rule that

$$\alpha_j = \frac{D_j}{g(y_1, \ldots, y_n)}, \tag{5.2}$$

where $g(y_1, \ldots, y_n)$ is the (non-zero) determinant of the coefficient matrix

$$\begin{pmatrix} \langle y_1, y_1 \rangle & \langle y_2, y_1 \rangle & \cdots & \langle y_n, y_1 \rangle \\ \vdots & \vdots & & \vdots \\ \langle y_1, y_n \rangle & \langle y_2, y_n \rangle & \cdots & \langle y_n, y_n \rangle \end{pmatrix}$$

and D_j is the determinant of the matrix which is obtained by replacing the jth column of the coefficient matrix by the column $(\langle y, y_i \rangle)_{i=1}^{n}$.

The determinant $g(y_1, \ldots, y_n)$ is called the *Gram determinant* corresponding to the n-tuple $(y_1, \ldots, y_n)$.

A neat way to represent $w = \sum_{i=1}^{n} \alpha_i y_i$ is by the determinant

$$w = -\frac{1}{g(y_1, \ldots, y_n)} \det \begin{pmatrix} \langle y_1, y_1 \rangle & \langle y_2, y_1 \rangle & \cdots & \langle y_n, y_1 \rangle & \langle y, y_1 \rangle \\ \vdots & \vdots & & \vdots & \vdots \\ \langle y_1, y_n \rangle & \langle y_2, y_n \rangle & \cdots & \langle y_n, y_n \rangle & \langle y, y_n \rangle \\ y_1 & y_2 & \cdots & y_n & 0 \end{pmatrix}.$$

(5.3)

This can be verified by expanding the determinant of the last row as minors and referring to (5.2).

We shall now obtain a formula for the distance $d = d(y, M)$.

Since $y - w \perp M$,

$$d^2 = \|y - w\|^2 = \langle y - w, y - w \rangle = \langle y, y \rangle - \langle w, y \rangle$$
$$= \langle y, y \rangle - \sum_{i=1}^{n} \alpha_i \langle y_i, y \rangle$$

or

$$\alpha_1 \langle y_1, y \rangle + \alpha_2 \langle y_2, y \rangle + \cdots + \alpha_n \langle y_n, y \rangle = \langle y, y \rangle - d^2. \qquad (5.4)$$

Attaching (5.4) to the system of equations (5.1), we obtain a non-trivial solution to the sytem of equations

$$\langle y_1, y_1 \rangle x_1 + \langle y_2, y_1 \rangle x_2 + \cdots + \langle y_n, y_1 \rangle x_n + \langle -y, y_1 \rangle x_{n+1} = 0$$
$$\vdots \qquad \vdots \qquad \vdots \qquad \vdots \qquad \vdots$$
$$\langle y_1, y_n \rangle x_1 + \langle y_2, y_n \rangle x_2 + \cdots + \langle y_n, y_n \rangle x_n + \langle -y, y_n \rangle x_{n+1} = 0$$
$$\langle y_1, y \rangle x_1 + \langle y_2, y \rangle x_2 + \cdots + \langle y_n, y \rangle x_n + (d^2 - \langle y, y \rangle) x_{n+1} = 0$$

A non-trivial solution, as seen from (5.1) and (5.4), is

$$x_i = \alpha_i, \quad 1 \le i \le n, \quad x_{n+1} = 1.$$

Hence, by Cramer's rule,

$$\det \begin{pmatrix} \langle y_1, y_1 \rangle & \langle y_2, y_1 \rangle & \cdots & \langle y_n, y_1 \rangle & \langle -y, y_1 \rangle \\ \vdots & \vdots & & \vdots & \vdots \\ \langle y_1, y_n \rangle & \langle y_2, y_n \rangle & \cdots & \langle y_n, y_n \rangle & \langle -y, y_n \rangle \\ \langle y_1, y \rangle & \langle y_2, y \rangle & \cdots & \langle y_n, y \rangle & d^2 - \langle y, y \rangle \end{pmatrix} = 0.$$

Let us expand the determinant by using the last column as minors. If C_j is the cofactor corresponding to the number in the jth row of the last column, then $C_{n+1} = g(y_1, \ldots, y_n)$ and

$$(d^2 - \langle y, y \rangle)g(y_1, \ldots, y_n) - \langle y, y_n \rangle C_n - \cdots - \langle y, y_1 \rangle C_1 = 0.$$

Thus

$$d^2 g(y_1, \ldots, y_n) = \langle y, y \rangle C_{n+1} + \sum_{k=1}^{n} \langle y, y_k \rangle C_k$$

$$= \det \begin{pmatrix} \langle y_1, y_1 \rangle & \cdots & \langle y_n, y_1 \rangle & \langle y, y_1 \rangle \\ \vdots & & \vdots & \vdots \\ \langle y_1, y_n \rangle & \cdots & \langle y_n, y_n \rangle & \langle y, y_n \rangle \\ \langle y_1, y \rangle & \cdots & \langle y_n, y \rangle & \langle y, y \rangle \end{pmatrix}$$

$$= g(y_1, \ldots, y_n, y). \tag{5.5}$$

Now $g(y_1) = \langle y_1, y_1 \rangle > 0$. Applying (5.5) to $M_k = \mathrm{sp}\{y_1, \ldots, y_k\}$ and $y = y_{k+1}$, and noting that $d(y_{k+1}, M_k) \leq \|y_{k+1}\|$, it follows by induction that

$$0 < g(y_1, \ldots, y_n) \leq g(y_1, \ldots, y_{n-1}) \|y_n\|^2$$
$$\leq g(y_1, \ldots, y_{n-2}) \|y_{n-1}\|^2 \|y_{n-2}\|^2 \leq \cdots \leq \|y_1\|^2 \|y_2\|^2 \ldots \|y_n\|^2. \tag{5.6}$$

To summarize, we obtain from (5.3), (5.5) and (5.6) the following theorem.

Theorem 5.1 *Let $\{y_1, \ldots, y_n\}$ be a basis for the subspace $M \subset E$. For $y \in E$,*

$$d(y, M) = \left(\frac{g(y_1, \ldots, y_n, y)}{g(y_1, \ldots, y_n)} \right)^{1/2}$$

and

$$0 < g(y_1, \ldots, y_n) \leq \|y_1\|^2 \|y_2\|^2 \ldots \|y_n\|^2,$$

where $g(x_1, \ldots, x_j) = \det(\langle x_i, x_j \rangle)$ is the Gram determinant corresponding to $(x_1, x_2, \ldots, x_j)$.

The vector $w \in M$ for which $d(y, M) = \|y - w\|$ is given by

$$w = -\frac{1}{g(y_1, \ldots, y_n)} \det \begin{pmatrix} \langle y_1, y_1 \rangle & \langle y_2, y_1 \rangle & \cdots & \langle y_n, y_1 \rangle & \langle y, y_1 \rangle \\ \vdots & \vdots & & \vdots & \vdots \\ \langle y_1, y_n \rangle & \langle y_2, y_n \rangle & \cdots & \langle y_n, y_n \rangle & \langle y, y_n \rangle \\ y_1 & y_2 & \cdots & y_n & 0 \end{pmatrix}.$$

There is a very nice geometric interpretation of the Gram determinant. For example, let y_1, y_2 and y_3 be linearly independent vectors in 3-space. The volume V of the parallelopiped in Figure 1 is

$$V = \|y_1\| d(y_2, \mathrm{sp}\{y_1\}) d(y_2, \mathrm{sp}\{y_1, y_2\}).$$

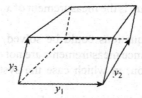

Figure 1

Thus by Theorem 5.1,

$$V^2 = \langle y_1, y_1 \rangle \frac{g(y_1, y_2)}{g(y_1)} \frac{g(y_1, y_2, y_3)}{g(y_1, y_2)} = g(y_1, y_2, y_3).$$

As an application of Theorem 5.1, we have the following result.

Hadamard's Inequality for a Determinant. *Given $A = (a_{ij})$, an $n \times n$ matrix of complex numbers,*

$$|\det A|^2 \le \prod_{i=1}^{n} \sum_{j=1}^{n} |a_{ij}|^2.$$

Proof: Let $y_i = (a_{i1}, \ldots, a_{in})$ be considered as a vector in the Hilbert space $\mathbb{C}^n$. Then

$$\langle y_i, y_j \rangle = \sum_{k=1}^{n} a_{ik} \bar{a}_{jk} \quad \text{and} \quad \|y_i\|^2 = \sum_{k=1}^{n} |a_{ik}|^2.$$

Hence if A^* is the conjugate transpose of A, then by Theorem 5.1,

$$\|y_1\|^2 \|y_2\|^2 \cdots \|y_n\|^2 \ge g(y_1, \ldots, y_n) = \det(\langle y_i, y_j \rangle) = \det AA^*$$
$$= \det A \, \overline{\det A} = |\det A|^2,$$

which establishes the desired inequality.   □

1.6 Incompatible Systems of Equations

Suppose that y is known to be a linear function of $x_1, x_2, \ldots, x_n$, i.e., $y = \lambda_1 x_1 + \lambda_2 x_2 + \cdots + \lambda_n x_n$, but the constants λ_i have not been determined. An experiment is devised in which the inputs $x_1, \ldots, x_n$ and the corresponding output y are measured. The experiment is repeated m times so as to provide a system of equations of the form

$$
\begin{aligned}
y_1 &= a_{11}\lambda_l + a_{12}\lambda_2 + \cdots + a_{1n}\lambda_n \\
&\vdots \qquad \vdots \qquad \vdots \qquad \qquad \vdots \\
y_m &= a_{m1}\lambda_1 + a_{m2}\lambda_2 + \cdots + a_{mn}\lambda_n,
\end{aligned}
\tag{6.1}
$$

where in the ith experiment, a_{ij} is the measurement of x_j and y_i is the measurement of y.

Even if this system of equations could be solved, the solutions would only approximate the desired λ_i since measurements are not exact. Moreover, if $m > n$, there might not be a solution, in which case the system of equations is called *incompatible*.

As a compromise, we seek to minimize the deviation between the right and left sides of each equation. To be specific, it is desired to find $\alpha_1, \alpha_2, \ldots, \alpha_n$ so that $f(\lambda_1, \ldots, \lambda_n) = |y_1 - \sum_{i=1}^n a_{1i}\lambda_i|^2 + \cdots + |y_m - \sum_{i=1}^n a_{mi}\lambda_i|^2$ attains it minimum at $(\alpha_1, \ldots, \alpha_n)$.

The solution to this minimization problem is obtained directly from Theorem 5.1 as follows.

Let

$$A_i = (a_{1i}, \ldots, a_{mi}) \in \mathbb{C}^m, \quad 1 \le i \le n, \quad Y = (y_1, \ldots, y_m) \in \mathbb{C}^m.$$

The components of A_i and Y correspond to the columns appearing in (6.1). Let $M = \mathrm{sp}\{A_i\} \subset \mathbb{C}^m$. Since, in practice, there are usually more experiments than there are variables, i.e., $m > n$, we assume that $A_1, \ldots, A_n$ are linearly independent in $\mathbb{C}^m$. In this setting we seek $(\alpha_1, \ldots, \alpha_n)$ so that for $w = \sum_{i=1}^n \alpha_i A_i$ and any $\lambda_1, \ldots, \lambda_n$ in $\mathbb{C}$,

$$f(\alpha_1, \ldots, \alpha_n) = \|Y - w\|^2 = d^2(Y, M) \le \left\| Y - \sum_{i=1}^n \lambda_i A_i \right\|^2$$

$$= f(\lambda_1, \ldots, \lambda_n).$$

Let

$$C_{ij} = \langle A_j, A_i \rangle = \sum_{k=1}^m a_{kj}\bar{a}_{ki}, \, 1 \le i, j \le n,$$

$$Y_i = \langle Y, A_i \rangle = \sum_{k=1}^m y_k \bar{a}_{ki}^2, \, 1 \le i \le n, \, Y_{n+1} = \sum_{k=1}^m |y_k|^2.$$

Then by Theorem 5.1, the desired w is given by

$$w = -\frac{1}{\det(C_{ij})} \det \begin{pmatrix} C_{11} & C_{12} & \cdots & C_{1n} & Y_1 \\ \vdots & \vdots & & \vdots & \vdots \\ C_{n1} & C_{n2} & \cdots & C_{nn} & Y_n \\ A_1 & A_2 & \cdots & A_n & 0 \end{pmatrix}$$

and

$$f(\alpha_1, \ldots, \alpha_n) = \|Y - w\|^2 = \frac{1}{\det(C_{ij})} \det \begin{pmatrix} C_{11} & C_{12} & \cdots & C_{1n} & Y_1 \\ \vdots & \vdots & & \vdots & \vdots \\ C_{n1} & C_{n2} & \cdots & C_{nn} & Y_n \\ \bar{Y}_1 & \bar{Y}_2 & \cdots & \bar{Y} & Y_{n+1} \end{pmatrix}$$

1.7 Least Squares Fit

Given points $t_1, \ldots, t_k$ in the interval $[a, b]$ and given complex numbers $y_1, \ldots, y_k$, let us first consider the problem to find a polynomial P of degree at most $k - 1$ such that $P(t_j) = y_j$, $1 \leq j \leq k$. This problem has a solution which is given by the Lagrange interpolation formula:

$$P(t) = \sum_{j=1}^{k} y_j P_j(t), \quad P_j(t) = \prod_{\substack{i=1 \\ i \neq j}}^{k} \frac{t - t_i}{t_j - t_i}.$$

In many cases, k is very large and it is reasonable to look for a polynomial P of degree $n < k - 1$ which has the properties $P(t_j) = y_j$, $1 \leq j \leq k$. It is clear that, in general, such a polynomial does not exist. As a compromise, we seek a polynomial P of degree at most n such that

$$S(P) = \sum_{i=1}^{k} |y_i - P(t_i)|^2$$

has the smallest value among all polynomials of degree at most n.

To solve this problem of "least squares," let E be the vector space of all complex valued functions defined on $t_1, \ldots, t_k$. Define $\langle g, h \rangle = \sum_{i=1}^{k} g(t_i) \overline{h(t_i)}$, $g, h \in E$. Clearly, $\langle , \rangle$ is an inner product on E. Let M be the subspace of E consisting of all polynomials of degree at most n. The polynomials $\{1, t, \ldots, t^n\}$ are linearly independent in M; for if $P(t) = \sum_{j=0}^{n} \alpha_j t^j = 0$, i.e., $P(t_j) = 0$, $1 \leq j \leq k$, then P is a polynomial of degree $n < k$ which has at least $n + 1$ zeros. Thus $\alpha_j = 0$, $1 \leq j \leq n$.

For $0 \leq i, j \leq n$, let

$$C_{ij} = \langle t^j, t^i \rangle = \sum_{m=1}^{k} t_m^{i+j}, \quad B_i = \sum_{m=1}^{k} y_m t_m^i, \quad B_{n+1} = \sum_{m=1}^{n} y_m^2.$$

It follows from Theorem 5.1 that the desired polynomial P is given by

$$P(t) = -\frac{1}{\det(C_{ij})} \det \begin{pmatrix} C_{00} & C_{01} & \cdots & C_{0n} & B_0 \\ \vdots & \vdots & & \vdots & \vdots \\ C_{n0} & C_{n1} & \cdots & C_{nn} & B_n \\ 1 & t & \cdots & t^n & 0 \end{pmatrix}$$

and

$$d^2(P, M) = \frac{1}{\det(C_{ij})} \det \begin{pmatrix} C_{00} & C_{01} & \cdots & C_{0n} & B_0 \\ \vdots & \vdots & & \vdots & \vdots \\ C_{n0} & C_{n1} & \cdots & C_{nn} & B_n \\ \overline{B}_0 & \overline{B}_1 & \cdots & \overline{B}_n & B_{n+1} \end{pmatrix}$$

In practice, some of the datapoints $y_1, \ldots, y_k$ are, for various reasons, more reliable than others. Therefore, certain "weights" δ_i are chosen and the corresponding least squares problem is to find the polynomial of degree at most n which minimizes

$$S(P) = \sum_{i=1}^{k} |y_i - P(t_i)|^2 \delta_i^2$$

among all polynomials of degree at most n. If we apply the above results to E with the inner product $\langle g, h \rangle = \sum_{i=1}^{k} g(t_i) \overline{h(t_i)} \delta_i^2$, the desired polynomial is obtained.

1.8 Distance to a Convex Set and Projections onto Subspaces

Throughout this chapter, $\mathcal{H}$ denotes a Hilbert space. It was shown in Theorem 4.2 that the distance from a vector to a finite dimensional subspace is always attained. This result can be extended to closed convex sets. However, a very different proof is needed.

Definition: A set $C \subset \mathcal{H}$ is *convex* if for any two vectors x and y in C, the set

$$\{tx + (1 - t)y : \quad 0 \leq t \leq 1\} \text{ is contained in } C.$$

Examples: 1. Any subspace of $\mathcal{H}$ is convex.
   2. If x any y are vectors in the plane or 3-space, then $\{tx + (1-t)y : 0 \leq t \leq 1\}$ is the line segment joining x to y. Thus a set C in the space is convex if and only if the line segment joining any two points in C lies in C.
   3. The r-ball $S_r(x_0) = \{x : \|x - x_0\| \leq r\}$ is convex. For if x and y are in $S_r(x_0)$, then for $0 \leq t \leq 1$,

$$\|tx + (1 - t)y - x_0\| \leq t\|x - x_0\| + (1 - t)\|y - x_0\| \leq r.$$

   Thus $tx + (1 - t)y$ is in $S_r(x_0)$.
   4. The set of all functions in $L_2([a, b])$ which are positive almost everywhere on $[a, b]$ is convex.

Definition: Given a set $S \subset \mathcal{H}$, the *closure* of S, written $\overline{S}$, is the set of those vectors in $\mathcal{H}$ which are limits of sequences of vectors in S, i.e., $x \in \overline{S}$ if $x_n \to x$ for some sequence $\{x_n\} \subset S$.
If $\overline{S} = S$, we call S a *closed* set.
   Every r-ball in $\mathcal{H}$ is a closed set.
   The remark following Theorem 18.1 shows that every finite dimensional subspace of $\mathcal{H}$ is closed.
   The subspace of ℓ_2 consisting of all sequences $(0, \xi_1, \xi_2, \ldots)$, where $\xi_k \neq 0$ for at most a finite number of k is not closed in ℓ_2.
   We are now ready for the generalization of Theorem 4.2.

Theorem 8.1 *Suppose M is a closed convex subset of $\mathcal{H}$. Given $y \in \mathcal{H}$, there exists a unique $w \in M$ such that $d(y, M) = \|y - w\|$.*

Proof: Let $d = \inf\{\|y - z\| : z \in M\}$. There exists a sequence $\{z_n\} \subset M$ such that $\|y - z_n\| \to d$. The idea of the proof is to show that $\{z_n\}$ converges and that its limit is the desired w.

By Theorem 3.1 applied to $y - z_n$ and $y - z_m$,

$$2(\|y - z_n\|^2 + \|y - z_m\|^2) = \|2y - (z_n + z_m)\|^2 + \|z_n - z_m\|^2. \qquad (8.1)$$

Since M is convex, $\frac{1}{2}(z_n + z_m)$ is in M and

$$\|2y - (z_n + z_m)\| = 2 \left\| y - \frac{1}{2}(z_n + z_m) \right\| \geq 2d. \qquad (8.2)$$

Combining (8.1) and (8.2), we get

$$\|z_n - z_m\|^2 \leq 2(\|y - z_n\|^2 + \|y - z_m\|^2) - 4d^2 \to 4d^2 - 4d^2 = 0$$

as $n, m \to \infty$. Since $\mathcal{H}$ is complete and M is closed, there exists a $w \in M$ such that $z_n \to w$. Thus $d = \lim_{n \to \infty} \|y - z_n\| = \|y - w\|$. Finally, suppose $z \in M$ and $d = \|y - z\|$. Computing the distance between y and the midpoint of the segment jointing z to w, namely $\frac{1}{2}(z + w) \in M$, we obtain

$$d^2 \leq \|y - \frac{1}{2}(z + w)\|^2 = \left\| \frac{1}{2}(y - z) + \frac{1}{2}(y - w) \right\|^2.$$

Hence, by the parallelogram law applied to $\frac{1}{2}(y - z)$ and $\frac{1}{2}(y - w)$,

$$d^2 \leq \left\| \frac{1}{2}(y - z) + \frac{1}{2}(y - w) \right\|^2$$

$$= 2 \left(\left\| \frac{1}{2}(y - z) \right\|^2 + \left\| \frac{1}{2}(y - w) \right\|^2 \right) - \left\| \frac{1}{2}(z - w) \right\|^2$$

$$= d^2 - \frac{1}{4}\|z - w\|^2.$$

Thus $z = w$ and the theorem is proved.   □

Since any r-ball S_r in $\mathcal{H}$ is closed and convex, we can apply the above theorem to S_r.

Definition: Given $S \subset \mathcal{H}$, the *orthogonal complement* $S^{\perp}$ of S is the set $\{x \in \mathcal{H} : x \perp S\}$.

Theorem 8.2 *Let M be a closed subspace of $\mathcal{H}$. Given $y \in \mathcal{H}$, there exists a unique $w \in M$ and a unique $v \in M^{\perp}$ such that $y = w + v$.*

Proof: By Theorems 8.1 and 4.3, there exists a unique $w \in M$ such that $v = y - w \in M^\perp$ and $y = w + v$. Suppose $y = w_1 + v_1$, $w_1 \in M$, $v_1 \in M^\perp$. Then $y - w_1 \in M^\perp$. Hence by the uniqueness of w, $w = w_1$ and therefore $v = v_1$. $\square$

Corollary 8.3 *If M is a closed subspace of $\mathcal{H}$, then $(M^\perp)^\perp = M$.*

Clearly, $M \subset (M^\perp)^\perp$. Suppose $y \in (M^\perp)^\perp$. Theorem 8.2 guarantees the existence of a $w \in M \subset (M^\perp)^\perp$ and a $v \in M^\perp$ such that $y = w + v$. Thus $v = y - w \in (M^\perp)^\perp \cap M^\perp = (0)$. Hence $y = w \in M$.

1.9 Orthonormal Systems

We recall from linear algebra that given linearly independent vectors $y_1, y_2, \ldots, y_n$ in $\mathcal{H}$, there exist an orthonormal set of vectors $\varphi_1, \varphi_2, \ldots, \varphi_n$ in $\mathcal{H}$ such that $\mathrm{sp}\{\varphi_i\}_{i=1}^k = \mathrm{sp}\{y_i\}_{i=1}^k$, $1 \le k \le n$. The φ_i are defined indirectively as follows:

$$\varphi = \frac{y_1}{\|y_1\|}, \quad \varphi_k = \frac{y_k - w_{k-1}}{\|y_k - w_{k-1}\|},$$

where,

$$w_{k-1} = \sum_{i=1}^{k-1} \langle y_k, \varphi_i \rangle \varphi_i$$

and

$$\|y_k - w_{k-1}\| = d(y_k, \mathrm{sp}\{y_1, \ldots, y_{k-1}\}).$$

The following result enables one to find φ_n without first determining $\varphi_1, \varphi_2, \ldots, \varphi_{n-1}$.

Theorem 9.1 *Let $y_1, y_2, \ldots$ be linearly independent in inner product space E and let $\varphi_1, \varphi_2, \ldots$ be the corresponding orthonormal set obtained by applying the Gram-Schmidt orthogonalization procedure to $y_1, y_2, \ldots$. Then*

$$\varphi_1 = \frac{y_1}{\|y_1\|}$$

$$\vdots$$

$$\varphi_{n+1} = (d_n d_{n+1})^{-\frac{1}{2}} \det \begin{pmatrix} \langle y_1, y_1 \rangle & \langle y_2, y_1 \rangle & \cdots & \langle y_{n+1}, y_1 \rangle \\ \vdots & \vdots & & \vdots \\ \langle y_1, y_n \rangle & \langle y_2, y_n \rangle & \cdots & \langle y_{n+1}, y_n \rangle \\ y_1 & y_2 & \cdots & y_{n+1} \end{pmatrix}, \quad n = 1, 2, \ldots$$

$$(9.1)$$

where

$$0 < d_k = g(y_1, \ldots, y_k), \quad k = n, n+1. \tag{9.2}$$

Proof: By the Gram-Schmidt procedure and Theorem 5.1,

$$\varphi_{n+1} = \frac{y_{n+1} - w}{\|y_{n+1} - w\|} = \left(\frac{d_n}{d_{n+1}}\right)^{1/2} (y_{n+1} - w),$$

where w is given in Theorem 5.1 with y replaced by y_{n+1}.

Hence

$$\varphi_{n+1} = \left(\frac{d_n}{d_{n+1}}\right)^{1/2} y_{n+1} + (d_n d_{n+1})^{-1/2}$$

$$\begin{pmatrix} \langle y_1, y_1 \rangle & \langle y_2, y_1 \rangle & \cdots & \langle y_n, y_1 \rangle & \langle y_{n+1}, y_1 \rangle \\ \vdots & & & \vdots & \vdots \\ \langle y_1, y_n \rangle & \langle y_2, y_n \rangle & & \langle y_n, y_n \rangle & \langle y_{n+1}, y_n \rangle \\ y_1 & y_2 & & y_n & 0 \end{pmatrix}$$

$$(9.3)$$

But this equality is precisely the right hand side of equality (9.1). This follow from the observation that the cofactors of $y_1, y_2, \ldots, y_n$ are the same in the determinants appearing in (9.1) and (9.3) while the cofactor of y_{n+1} in the determinant in (9.1) is d_n. $\square$

It is obvious from the theorem above that every infinite dimensional Hilbert space contains an infinite orthonormal set.

Examples: 1. The *standard basis* $e_1 = (1, 0, 0, \ldots)$, $e_2 = (0, 1, 0, \ldots)$, $\ldots$ is an infinite orthonormal system in ℓ_2.

2. $\left\{ \frac{1}{\sqrt{2\pi}} e^{\mathrm{int}} \right\}_{n=-\infty}^{\infty}$ is an infinite orthonormal system in $L_2([-\pi, \pi])$.

3. $\frac{1}{\sqrt{2\pi}}, \frac{\cos x}{\sqrt{\pi}}, \frac{\sin x}{\sqrt{\pi}}, \ldots, \frac{\cos nx}{\sqrt{\pi}}, \frac{\sin nx}{\sqrt{\pi}}, \ldots$ is an infinite orthonormal system in $L_2([\pi, \pi])$.

1.10 Szegö Polynomials

Let w be a strictly positive Lebesgue integrable function defined on the unit circle. Define on the vector space P of polynomials with complex coefficients on inner product

$$\langle p, q \rangle_w = \int_{-\pi}^{\pi} p(e^{i\theta}) \overline{q(e^{i\theta})} \, w(e^{i\theta}) \, d\theta. \qquad (10.1)$$

Applying the Gram-Schmidt process to the polynomials $1, z, z^2, \ldots$, we obtain polynomials $\varphi_0, \varphi_1, \ldots$ which are orthonormal relative to the inner product $\langle \, . \, , \, . \, \rangle_w$. These polynomials $\varphi_0, \varphi_1, \ldots$ are called the *Szegö polynomials* for the weight w.

If w is the constant function $1/2\pi$, then the corresponding Szegö polynomials are $1, z, z^2, \ldots$.

In this section we prove three theorems. The first two describe in two different ways the Szegö polynomials. The third is Szegö's theorem which states that all the zeros of the Szegö polynomials lie inside the unit circle.

Theorem 10.1 *The Szegö polynomials* $\varphi_0, \varphi_1, \ldots$ *corresponding to the strictly positive weight* w *are given by*

$$\varphi_0(z) = a_0^{-1/2}$$

$$\varphi_n(z) = (d_{n-1}, d_n)^{-1/2} \det \begin{pmatrix} a_0 & a_{-1} & \cdots & a_{-n} \\ a_1 & a_0 & \cdots & a_{-n+1} \\ \vdots & \vdots & & \vdots \\ a_{n-1} & a_{n-2} & \cdots & a_{-1} \\ 1 & z & & z^n \end{pmatrix}, n \geq 1, \qquad (10.2)$$

where

$$a_m = \int_{-\pi}^{\pi} e^{-im\theta} w(e^{i\theta}) d\theta, m \in \mathbb{Z}, \qquad (10.3)$$

$$d_k = \det T_k, \quad k = 0, 1, 2, \ldots, \qquad (10.4)$$

and T_k is the Toeplitz matrix

$$T_k = \begin{pmatrix} a_0 & a_{-1} & \cdots & a_{-k} \\ a_1 & a_0 & \cdots & a_{-k+1} \\ \vdots & \vdots & \ddots & \vdots \\ a_k & a_{k-1} & \cdots & a_0 \end{pmatrix}. \qquad (10.5)$$

Proof: Since

$$\langle z^j, z^k \rangle_w = \int_{-\pi}^{\pi} e^{-i(k-j)\theta} w(e^{i\theta}) = a_{k-j}, k, j = 0, 1, 2, \ldots, \qquad (10.6)$$

we have

$$0 < g(1, z, \ldots, z^m) := \det(\langle z^j, z^k \rangle_w)_{j,k=0}^m = \det T_m = d_m. \qquad (10.7)$$

Equality (10.2) now follows directly from (10.6), (10.7) and Theorem 9.1 applied to the polynomials $1, z, \ldots, z^n$. $\square$

Let T_n be the Toeplitz matrix described in (10.5) with $k = n$. The matrix T_n is invertible since det $T_n > 0$ by (10.7). Therefore there exist numbers $\beta_0, \beta_1, \ldots, \beta_n$ such that

$$T_n \begin{pmatrix} \beta_0 \\ \vdots \\ \beta_{n-1} \\ \beta_n \end{pmatrix} = \begin{pmatrix} 0 \\ \vdots \\ 0 \\ 1 \end{pmatrix}. \tag{10.8}$$

These numbers can be used to give an alternative description of the n-th Szegő polynomial.

Theorem 10.2 *Let $\beta_0, \beta_1, \ldots, \beta_n$ be numbers such that (10.8) holds, and put*

$$p_n(z) = \beta_0 + \beta_1 z + \cdots + \beta_n z^n \tag{10.9}$$

Then the n-th Szegő polynomial φ_n is given by

$$\varphi_n(z) = \left(\frac{d_n}{d_{n-1}} \right)^{1/2} p_n(z), n = 0, 1, 2, \ldots, \tag{10.10}$$

where d_k is defined by (10.4) and $d_{-1} = 1$.

Proof: Notice that $1 = T_0 \beta_0 = a_0 \beta_0$. Therefore $\beta_0 = a_0^{-1}$, and thus

$$d_0^{1/2} p_0(z) = a_0^{1/2} a_0^{-1} = a_0^{-1/2} = \varphi_0(z),$$

which proves (10.10) for $n = 0$.

For $n \geq 1$, we find the jth coefficient β_j of $p_n(z)$ by applying Cramer's rule to the system of equations corresponding to the matrix equation (10.8). We get

$$\beta_j = (-1)^{n+j+2} \frac{\Delta_j}{d_n}, \tag{10.11}$$

where Δ_j is the determinant of the matrix obtained by deleting the $(k+1)$st column of the matrix

$$\begin{pmatrix} a_0 & a_{-1} & \cdots & a_{-n} \\ a_1 & a_0 & & a_{-n+1} \\ \vdots & \vdots & & \vdots \\ a_{n-1} & a_{n-2} & & a_{-1} \end{pmatrix}.$$

Now from (10.2) and (10.9) we have that the coefficient of z^j of $\varphi(z)$ is

$$(d_{n-1} d_n)^{-1/2} (-1)^{n+k+2} \Delta_j = (-1)^{n+j+2} (d_{n-1} d_n)^{-1/2} d_n \beta_j = \left(\frac{d_n}{d_{n-1}} \right)^{-1/2} \beta_j.$$

Thus formula (10.10) follows.   □

We are now prepared to prove the result about the location of the zeros of the Szegő polynomials. The proof given here is due to H. Landau.

Theorem 10.3 (Szegö's Theorem). *All the zeros of the Szegö polynomials for the positive weight w are contained inside the unit circle.*

Proof: From equality (10.10) we see that it suffices to prove the theorem for $p_n(z)$, where $p_n(z)$ is defined by (10.8) and (10.9). Let λ be any zero of p_n. Then

$$p_n(z) = (z - \lambda)q(z), \qquad (10.12)$$

where $q(z)$ is a polynomial of degree $n - 1$. Since φ_n is orthogonal to $\mathrm{sp}\{\varphi_0, \ldots, \varphi_{n-1}\} = \mathrm{sp}\{1, z, \ldots, z^{n-1}\}$, and p_n is a scalar multiple of φ_n, if follows that p_n is orthogonal to q. Now

$$p_n(z) + \lambda q(z) = zq(z),$$

and from the definition of the inner product $\langle .,. \rangle_w$ we get

$$\langle q, q \rangle_w = \langle zq, zq \rangle_w = \langle p_n + \lambda q, p_n + \lambda q \rangle_w = \langle p_n, p_n \rangle_w + |\lambda|^2 \langle q, q \rangle.$$

Thus

$$(1 - |\lambda|^2)\langle q, q \rangle_w = \langle p_n, p_n \rangle_w > 0,$$

which implies $|\lambda| < 1$. $\qquad\qquad\qquad\qquad\qquad\qquad\qquad\qquad\qquad\qquad\square$

Examples: 1) Let w_1 be the weight given by

$$w_1(z) = \frac{-3z}{2z^2 - 5z + 2} \qquad (|z| = 1).$$

From $2z^2 - 5z + 2 = (2z - 1)(z - 2)$ and $\bar{z} = z^{-1}$ for $|z| = 1$, we see that

$$w_1(z) = \frac{3}{(2z - 1)(2z^{-1} - 1)} = \frac{3}{|2z - 1|^2}, \qquad |z| = 1.$$

Thus the weight w_1 is strictly positive on the unit circle. Also, it is easily seen that

$$w_1(z) = -1 + \frac{1}{1 - \frac{1}{2}z} + \frac{1}{1 - \frac{1}{2}z^{-1}} = \sum_{j=-\infty}^{\infty} 2^{-|j|}z^j \qquad (|z| = 1).$$

Now let us use Theorem 10.2 to find the corresponding n-th orthogonal polynomial φ_n. This requires to solve the following system of equations:

$$\begin{cases} \beta_0 + 2\beta_1 + \cdots + 2^n\beta_n = 0 \\ 2\beta_0 + \beta_1 + \cdots + 2^{n-1}\beta_{n-1} = 0 \\ \vdots \\ 2^n\beta_0 + 2^{n-1}\beta_1 + \cdots + \beta_n = 1 \end{cases} \qquad (10.13)$$

Solving this system (for instance by elimination) we obtain the following solution:

$$\beta_0 = 0, \ldots, \beta_{n-2} = 0, \beta_{n-1} = \frac{2}{3}, \beta_n = -\frac{1}{3}.$$

From Theorem 10.2 we conclude that the n-th Szegö polynomial for the weight w_1 is given by

$$\varphi_n(z) = \left(\frac{d_n}{d_{n-1}}\right)^{1/2} \left(\frac{2}{3} z^{n-1} - \frac{1}{3} z^n\right),$$

where $d_{-1} = 1$ and d_n is the determinant of the coefficient matrix of (10.13). Notice that the zeros of φ_n are given by $z = 0$ and $z = \frac{1}{2}$ which is in agreement with Theorem 10.3.

2) In this example we take as a weight the function

$$w_2(z) = -2z^{-1} + 5 - 2z = \left| z - \frac{1}{2} \right|^2 > 0 \quad (|z| = 1).$$

The Toeplitz matrix that we are interested in now is the following $(n+1) \times (n+1)$ tridiagonal diagonal matrix:

$$T_n = \begin{pmatrix} 5 & -2 & 0 & 0 & \ldots & 0 & 0 \\ -2 & 5 & -2 & 0 & \ldots & 0 & 0 \\ 0 & -2 & 5 & -2 & & 0 & 0 \\ \vdots & & & & & & \\ 0 & 0 & 0 & 0 & & 5 & 2 \\ 0 & 0 & 0 & 0 & & -2 & 5 \end{pmatrix}. \tag{10.14}$$

Put $d_n = \det T_n$, $n = 0, 1, 2, \ldots$, and set $d_{-1} = 1$. From the tridiagonal structure of the matrix it follows that the determinants d_n satisfy the following recurrence relation:

$$d_k = 5\,d_{k-1} - 4\,d_{k-2}, k = 1, 2, \ldots.$$

This together with $d_{-1} = 1$ and $d_0 = 5$ yields

$$d_k = 5^{k+1} \sum_{j=0}^{\gamma} (-1)^j \binom{k+i-j}{j} \left(\frac{2}{5}\right)^j,$$

where $\gamma = [(k+1)/2]$ denotes the largest integer which does not exceed $(k+1)/2$.

The above formula for d_k is proved by induction using the above recurrence relations.

Using Cramer's rule it is straight forward to check that for this example the solution $\beta_0, \ldots, \beta_n$ of equation (10.8) is given by

$$\beta_k = (-2)^{n-k} \frac{d_{k-1}}{d_n}, \quad k = 0, 1, 2, \ldots, n.$$

Thus the n-th Szegö polynomial for the weight w_2 has the form

$$\varphi_n(z) = \frac{(-2)^n}{\sqrt{d_{n-1}d_n}} \sum_{k=0}^{n}(-2)^{-k}d_{k-1}\,z^k,$$

where $d_{-1} = 1$ and d_n is the determinant of the Toeplitz matrix T_n in (10.14).

1.11 Legendre Polynomials

Let $\mathcal{H} = L_2([-1, 1])$ and let $u_n(x) = x^n$, $n = 0, 1, \ldots$. Using the Gram-Schmidt orthogonalization procedure, we shall exhibit an orthonormal system whose span coincides with the span of $u_1, u_2, \ldots$.

Let

$$\varphi_0 = \frac{u_0}{\|u_0\|}; \quad \varphi_0(x) = \frac{1}{\sqrt{2}}$$

$$v_1 = u_1 - \langle u_1, \varphi_0\rangle\varphi_0$$

$$\varphi_1 = \frac{v_1}{\|v_1\|}; \quad \varphi_1(x) = \sqrt{\frac{3}{2}}\,x$$

$$v_2 = u_2 - \langle u_2, \varphi_0\rangle\varphi_0 - \langle u_2, \varphi_1\rangle\varphi_1$$

$$\varphi_2 = \frac{v_2}{\|v_2\|}; \quad \varphi_2(x) = \sqrt{\frac{5}{2}}\cdot\frac{1}{2}(3x^2 - 1).$$

Continuing in this manner,

$$\varphi_n(x) = \sqrt{\frac{2n+1}{2}}\,\frac{1}{2^n n!}\,\frac{d^n}{dx^n}(x^2 - 1)^n, \quad n = 0, 1, \ldots. \tag{11.1}$$

The proof of the formula is as follows.

An inspection of the Gram-Schmidt procedure verifies that each φ_k is a polynomial of degree k and the coefficient of x^k is positive. Thus, since $\text{sp}\{\varphi_0, \ldots, \varphi_k\} = \text{sp}\{1, x, \ldots, x^k\}$.

$$\varphi_k(x) = \sum_{j=0}^{k} a_{kj}x^j, \quad a_{kk} > 0 \tag{11.2}$$

and

$$x^k = \sum_{j=0}^{k} b_{kj}\varphi_j(x), \quad b_{kk} > 0. \tag{11.3}$$

We observe that if P_n is a polynominal of degree n and $P_n \perp x^j$, $j = 0, 1, \ldots,$ $n - 1$, then $P_n = C_n \varphi_n$ for some number C_n. To see this, we know from (11.3) that $P_n = \sum_{i=1}^{n} C_i \varphi_i$ and therefore

$$\langle P_n, \varphi_k \rangle = \sum_{i=1}^{n} C_i \langle \varphi_i, \varphi_k \rangle = C_k.$$

But, by (11.2),

$$\langle P_n, \varphi_k \rangle = \sum_{j=0}^{k} \bar{a}_{kj} \langle P_n, x^j \rangle = 0, \quad 0 \leq k < n.$$

Hence $P_n = C_n \varphi_n$.

Let $P_n(x) = \frac{d^n}{dx^n}(x^2 - 1)^n$, which is a polynomial of degree n. To establish (11.1), we show $P_n \perp x^k$, $k < n$, which implies that $P_n = C_n \varphi_n$. The constant C_n is computed and (11.1) is verified.

Since $Q_n^{(k)}(x) = \frac{d^k}{dx^k}(x^2 - 1)^n$ vanishes at 1 and -1 when $0 \leq k < n$, repeated integration by parts yields for $k < n$,

$$\langle P_n, x^k \rangle = \int_{-1}^{1} Q_n^{(n)}(x) x^k dx = (-1)^k k! \int_{-1}^{1} Q_n^{(n-k)}(x) dx$$

$$= (-1)^k k! Q_n^{(n-k-1)}(x)|_{-1}^{1} = 0.$$

Hence by the observation above,

$$P_n = C_n \varphi_n \text{ for some } C_n \in \mathbb{C}. \tag{11.4}$$

Since the leading coefficients of P_n and φ_n are positive, $C_n > 0$. Thus

$$\|P_n\| = \|C_n \varphi_n\| = C_n. \tag{11.5}$$

Integrating by parts, we get

$$\langle P_n, P_n \rangle = \int_{-1}^{1} Q_n^{(n)}(x) \, Q_n^{(n)}(x) \, dx = - \int_{-1}^{1} Q_n^{(n-1)} Q_n^{(n+1)}(x) \, dx$$

$$= \cdots = (-1)^n \int_{-1}^{1} Q_n(x) \, Q_n^{(2n)}(x) \, dx.$$

Since $Q_n^{2n}(x) = \frac{d^{2n}}{dx^{2n}}(x^2 - 1)^n = (2n)!$,

$$\langle P_n, P_n \rangle = (-1)^n (2n)! \int_{-1}^{1} (x^2 - 1)^n \, dx$$

$$= (2n)! \int_{-1}^{1} (1 - x)^n (1 + x)^n \, dx \tag{11.6}$$

and

$$\int_{-1}^{1} (1-x)^n (1+x)^n \, dx = \frac{n}{n+1} \int_{-1}^{1} (1-x)^{n-1}(1+x)^{n+1} \, ds$$

$$= \cdots = \frac{n!}{(n+1)(n+2)\ldots 2n} \int_{-1}^{1} (1+x)^{2n} \, dx$$

$$= \frac{(n!)^2 2^{2n+1}}{(2n)!(2n+1)}.$$

Thus by (11.4), (11.5) and (11.6),

$$\varphi_n = \frac{P_n}{\|P_n\|} = \sqrt{\frac{2n+1}{2}} \frac{1}{2^n n!} \frac{d^n}{dx^n} (x^2 - 1)^n.$$

The polynomial $\frac{1}{2^n n!} \frac{d^n}{dx^n} (x^2 - 1)^n$ is called the *Legendre polynomial* of degree n. We shall refer to φ_n as the *normalized* Legendre polynomial of degree n. This polynomial φ_n has the following interesting property.

Let $a_n \in \mathbb{C}$ be chosen so that $\widetilde{\varphi}_n = a_n \varphi_n$ has a leading coefficient 1, i.e., $a_n = \frac{2^n (n!)^2}{(2n)!} \sqrt{\frac{2}{2n+1}}$. For any polynomial of degree n with leading coefficient 1, (11.3) implies that for some $a_k \in \mathbb{C}, 0 \leq k \leq n$, $Q = \sum_{k=0}^{n} a_k \varphi_k$. Hence

$$\int_{-1}^{1} |Q(x)|^2 \, dx = \langle Q, Q \rangle = \sum_{k=0}^{n} |a_k|^2 \geq |a_n|^2$$

$$= \int_{-1}^{1} |\widetilde{\varphi}_n (x)|^2 \, dx = \left(\frac{2^n (n!)^2}{(2n)!} \sqrt{\frac{2}{2n+1}} \right)^2.$$

1.12 Orthonormal Bases

Now that we know that every vector in a Hilbert space has a closest vector w in a closed subspace M, it remains to find a representation of w. It turns out that $w = \sum_k \langle w, \varphi_k \rangle \varphi_k$, where $\{\varphi_k\}$ is a certain orthonormal system in M. It is therefore necessary to concern ourselves first with the convergence of the series.

Definition: A series $\sum_{k=1}^{\infty} x_k$ of vectors $x_k \in \mathcal{H}$ converges to $x \in \mathcal{H}$, written $x = \sum_{k=1}^{\infty} x_k$, provided $s_n \to x$, where $s_n = \sum_{k=1}^{n} x_k$.

For example, if $\{e_k\}$ is the standard basis for ℓ_2, and $x = (\alpha_1, \alpha_2, \ldots) \in \ell_2$, then $x = \sum_{k=1}^{\infty} \alpha_k e_k$ since

$$\left\| x - \sum_{k=1}^{n} \alpha_k e_k \right\|^2 = \sum_{k=n+1}^{\infty} |\alpha_k|^2 \to 0 \quad \text{as } n \to \infty.$$

Lemma 12.1 *The inner product is continuous on* $\mathcal{H} \times \mathcal{H}$, *i.e., if* $x_n \to x$ *and* $y_n \to y$ *in* $\mathcal{H}$, *then*

$$\langle x_n, y_n \rangle \to \langle x, y \rangle.$$

Proof: By Schwarz's inequality,

$$|\langle x_n, y_n \rangle - \langle x, y \rangle| \leq |\langle x_n, y_n - y \rangle| + |\langle x_n - x, y \rangle|$$
$$\leq \|x_n\| \|y_n - y\| + \|x_n - x\| \|y\| \to 0,$$

since $\|x_n\| \to \|x\|$, $y_n \to y$ and $x_n \to x$. $\qquad\qquad\square$

Theorem 12.2 *If* $\{\varphi_1, \varphi_2, \ldots\}$ *is an orthonormal system in* $\mathcal{H}$, *then for every* $x \in \mathcal{H}$

   (i) $\sum_k |\langle x, \varphi_k \rangle|^2 \leq \|x\|^2$ *(Bessel's inequality).*

   (ii) $\sum_k \langle x, \varphi_k \rangle \varphi_k$ *converges.*

   (iii) $\sum_k \alpha_k \varphi_k$ *converges if and only if* $\{a_k\} \in \ell_2$.

   (iv) *If* $y = \sum_k \alpha_k \varphi_k$, *then* $\alpha_k = \langle y, \varphi_k \rangle$.

Proof: (i) $0 \leq \langle x - \sum_{k=1}^n \langle x, \varphi_k \rangle \varphi_k, x - \sum_{k=1}^n \langle x, \varphi_k \rangle \varphi_k \rangle$

$$= \|x\|^2 - 2 \sum_{k=1}^n |\langle x, \varphi_k \rangle|^2 + \sum_{k=1}^n |\langle x, \varphi_k \rangle|^2.$$

Thus for all n,

$$\sum_{k=1}^n |\langle x, \varphi_k \rangle|^2 \leq \|x\|^2$$

which implies (i).

(ii) For $s_n = \sum_{k=1}^n \langle x, \varphi_k \rangle \varphi_k$, it follows from (i) that for $n > m$,

$$\|s_n - s_m\|^2 = \langle \sum_{k=m+1}^n \langle x, \varphi_k \rangle \varphi_k, \sum_{k=m+1}^n \langle x, \varphi_k \rangle \varphi_k \rangle$$
$$= \sum_{k=m+1}^n |\langle x, \varphi_k \rangle|^2 \to 0 \quad \text{as} \quad n, m \to \infty.$$

Thus $\{s_n\}$ converges since $\mathcal{H}$ is complete, i.e., $\sum_k \langle x, \varphi_k \rangle \varphi_k$ converges.

(iii) Let $s_n = \sum_{k=1}^n \alpha_k \varphi_k$, $S_n = \sum_{k=1}^n |\alpha_k|^2$. Then for $n > m$,

$$\|s_n - s_m\|^2 = \langle \sum_{k=m+1}^n \alpha_k \varphi_k, \sum_{k=m+1}^n \alpha_k \varphi_k \rangle = \sum_{k=m+1}^n |\alpha_k|^2 = S_n - S_m.$$

Thus $\{s_n\}$ is a Cauchy sequence if and only if $\{S_n\}$ is a Cauchy sequence. Therefore $\{s_n\}$ converges if and only if $\{S_n\}$ converges, which implies (iii).

(iv) Suppose $y = \sum_k \alpha_k \varphi_k$. Then by Lemma 12.1,

$$\langle y, \varphi_j \rangle = \lim_{n \to \infty} \langle \sum_{k=1}^n \alpha_k \varphi_k, \varphi_j \rangle = \alpha_j.$$

$\qquad\qquad\qquad\qquad\qquad\qquad\qquad\qquad\qquad\qquad\qquad\qquad\qquad\square$

Definition: An orthonormal system $\{\varphi_1, \varphi_2, \ldots\}$ is called an *orthonormal basis* for $\mathcal{H}$ if for every $v \in \mathcal{H}$, $v = \sum_k \alpha_k \varphi_k$ for some $\alpha_1, \alpha_2, \ldots$ in $\mathbb{C}$.

By Theorem 12.2 (iv), $\alpha_k = \langle y, \varphi_k \rangle$. Each $\langle y, \varphi_k \rangle$ is called a *Fourier coefficient* of y.

Examples: 1. The standard basis $\{e_k\}$ is an orthonormal basis for ℓ_2. Also $\ell_2(\mathbb{Z})$ has a natural orthonormal basis. Indeed, put $e_k = (\delta_{jk})_{j \in \mathbb{Z}}$, where δ_{jk} is the Kronecker delta. Then $e_k \in \ell_2(\mathbb{Z})$, and

$$\langle e_k, e_\ell \rangle = \begin{cases} 0 & \text{if } k \neq \ell, \\ 1 & \text{if } k = \ell. \end{cases}$$

The vectors e_k, $k \in \mathbb{Z}$, have norm one, are mutually orthogonal, and each $x = (\xi_j)_{j \in \mathbb{Z}}$ can be written as $x = \sum_{j=-\infty}^{\infty} \xi_j e_j$ with the series converging in the norm of $\ell_2(\mathbb{Z})$. Thus $\ldots, e_{-1}, e_0, e_1, \ldots$ is an orthonormal basis of $\ell_2(\mathbb{Z})$. We shall refer to this basis as the *standard* orthonormal basis of $\ell_2(\mathbb{Z})$.

We shall prove the next assertions 2, 3 and 4 in Sections 13 and 14.

2. $\frac{1}{\sqrt{2\pi}} e^{inx}$, $n = 0, \pm 1, \ldots$ is an orthonormal basis for $L_2([-\pi, \pi])$.

3. $\{\frac{1}{\sqrt{2\pi}}, \frac{\cos nx}{\sqrt{\pi}}, \frac{\sin nx}{\sqrt{\pi}}\}_{n=1}^{\infty}$ is an orthonormal basis for $L_2([-\pi, \pi])$.

4. The normalized Legendre polynomials

$$\varphi_n(x) = \sqrt{\frac{2n+1}{n}} \frac{1}{2^n n!} \frac{d^n}{dx^n} (x^2 - 1)^n, \quad n = 0, 1, \ldots$$

is an orthonormal basis for $L_2([-1, 1])$.

Suppose $\dim \mathcal{H} = n$ and $\{\varphi_1, \ldots, \varphi_k\}$ are orthonormal in $\mathcal{H}$. Since orthonormal vectors are linearly independent, $\{\varphi_1, \ldots, \varphi_k\}$ is an orthonormal basis for $\mathcal{H}$ if and only if $k = n$.

If we have an infinite orthonormal system, then it need not be an orthonormal basis. For example, if one of the vectors from an orthonormal basis is deleted, then the remaining set is no longer an orthonormal basis. The following theorem provides very useful way to check if an orthonormal system is an orthonormal basis.

Theorem 12.3 *Let $\{\varphi_1, \varphi_2, \ldots\}$ be an orthonormal system in $\mathcal{H}$. The following statements are equivalent.*

(i) $\{\varphi_1, \varphi_2, \ldots\}$ *is an orthonormal basis for $\mathcal{H}$.*

(ii) *If $\langle x, \varphi_k \rangle = 0$ for $k = 1, 2, \ldots$, then $x = 0$.*

(iii) $\mathrm{sp}\{\varphi_k\}$ *is dense in $\mathcal{H}$, i.e., every vector in $\mathcal{H}$ is a limit of a sequence of vector in $\mathrm{sp}\{\varphi_k\}$.*

(iv) *For every* $X \in \mathcal{H}$,

$$\|x\|^2 = \sum_k |\langle x, \varphi_k \rangle|^2 \quad (Parseval's \ equality).$$

(v) *For all x and y in* $\mathcal{H}$,

$$\langle x, y \rangle = \sum_k \langle x, \varphi_k \rangle \overline{\langle y, \varphi_k \rangle}.$$

Proof: (i) implies (v). Let $s_n = \sum_{k=1}^n \langle x, \varphi_k \rangle \varphi_k$, $S_n = \sum_{k=1}^n \langle y, \varphi_k \rangle \varphi_k$. Then

$$\langle x, y \rangle = \lim_{n \to \infty} \langle s_n, S_n \rangle = \lim_{n \to \infty} \sum_{k=1}^n \langle x, \varphi_k \rangle \overline{\langle y, \varphi_k \rangle}.$$

(v) implies (iv). Take $x = y$ in (v).
(iv) implies (iii). Given $x \in \mathcal{H}$,

$$\left\| x - \sum_{k=1}^n \langle x, \varphi_k \rangle \varphi_k \right\|^2 = \|x\|^2 - \sum_{k=1}^n |\langle x, \varphi_k \rangle|^2 \to 0$$

as $n \to \infty$.

(iii) implies (ii). If $\langle x, \varphi_k \rangle = 0$, $k = 1, 2, \ldots$, then clearly $x \perp \text{sp}\{\varphi_k\}$. Therefore, by the continuity of the inner product, x is orthogonal to the closure of $\text{sp}\{\varphi_k\}$, which is $\mathcal{H}$. In particular, $x \perp x$ and $x = 0$.

(ii) implies (i). For any $z \in \mathcal{H}$, $w = \sum_k \langle z, \varphi_k \rangle \varphi_k$ converges by Theorem 11.2. Thus for each j,

$$\langle z - w, \varphi_j \rangle = \langle z, \varphi_j \rangle - \lim_{n \to \infty} \left\langle \sum_{k=1}^n \langle z, \varphi_k \rangle \varphi_k, \varphi_j \right\rangle$$

$$= \langle z, \varphi_j \rangle - \langle z, \varphi_j \rangle = 0.$$

Assumption (ii) assures us that $z - w = 0$ which establishes (i).
An orthonormal basis for $\mathcal{H}$ is also called a *complete orthonormal* system. □

1.13 Fourier Series

The proofs of the assertions 2, 3, and 4 preceding Theorem 12.3 rely on the following two approximation theorems ([R], pp. 174-5).

Weierstrass Approximation Theorem. *If f is a complex valued function which is continuous on* [a, b], *then for every* $\varepsilon > 0$ *there exists a polynominal P such that*

$$|f(x) - P(x)| < \varepsilon \quad for \ all \quad x \in [a, b].$$

Weierstrass Second Approximation Theorem. *If f is a complex valued function which is continuous on $[-\pi, \pi]$ and $f(-\pi) = f(\pi)$, then for every $\varepsilon > 0$ there exists a trigonometric polynomial*

$$T_n(x) = \sum_{j=0}^{n} (a_j \cos jx + b_j \sin jx)$$

such that

$$|f(x) - T_n(x)| < \varepsilon \quad \text{for all} \quad x \in [-\pi, \pi].$$

1. To prove that the orthonormal system

$$S = \left\{ \frac{1}{\sqrt{2\pi}}, \frac{\cos nx}{\sqrt{\pi}}, \frac{\sin nx}{\sqrt{\pi}} : \quad n = 1, 2, \ldots \right\}$$

is an orthonormal basis for $L_2([-\pi, \pi])$, it suffices, by Theorem 12.3, to show that the span of S is dense in the space.

Suppose f is a real valued function in $L_2([-\pi, \pi])$. Given $\varepsilon > 0$, there exists a real valued function g which is continuous on $[-\pi, \pi]$ such that

$$\|f - g\| < \varepsilon/3 \tag{13.1}$$

(cf. [12], p. 90). Next we choose a function h which is continuous on $[-\pi, \pi]$ so that $h(-\pi) = h(\pi)$ and

$$\|g - h\| < \varepsilon/3. \tag{13.2}$$

By the second Weierstrass approximation theorem, there exists a trigonometric polynomial T_n such that

$$|h(x) - T_n(x)| < \frac{\varepsilon}{10}, \quad x \in [-\pi, \pi].$$

Thus

$$\|h - T_n\|^2 = \int_{-\pi}^{\pi} |h(x) - T_n(x)|^2 \, dx < \frac{\varepsilon^2}{9}. \tag{13.3}$$

Combining (13.1), (13.2) and (13.3) we have

$$\|f - T_n\| < \varepsilon. \tag{13.4}$$

If f is complex valued, it follows from (13.4) applied to $\Re\, f$ and $\Im\, f$ that the span of S is dense in $L_2([-\pi, \pi])$.

Now that we know that S is an orthonormal basis for $L_2([-\pi, \pi])$, we may conclude from Theorem 12.3 that given $f \in L_2([-\pi, \pi])$, the series

$$\frac{a_0}{2} + \sum_{n=1}^{\infty} (a_n \cos nx + b_n \sin nx),$$

where

$$a_n = \frac{1}{\pi} \int_{-\pi}^{\pi} f(x) \cos nx \, dx, \quad b_n = \frac{1}{\pi} \int_{-\pi}^{\pi} f(x) \sin nx \, dx,$$

$n = 0, 1, \ldots,$ converges in $L_2([-\pi, \pi])$ to f, i.e.,

$$\lim_{n \to \infty} \int_{-\pi}^{\pi} \left| f(x) - \frac{a_0}{2} - \sum_{k=1}^{n} (a_k \cos kx + b_k \sin kx) \right|^2 dx = 0.$$

The series is called the *Fourier series* of f; a_n, b_n are called the *Fourier coefficients* of f.

By Parseval's equality (Theorem 12.3),

$$\int_{-\pi}^{\pi} |f(x)|^2 \, dx = \|f\|^2 = \pi \left[\frac{|a_0|^2}{2} + \sum_{n=1}^{\infty} (|a_n|^2 + |b_n|^2) \right].$$

2. Since $\cos nx = \frac{e^{inx} + e^{-inx}}{2}$ and $\sin nx = \frac{e^{inx} - e^{-inx}}{2i}$, it follows from the above result that sp$\{e^{inx} : n = 0, \pm 1, \ldots\}$ is dense in $L_2([-\pi, \pi])$. Hence $\{\frac{1}{\sqrt{2\pi}} e^{inx} : n = 0, \pm 1, \ldots\}$ is an orthonormal basis for $L_2([-\pi, \pi])$. Therefore, given $f \in L_2([-\pi, \pi])$, the Fourier series $\sum_{n=-\infty}^{\infty} c_n e^{inx}$, where

$$c_n = \frac{1}{2\pi} \int_{-\pi}^{\pi} f(x) e^{-inx} \, dx, \quad n = 0, \pm 1, \ldots$$

converges in $L_2([-\pi, \pi])$ to f, and by Parseval's equality,

$$\int_{-\pi}^{\pi} |f(x)|^2 = 2\pi \left(\sum_{n=-\infty}^{\infty} |c_n|^2 \right).$$

1.14 Completeness of the Legendre Polynomials

It was shown in Section 11 that the set of normalized Legendre polynomials

$$\varphi_n(x) = \sqrt{\frac{2n+1}{2}} \frac{1}{2^n n!} \frac{d^n}{dx^n} (x^2 - 1)^n, \quad n = 0, 1, \ldots$$

is orthonormal in $L_2([-1, 1])$ and

$$\text{sp}\{1, x, \ldots, x^k\} = \text{sp}\{\varphi_0, \varphi_1, \ldots, \varphi_k\}, \quad k = 0, 1, \ldots. \tag{14.1}$$

We now proceed to prove that $\{\varphi_0, \varphi_1, \ldots\}$ is an orthonormal basis for $L_2([-1, 1])$ by showing that (iii) of Theorem 12.3 holds.

Since the set of all complex valued functions which are continuous on $[-1, 1]$ is dense in $L_2([-1, 1])$, it follows from the Weierstrass approximation theorem that the set of all polynomials is also dense in $L_2([-1, 1])$. Thus from (14.1), the span of $\varphi_1, \varphi_2, \ldots$ is dense in $L_2([-\pi, \pi])$. This proves that $\{\varphi_0, \varphi_1, \ldots\}$ is an orthonormal basis. Consequently

$$\lim_{k \to \infty} \int_{-1}^{1} \left| f(x) - \sum_{n=0}^{k} a_n \varphi_n(x) \right|^2 dx = 0,$$

where

$$a_n = \langle f, \varphi_n \rangle = \sqrt{\frac{2n+1}{2}} \frac{1}{2^n n!} \int_{-1}^{1} f(x) \frac{d^n}{dx^n} (x^2 - 1)^n \, dx.$$

By Parseval's equality,

$$\int_{-1}^{1} |f(x)|^2 \, dx = \sum_{n=0}^{\infty} |a_n|^2.$$

1.15 Bases for the Hilbert Space of Functions on a Square

Let $\mathcal{H}_0 = L_2([a, b] \times [a, b])$ be the vector space of all those complex valued functions f which are Lebesgue measurable on the square $[a, b] \times [a, b]$ and for which

$$\int_{a}^{b} \int_{a}^{b} |f(t, s)|^2 \, ds \, dt < \infty.$$

By identifying functions which differ at most on a set of Lebesgue measure zero on $[a, b] \times [a, b]$,

$$\langle h, g \rangle = \int_{a}^{b} \int_{a}^{b} h(t, s) \overline{g(t, s)} \, ds \, dt$$

defines the inner product. With this inner product, $\mathcal{H}_0$ is a Hilbert space.

The reader in referred to [R] for a treatment of $\mathcal{H}_0$.

An orthonormal basis for $\mathcal{H}_0$ can be constructed from an orthonormal basis for $L_2([a, b])$ as follows.

Theorem 15.1 *If $\varphi_1, \varphi_2, \ldots$ is an orthonormal basis for $L_2([a, b])$, then $\Phi_{ij}(s, t) = \varphi_i(s)\varphi_j(t)$, $1 \leq i, j$, forms an orthonormal basis for $L_2([a, b] \times [a, b])$.*

Proof: The set Φ_{ij} is orthonormal since

$$\langle \Phi_{jk}, \Phi_{mn} \rangle = \int_a^b \int_a^b \varphi_j(s)\varphi_k(t)\overline{\varphi_m(s)}\,\overline{\varphi_n(t)}\,ds\,dt$$

$$= \int_a^b \varphi_j(s)\overline{\varphi_m(s)}\,ds \int_a^b \varphi_k(t)\overline{\varphi_n(t)}\,dt = \delta_{jm}\,\delta_{kn},$$

where δ_{jm}, called the *Kronecker delta*, is defined to be 1 if $j = m$ and 0 if $j \neq m$.

To prove that the system $\{\Phi_{jk}\}$ is an orthonormal basis, we show that if $g \in L_2([a, b] \times [a, b])$ and $\langle g, \Phi_{jk} \rangle = 0$, then $g = 0$ a.e. Now

$$0 = \langle g, \Phi_{jk} \rangle = \int_a^b \int_a^b g(s, t)\overline{\varphi_j(s)}\,\overline{\varphi_k(t)}\,ds\,dt$$

$$= \int_a^b \overline{\varphi_j(s)} \left\{ \int_a^b g(s, t)\overline{\varphi_k(t)}\,dt \right\}\,ds.$$

Thus

$$h(s) = \int_a^b g(s, t)\overline{\varphi_k(t)}\,dt$$

is orthogonal to each φ_j which implies $h = 0$ a.e. Hence there exists a set Z of Lebesgue measure zero such that for each $s \notin Z$, $g(s,)$ is orthogonal to φ_k, $k = 1, 2, \ldots$. Consequently, $g(s, t) = 0$ for almost every t and each $s \notin Z$. It therefore follows that

$$\int_a^b \int_a^b |g(s, t)|^2 \,dt\,ds = 0,$$

from which we may conclude that $g = 0$ a.e.

The argument above shows that if $\{\varphi_1, \varphi_2, \ldots\}$ and $\{\psi_1, \psi_2, \ldots\}$ are orthonormal bases for $L_2([a, b])$, then $F_{jk}(s, t) = \varphi_j(s)\psi_k(t)$, $1 \leq j, k$ is an orthonormal basis for $L_2([a, b] \times [a, b])$. We can write $\{F_{jk}\}$ as a sequence $F_{11}, F_{12}, F_{21}, F_{22}, \ldots$ (cf. Appendix 1).

Since $\{\frac{1}{\sqrt{2\pi}}e^{int}, n = 0, \pm 1, \ldots\}$ is an orthonormal basis for $L_2([-\pi, \pi])$, Theorem 15.1 shows that $\{\frac{1}{\sqrt{2\pi}}e^{i(nt+ms)} : m, n = 0, \pm 1, \ldots\}$ is an orthonormal basis for $L_2([-\pi, \pi] \times [-\pi, \pi])$. Hence, for f in this space,

$$\int_{-\pi}^{\pi} \int_{-\pi}^{\pi} \left| f(t, s) - \sum_{n=p}^{q} \sum_{m=j}^{k} a_{nm}e^{i(nt+ms)} \right|^2 \,ds\,dt$$

converges to zero as $j, p \to -\infty$ and $k, q \to \infty$, where

$$a_{nm} = \frac{1}{2\pi} \int_{-\pi}^{\pi} \int_{-\pi}^{\pi} f(t, s)e^{-i(nt+ms)} \,ds\,dt.$$

$\square$

1.16 Stability of Orthonormal Bases

The following theorem shows that an orthonormal system which is "close enough" to an orthonormal basis is also an orthonormal basis.

Lemma 16.1 *If M and N are subspaces of an inner product space and $\dim M < \dim N$, then $M^{\perp} \cap N \neq (0)$*

Proof: Let $\varphi_1, \ldots, \varphi_n$ be a basis for M and let $\psi_1, \ldots, \psi_{n+1}$ be linearly independent in N. The lemma is proved once it is established that there exist numbers $\beta_1, \ldots, \beta_{n+1}$, not all zero, such that $\sum_{i=1}^{n+1} \beta_i \psi_i \perp \varphi_k, k = 1, 2, \ldots, n$, or

$$\sum_{i=1}^{n+1} \beta_i \langle \psi_i, \varphi_k \rangle = 0, \quad k = 1, 2, \ldots, n.$$

But, from linear algebra, this system of n equations in $n + 1$ unknowns has a non-trivial solution.   □

Theorem 16.2 *Let $\varphi_1, \varphi_2, \ldots$ be an orthonormal basis for a Hilbert space $\mathcal{H}$. If $\psi_1, \psi_2, \ldots,$ is an orthonormal system in $\mathcal{H}$ such that $\sum_{n=1}^{\infty} \|\varphi_n - \psi_n\|^2 < \infty$, then $\psi_1, \psi_2, \ldots$ is an orthonormal basis for $\mathcal{H}$.*

Proof: Assume $\{\psi_n\}_{n=1}^{\infty}$ is not an orthonormal basis for $\mathcal{H}$. Then by Theorem 12.3, there exists a $\psi_0 \in \mathcal{H}$ such that $\|\psi_0\| = 1$ and $\langle \psi_0, \psi_j \rangle = 0, j = 1, 2, \ldots$. Choose an integer N so that

$$\sum_{n=N+1}^{\infty} \|\varphi_n - \psi_n\|^2 < 1. \tag{16.1}$$

Lemma 16.1 guarantees the existence of a $w \neq 0$ in $\mathrm{sp}\{\psi_0, \ldots, \psi_N\}$ such that $w \perp \varphi_j, 1 \leq j \leq N$. Since $\{\varphi_n\}_{n=1}^{\infty}$ is an orthonormal basis for $\mathcal{H}$ and $w \perp \psi_n$, $n > N$,

$$0 < \|w\|^2 = \sum_{n=1}^{\infty} |\langle w, \varphi_n \rangle|^2 = \sum_{n=N+1}^{\infty} |\langle w, \varphi_n \rangle|^2$$

$$= \sum_{n=N+1}^{\infty} |\langle w, \varphi_n - \psi_n \rangle|^2 \leq \|w\|^2 \sum_{n=N+1}^{\infty} \|\varphi_n - \psi_n\|^2,$$

which contradicts (16.1). Thus $\{\psi_n\}_{n=1}^{\infty}$ is an orthonormal basis for $\mathcal{H}$.   □

1.17 Separable Spaces

Now that we have studied orthonormal bases in considerable detail, it is natural to determine which Hilbert spaces possess an orthonormal basis.

Definition: A Hilbert space $\mathcal{H}$ is called *separable* if there exist vectors $v_1, v_2, \ldots$ which span a subspace dense in $\mathcal{H}$.

An equivalent definition appears in Appendix 1.

Every Hilbert space $\mathcal{H}$ which has an orthonormal basis is separable since the span of the basis is dense in $\mathcal{H}$ by Theorem 12.3(iii). In particular, ℓ_2 and $L_2([a, b])$ are separable.

It turns out that only separable Hilbert spaces have an orthonormal basis.

Theorem 17.1 *A Hilbert space contains an orthonormal basis if and only if it is separable.*

Proof: Suppose $\mathcal{H}$ is a separable Hilbert space and $\text{sp}\{w_1, w_2, \ldots\}$ is dense in $\mathcal{H}$. By discarding those w_k which are linear combinations of $w_1, \ldots, w_{k-1}$ it follows that linear independent vectors $\{v_1, v_2, \ldots\}$ can be found such that $\text{sp}\{v_1, v_2, \ldots\} = \text{sp}\{w_1, w_2, \ldots\}$. Hence there exists, by Theorem 9.1, orthonormal vectors $\{\varphi_1, \varphi_2, \ldots\}$ such that $\text{sp}\{\varphi_i\} = \text{sp}\{v_i\}$ which is dense in $\mathcal{H}$. Thus $\{\varphi_1, \varphi_2, \ldots\}$ is an orthonormal basis for $\mathcal{H}$ by Theorem 12.3.

Conversely, if $\varphi_1, \varphi_2, \ldots$ is an orthonormal basis for $\mathcal{H}$, then $\text{sp}\{\varphi_1, \varphi_2, \ldots\}$ is dense in $\mathcal{H}$. $\qquad\square$

Theorem 17.2 *A closed subspace of a separable Hilbert space is separable.*

Proof: Let $\text{sp}\{v_1, v_2, \ldots\}$ be dense in the Hilbert space $\mathcal{H}$ and let M be a closed subspace of $\mathcal{H}$. For each k there exists a unique vector $w_k \in M$ such that $v_k - w_k \perp M$. We shall show that $\text{sp}\{w_1, w_2, \ldots\}$ is dense in M which proves that M is separable.

Given $w \in M$ and given $\varepsilon > 0$, there exists a vector $\sum_{j=1}^{n} \beta_j v_j$ such that

$$\left\| w - \sum_{j=1}^{n} \beta_j v_j \right\| < \varepsilon. \tag{17.1}$$

Since $\sum_{j=1}^{n} \beta_j (v_j - w_j)$ is orthogonal to M, (17.1) and Theorem 4.3 imply

$$\varepsilon > d\left(\sum_{j=1}^{n} \beta_j v_j, M \right) = \left\| \sum_{j=1}^{n} \beta_j v_j - \sum_{j=1}^{n} \beta_j w_j \right\|. \tag{17.2}$$

Thus, from (17.1) and (17.2),

$$\left\| w - \sum_{j=1}^{n} \beta_j w_j \right\| \le \left\| w - \sum_{j=1}^{n} \beta_j v_j \right\| + \left\| \sum_{j=1}^{n} \beta_j v_j - \sum_{j=1}^{n} \beta_j w_j \right\| < 2\varepsilon.$$

Therefore, $\mathrm{sp}\{w_1, w_2, \ldots\}$ is dense in M.

We can now give a representation of the projection of a given vector into a closed, separable subspace of a Hilbert space. $\qquad\square$

Theorem 17.3 *Suppose M is a closed, separable subspace of a Hilbert space $\mathcal{H}$. Then M has an orthonormal basis $\{\varphi_1, \varphi_2, \ldots\}$. Given $y \in \mathcal{H}$, the vector $w = \sum_k \langle y, \varphi_k \rangle \varphi_k$ is the projection of y into M, i.e., $y - w \perp M$.*

Proof: Since M is closed and $\mathcal{H}$ is complete, it follows that M, with the inner product inherited from $\mathcal{H}$, is a separable Hilbert space. Hence M contains an orthonormal basis $\{\varphi_1, \varphi_2, \ldots\}$. Now $w = \sum_k \langle y, \varphi_k \rangle \varphi_k$ converges by Theorem 12.3 and for each j,

$$\langle y - w, \varphi_j \rangle = \langle y, \varphi_j \rangle - \lim_{n \to \infty} \left\langle \sum_{k=1}^{n} \langle y, \varphi_k \rangle \varphi_k, \varphi_j \right\rangle = \langle y, \varphi_j \rangle - \langle y, \varphi_j \rangle = 0.$$

Thus $(y - w) \perp \mathrm{sp}\{\varphi_1, \varphi_2, \ldots\}$ and since $\mathrm{sp}\{\varphi_1, \varphi_2, \ldots\}$ is dense in M, the continuity of the inner product implies $(y - w) \perp M$. $\qquad\square$

1.18 Isometry of Hilbert Spaces

The existence of orthonormal bases for separable Hilbert spaces enables one to identify all separable Hilbert spaces as follows.

Definition: Inner product spaces E and F are *linearly isometric* if there exists a function A which maps E *onto* F such that for all u, v in E and α, $\beta \in \mathbb{C}$,

    (i) $A(\alpha u + \beta v) = \alpha A u + \beta A v$;

    (ii) $\|Au\| = \|u\|$.

The operator A is called a *linear isometry*.

Theorem 18.1 *Any two infinite dimensional separable Hilbert spaces (over $\mathbb{C}$) are linearly isometric.*

Proof: Suppose $\mathcal{H}_1$ and $\mathcal{H}_2$ are infinite dimensional separable Hilbert spaces with orthonormal bases $\{\varphi_1, \varphi_2, \ldots\}$ and $\{\psi_1, \psi_2, \ldots\}$, respectively. For each $u \in \mathcal{H}_1$, define $Au = \sum_k \langle u, \varphi_k \rangle \psi_k$. The series converges by Theorem 12.3. Clearly, A has property (i). By Parseval's equality

$$\|Au\|^2 = \sum_k |\langle u, \varphi_k \rangle|^2 = \|u\|^2.$$

Finally, A maps $\mathcal{H}_1$ onto $\mathcal{H}_2$; for if $y \in \mathcal{H}_2$, then

$$y = \sum_k \langle y, \psi_k \rangle \psi_k = A \left(\sum_k \langle y, \psi_k \rangle \varphi_k \right).$$

The same proof shows that every n-dimensional Hilbert space (over $\mathbb{C}$) is linear isometric to $\mathbb{C}^n$. Thus every finite dimensional subspace of a Hilbert space is closed. $\qquad \square$

Corollary 18.2 *Any infinite dimensional separable Hilbert space is linearly isometric to ℓ_2.*

It is easy to verify that the operator A defined in the proof of Theorem 18.1 has the property that for all u, v in $\mathcal{H}_1$.

$$\langle Au, Av \rangle_{\mathcal{H}_2} = \langle u, v \rangle_{\mathcal{H}_1},$$

which means that the spaces $\mathcal{H}_1$, $\mathcal{H}_2$ from the point of view of Hilbert spaces, practically do not differ.

Often one can give an explicit description of the operator that establishes the linear isometry between two Hilbert spaces. To illustrate this let us consider the spaces $\ell_2(\mathbb{Z})$ and $L_2([-\pi, \pi])$. Both are infinite dimensional separable Hilbert spaces, and hence by Theorem 18.1 they are linearly isometric. To make this connection explicit, let $\mathcal{F}$ be the map that assigns to each function f in $L_2([-\pi, \pi])$ the sequence of the Fourier coefficients of f with respect to the orthonormal basis $\varphi_n(t) = \frac{1}{\sqrt{2\pi}} e^{int}, n = 0, \pm 1, \pm 2, \ldots$. Thus

$$\mathcal{F}f = (f_n)_{n \in \mathbb{Z}}, \quad f_n = \frac{1}{\sqrt{2\pi}} \int_{-\pi}^{\pi} f(t) e^{-int} \, dt.$$

By Parseval's equality, the sequence $\mathcal{F}f$ is square summable, that is $\mathcal{F}f \in \ell_2(\mathbb{Z})$, and

$$\|f\| = \left(\sum_{n=-\infty}^{\infty} |f_n|^2 \right)^{1/2} = \|\mathcal{F}f\|.$$

It follows that the map $f \mapsto \mathcal{F}f$ is a linear operator which maps $L_2([-\pi, \pi])$ in a one to one way onto $\ell_2(\mathbb{Z})$ and preserves the norm. In the sequel we refer to $\mathcal{F}$ as the *Fourier transform* on $L_2([-\pi, \pi])$. Notice that $\mathcal{F}$ transforms the orthonormal basis $\varphi_n(t) = \frac{1}{\sqrt{2\pi}} e^{int} n = 0, \pm 1, \pm 2, \ldots$ of $L_2([-\pi, \pi])$ into the standard orthonormal basis of $\ell_2(\mathbb{Z})$.

1.19 Example of a Non Separable Space

The Hilbert spaces which are usually encountered in applications are separable. However, there is a very important non-separable inner product space which has been a source of active research for more than fifty years. This is the space of almost periodic functions.

Definition: A complex valued function which is continuous on $R = (-\infty, \infty)$ is *almost periodic* if it is the uniform limit on R of a sequence of trigonometric polynomials of the form $\sum_{k=1}^{n} a_k e^{i\lambda_k t}$, λ_k real.

Let E be the set of almost periodic functions. Under the usual operations of addition and scalar multiplication, E is a vector space over $\mathbb{C}$.

It can be shown that if f and g are in E, then

$$\langle f, g \rangle = \lim_{T \to \infty} \frac{1}{T} \int_0^T f(x) \bar{g}(x) \, dx$$

exists and defines an inner product on E. Since $\{e^{i\lambda t} : \lambda \text{ real}\}$ is an uncountable orthonormal set, E is not separable (cf. Appendix 1).

Exercises I

1. For which $\alpha \in \mathbb{R}$ (the set of real numbers) is $\{\frac{1}{n^\alpha}\}_{n=1}^{\infty}$ in ℓ_2?

2. Determine which of the following functions are in $L_2[0, 1]$. Then compute their norms.

   (a) $\dfrac{1}{\sqrt{t^2+1}}$

   (b) $\dfrac{\sqrt{\alpha}}{\cos \alpha t}$, $\alpha \in \mathbb{R}$

   (c) $\frac{1}{t^\alpha}$, $\alpha \in \mathbb{R}$

   (d) $t^n e^{\alpha t}$, $\alpha \in \mathbb{R}$, $n \in \mathbb{N}$ (the set of positive integers).

3. Which of the functions in problem 2 are in $L_2[1, \infty)$?

4. Compute the ℓ_2-inner product of $\xi = \{(\frac{i}{2})^n\}_{n=1}^{\infty}$ with $\eta = \{(\frac{1}{4}+\frac{1}{4}i\sqrt{2})^n\}_{n=1}^{\infty}$.

5. Check if $\|x + y\| = \|x\| + \|y\|$ where x, y in ℓ_2 are given by

   (a) $x = (0, \frac{1}{2}, \frac{1}{4}, \ldots)$; $y = (1, 0, 0, \ldots)$.

   (b) $x = (1, \frac{1}{3}, \frac{1}{9}, \ldots)$; $y = 3x$.

6. Check if $\|x + y\|^2 = \|x\|^2 + \|y\|^2$ for the following pairs x, y in $L_2[0, 1]$.

   (a) $x(t) = t$; $y(t) = -t$

   (b) $x(t) = \begin{cases} 1; & 0 \le t \le \frac{1}{2} \\ 0; & \frac{1}{2} < t \le 1 \end{cases}$; $y(t) = \begin{cases} 0; & 0 \le t \le \frac{1}{2} \\ 1; & \frac{1}{2} < t \le 1 \end{cases}$.

7. Let $\mathcal{H}$ be a Hilbert space.

   (a) Describe all pairs of vectors x, y for which

   $$\|x + y\| = \|x\| + \|y\|.$$

   (b) Describe all pairs of vectors x, y for which

   $$\|x + y\|^2 = \|x\|^2 + \|y\|^2.$$

8. Let $C[a, b]$ be the vector space of all continuous complex valued functions on $[a, b]$. Intoduce a norm $\|\cdot\|$ on $C[a, b]$ by $\|f\| = \max_{t\in[a,b]} |f(t)|$. Show that it is impossible to define an inner product on $C[a, b]$ such that the norm it induces is the same as the given norm.

9. Find the intersection of the ball $\|z - \xi_0\| \leq 1$ with the line $z = \lambda\xi_1$, where $\xi_0 = (1, \frac{1}{2}, \frac{1}{4}, \frac{1}{8}, \ldots), \xi_1 = (1, 0, \ldots)$.

10. Find the intersection of the line $z = \lambda\varphi$ with the sphere $\|z - \varphi_0\| = 1$ in $L_2[0, 1]$, where $\varphi_0(t) = 2t^2$, $\varphi(t) = t$.

11. Find the intersection of the ball $\|z - \xi_0\| \leq 1$ with the plane $z = \alpha\xi_1 + \beta\xi_2$, where

$$\xi_0 = \left(1, \frac{1}{2}, \frac{1}{4}, \ldots\right), \quad \xi_1 = (1, 0, 0, \ldots) \quad \text{and} \quad \xi_2 = (0, 1, 0, \ldots).$$

12. Find the intersection of the plane $z = \alpha\varphi_1 + \beta\varphi_2$ with the sphere $\|z - \varphi_0\| = 1$ in $L_2[0, 1]$, where

$$\varphi_0(t) \equiv \gamma \in \mathbb{C}, \quad \varphi_1(t) = \begin{cases} 1; & 0 \leq t \leq \frac{1}{2} \\ -1; & \frac{1}{2} < t \leq 1 \end{cases}$$

$$\varphi_2(t) = \begin{cases} \frac{1}{2}; & 0 \leq t \leq \frac{1}{4}, \frac{1}{2} \leq t \leq \frac{3}{4} \\ -\frac{1}{2}; & \frac{1}{4} < t < \frac{1}{2}, \frac{3}{4} < t \leq 1 \end{cases}$$

13. Is the intersection of the two balls $\|\xi_1 - z\| \leq R_1$ and $\|\xi_2 - z\| \leq R_2$ empty?

    (a) $\xi_1 = (\frac{1}{\sqrt{(n-1)!}})_{n\in\mathbb{N}}, R_1 = \frac{1}{2}; \xi_2 = (1, 0, 0, \ldots), R_2 = \frac{1}{2}$.

    (b) $\xi_1 = (\frac{1}{\sqrt{n+1}\sqrt{n}})_{n\in\mathbb{N}}, R_1 = \frac{1}{4}; \xi_2 = (\frac{1}{\sqrt{2}}, 0, 0, \ldots), R_2 = \frac{1}{2}$.

    (c) $\xi_1 = (1, 0, 0, 0, \frac{1}{81}, \frac{1}{243}, \frac{1}{729}, \ldots), R_1 = \frac{1}{2}; \xi_2 = (0, \frac{1}{3}, \frac{1}{9}, \frac{1}{27}, 0, 0, \ldots), R_2 = \frac{3-\sqrt{2}}{2\sqrt{2}}$.

14. (a) Show that the system $\{z_j\}_1^\infty$ is linearly independent in $L_2[0, 1]$, where

$$z_j(t) = \begin{cases} 0; & 0 \le t < \frac{1}{j+1} \\ 1; & \frac{1}{j+1} < t \le 1 \end{cases}.$$

    (b) Let $z_j(t) = t^j e^{-\alpha t}$, $j = 1, 2, \ldots$. Show that the system is linearly independent in $L_2[0, \infty)$.

15. Let $\omega = (\omega_1, \omega_2, \ldots)$, where $\omega_j > 0$. Define $\ell_2(\omega)$ to be the set of all sequences $\xi = (\xi_1, \xi_2, \ldots)$ of complex numbers with $\sum_{j=1}^\infty \omega_j |\xi_j|^2 < \infty$. Define an inner product on $\ell_2(\omega)$ by

$$\langle \xi, \eta \rangle = \sum_{j=1}^\infty \omega_j \xi_j \bar{\eta}_j.$$

    Show that $\ell_2(\omega)$ is a Hilbert space.

16. (a) Find a vector ω such that $(1, \frac{1}{2^2}, \frac{1}{3^3}, \ldots) \notin \ell_2(\omega)$.

    (b) Find a vector ω such that the set of all $\xi = (\xi_1, \xi_2, \ldots)$ with $|\xi_n| < n^n$ is in $\ell_2(\omega)$.

17. Let $g(t)$ be continuous and strictly positive on $[a, b]$. Give an inner product on $L_2[a, b]$ such that

$$\| f \| = \left(\int_a^b g(t)|f(t)|^2 \, dt \right)^{1/2}.$$

18. Let H be a Hilbert space. Denote by $\ell_2(H)$ the space of all sequences $(x_j)_{j \in \mathbb{N}}$ of vectors in H with $\sum_{j=1}^\infty \|x_j\|^2 < \infty$. Define the inner product $(x, y) = \sum_{j=1}^\infty \langle x_j, y_j \rangle$. Show that $\ell_2(H)$ is Hilbert space.

19. Let $z^{(1)} = (z_1^{(1)}, z_2^{(2)}, \ldots), z^{(2)} = (z_1^{(2)}, z_2^{(2)}, \ldots), \ldots, z^{(n)} = (z_1^{(n)}, z_2^{(n)}, \ldots)$ be any n vectors in ℓ_2. Take $z^{(i)}(k) = (z_1^{(i)}, \ldots, z_k^{(i)}, 0, 0, \ldots)$, for $i = 1, 2, \ldots, n$.

    (a) If for some k the vectors $\{z^{(1)}(k), \ldots, z^{(n)}(k)\}$ are linearly independent, show that $\{z^{(1)}, \ldots, z^{(n)}\}$ are linearly independent too.

    (b) Check if the converse statement is true.

20. Check that the following sets are closed subspaces of ℓ_2.

    (a) $A = \{(\xi_1, 2\xi_1, \xi_3, 4\xi_3, \xi_5, \xi_6, \xi_7, \ldots) | \sum_{j=1}^\infty |\xi_j|^2 < \infty\}$.

    (b) $B = \{(\xi_1, 0, \xi_3, 0, \xi_5, 0, \ldots) | \sum_{j=1}^\infty |\xi_{2j-1}|^2 < \infty\}$.

    (c) $C = \{(0, \xi_2, 0, \xi_4, 0, \xi_6, \ldots) | \sum_{j=1}^\infty |\xi_{2j}|^2 < \infty\}$.

21. Find the pairwise intersections between the subspaces defined problem 20.

22. Let $r_1, r_2, r_3 \in \mathbb{C}$ be such that $|r_i| < 1$, $i = 1, 2, 3$; and $r_j \neq r_i$. Let $x_i = (1, r_i, r_i^2, \ldots)$. Prove that $\xi = (\xi_1, \xi_2, \xi_3, \ldots)$ is in the subspace spanned by x_1, x_2 and x_3 if and only if $\xi_{n+3} - (r_1 + r_2 + r_3)\xi_{n+2} + (r_1 r_2 + r_1 r_3 + r_2 r_3)\xi_{n+1} - r_1 r_2 r_3 \xi_n = 0$.

23. Let $x = (1, \frac{1}{2}, \frac{1}{4}, \frac{1}{8}, \frac{1}{16}, \ldots)$, $y = (1, 2.\frac{1}{2}, 3.\frac{1}{4}, 4.\frac{1}{8}, \ldots)$, $z = (2, 3.2.\frac{1}{2},$ $4.3.\frac{1}{4}, 5.4.\frac{1}{8}, \ldots)$ be vectors in ℓ_2. Prove that $\xi = (\xi_1, \xi_2, \ldots)$ is in the subspace spanned by x, y and z if and only if

$$\xi_{n+3} - \frac{3}{2}\xi_{n+2} + \frac{3}{4}\xi_{n+1} - \frac{1}{8}\xi_n = 0.$$

24. Let $f_1(t) = e^t$, $f_2(t) = e^{it}$ and $f_3(t) = e^{-it}$. Prove that a vector $y \in L_2[a, b]$ is in the subspace spanned by f_1, f_2 and f_3 if and only if y satisfies the differential equation

$$y^{(3)} - y^{(2)} + y' - y = 0.$$

25. (a) Prove that a vector $y \in L_2[a, b]$ is in the subspace spanned by $\{e^t, te^t,$ $e^{-t}\}$ if and only if y satisfies the differential equation $y^{(3)} - y^{(2)} - y' + y = 0.$

    (b) What is the dimension of this subspace?

26. (a) Let y_1 and y_2 be twice continuously differentiable functions in $L_2[a, b]$ with $\det\begin{pmatrix} y_1 & y_2 \\ y_1' & y_2' \end{pmatrix} \neq 0$. Prove that y is an element of the subspace spanned by y_1 and y_2 if and only if

$$\det\begin{pmatrix} y_1 & y_2 & y \\ y_1' & y_2' & y' \\ y_1'' & y_2'' & y'' \end{pmatrix} \equiv 0.$$

    (b) What is the dimension of that subspace?

27. Show that the set of all vectors in ℓ_2 of the form $\xi = (\xi_1, \xi_1, \xi_3, \xi_4, \ldots)$ is the orthogonal complement of the subspace spanned by $(1, -1, 0, 0, \ldots)$. Show that any vector $x \in \ell_2$ can be represented in the form $x = x_1 + x_2$, where x_1 is in $\mathrm{sp}(1, -1, 0, 0, \ldots)$ and x_2 is in the orthogonal complement. Show that this representation is unique.

28. Show that the set of all vectors in ℓ_2 of the form $(\xi_1, \xi_1, \xi_1, \xi_4, \xi_5, \ldots)$ is the orthogonal complement of the subspace spanned by $(1, -1, 0, 0, \ldots)$ and $(0, 1, -1, 0, \ldots)$. Show that any $x \in \ell_2$ can be represented uniquely as $x = x_1 + x_2$, where x_1 is in the span of $(1, -1, 0, 0, \ldots)$ and $(0, 1, -1, 0, \ldots)$ and x_2 is in its orthogonal complement.

29. Let $x_1 = (1, -5, 6, 0, \ldots)$, $x_2 = (0, 1, -5, 6, 0, \ldots)$, $\ldots$, $x_j = (\underbrace{0, 0, 0, 0,}_{j-1 \text{ times}}$
    $1, -5, 6, 0, \ldots)$, $\ldots$. Prove that the following three statements are equivalent:

1. $\xi = (\xi_1, \xi_2, \ldots)$ is a vector in the orthogonal complement (in ℓ_2) of the subspace $\text{sp}\{x_1, x_2, \ldots\}$.

2. For all $n \in \mathbb{N}$, $6\xi_{n+2} - 5\xi_{n+1} + \xi_n = 0$.

3. ξ is a vector of the subspace spanned by $(1, \frac{1}{2}, \frac{1}{4}, \frac{1}{8}, \ldots)$ and $(1, \frac{1}{3}, \frac{1}{9}, \frac{1}{27}, \ldots)$.

30. Find in ℓ_2 the orthogonal complement of the subspace

$$\text{sp}\left(1, \frac{1}{2}, \frac{1}{4}, \frac{1}{8}, \ldots\right), \left(1, \frac{1}{4}, \frac{1}{16}, \frac{1}{64}, \ldots\right).$$

31. Let $\omega = (1, \alpha, \alpha^2, \ldots)$ for $\alpha > 1$. Denote by $\ell_2(\omega)$ the Hilbert space consisting of all sequences $\xi = (\xi_1, \xi_2, \ldots)$ such that $\sum_{j=1}^{\infty} \alpha^{j-1} |\xi_j|^2 < \infty$. Prove that $\xi = (1, \frac{1}{\alpha}, \frac{1}{\alpha^2}, \ldots) \in \ell_2$ and find the orthogonal complement in $\ell_2(\omega)$ of $\text{sp}\{\xi\}$.

32. Let $x_1 = (1, 2, 0, \ldots),\qquad y_1 = (1, 0, 0, \ldots);$
$x_2 = (0, 1, 2, 0, \ldots),\qquad y_2 = (1, 1, 0, 0, \ldots);$
$x_3 = (0, 0, 1, 2, 0, \ldots),\quad y_3 = (1, 1, 1, 0, 0, \ldots);$
$\vdots \qquad\qquad \vdots\qquad\qquad \vdots\qquad\qquad \vdots$

be two given systems in ℓ_2. Prove that for all j, $y_j \notin \overline{\text{sp}}\{x_1, x_2, \ldots\}$.

33. (a) Prove that the orthogonal complement in ℓ_2 of $\text{sp}\{y\}$, where $y = (1, -\frac{1}{3}, \frac{1}{9}, -\frac{1}{27}, \ldots)$ is $\overline{\text{sp}}\{x_1, x_2, \ldots)\}$, where

$$x_j = (\underbrace{0, \ldots, 0}_{j-1 \text{ times}}, 1, 3, 0, \ldots).$$

(b) Decompose each vector with respect to $\text{sp}\{x\}$ and its orthogonal complement.

34. Prove that $\xi \in \overline{\text{sp}}\{x_1, x_2, \ldots\}$, where $x_1, x_2, \ldots$ are as in problem 32 if and only if $\xi_1 = -\sum_{j=1}^{\infty}(-\frac{1}{2})^j \xi_{j+1}$. Therefore ξ is of the form $(\frac{1}{2}\xi_2 - \frac{1}{4}\xi_3 + \frac{1}{8}\xi_4 - \ldots, \xi_2, \xi_3, \ldots)$.

35. Let $x_1 = (\alpha, \beta, 0, 0, \ldots)$, $x_2 = (0, \alpha, \beta, 0, \ldots)$, $x_3 = (0, 0, \alpha, \beta, \ldots), \ldots$, where $|\frac{\beta}{\alpha}| > 1$. Prove that for all j, the vectors y_j from problem 32 are not in the subspace $\overline{\text{sp}}\{x_1, x_2, \ldots\}$ of ℓ_2.

36. Let $x_1, x_2, x_3, \ldots$ be as in problem 35. Prove that the orthogonal complement in ℓ_2 to $\text{sp}\{(1, -\frac{\alpha}{\beta}, \frac{\alpha^2}{\beta^2}, -\frac{\alpha^3}{\beta^3}, \ldots)\}$ is $\overline{\text{sp}}\{x_1, x_2, \ldots\}$.

37. Let $x_1, x_2, x_3, \ldots$ be the vectors in ℓ_2 as in problem 35. Prove that $\xi \in \overline{\text{sp}}\{x_1, x_2, \ldots\}$ if and only if $\xi_1 = -\sum_{j=1}^{\infty}(-\frac{\alpha}{\beta})^j \xi_{j+1}$ so that ξ is of the form

$$\xi = \left(\frac{\alpha}{\beta}\xi_2 - \frac{\alpha^2}{\beta^2}\xi_3 + \frac{\alpha^3}{\beta^3}\xi_4 - \ldots, \xi_2, \xi_3, \xi_4, \ldots\right).$$

38. Let $x_1 = (\alpha, \beta, 0, 0, \ldots)$, $x_2 = (0, \alpha, \beta, 0, \ldots)$, $x_3 = (0, 0, \alpha, \beta, 0, \ldots)$, $\ldots$, where $|\frac{\beta}{\alpha}| \leq 1$. Check that $\overline{\mathrm{sp}}\{x_1, x_2, \ldots\} = \ell_2$.

39. Let $x_1 = (1, 0, 0, \ldots)$, $x_2 = (\alpha, \beta, 0, \ldots)$, $x_3 = (0, \alpha, \beta, 0, \ldots)$, $\ldots$, where $|\frac{\alpha}{\beta}| > 1$.

    (a) Check that $\overline{\mathrm{sp}}\{x_1, x_2, \ldots\} = \ell_2$.
    (b) Show that any finite system of these vectors is linearly independent.
    (c) Find $\alpha_1, \alpha_2, \ldots$ in $\mathbb{C}$, not all zero, such that $\sum_{j=1}^{\infty} \alpha_j x_j$ converges to zero.

40. Let $x_1 = (1, 0, 0, \ldots)$, $x_2 = (\alpha, \beta, 0, \ldots)$, $x_3 = (0, \alpha, \beta, 0, \ldots)$, $\ldots$, where $|\frac{\alpha}{\beta}| \leq 1$.

    (a) Check that $\overline{\mathrm{sp}}\{x_1, x_2, \ldots\} = \ell_2$.
    (b) Show that each finite system of these vectors is linearly independent.
    (c) Show that one cannot find $\alpha_1, \alpha_2, \ldots$, not all zero, such that $\sum_{j=1}^{\infty} \alpha_j x_j$ converges to zero.

41. Let $\chi_\tau(t) = \begin{cases} 0 & \text{for } \tau \leq t < 1 \\ 1 & \text{for } 0 < t < \tau \end{cases}$

    (a) Prove that $\overline{\mathrm{sp}}\{\chi_\tau\}_{\tau \in [0,1]} = L_2(0, 1)$.
    (b) Prove that any finite number of these functions with different τ are linearly independent.
    (c) Prove that $\overline{\mathrm{sp}}\{\chi_\tau\}_{\tau \in Q \cap [0,1]} = L_2(0, 1)$, where Q is the set of rational numbers.

42. Let $\{x_i\}$ be the vectors in problem 32. For every $\xi \in \ell_2$, find the projection of ξ into $\overline{\mathrm{sp}}\{x_1, x_2, \ldots\}$ and find the distance from ξ to this subspace.

43. Does the line $z = \lambda x$ intersect the ball $\|z - \alpha_0\| \leq R_0$?

    (a) $x = (1, \frac{1}{\sqrt{3}}, \frac{1}{\sqrt{9}}, \frac{1}{\sqrt{27}}, \ldots)$, $\alpha_0 = (0, 1, 0, \ldots)$, $R_0 = \frac{1}{2}$.
    (b) $x = (1, 0, \frac{1}{2}, 0, \frac{1}{4}, 0, \frac{1}{8}, \ldots)$, $\alpha_0 = (0, 1, 0, \frac{1}{2}, 0, \frac{1}{4}, \ldots)$, $R_0 = \frac{4}{3}$.

44. Suppose $\varphi_1, \ldots, \varphi_n$ is an orthonormal system in a Hilbert space $\mathcal{H}$. Let S be the ball $\{z : \|z - z_0\| \leq R_0\}$, and let $M = \mathrm{sp}\{\varphi_1, \ldots, \varphi_n\}$. The distance d between S and M is defined by $d = \inf\{\|u - v\| : u \in S, v \in M\}$. Prove that

$$d = \max\left\{ \left(\|z_0\|^2 - \sum_{j=1}^{n} |\langle z_0, \varphi_j \rangle|^2 \right)^{1/2} - R_0, 0 \right\}.$$

45 Let S and M be as in problem 44. Prove that the intersection of S with M is a ball in M with $\sum_{j=1}^{n} \langle z_0, \varphi_j \rangle \varphi_j$ as the center and $(R_0^2 - \|z_0\|^2 + \sum_{j=1}^{n} |\langle z_0, \varphi_j \rangle|^2)^{1/2}$ as the radius.

46. Let $\|z - z_1\| = R$ and $\|z - z_2\| = R$ be two spheres with $\|z_1\| = \|z_2\|$. Show that the intersection of the two spheres is a sphere in the subspace orthogonal to the vector $z_2 - z_1$. Find its radius and its center.

47. Prove that $|\det(a_{jk})|^2 = \prod_{j=1}^{n} \sum_{k=1}^{n} |a_{jk}|^2$ if and only if the vectors $y_1 = (a_{11}, a_{12}, \ldots, a_{1n}), y_2 = (a_{21}, a_{22}, \ldots, a_{2n}), \ldots, y_n = (a_{n1}, a_{n2}, \ldots, a_{nn})$ are orthogonal to each other.

48. Consider the incompatible systems of equations:

$$\begin{aligned}
\text{(a)} \quad 1 &= 2x_1 + 3x_2 \\
0 &= x_1 + x_2 \\
2 &= 3x_1 - x_2
\end{aligned}
\qquad
\begin{aligned}
\text{(b)} \quad 1 &= x_1 - x_2 \\
-1 &= 2x_1 + x_2 \\
-1\tfrac{7}{8} &= x_1 + 2x_2
\end{aligned}$$

Minimize the deviation between the right and left sides.

49. Let the points $(0, 0)$, $(1, 1)$, $(2, 1)$ be given. Find the polynomial $P(t)$ of degree 1 with the least squares fit to these three points.

50. Let $z_1, \ldots, z_n$ be vectors in a Hilbert space $\mathcal{H}$.

    (a) Prove that if $z_1, \ldots, z_n$ are linearly independent, then there exists an $\varepsilon > 0$ such that any vectors $y_1, \ldots, y_n$ in $\mathcal{H}$ which satisfy

    $$\|z_i - y_i\| < \varepsilon, \quad i = 1, \ldots, n,$$

    are also linearly independent. Hint: Consider the Gram determinant.

    (b) Let $z_1, \ldots, z_n$ be linearly dependent. Determine whether or not there exists a $\delta > 0$ such that if $y_1, \ldots, y_n$ satisfy $\|z_i - y_i\| < \delta$, then the system $y_1, \ldots, y_n$ is linearly dependent too.

51. Let $A = \{a_1, \ldots, a_n\}$ be a system of vectors in a Hilbert space $\mathcal{H}$. For any $y \in \mathcal{H}$, let y_A denote the projection of y into the subspace $\mathrm{sp}\{a_1, \ldots, a_n\}$. Prove that for any $\varepsilon > 0$, there exists a $\delta > 0$ such that for any system of vectors $B = \{b_1, \ldots, b_n\}$ with the property $\|a_j - b_j\| < \delta, 1 \leq j \leq n$, the inequality

$$\|y_A - y_B\| \leq \varepsilon \|y\|$$

holds for any $y \in \mathcal{H}$. Hint: Use the formula for y_A.

52. Let

$$y_1 = a_{11}\lambda_1 + \cdots + a_{1n}\lambda_n$$
$$\vdots \qquad \vdots \qquad \vdots$$
$$y_m = a_{m1}\lambda_1 + \cdots + a_{mn}\lambda_n$$

where $m > n$, be an incompatible system. Let $\alpha_A = (\alpha_1, \ldots, \alpha_n)$ be its solution as defined in Section 6. Prove that for any $\varepsilon > 0$, there exists a $\delta > 0$ such that if $z_1, \ldots, z_n$ and $\{b_{ij}\}_{\substack{i=1,\ldots,n \\ j=1,\ldots,m}}$ satisfy $\|z_k - y_k\} < \delta$,

$1 \leq k \leq m$, $\|a_{ij} - b_{ij}\| < \delta$, $1 \leq i \leq m$, $1 \leq j \leq n$, then the solution $\beta_B = (\beta_1, \ldots, \beta_n)$ of the incompatible system

$$z_1 = b_{11}\lambda_1 + \cdots + b_{1n}\lambda_n$$
$$\vdots \qquad \vdots \qquad \vdots$$
$$z_m = b_{m1}\lambda_1 + \cdots + b_{mn}\lambda_n$$

satisfies

$$\left(\sum_{i=1}^{n} |\alpha_i - \beta_i|^2 \right)^{1/2} < \varepsilon.$$

53. Let $t_1, \ldots, t_k$ be k points in $[0, 1]$ and let $y_1, \ldots, y_k$ be in $\mathbb{C}$. Let $P(t)$ be their least square fit polynomial of degree $n < k$. Prove that for any $\varepsilon > 0$, there exists a $\delta > 0$ such that if $s_1, \ldots, s_k$ in $[0, 1]$ and $z_1, \ldots, z_n \in \mathbb{C}$ satisfy $|s_i - t_i| < \delta$, $i = 1, \ldots, k$; $|z_i - y_i| < \delta$, $i = 1, \ldots, k$, then their least square fit polynomial $Q(t)$ (degree $< k$) satisfies

$$\left[\int_0^1 |P(t) - Q(t)|^2 dt \right]^{1/2} < \varepsilon.$$

54. Let $L_0 = \{\varphi \in L_2[-a, a] \mid \varphi(t) = -\varphi(-t) \text{ a.e.}\}$
    $L_E = \{\varphi \in L_2[-a, a] \mid \varphi(t) = \varphi(-t) \text{ a.e.}\}$.

    (a) Show that both sets are closed infinite dimensional subspaces of $L_2[-a, a]$.
    (b) Show that L_0 and L_E are orthogonal.
    (c) Show that L_E is the orthogonal complement of L_0.
    (d) For $f \in L_2[-a, a]$, find its projections into L_0 and L_E.
    (e) Find the distances from $f(t) = t^2 + t$ to L_0 and to L_E. Find the distances from any $f \in L_2[-a, a]$ to L_0 and L_E.

55. Let $N_1 = \{(\xi_1, \xi_1, \xi_1, \xi_1, \xi_2, \xi_2, \xi_2, \xi_2, \ldots)\}$
    $N_2 = \{(\xi_1, i\xi_1, -\xi_1, -i\xi_1, \xi_2, i\xi_2, -\xi_2, -i\xi_2, \ldots)\}$
    $N_3 = \{(\xi_1, -\xi_1, \xi_1, -\xi_1, \xi_2, -\xi_2, \xi_2, -\xi_2, \ldots)\}$
    $N_4 = \{(\xi_1, -i\xi_1, -\xi_1, i\xi_1, \xi_2, -i\xi_2, -\xi_2, i\xi_2, \ldots)\}$

    (a) Prove that they are closed infinite dimensional subspaces of ℓ_2.
    (a) Check that they are mutually orthogonal.
    (c) Show that $N_1 \oplus N_2 \oplus N_3 \oplus N_4 = \ell_2$.

56. Let $\{\varphi_1, \varphi_2, \ldots\}$ be a set of vectors and let $\{\omega_1, \omega_2, \ldots\}$ be an orthognal system. We call $\{\omega_1, \omega_2, \ldots\}$ a *backward orthognalization* of $\{\varphi_1, \varphi_2, \ldots\}$ if $\overline{sp}\{\varphi_j, \varphi_{j+1}, \ldots\} = \overline{sp}\{\omega_j, \omega_{j+1}, \ldots\}$ for $j = 1, 2, \ldots$. Prove that there exists a backward orthogonalization for $\{\varphi_1, \varphi_2, \ldots\}$ if and only if for every j we have $\varphi_j \notin \overline{sp}\{\varphi_{j+1}, \varphi_{j+2}, \ldots\}$.

57. (a) Let $\varphi_1 = (1, 2, 0, 0, \ldots)$, $\varphi_2 = (-1, 2, 0, 0, \ldots)$; $\varphi_j = e_j$ for $j = 3, 4, 5, \ldots$, where $\{e_j\}$ is the standard basis for ℓ_2. Construct a backward orthongonalization in ℓ_2 for $\{\varphi_j\}_{j \in \mathbb{N}}$.

(b) Let $x_1, \ldots, x_n$ be in $\mathbb{C}^n$. Define $(x_i)_j$ to be the jth component of x_i.
Set

$$\varphi_1 = ((x_1)_1, (x_1)_2, \ldots, (x_1)_n, 0, \ldots)$$
$$\varphi_2 = ((x_2)_1, (x_2)_2, \ldots, (x_2)_n, 0, \ldots)$$

$$\vdots \qquad \vdots \qquad \vdots \qquad \qquad \vdots$$

$$\varphi_n = ((x_n)_1, (x_n)_2, \ldots, (x_n)_n, 0, \ldots)$$
$$\varphi_j = e_j \quad \text{for} \quad j > n.$$

(i) Which conditions should be imposed on $x_1, \ldots, x_n$, so that there exists a backward orthogonalization for $\{\varphi_j\}_{j \in \mathbb{N}}$?

(ii) Construct, in this case, the backward orthogonalization.

58. Find the Fourier coefficients of the following functions:

(a) $f(t) = t$

(b) $f(t) = t^2$

(c) $\cos at$, $a \in \mathbb{R} \backslash \mathbb{Z}$ ($\mathbb{Z}$ is the set of integer)

(d) $f(t) = \begin{cases} 1; & t \geq 0 \\ -1; & t < 0 \end{cases}$

(e) $f(t) = |t|$

(f) Use the Parseval equality to prove that $\sum_{n=1}^{\infty} \frac{1}{n^2} = \frac{\pi^2}{6}$. (Hint: Consider t.)

(g) Use the Fourier expansion of $\cos \frac{1}{2} t$ to prove that $2 + 4 \sum_{n=1}^{\infty} \frac{(-1)^n}{1-4n^2} = \pi$.

59. For $f \in L_2[-\pi, \pi]$, find the projection of f into $\text{sp}\{e^{-int}, \ldots, e^{int}\}$ and find the distance from f to that subspace.

60. (a) Find the Legendre coefficients a_0, a_1, a_2 (i.e., the Fourier coefficients with respect to the normalized Legendre polynomials) of the following functions

(i) $f(t) = t^2$

(ii) $\cos \frac{\pi t}{2}$

(b) Prove that for every m-times differentiable function f, the Legendre coefficients a_k, $k \leq m$, are

$$a_k = \sqrt{\frac{2k+1}{k}} \frac{1}{2^k k!} \int_{-1}^{1} f^{(k)}(t)(1 - t^2)^k \, dt.$$

61. (a) In $L_2[-1, 1]$, find the projection of x^n into $\text{sp}\{x^{n-1}, x^{n-2}, \ldots, 1\}$.

(b) Express $x^n = \varphi(x) + \psi(x)$, where

$$\varphi(x) \in \text{sp}\{x^{n-1}, \ldots, 1\} \text{ and } \psi(x) \in \text{sp}\{x^{n-1}, \ldots, 1\}^{\perp}.$$

62. Consider the two vectors $\cos t$ and $\cos t + \sin t$ in $L_2[-\pi, \pi]$. Change the inner product on $L_2[-\pi, \pi]$ in such a way that it remains a Hilbert space and these two vectors become orthogonal.

63. Consider the vectors $(1, 2, 0, 0, \ldots)$ and $(1, 1, 1, 0, 0, \ldots)$ in ℓ_2. Change the inner product on ℓ_2 such that it remains a Hilbert space and these two vectors become orthogonal.

64. In general, given linearly independent vectors $\varphi_1, \ldots \varphi_n$ in a Hilbert space $\mathcal{H}$, change the inner product on $\mathcal{H}$ such that it remains a Hilbert space and $\varphi_1, \ldots, \varphi_n$ became orthogonal.

65. Let L be a closed subspace of a Hilbert space $\mathcal{H}$. Given $g \in L$ and $f \in \mathcal{H}$, denote by $P_L f$ the projection of f into L. Prove that $g \perp P_L f$ if and only if $g \perp f$.

66. Prove that for any two subspaces of a Hilbert space $\mathcal{H}$,

    (a) $(L_1 + L_2)^{\perp} = L_1^{\perp} \cap L_2^{\perp}$

    (b) $(L_1 \cap L_2)^{\perp} = \overline{L_1^{\perp} + L_2^{\perp}}$

67. Generalize problem 66 to the case of n subspaces.

68. Set $L_1 = \mathrm{sp}\{1, 2, 0, \ldots), (0, 1, 2, 0, \ldots), (0, 0, 1, 2, 0, \ldots), \ldots\}$
    $L_2 = \mathrm{sp}\{(1, 0, 0, \ldots)\}$.

    Prove that $L_1 + L_2$ is dense in ℓ_2.

69. Let $L_1 = \overline{\mathrm{sp}}\{1, 2, 0, \ldots), (0, 1, 2, 0, \ldots), (0, 0, 1, 2, 0, \ldots), \ldots\}$
    $L_2 = \overline{\mathrm{sp}}\{1, 3, 0, \ldots), (0, 1, 3, 0, \ldots), (0, 0, 1, 3, 0, \ldots), \ldots\}$

    be two subspaces in ℓ_2.

    (a) Prove that $\xi \in L_1 \cap L_2$ if and only if ξ is orthogonal to all vectors $(\eta_1, \eta_2, \ldots) \in \ell_2$, where $6\eta_{k+2} + 5\eta_{k+1} + \eta_k = 0$.

    (b) Prove that $L_1 \cap L_2 = \overline{\mathrm{sp}}\{(1, 5, 6, 0, \ldots), (0, 1, 5, 6, 0, \ldots), \ldots\}$.

70. Define $\ell_2(\mathbb{N} \times \mathbb{N})$ to be the set of all double sequences $\{\xi_{jk}\}$ with $\sum_{j,k=1}^{\infty} |\xi_{jk}|^2 < \infty$, and an inner product defined by

$$\langle \xi, \eta \rangle = \sum_{j,k=1}^{\infty} \xi_{jk} \overline{\eta}_{jk}.$$

    (a) Prove that it is a Hilbert space.

    (b) Let $\varphi^{(1)}, \varphi^{(2)}, \ldots$ be an orthonormal basis in ℓ_2, $\varphi^{(j)} = (x_k^{(j)})_{k=1}^{\infty}$. Set

$$\varphi^{(ij)} = (x_k^{(i)} \overline{x}_p^{(j)})_{k,p=1}^{\infty},$$

    Prove that $\{\varphi^{(ij)}\}_{i,j=1}^{\infty}$ is an orthonormal basis for $\ell_2(\mathbb{N} \times \mathbb{N})$.

71. Determine which of the following systems are orthogonal bases in ℓ_2 and which are not.

    (a) $(1, 2, 0, 0, \ldots), (0, 0, 1, 2, 0, \ldots), (0, 0, 0, 0, 1, 2, 0, \ldots), \ldots$

(b) $(1, -1, 0, 0, \ldots)$, $(1, 1, 0, 0, \ldots)$, $(0, 0, 1, -1, 0, 0, \ldots)$,

$(0, 0, 1, 1, 0, 0, \ldots)$, $\ldots$.

72. Given vectors $\varphi_1, \ldots, \varphi_n$ in a Hilbert space $\mathcal{H}$, when is it possible to find a vector $\chi_0 \in \mathcal{H}$ such that $\langle \chi_0, \varphi_1 \rangle = 1$ and $\langle \chi_0, \varphi_j \rangle = 0$ for $j > 1$? If it exists, find such a χ_0.

73. Given vectors $\varphi_1, \ldots, \varphi_n$ in a Hilbert space $\mathcal{H}$,

   (a) prove that there exist vectors $\chi_1, \ldots, \chi_n$ with $\langle \chi_j, \varphi_k \rangle = \delta_{jk}$, if and only if $\varphi_1, \ldots, \varphi_n$ are linearly independent. Such a system $\chi_1, \ldots, \chi_n$ is called a *biorthogonal system*.

   (b) Let $\widetilde{\varphi}_j$ be the projection of φ_j into $\mathrm{sp}\{\varphi_1, \ldots, \varphi_{j-1}, \varphi_{j+1}, \ldots, \varphi_n\}$. Prove that the system $\{\chi_j\}_{j=1}^n$, where $\chi_j = \frac{1}{\|\varphi_j\|^2 - \|\widetilde{\varphi}_j\|^2} (\varphi_j - \widetilde{\varphi}_j)$ is a biorthogonal system.

   (c) How many biorthogonal systems corresponding to $\varphi_1, \ldots, \varphi_n$ exist?

74. Given a system of vectors $\{\varphi_j\}_{j=1}^\infty$ in a Hilbert space $\mathcal{H}$,

   (a) prove that there exists a system $\{\chi_j\}_{j=1}^\infty$ with $\langle \chi_j, \varphi_k \rangle = \delta_{jk}$ if and only if

   $$\varphi_j \notin \overline{\mathrm{sp}}\{\varphi_i\}_{i \neq j}.$$

   (b) Prove that the formula in 73b gives a biorthogonal system in this case too.

75. Let $A_n = (a_{jk})_{j,k=0}^n$ be a positive definite $(n+1) \times (n+1)$ matrix with complex entries. Show that

   $$\langle x, y \rangle_{A_n} = \sum_{j,k=0}^n a_{jk} x_k \bar{y}_j.$$

   defines an inner product on $\mathbb{C}^{n+1}$. Denote by $\mathbb{C}_{A_n}$ the linear space of all vectors in $\mathbb{C}^{n+1}$ endowed with this inner product. Prove that $\mathbb{C}_{A_n}$ is a Hilbert space.

76. Consider the vectors $e_k = (\delta_{kj})_{j=0}^n$, $k = 0, \ldots, n$, in $\mathbb{C}_{A_n}$. Let $x_0, x_1, \ldots, x_n$ be the vectors in $\mathbb{C}_{A_n}$ obtained from $e_0, \ldots, e_n$ by the Gram-Schmidt orthogonalization procedure in the inner product $\langle ., . \rangle_{A_n}$; see the previous exercise. Show that for each k the vector x_k has the form

   $$x_k = c_k(\zeta_0^{(k)}, \ldots, \zeta_k^{(k)}, 0, 0, \ldots, 0),$$

   where $c_k \in \mathbb{C}$ and

   $$A_k \begin{pmatrix} \zeta_0^{(k)} \\ \vdots \\ \zeta_k^{(k)} \end{pmatrix} = \begin{pmatrix} \delta_{0k} \\ \vdots \\ \delta_{kk} \end{pmatrix}. \qquad (*)$$

77. Let the matrix A_n in the previous exercise be replaced by the positive definite Toeplitz matrix $T_n = (t_{j-k})_{j,k=0}^n$. Prove that the polynomial

$$q_k(z) = \zeta_0^{(k)} + \zeta_1^{(k)} z + \cdots + \zeta_k^{(k)} z^k,$$

with the coefficients given by $(*)$ in the previous exercise, has all its zeros inside the unit disc.

78. Let T_n be as in the previous exercise, and

$$T_n^{-1} = (s_{jk})_{j,k=0}^n.$$

Prove that the polynomial

$$s_{n0} + s_{n1} z + \cdots + s_{nn} z^n$$

has all its zeros inside the unit circle.

79. Let $\alpha \in \mathbb{C}$ and $|\alpha| \neq 1$. Prove that the weight

$$w(\lambda) = \alpha\lambda + (1 + |\alpha|^2) + \bar{\alpha}\lambda^{-1}$$

is strictly positive on $\mathbb{T}$, the unit circle in the complex plane, and determine the Szegö polynomials for this weight.

80. Consider the matrix

$$A_n = \begin{pmatrix} 1 & a & a^2 & \cdots & a^n \\ b & 1 & a & \cdots & a^{n-1} \\ \vdots & & & & \vdots \\ b^n & b^{n-1} & b^{n-2} & \cdots & 1 \end{pmatrix}.$$

(a) Prove that A_n is invertible if and only, if $1 - ab \neq 0$.

(b) Assume $1 - ab \neq 0$. Prove that $A^{-1} = (f_{jk})_{j,k=0}^n$ where

$$\begin{aligned}
f_{00} &= (1 - ab)^{-1}, \ f_{nn} = (1 - ab)^{-1}, \\
f_{kk} &= (1 + ab)(1 - ab)^{-1}, k = 2, 3, \ldots, n - 1, \\
f_{k-1,k} &= -(1 - ab)^{-1}a, k = 1, \ldots, n, \\
f_{k,k-1} &= -(1 - ab)^{-1}b, k = 1, \ldots, n,
\end{aligned}$$

and $f_{jk} = 0$ for the indices $|j - k| \geq 2$. In particular, A_n^{-1} is a tridiagonal matrix.

81. Fix $a \in \mathbb{C}$, $|a| < 1$. Consider on $\mathbb{T}$ the weight

$$\omega(\lambda) = \frac{-1 - |a|^2}{-1 + |a|^2 - a\lambda - \bar{a}\lambda^{-1}}.$$

(a) Prove that $w(\lambda) > 0$ for each $\lambda \in \mathbb{T}$ and that

$$w(e^{it}) = 1 + \sum_{j=1}^{\infty} a^j e^{ijt} + \sum_{j=1}^{\infty} \bar{a}^j e^{-ijt}.$$

(b) Construct the Szegö polynomials for the weight w (hint: use the previous exercise).

82. Let the weight w be given by

$$w(\lambda) = |q_n(\lambda)|^{-2},$$

where $q_n(\lambda) = \lambda^n + a_{n-1}\lambda^{n-1} + \cdots + a_0$ with $q_n(\lambda) \neq 0$ for $|\lambda| \geq 1$. Prove that up to a constant the $(n+r)$-th Szegö polynomial for w is equal to $\lambda^r q_n(\lambda)$.

83. Let $|a| < 1$, and consider the weight

$$w(\lambda) = \left| \sum_{j=0}^{\infty} a^j \lambda^{-j} \right|^2.$$

Compute up to a constant the Szegö polynomials for this weight.

Chapter II
Bounded Linear Operators on Hilbert Spaces

In this chapter, continuous linear functions defined on a Hilbert space are introduced and studied. These functions are described by infinite matrices in the same way as linear transformations on $\mathbb{C}^n$ are represented by finite matrices. In this way the chapter may be viewed as a beginning of a theory of infinite matrices. As may be expected, analysis plays a very important role.

Throughout this chapter, $\mathcal{H}$, $\mathcal{H}_1$ and $\mathcal{H}_2$ denote Hilbert spaces over the complex numbers $\mathbb{C}$.

2.1 Properties of Bounded Linear Operators

Definition: A function A which maps $\mathcal{H}_1$ into $\mathcal{H}_2$ is called a *linear operator* if for all x, y in $\mathcal{H}_1$ and $\alpha \in \mathbb{C}$,

(i) $A(x + y) = A(x) + A(y)$;

(ii) $A(\alpha x) = \alpha A(x)$.

For convenience, we write Ax instead of $A(x)$.

Taking $\alpha = 0$ in (ii) gives $A0 = 0$.

For example, with each $n \times n$ matrix of complex numbers there is a natural way to associate a linear operator A mapping $\mathbb{C}^n$ into $\mathbb{C}^n$, namely

$$A(\alpha_1, \ldots, \alpha_n) = (\beta_1, \ldots, \beta_n)$$

where

$$(a_{ij}) \begin{pmatrix} \alpha_1 \\ \vdots \\ \alpha_n \end{pmatrix} = \begin{pmatrix} \beta_1 \\ \vdots \\ \beta_n \end{pmatrix} \text{ i.e., } \beta_i = \sum_{j=1}^{n} a_{ij}\alpha_j.$$

Definition: The linear operator $A : \mathcal{H}_1 \to \mathcal{H}_2$ is called *bounded* if

$$\sup_{\|x\| \le 1} \|Ax\| < \infty.$$

The *norm* of A, written $\|A\|$, is given by

$$\|A\| = \sup_{\|x\| \le 1} \|Ax\|.$$

When there is no cause for confusion, we use the same notation for the norms and inner products on $\mathcal{H}_1$ and $\mathcal{H}_2$.

An operator A is bounded if and only if it takes the 1-ball S_1 with center zero in $\mathcal{H}_1$ into some r-ball in $\mathcal{H}_2$. The smallest ball in $\mathcal{H}_2$ which contains AS_1 has radius $\|A\|$.

The identity operator $I : \mathcal{H}_1 \to \mathcal{H}_1$ defined by $Ix = x$ is obviously a bounded linear operator of norm 1.

The operator $A : \mathbb{C}^n \to \mathbb{C}^n$ defined above is bounded; for if $x = (\alpha_1, \ldots, \alpha_n)$, then by (1.3) in Section I.1

$$\|Ax\|^2 = \sum_{i=1}^n |\beta_i|^2 \leq \left(\sum_{i,j=1}^n |a_{ij}|^2 \right) \left(\sum_{j=1}^n |\alpha_j|^2 \right) = \|x\|^2 \sum_{i,j=1}^n |a_{ij}|^2,$$

from which it follows that $\|A\| \leq (\sum_{i,j=1}^n |a_{ij}|^2)^{1/2}$.

It is easy to verify the next three properties of $\|A\|$.

(a) $\|A\| = \sup_{x \neq 0} \frac{\|Ax\|}{\|x\|} = \sup_{\|x\|=1} \|Ax\|$.

(b) $\|A\| = \sup_{\|x\|=\|y\|=1} |\langle Ax, y \rangle| = \sup_{\|x\|\leq 1, \|y\|\leq 1} |\langle Ax, y \rangle|$.   (1.1)

(c) If $\|Ax\| \leq C\|x\|$ for all $x \in \mathcal{H}_1$, then A is bounded and $\|A\| \leq C$.

The set of bounded linear operators whick map $\mathcal{H}_1$ into $\mathcal{H}_2$ is denoted by $\mathcal{L}(\mathcal{H}_1, \mathcal{H}_2)$. If $\mathcal{H}_1 = \mathcal{H}_2$, we write $\mathcal{L}(\mathcal{H}_1)$ instead of $\mathcal{L}(\mathcal{H}_1, \mathcal{H}_1)$.

If A and B are in $\mathcal{L}(\mathcal{H}_1, \mathcal{H}_2)$, it is a simple matter to verify (i)–(v).

(i) $\alpha A + \beta B \in \mathcal{L}(\mathcal{H}_1, \mathcal{H}_2)$; $\alpha, \beta \in \mathbb{C}$.

(ii) $\|\alpha A\| = |\alpha| \|A\|$; $\alpha \in \mathbb{C}$.

(iii) $\|A + B\| \leq \|A\| + \|B\|$.

(iv) $\|A - B\| \geq |\|A\| - \|B\||$.

(v) If $C \in \mathcal{L}(\mathcal{H}_2, \mathcal{H}_3)$, where $\mathcal{H}_3$ is a Hilbert space, then defining CA by $CAx = C(Ax)$, it follows that CA is in $\mathcal{L}(\mathcal{H}_1, \mathcal{H}_3)$ and $\|CA\| \leq \|C\| \|A\|$.

2.2 Examples of Bounded Linear Operators with Estimates of Norms

While it is often easy to verify that a linear operator is bounded, it may be very difficult to determine its norm. A few examples are discussed in this section.

1. Every linear operator which maps a finite dimensional Hilbert space into a Hilbert space is bounded. Indeed, suppose A is a linear map from $\mathcal{H}_1$ into $\mathcal{H}_2$, and dim $\mathcal{H}_1 < \infty$. Let $\varphi_1, \ldots, \varphi_n$ be an orthonormal basis for $\mathcal{H}_1$. Then for any $x \in \mathcal{H}_1$,

$$x = \sum_{i=1}^n \langle x, \varphi_k \rangle \varphi_k \quad \text{and} \quad Ax = \sum_{k=1}^n \langle x, \varphi_k \rangle A\varphi_k.$$

Hence

$$\|Ax\| \le \sum_{k=1}^{n} |\langle x, \varphi_k \rangle| \|A\varphi_k\| \le \left(\sum_{k=1}^{n} |\langle x, \varphi_k \rangle|^2 \right)^{1/2} \left(\sum_{k=1}^{n} \|A\varphi_k\|^2 \right)^{1/2}$$

$$= \|x\| \left(\sum_{k=1}^{n} \|A\varphi_k\|^2 \right)^{1/2},$$

which implies that

$$\|A\| \le \left(\sum_{k=1}^{n} \|A\varphi_k\|^2 \right)^{1/2}.$$

Suppose, in addition, that $\mathcal{H}_1 = \mathcal{H}_2$ and $A\varphi_k = \lambda_k \varphi_k$, $1 \le k \le n$. Then

$$\|Ax\|^2 = \left\langle \sum_{k=1}^{n} \langle x, \varphi_k \rangle A\varphi_k, \sum_{k=1}^{n} \langle x, \varphi_k \rangle A\varphi_k \right\rangle$$

$$= \sum_{k=1}^{n} |\lambda_k|^2 |\langle x, \varphi_k \rangle|^2 \le M^2 \|x\|^2,$$

where $M = \max_k |\lambda_k|$. Hence $\|A\| \le M$. On the other hand, suppose $M = |\lambda_j|$. Since $\|\varphi_j\| = 1$,

$$\|A\| \ge \|A\varphi_j\| = |\lambda_j| = M.$$

Thus

$$\|A\| = \max_k |\lambda_k|.$$

2. Let $\{\varphi_1, \varphi_2, \ldots\}$ be an orthonormal basis for the Hilbert space $\mathcal{H}$ and let $\{\lambda_1, \lambda_2, \ldots\}$ be a bounded sequence of complex numbers. For each $x \in \mathcal{H}$, define

$$Ax = \sum_k \lambda_k \langle x, \varphi_k \rangle \varphi_k.$$

A is a linear operator on $\mathcal{H}$ and by Bessel's inequality,

$$\|Ax\|^2 = \sum_k |\lambda_k|^2 |\langle x, \varphi_k \rangle|^2 \le m^2 \|x\|^2,$$

where $m = \sup_k |\lambda_k|$. Thus $\|A\| \le m$. On the other hand, given $\varepsilon > 0$, there exists a λ_j such that $|\lambda_j| > m - \varepsilon$. Hence

$$\|A\| \ge \|A\varphi_j\| = \|\lambda_j \varphi_j\| = |\lambda_j| > m - \varepsilon.$$

Since ε was arbitrary, $\|A\| \ge m$. Therefore $\|A\| = \sup_k |\lambda_k|$.

We shall see in Chapter IV that there is a large class of operators A of the form given above.

3. Certain infinite matrices give rise to bounded linear operators on ℓ_2 as follows: Given an infinite matrix $(a_{ij})_{i,j=1}^{\infty}$, where

$$\sum_{i=1}^{\infty}\sum_{j=1}^{\infty}|a_{ij}|^2 < \infty, \qquad (2.1)$$

define $A: \ell_2 \to \ell_2$ by

$$A(\alpha_1, \alpha_2, \ldots) = (\beta_1, \beta_2, \ldots),$$

where

$$(a_{ij})\begin{pmatrix}\alpha_1\\\alpha_2\\\vdots\end{pmatrix} = \begin{pmatrix}\beta_1\\\beta_2\\\vdots\end{pmatrix}; \text{ i.e., } \beta_i = \sum_{j=1}^{\infty}a_{ij}\,\alpha_j.$$

The operator A is a bounded linear operator on ℓ_2 and

$$\|A\|^2 \le \sum_{i=1}^{\infty}\sum_{j=1}^{\infty}|a_{ij}|^2$$

since

$$|\beta_i| \le \sum_{j=1}^{\infty}|a_{ij}\,\alpha_j| \le \left(\sum_{j=1}^{\infty}|a_{ij}|^2\right)^{1/2}\left(\sum_{j=1}^{\infty}|\alpha_j|^2\right)^{1/2}$$

implies that for $x = (\alpha_1, \alpha_2, \ldots)$,

$$\|Ax\|^2 = \sum_{i=1}^{\infty}|\beta_i|^2 \le \|x\|^2\sum_{i=1}^{\infty}\sum_{j=1}^{\infty}|a_{ij}|^2.$$

Condition (2.1) is not a necessary condition for A to be bounded since the identity matrix $(a_{ij}) = (\delta_{ij})$ does not satisfy (2.1), yet $A = I$.

No conditions on the matrix entries a_{ij} have been found which are necessary and sufficient for A to be bounded, nor has $\|A\|$ been determined in the general case.

4. Let $\mathcal{H} = L_2([a, b])$ and let $a(t)$ be a complex valued function which is continuous on $[a, b]$. Define $A : \mathcal{H} \to \mathcal{H}$ by

$$(Af)(t) = a(t)f(t).$$

A is linear and for $M = \max_{a \leq t \leq b} |a(t)|$,

$$\|Af\|^2 = \int_a^b |a(t)f(t)|^2 dt \leq M^2 \|f\|^2.$$

Thus $\|A\| \leq M$. To show that $\|A\| = M$, suppose $M = |a(t_0)|$. Define a sequence $\{\varphi_n\}$ in $\mathcal{H}$ by

$$\varphi_n(t) = \begin{cases} \sqrt{\frac{n}{2}}; & t \in [t_0 - \frac{1}{n}, t_0 + \frac{1}{n}] \\ 0; & \text{otherwise.} \end{cases}$$

Now

$$\|\varphi_n\|^2 = \int_{t_0 - \frac{1}{n}}^{t_0 + \frac{1}{n}} \frac{n}{2} dt = 1,$$

and from the continuity of $a(t)$ at t_0, it follows that

$$\|A\|^2 \geq \|A\varphi_n\|^2 = \frac{n}{2} \int_{t_0 - \frac{1}{n}}^{t_0 + \frac{1}{n}} |a(t)|^2 dt \rightarrow |a(t_0)|^2.$$

Hence

$$\|A\| \geq |a(t_0)| = M.$$

5. Let k be a complex valued Lebesgue measurable function on $[a, b] \times [a, b]$ such that

$$\int_a^b \int_a^b |k(t, s)|^2 dsdt < \infty.$$

Define the operator $K : L_2[a, b] \rightarrow L_2([a, b])$ by

$$(Kf)(t) = \int_a^b k(t, s)f(s)ds.$$

The integral exists since for almost every t, $k(t, s)f(s)$ is Lebesgue measurable on $[a, b]$ and by Schwarz's inequality,

$$\int_a^b |k(t, s)f(s)|ds \leq \left(\int_a^b |k(t, s)|^2 ds \right)^{1/2} \left(\int_a^b |f(s)|^2 ds \right)^{1/2}.$$

Thus

$$\|Kf\|^2 \leq \int_a^b \left(\int_a^b |k(t, s)f(s)|ds \right)^2 dt \leq \|f\|^2 \int_a^b \int_a^b |k(t, s)|^2 dsdt$$

Hence

$$\|K\|^2 \leq \int_a^b \int_a^b |k(t, s)|^2 dsdt.$$

Clearly, K is linear.

The operator K is called an *integral operator* and $k(t, s)$ is called the *kernel function* corresponding to K.

6. Let $S_r: \ell_2 \to \ell_2$ be defined by

$$S_r(\alpha_1, \alpha_2, \ldots) = (0, \alpha_1, \alpha_2, \ldots).$$

S_r is called the *forward shift* operator. Obviously, S_r is linear and $\|S_r x\| = \|x\|, x \in \ell_2$. In particular, $\|S_r\| =]$

7. Let $S_\ell: \ell_2 \to \ell_2$ be defined by

$$S_\ell(\alpha_1, \alpha_2, \ldots) = (\alpha_2, \alpha_3, \ldots).$$

S_ℓ is called the *backward shift* operator. Obviously, S_ℓ is linear and $\|S_\ell\| = 1$.

Some authors call a forward shift a right shift and a backward shift a left shift.

8. Let $V : \ell_2(\mathbb{Z}) \to \ell_2(\mathbb{Z})$ be defined by

$$V(\ldots, \xi_{-1}, \boxed{\xi_0}, \xi_{-1}, \ldots) = (\ldots, \xi_{-2}, \boxed{\xi_{-1}}, \xi_0, \ldots).$$

Here the box "indicates the zero position in the sequence. The operator V is called the *bilateral (forward) shift* on $\ell_2(\mathbb{Z})$. Notice that $\|V x\| = \|x\|$ for each x in $\ell_2(\mathbb{Z})$, and V maps $\ell_2(\mathbb{Z})$ in a one-one way onto $\ell_2(\mathbb{Z})$. In fact, the inverse map is given by

$$V^{-1}(\ldots, x_{-1}, \boxed{x_0}, x_1, \ldots) = (\ldots, x_0, \boxed{x_1}, x_2, \ldots),$$

which is referred to as the *bilateral backward shift* on $\ell_2(\mathbb{Z})$.

Linear operators which stem from differential equations are neither defined on all of $L_2([a, b])$ nor are bounded on those subspaces of $L_2([a, b])$ on which they are defined. For example, the operator $Af = \frac{d}{dx}f$ is not defined for all $f \in L_2([-\pi, \pi])$. If we want A to map a subspace $\mathcal{D}(A)$ of $L_2([-\pi, \pi])$ into $L_2([-\pi, \pi])$, we could take

$$\mathcal{D}(A) = \left\{ f \in L_2([-\pi, \pi]) : \frac{d}{dx}f \in L_2([-\pi, \pi]) \right\}, Af = \frac{d}{dx}f.$$

However, A is unbounded on $\mathcal{D}(A)$; for if $\varphi_n(x) = \frac{1}{\sqrt{2\pi}}e^{inx}, n = 1, 2, \ldots$, then $\|\varphi_n\| = 1$ but $\|A\varphi_n\| = n\|\varphi_n\| = n$.

Fortunately, there is a well developed theory of unbounded linear operators which enables one to treat large classes of differential operators. The reader is referred to [DS2], [G] and [GGK1] for an extensive treatment of the subject.

2.3 Continuity of a Linear Operator

Whereas a bounded real valued function can be discontinuous at every point, the situation for bounded linear operators is vastly different as the next theorem shows.

An operator A mapping $\mathcal{H}_1$ into $\mathcal{H}_2$ is said to be *continuous* at $x_0 \in \mathcal{H}_1$ if $x_n \to x_0$ implies $Ax_n \to Ax_0$.

Thus A is continuous at x_0 if and only if for every $\varepsilon > 0$, there exists a $\delta > 0$ such that $\|x - x_0\| < \delta$ implies $\|Ax - Ax_0\| < \varepsilon$.

Theorem 3.1 *Let A be a linear operator which maps $\mathcal{H}_1$ into $\mathcal{H}_2$. The following statements are equivalent.*

   (i) *A is continuous at a point.*

  (ii) *A is uniformly continuous on $\mathcal{H}_1$.*

 (iii) *A is bounded.*

Proof: (i) implies (ii). Suppose A is continuous at x_0. Given $\varepsilon > 0$, there exists a $\delta > 0$ such that $\|x - x_0\| < \delta$ implies $\|Ax - Ax_0\| < \varepsilon$. Thus if $\|w - z\| < \delta$, w, z in $\mathcal{H}_1$, then $\|x_0 - (x_0 + z - w)\| < \delta$. Therefore,

$$\|Aw - Az\| = \|Ax_0 - A(x_0 + z - w)\| < \varepsilon.$$

(ii) implies (iii). There exists a $\delta > 0$ such that

$$\|Az\| = \|Az - A0\| \le 1, \|z\| \le \delta. \tag{3.1}$$

Suppose $\|x\| = 1$, $x \in \mathcal{H}_1$. Then $\|\delta x\| = \delta$ and by (3.1),

$$1 \ge \|A(\delta x)\| = \delta \|Ax\|,$$

or $\|Ax\| \le \frac{1}{\delta}$. Hence $\|A\| \le \frac{1}{\delta}$.

(iii) implies (i). Given $\varepsilon > 0$ and $x_0 \in \mathcal{H}_1$, if $\|x - x_0\| < \frac{\varepsilon}{1+\|A\|}$, then

$$\|Ax - Ax_0\| \le \|A\| \|x - x_0\| < \varepsilon.$$

$\square$

2.4 Matrix Representations of Bounded Linear Operators

We have seen in Section 2 Example 3 how certain matrices give rise to bounded linear operators. In this section it is shown how to associate a matrix with a given bounded linear operator on a separable Hilbert space.

Suppose $A \in \mathcal{L}(H)$, where $\mathcal{H}$ is a Hilbert space with orthonormal basis φ_1, $\varphi_2, \ldots$. Then for $x \in \mathcal{H}$, $x = \sum_j \langle x, \varphi_j \rangle \varphi_j$. It follows from the linearity and continuity of A, applied to the sequence of partial sums of this series, that

$$Ax = \sum_j \langle x, \varphi_j \rangle A\varphi_j. \tag{4.1}$$

Now

$$A\varphi_j = \sum_k \langle A\varphi_j, \varphi_k \rangle \varphi_k. \tag{4.2}$$

Combining (4.1) and (4.2) gives

$$Ax = \sum_j \sum_k \langle x, \varphi_j \rangle \langle A\varphi_j, \varphi_k \rangle \varphi_k = \sum_k \sum_j \langle x, \varphi_j \rangle \langle A\varphi_j, \varphi_k \rangle \varphi_k. \tag{4.3}$$

Thus if $Ax = y$, then (4.3) implies

$$\begin{pmatrix} \langle A\varphi_1, \varphi_1 \rangle & \langle A\varphi_2, \varphi_1 \rangle & \cdots \\ \langle A\varphi_1, \varphi_2 \rangle & \langle A\varphi_2, \varphi_2 \rangle & \cdots \\ \vdots & \vdots & \end{pmatrix} \begin{pmatrix} \langle x, \varphi_1 \rangle \\ \langle x, \varphi_2 \rangle \\ \vdots \end{pmatrix} = \begin{pmatrix} \langle y, \varphi_1 \rangle \\ \langle y, \varphi_2 \rangle \\ \vdots \end{pmatrix}. \tag{4.4}$$

This matrix equation leads to the following definition.

Definition: Let $\varphi_1, \varphi_2, \ldots$ be an orthonormal basis for a Hilbert space $\mathcal{H}$ and let A be in $\mathcal{L}(\mathcal{H})$. The matrix (a_{ij}) corresponding to A and $\varphi_1, \varphi_2, \ldots$ is defined by $a_{ij} = \langle A\varphi_j, \varphi_i \rangle$.

Examples: 1. Let $S_r \in \mathcal{L}(\ell_2)$ be the forward shift given by

$$S_r(\alpha_1, \alpha_2, \ldots) = (0, \alpha_1, \alpha_2, \ldots).$$

If $e_1, e_2, \ldots$ is the standard orthonormal basis for ℓ_2, then

$$\langle S_r e_j, e_i \rangle = \langle e_{j+1}, e_i \rangle = \delta_{j+1,i}.$$

Thus the matrix corresponding to S_r and $e_1, e_2, \ldots$ is the lower diagonal matrix

$$\begin{pmatrix} 0 & 0 & 0 & \cdots \\ 1 & 0 & 0 & \cdots \\ 0 & 1 & 0 & \cdots \\ \cdots & \cdots & \cdots & \cdots \end{pmatrix}$$

2. Let $\mathcal{H} = L_2([-\pi, \pi])$ and let $a(t)$ be a bounded complex valued Lebesgue measurable function on $[-\pi, \pi]$.
    Define $A \in \mathcal{L}(\mathcal{H})$ by

$$(Af)(t) = a(t)f(t).$$

The "doubly infinite" matrix $\{a_{jk}\}_{j,k=-\infty}^{\infty}$ corresponding to A and the orthonormal basis $\varphi_n(t) = \frac{1}{\sqrt{2\pi}} e^{int}$, $n = 0, \pm 1, \pm 2, \ldots$ is obtained as follows. Let

$$a_n = \frac{1}{2\pi} \int_{-\pi}^{\pi} a(t)e^{-int}\, dt.$$

Then

$$a_{jk} = \langle A\varphi_k, \varphi_j \rangle = \frac{1}{2\pi} \int_{-\pi}^{\pi} a(t) e^{i(k-j)t} dt = a_{j-k}.$$

Thus the matrix is

$$\begin{pmatrix} \ddots & & \ddots & & \ddots & & \\ \cdots & a_1 & a_0 & a_{-1} & & \cdots & \\ \cdots & a_1 & \boxed{a_0} & a_{-1} & & \cdots & \\ \cdots & & a_1 & a_0 & a_{-1} & \cdots & \\ & & & \ddots & & \ddots & & \ddots \end{pmatrix}$$

which is called *a Toeplitz matrix*. The entry $\boxed{a_0}$ is located in row zero column zero.

3. Let K be the integral oprator on $L_2([a,b])$ with kernel function $k \in L_2([a,b] \times [a,b])$ as in Section 2, Example 5. Given an orthonormal basis $\{\varphi_j\}$ for $L_2([a,b])$, it follows from the remark at the end of the proof of Theorem I.15.1 that $\Phi_{ij}(s,t) = \varphi_i(s)\varphi_j(t)$, $1 \le i, j$, forms an orthonormal basis for $L_2([a,b] \times [a,b])$. Now the matrix (a_{ij}) corresponding to K and $\{\varphi_i\}$ is given by

$$a_{ij} = \langle K\varphi_j, \varphi_i \rangle = \int_a^b \int_a^b k(t,s)\varphi_j(s)\bar{\varphi}_i(t) ds\, dt = \langle k, \Phi_{ji} \rangle.$$

Thus we see that each a_{ij} is a Fourier coefficient of k with respect to $\{\Phi_{ij}\}$. Consequently, Parseval's equality gives

$$\sum_{i,j} |a_{ij}|^2 = \sum_{i,j} |\langle k, \Phi_{ij} \rangle|^2 = \|k\|^2.$$

So far we started with a bounded linear operator and considered its matrix representation relative to an orthonormal basis. In some problems the reverse direction is important: an infinite matrix is given and one looks for a bounded linear operator for which the given matrix is the matrix of the operator relative to some orthonormal basis. In the sequel we shall use the following definition.

Let $(a_{jk})_{j,k=1}^{\infty}$ be an infinite matrix. We say that this matrix *induces a bounded linear operator* A on ℓ_2 if $(a_{jk})_{j,k=1}^{\infty}$ is the matrix of A with respect to the standard orthonormal basis $e_1, e_2, \ldots$ of ℓ_2, that is,

$$a_{jk} = \langle Ae_k, e_j \rangle, \; j, k = 1, 2, \ldots.$$

In this case we simply write

$$A = \begin{bmatrix} a_{11} & a_{12} & a_{13} & \cdots \\ a_{21} & a_{22} & a_{23} & \cdots \\ a_{31} & a_{32} & a_{33} & \\ \vdots & & & \ddots \end{bmatrix} \tag{4.5}$$

As a first example consider a diagonal matrix

$$\text{diag}(w_1, w_2, w_3, \ldots) = (\delta_{jk} w_k)_{j,k=1}^{\infty}.$$

Here δ_{jk} is the Kroneclan delta and the diagonal entries are $w_1, w_2, w_3, \ldots$. This matrix induces a bounded linear operator $\mathcal{D}$ on ℓ_2 if and only if $\sup_{j \geq 1} |w_j| < \infty$, and in this case

$$\|\mathcal{D}\| = \sup_{j \geq 1} |w_j|. \tag{4.6}$$

Indeed, if the matrix $\text{diag}(w_j)_{j=1}^{\infty}$ induces a bounded linear operator $\mathcal{D}$ on ℓ_2, then

$$|w_j| = |\langle \mathcal{D}e_j, e_j \rangle| \leq \|\mathcal{D}\|. \quad j = 1, 2, \ldots,$$

and hence $\sup_{j \geq 1} |w_j| \leq \|\mathcal{D}\| < \infty$. Conversely, if $\sup_{j \geq 1} |w_j| < \infty$, then

$$\mathcal{D}(x_1, x_2, \ldots) = (w_1 x_1, w_2 x_2, \ldots)$$

is a bounded linear operator on ℓ_2 and the matrix of $\mathcal{D}$ with respect to the standard orthonormal basis of ℓ_2 is the diagonal matrix $\text{diag}(w_j)_{j=1}^{\infty}$. Moreover,

$$\|\mathcal{D}x\| = \left(\sum_{j=1}^{\infty} |w_j x_j|^2 \right)^{1/2} \leq \left(\sup_{j \geq 1} |\omega_j| \right) \cdot \|x\|,$$

and hence (4.6) holds.

Another example appears in parts of Section II.2. Indeed, if

$$\sum_{i=1}^{\infty} \sum_{j=1}^{\infty} |a_{ij}|^2 < \infty,$$

then the matrix with respect to the standard orthonormal basis of ℓ_2 of the operator A defined in Part 3 of Section II.2 is precisely equal to $(a_{ij})_{ij=1}^{\infty}$.

2.5 Bounded Linear Functionals

Definition: A function f which maps a Hilbert space $\mathcal{H}$ into $\mathbb{C}$ is called a *functional*. If f is in $\mathcal{L}(\mathcal{H}, \mathbb{C})$, we refer to f as a *bounded linear functional* on $\mathcal{H}$.

The inner product on $\mathcal{H}$ gives rise to bounded linear functionals on $\mathcal{H}$ as follows. For each $y \in \mathcal{H}$, the functional f_y defined on $\mathcal{H}$ by $f_y(x) = \langle x, y \rangle$ is linear. Moreover,

$$|f_y(x)| = |\langle x, y \rangle| \leq \|x\| \|y\|.$$

Thus f_y is bounded and $\|f_y\| \le \|y\|$. Since

$$\|f_y\|\|y\| \ge |f_y(y)| = \langle y, y \rangle = \|y\|^2,$$

we have $\|f_y\| \ge \|y\|$ and therefore

$$\|f_y\| = \|y\|, \, y \in \mathcal{H}. \tag{5.1}$$

In this section we prove the very useful result that every bounded linear functional on $\mathcal{H}$ is an f_y.

The motivation for the proof is based on the following observations.

Suppose f is a bounded linear functional on $\mathcal{H}$ and suppose there exists a $y \in \mathcal{H}$ such that $f(x) = \langle x, y \rangle$ for all $x \in \mathcal{H}$. In order to find this y, let Ker $f = \{x : f(x) = 0\}$. Then $0 = f(x) = \langle x, y \rangle$ for all $x \in$ Ker f, i.e., $y \perp$ Ker f. It follows readily from the assumption that f is bounded and linear that Ker f is a closed subspace of $\mathcal{H}$. The first step in finding y is to choose $v \ne 0$ which is orthogonal to Ker f (assuming $f \ne 0$). The existence of such a v is assured by Theorem I.8.2. But for any $\alpha \in \mathbb{C}$, $y = \alpha v$ is also orthogonal to Ker f. In order to determine which α to choose, we note that

$$\alpha f(v) = f(\alpha v) = f(y) = \langle y, y \rangle = \alpha \bar{\alpha} \|v\|^2.$$

Thus we choose $\alpha = \overline{\frac{f(v)}{\|v\|^2}}$, and $y = \alpha v$ is our candidate

Lemma 5.1 *If f is a linear functional on $\mathcal{H}$ and $f(x_0) \ne 0$ for some $x_0 \in \mathcal{H}$, then every $x \in \mathcal{H}$ has the form*

$$x = \beta x_0 + z, \, \beta \in \mathbb{C}, \, z \in \text{Ker} f.$$

Proof: Take $\beta = \frac{f(x)}{f(x_0)}$ and $z = x - \beta x_0$.   □

Riesz Representation Theorem 5.2 If f is a bounded linear functional on a Hilbert space $\mathcal{H}$, then there exists a unique $y \in \mathcal{H}$ such that for all $x \in \mathcal{H}$,

$$f(x) = \langle x, y \rangle.$$

Moreover, $\|f\| = \|y\|$.

Proof: If $f = 0$, take $y = 0$. Suppose $f \ne 0$. Then Ker f is a proper closed subspace of $\mathcal{H}$. Hence there exists a $v \ne 0$ in $(\text{Ker} f)^\perp$. Let $y = \alpha v$, where $\alpha = \overline{\frac{f(v)}{\|v\|^2}}$. Then

   (i) $y \perp$ Ker f.

   (ii) $f(y) = \langle y, y \rangle$.

Given $x \in \mathcal{H}$, we know from Lemma 5.1 that there exists a $\beta \in \mathbb{C}$ and a $z \in \text{Ker}$ f such that $x = \beta y + z$. From (i) and (ii) we get,

$$f(x) = f(\beta y) = \beta f(y) = \beta \langle y, y \rangle = \langle \beta y + z, y \rangle = \langle x, y \rangle.$$

To show that y is unique, suppose there exists a $w \in \mathcal{H}$ such that $f(x) = \langle x, w \rangle$ for all $x \in \mathcal{H}$. Then

$$0 = f(x) - f(x) = \langle x, y - w \rangle$$

for all $x \in \mathcal{H}$. In particular, $\langle y - w, y - w \rangle = 0$. Hence $y = w$. Equation (5.1) established that $\| f \| = \| y \|$. $\square$

If $\varphi_1, \varphi_2, \ldots$ is an orthonormal basis for $\mathcal{H}$, then the vector y corresponding to the functional f in the Riesz representation theorem is given by

$$y = \sum_k \overline{f(\varphi_k)} \varphi_k.$$

Indeed, since $f(\varphi_k) = \langle \varphi_k, y \rangle$,

$$y = \sum_k \langle y, \varphi_k \rangle \varphi_k = \sum_k \overline{f(\varphi_k)} \varphi_k.$$

Examples: 1. A functional F on $L_2([a, b])$ is bounded and linear if and only if there exists a $g \in L_2([a, b])$ such that

$$F(f) = \int_a^b f(t) \bar{g}(t) dt$$

for all $f \in L_2([a, b])$. In this case, $\| F \| = \| g \|$.

2. A functional f on ℓ_2 is bounded and linear if and only if there exists a $y = (\beta_1, \beta_2, \ldots) \in \ell_2$ such that for all $x = (\alpha_1, \alpha_2, \ldots) \in \ell_2$,

$$f(x) = \sum_{k=1}^{\infty} \alpha_k \bar{\beta}_k.$$

3. Let $\{\lambda_k\}$ be a bounded sequence of positive numbers. For $x = (\alpha_1, \alpha_2, \ldots)$ and $y = (\beta_1, \beta_2, \ldots)$, define a "weighted" inner product $\langle \, , \, \rangle_w$ by

$$\langle x, y \rangle_w = \sum_{k=1}^{\infty} \lambda_k \alpha_k \bar{\beta}_k.$$

It is clear that $\langle \, , \, \rangle_w$ is an inner product on ℓ_2. Furthermore, ℓ_2 is complete with respect to $\langle \, , \, \rangle_w$ if $\inf \{\lambda_k\} > 0$. The proof is essentially the same as the proof of the completeness of ℓ_2. Hence a functional f is bounded and linear on $(\ell_2, \langle \, , \, \rangle_w)$ if and only if there exists a $y = (\beta_1, \beta_2, \ldots) \in \ell_2$ such that

$$f(x) = \langle x, y \rangle_w = \sum_k \lambda_k \alpha_k \bar{\beta}_k$$

for all $x = (\alpha_1, \alpha_2, \ldots) \in \ell_2$. Also, $\| f \| = \sum_{k=1}^{\infty} \lambda_k |\beta_k|^2$.

2.6 Operators of Finite Rank

Definition: The *image* or *range* of $A \in \mathcal{L}(\mathcal{H}_1, \mathcal{H}_2)$, written Im A, is the subspace $A\mathcal{H}_1 = \{Ax : x \in \mathcal{H}_1\}$.

If Im A is finite dimensional, A is called an operator of *finite rank* and dim Im A is the *rank* of A.

Let $\{v_1, \ldots, v_n\}$ and $\{w_1, \ldots, w_n\}$ be two systems of vectors in $\mathcal{H}_1$ and $\mathcal{H}_2$, respectively. Define $K\colon \mathcal{H}_1 \to \mathcal{H}_2$ by

$$Kx = \sum_{i=1}^{n} \langle x, v_i \rangle w_i. \qquad (6.1)$$

K is linear and Im $K \subset$ sp $\{w_1, \ldots, w_n\}$. Thus the rank of K is at most n. Moreover, $\|K\| \leq \sum_{i=1}^{n} \|v_i\| \|w_i\|$ since

$$\|Kx\| \leq \sum_{n=1}^{n} \|\langle x, v_i \rangle w_i\| \leq \|x\| \sum_{i=1}^{n} \|v_i\| \|w_i\|.$$

It turns out that every $K \in \mathcal{L}(\mathcal{H}_1, \mathcal{H}_2)$ of finite rank has the form given in (6.1).

Theorem 6.1 *Suppose $K \in \mathcal{L}(\mathcal{H}_1, \mathcal{H}_2)$ is of rank n. There exist vectors $v_1, \ldots, v_n$ in $\mathcal{H}_1$ and vectors $\varphi_1, \ldots, \varphi_n$ in $\mathcal{H}_2$ such that for every $x \in \mathcal{H}_1$,*

$$Kx = \sum_{i=1}^{n} \langle x, v_i \rangle \varphi_i.$$

The vectors $\varphi_1, \ldots, \varphi_n$ may be chosen to be any orthonormal basis for Im K.

Proof: Let $\varphi_1, \ldots, \varphi_n$ be an orthonormal basis for Im K. Then for every $x \in \mathcal{H}_1$,

$$Kx = \sum_{i=1}^{n} \langle Kx, \varphi_i \rangle \varphi_i. \qquad (6.2)$$

Now for each i, $f_i(x) = \langle Kx, \varphi_i \rangle$ is a bounded linear functional on $\mathcal{H}_1$. Therefore the Riesz representation theorem guarantees the existence of a $v_i \in \mathcal{H}_1$ such that for all $x \in \mathcal{H}_1$,

$$\langle Kx, \varphi_i \rangle = f_i(x) = \langle x, v_i \rangle, \quad 1 \leq i \leq n.$$

Thus from Equation (6.2) we obtain

$$Kx = \sum_{i=1}^{n} \langle x, v_i \rangle \varphi_i.$$

It is clear that the representation of K is not unique. $\qquad \square$

Examples: 1. Let $\mathcal{H} = L_2([a, b])$. The general form of an operator $K \in \mathcal{L}(\mathcal{H})$ of finite rank is given by

$$(Kf)(t) = \sum_{j=1}^{n} w_j(t) \int_a^b f(s)\bar{v}_j(s)ds,$$

where $\{v_1, \ldots, v_n\}$ and $\{w_1, \ldots, w_n\}$ are arbitrary systems of functions in $L_2([a, b])$.

2. Let $K \in \mathcal{L}(\ell_2)$ be of rank 1. There exist $v = (\beta_1, \beta_2, \ldots)$ and $w = (\alpha_1, \alpha_2, \ldots)$ in ℓ_2 such that for all $x = (\xi_1, \xi_2, \ldots) \in \ell_2$,

$$Kx = \langle x, v \rangle w = (c\alpha_1, c\alpha_2, \ldots),$$

where $c = \sum_{i=1}^{\infty} \xi_i \bar{\beta}_i$. The vector Kx, considered as a column matrix, can be written

$$Kx = \begin{pmatrix} \alpha_1\bar{\beta}_1 & \alpha_1\bar{\beta}_2 & \cdots \\ \alpha_2\bar{\beta}_1 & \alpha_2\bar{\beta}_2 & \cdots \\ \vdots & \vdots & \end{pmatrix} \begin{pmatrix} \xi_1 \\ \xi_2 \\ \vdots \end{pmatrix}.$$

2.7 Invertible Operators

One of the fundamental problems in operator theory is to solve, if possible, the equation $Ax = y$, where A is a given linear operator and y is some vector. We shall now consider this problem for some simple, but important, operators.

Definition: An operator $A \in \mathcal{L}(\mathcal{H}_1, \mathcal{H}_2)$ is called *invertible if* there exists an operator $A^{-1} \in \mathcal{L}(\mathcal{H}_2, \mathcal{H}_1)$ such that

$$A^{-1}Ax = x \text{ for every } x \in \mathcal{H}_1$$

and

$$AA^{-1}y = y \text{ for every } y \in \mathcal{H}_2.$$

The operator A^{-1} is called the *inverse* of A.

Clearly, A has at most one inverse. If A is invertible, then for each $y \in \mathcal{H}_2$ there exists one and only one $x \in \mathcal{H}_1$ such that $Ax = y$, namely $x = A^{-1}y$. Conversely, if for each $y \in \mathcal{H}_2$ there exists a unique $x \in \mathcal{H}_1$ such that $Ax = y$, then A is invertible. The proof that A^{-1} is bounded is given in Theorem XII.4.1.

It is very useful to know that A^{-1} is continuous. For example, suppose that the equation $Ax = y$ has a unique solution in $\mathcal{H}_1$ for every $y \in \mathcal{H}_2$. It might very well be that this equation is too difficult to solve whereas $Ax = \tilde{y}$ can be solved rather easily for some $\tilde{y}$ "close" to y. In this case the solution $\tilde{x}$ of this equation is "close" to the solution x of the original equation if A^{-1} is bounded since

$$\|x - \tilde{x}\| = \|A^{-1}y - A^{-1}\tilde{y}\| \leq \|A^{-1}\| \|y - \tilde{y}\|.$$

Definition: The *kernel* of $A \in \mathcal{L}(\mathcal{H}_1, \mathcal{H}_2)$, written Ker A, is the closed subspace $\{x \in H_1 : Ax = 0\}$.

A is called *injective* or *one-one* if Ker $A = (0)$.

A is injective if and only if $Ax \neq Az$ whenever $x \neq z$. Thus, if A is injective, then the equation $Ax = y$ has at most one solution.

Suppose A is in $\mathcal{L}(\mathbb{C}^m)$ and (a_{ij}) is the $n \times n$ matrix corresponding to A and the standard basis for $\mathbb{C}^n$. Then A is invertible if and only if $\det(a_{ij}) \neq 0$, in which case the matrix corresponding to A^{-1} and the standard basis is

$$\frac{1}{\det(a_{ij})}(A_{ij})^T,$$

where A_{ij} is the cofactor corresponding to a_{ij} (cf. [S], p. 170).

Theorem 7.1 *Suppose $K \in L(\mathcal{H})$ is of finite rank, say*

$$Kx = \sum_{j=1}^{n} \langle x, \varphi_j \rangle \psi_j.$$

The operator $I - K$ is invertible if and only if

$$\det(\delta_{ij} - \langle \psi_j, \varphi_i \rangle)_{i,j=1}^n \neq 0.$$

In this case, for every $y \in \mathcal{H}$,

$$(I - K)^{-1}y = y - \frac{1}{\det(a_{ij})}\det \begin{pmatrix} a_{11} & a_{12} & \cdots & a_{1n} & \langle y, \varphi_1 \rangle \\ a_{21} & a_{22} & \cdots & a_{2n} & \langle y, \varphi_2 \rangle \\ \vdots & \vdots & & \vdots & \vdots \\ a_{n1} & a_{n2} & \cdots & a_{nn} & \langle y, \varphi_n \rangle \\ \psi_1 & \psi_2 & \cdots & \psi_n & 0 \end{pmatrix} \quad (7.1)$$

where $a_{ij} = \delta_{ij} - \langle \psi_j, \varphi_i \rangle$. Thus $I - (I - K)^{-1}$ is of finite rank.

The kernel of $I - K$ consists of the set of all vectors of the form $\sum_{i=1}^{n} \alpha_j \psi_j$, where

$$(a_{ij}) \begin{pmatrix} \alpha_1 \\ \alpha_2 \\ \vdots \\ \alpha_n \end{pmatrix} = \begin{pmatrix} 0 \\ 0 \\ \vdots \\ 0 \end{pmatrix}. \quad (7.2)$$

Proof: Given $y \in \mathcal{H}$, suppose $(I - K)x = y$. Then

$$x - \sum_{j=1}^{n} \langle x, \varphi_j \rangle \psi_j = y. \quad (7.3)$$

Taking the inner product of both sides with φ_k yields

$$\langle x, \varphi_k \rangle - \sum_{j=1}^{n} \langle x, \varphi_j \rangle \langle \psi_j, \varphi_k \rangle = \langle y, \varphi_k \rangle.$$

Hence the system of equations

$$\alpha_k - \sum_{j=1}^{n} \alpha_j \langle \psi_j, \varphi_k \rangle = \langle y, \varphi_k \rangle, \, 1 \le k \le n, \tag{7.4}$$

has a solution, namely $\alpha_j = \langle x, \varphi_j \rangle, \, 1 \le j \le n$.

Conversely, if $\alpha_1, \ldots, \alpha_n$ satisfies (7.4), then guided by (7.3), we let

$$x = y + \sum_{j=1}^{n} \alpha_j \psi_j. \tag{7.5}$$

Then

$$\langle x, \varphi_k \rangle = \langle y, \varphi_k \rangle + \sum_{j=1}^{n} \alpha_j \langle \psi_j, \varphi_k \rangle = \alpha_k.$$

Hence

$$(I - K)x = x - \sum_{k=1}^{n} \langle x, \varphi_k \rangle \psi_k = x - \sum_{k=1}^{n} \alpha_k \psi_k = y.$$

To summarize, given $y \in \mathcal{H}$, the equation $(I - K)x = y$ has a solution if and only if the system of equations (7.4) has a solution. Thus if $\det(a_{ij}) = D \ne 0$, where $a_{ij} = \delta_{ij} - \langle \psi_j, \varphi_i \rangle$, then by Cramer's rule, there exists a unique solution to the system of equations (7.4) given by $\alpha_j = \frac{D_j}{D}$, where D_j is the determinant of the matrix which is obtained by replacing the jth column of (a_{ij}) by the column

$$\begin{pmatrix} \langle y, \varphi_1 \rangle \\ \vdots \\ \langle y, \varphi_n \rangle \end{pmatrix}.$$

Once this unique solution $\alpha_1, \ldots, \alpha_n$ is obtained, it follows from what we have shown that $(I - K)x = y$ has a unique solution and this solution is given by (7.5). Thus

$$x = y + \frac{1}{D} \sum_{j=1}^{n} D_j \psi_j, \tag{7.6}$$

and defining $(I - K)^{-1}y = x$, a straightforward computation verifies that $(I - K)^{-1}$ is bounded and has the representation (7.1).

If $\det(a_{ij}) = 0$, then taking $y = 0$ in (7.4), we have by Cramer's rule, numbers $\alpha_1, \ldots, \alpha_n$, not all zero, such that (7.4) holds and by (7.5), $x = \sum_{j=1}^{m} \alpha_j \psi_j$ is in $\text{Ker}(I - K)$. Now $x \neq 0$, otherwise

$$0 = \langle x, \varphi_k \rangle = \sum_{j=1}^{n} \alpha_j \langle \psi_j, \varphi_k \rangle = \alpha_k, \quad 1 \leq k \leq n.$$

Hence $I - K$ is not invertible.

It also follows from the above argument that if x is in $\text{Ker}(I - K)$, then $\alpha_k = \langle x, \varphi_k \rangle$, $1 \leq k \leq n$, satisfies the matrix Equation (7.2). $\qquad\square$

Examples 1. Given the integral equation

$$f(t) - \lambda \int_0^{2\pi} f(s)\sin(s + t)ds = g(t), \tag{7.7}$$

we shall

(a) determine those λ such that for each $g \in L_2([0, 2\pi])$, there exists a solution in $L_2([0, 2\pi])$ to Equation (7.7).

(b) find the solutions.

Define $K: L_2([0, 2\pi]) \to L_2([0, 2\pi])$ by

$$(Kf)(t) = \int_0^{2\pi} f(s)\sin(s + t)ds.$$

Example 5 in Section 2 shows that K is bounded and linear. From the trigonometric identity $\sin(\alpha + \beta) = \sin\alpha \cos\beta + \cos\alpha \sin\beta$, it follows that

$$\lambda Kf = \langle f, \varphi_1 \rangle \psi_1 + \langle f, \varphi_2 \rangle \psi_2,$$

where

$$\psi_1(t) = \lambda\cos t, \quad \psi_2(t) = \lambda\sin t,$$
$$\varphi_1(t) = \sin t, \quad \varphi_2(t) = \cos t.$$

Thus λK has rank 2 if $\lambda \neq 0$. Using the same notation as in Theorem 7.1, we get

$$a_{11} = 1 - \langle \psi_1, \varphi_1 \rangle = 1, \qquad a_{12} = -\langle \psi_2, \varphi_1 \rangle = -\lambda\pi,$$
$$a_{21} = -\langle \psi_1, \varphi_2 \rangle = -\lambda\pi, \qquad a_{22} = 1 - \langle \psi_2, \varphi_2 \rangle = 1,$$
$$\det(a_{ij}) = 1 - \lambda^2\pi^2.$$

Hence $I - \lambda K$ is invertible if and only if $\lambda \neq \pm\frac{1}{\pi}$, in which case for each $g \in L_2([0, 2\pi])$ there exists a unique $f \in L_2([0, 2\pi])$ (identifying functions which are equal almost everywhere) which satisfies (7.7) a.e. The solution is

$$f(t) = [(1 - \lambda K)^{-1}g](t) = g(t) - \frac{1}{1 - \lambda^2\pi^2}\det\begin{pmatrix} 1 & -\lambda\pi & b_1 \\ -\lambda\pi & 1 & b_2 \\ \lambda\cos t & \lambda\sin t & 0 \end{pmatrix}$$

$$= g(t) + \frac{\lambda}{1 - \lambda^2\pi^2}[(b_1 + \lambda\pi b_2)\cos t + (b_2 + \lambda\pi b_1)\sin t],$$

where

$$b_1 = \langle g, \varphi_1 \rangle = \int_0^{2\pi} g(t) \sin t \, dt,$$

$$b_2 = \langle g, \varphi_2 \rangle = \int_0^{2\pi} g(t) \cos t \, dt.$$

When $\lambda = \pm \frac{1}{\pi}$, Ker $(I - \lambda K)$ consists of vectors of the form $\alpha_1 \psi_1 + \alpha_2 \psi_2$, where

$$\begin{pmatrix} 1 & -\lambda\pi \\ -\lambda\pi & 1 \end{pmatrix} \begin{pmatrix} \alpha_1 \\ \alpha_2 \end{pmatrix} = \begin{pmatrix} 0 \\ 0 \end{pmatrix}$$

or

$$\begin{aligned} \alpha_1 - \lambda\pi\alpha_2 &= 0 \\ -\lambda\pi\alpha_1 + \alpha_2 &= 0. \end{aligned}$$

Thus if $\lambda = \frac{1}{\pi}$, then $\alpha_1 = \alpha_2$. If $\lambda = -\frac{1}{\pi}$, then $\alpha_1 = -\alpha_2$. Hence

$$\mathrm{Ker} \left(I - \frac{1}{\pi} K \right) = \mathrm{sp}\{\cos t + \sin t\}.$$

$$\mathrm{Ker} \left(I + \frac{1}{\pi} K \right) = \mathrm{sp}\{\cos t - \sin t\}.$$

Therefore, the general solution to the homogeneous equation

$$f(t) - \frac{1}{\pi} \int_0^{2\pi} f(s) \sin(s + t) \, ds = 0$$

is $f(t) = c(\cos t + \sin t)$, c arbitrary in $\mathbb{C}$, and the general solution to the homogeneous equation

$$f(t) + \frac{1}{\pi} \int_0^{2\pi} f(s) \sin(s + t) \, ds = 0$$

is

$$f(t) = c(\cos t - \sin t), \ c \text{ arbitrary in } \mathbb{C}.$$

2. Let $a(t)$ be a continuous complex valued function defined on the closed interval $[a, b]$. Define a bounded linear operator A on $\mathcal{H} = L_2([a, b])$ by

$$(Af)(t) = a(t)f(t).$$

In Example 4 in Section 2, we saw that A is a bounded linear operator with $\|A\| = \max_{t \in [a,b]} |a(t)|$. If $a(t) \neq 0$ for all $t \in [a, b]$, then A is invertible,

$$(A^{-1}g)(t) = \frac{1}{a(t)} g(t).$$

and

$$\|A^{-1}\| = \max_{t \in [a,b]} \frac{1}{|a(t)|}. \tag{7.8}$$

Conversely, suppose A is invertible. We show that $a(t) \neq 0$ for all $t \in [a, b]$. For $f \in \mathcal{H}$,

$$\|Af\| \geq \frac{1}{\|A^{-1}\|} \|f\| \tag{7.9}$$

Suppose $a(t_0) = 0$ for some $t_0 \in [c, d]$. Then given $\varepsilon > 0$, there exists a $\delta > 0$ such that $|a(t)| < \varepsilon$ if $|t - t_0| < \delta$. Define

$$f(t) = \left\{ \begin{array}{l} 1, |t - t_0| \leq \delta \\ 0 \ |t - t_0| > 0 \end{array} \right\}.$$

Then

$$\|Af\|^2 = \int_{t_0 - \delta}^{t_0 + \delta} |a(t)|^2 \, |f(t)|^2 \leq \varepsilon^2 \|f\|^2,$$

which contradicts (7.9) if $\varepsilon < \frac{1}{\|A^{-1}\|}$.

If A and B are invertible operators in $\mathcal{L}(\mathcal{H})$, then AB is invertible and $(AB)^{-1} = B^{-1}A^{-1}$.

2.8 Inversion of Operators by the Iterative Method

The next theorem, which is rather easy to prove, is nevertheless very useful. The formula for $(I - A)^{-1}$ is suggested by the geometric series

$$1 + a + a^2 + \cdots = (1 - a)^{-1}, |a| < 1.$$

Theorem 8.1 *Suppose $A \in \mathcal{L}(\mathcal{H})$ and $\|A\| < 1$. Then $I - A$ is invertible, and for every $y \in \mathcal{H}$,*

$$(I - A)^{-1} y = \sum_{k=0}^{\infty} A^k y \ (A^0 = I).$$

Moreover,

$$\left\| (I - A)^{-1} - \sum_{k=0}^{n} A^k \right\| \to 0 \ as \ n \to \infty,$$

i.e.,

$$(I - A)^{-1} = \sum_{k=0}^{\infty} A^k$$

and

$$\|(I - A)^{-1}\| \le \frac{1}{1 - \|A\|}.$$

Proof: Given $y \in \mathcal{H}$, the series $\sum_{k=0}^{\infty} A^k y$ converges. Indeed, let $s_n = \sum_{k=0}^{n} A^k y$. Then for $n > m$,

$$\|s_n - s_m\| \le \sum_{k=m+1}^{n} \|A^k y\| \le \|y\| \sum_{k=m+1}^{n} \|A\|^k \to 0.$$

as $m, n \to \infty$ since $\sum_{k=0}^{\infty} \|A\|^k = \frac{1}{1-\|A\|}$. The completeness of $\mathcal{H}$ ensures the convergence of $\{s_n\}$, i.e., $\sum_{k=0}^{\infty} A^k y$ converges. Define $B : \mathcal{H} \to \mathcal{H}$ by $By = \sum_{k=0}^{\infty} A^k y$. B is linear and

$$\|By\| \le \sum_{k=0}^{\infty} \|A\|^k \|y\| = \left(\frac{1}{1 - \|A\|}\right) \|y\|.$$

Hence $\|B\| \le \frac{1}{1-\|A\|}$ and

$$(I - A)By = (I - A) \sum_{k=0}^{\infty} A^k y = \sum_{k=0}^{\infty} (I - A) A^k y$$

$$= \sum_{k=0}^{\infty} A^k (I - A) y = B(I - A)y$$

$$= \sum_{k=0}^{\infty} A^k y - \sum_{k=0}^{\infty} A^{k+1} y = y.$$

Therefore, $I - A$ is invertible and $(I - A)^{-1} = B$. Finally,

$$\left\| (I - A)^{-1} - \sum_{k=0}^{n} A^k \right\| = \sup_{\|y\|=1} \left\| \sum_{k=n+1}^{\infty} A^k y \right\| \le \sum_{k=n}^{\infty} \|A\|^k \to 0$$

as $n \to \infty$. $\qquad\qquad\qquad\qquad\qquad\qquad\qquad\qquad\qquad\qquad\qquad\square$

Corollary 8.2 *Let $A \in \mathcal{L}(\mathcal{H})$ be invertible. Suppose $B \in \mathcal{L}(\mathcal{H})$ and $\|A - B\| < \frac{1}{\|A^{-1}\|}$. Then B is invertible,*

$$B^{-1} y = \sum_{k=0}^{\infty} [A^{-1}(A - B)]^k A^{-1} y$$

and

$$\|A^{-1} - B^{-1}\| \leq \frac{\|A^{-1}\|^2 \|A - B\|}{1 - \|A^{-1}\| \|A - B\|}.$$

Proof: Since $B = A - (A - B) = A[I - A^{-1}(A - B)]$ and $\|A^{-1}(A - B)\| \leq \|A^{-1}\| \|A - B\| < 1$, it follows from Theorem 8.1 that B is invertible and

$$B^{-1} = [I - A^{-1}(A - B)]^{-1}A^{-1} = \sum_{k=0}^{\infty} [A^{-1}(A - B)]^k A^{-1}.$$

Hence

$$\|A^{-1} - B^{-1}\| \leq \|A^{-1}\| \sum_{k=1}^{\infty} \|A^{-1}\|^k \|A - B\|^k = \frac{\|A^{-1}\|^2 \|A - B\|}{1 - \|A^{-1}\| \|A - B\|}.$$

□

Corollary 8.3 *Let* $\{A_n\}$ *be a sequence of operators in* $\mathcal{L}(\mathcal{H}_1, \mathcal{H}_2)$ *which converges in norm to* $A \in \mathcal{L}(\mathcal{H}_1, \mathcal{H}_2)$. *Then* A *is invertible if and only if there exists an integer* N *such that* A_n *is invertible for* $n \geq N$ *and* $\sup_{n \geq N} \|A_n^{-1}\| < \infty$. *In this case,* $\|A_n^{-1} - A^{-1}\| \to 0$.

Proof: Suppose A is invertible, then if follows from Corollary 8.2 that there exists an N such that A_n is invertible for all $n \geq N$ and $\|A_n^{-1} - A^{-1}\| \to 0$. Thus $\sup_{n \geq N} \|A_n^{-1}\| < \infty$.

Conversely, suppose that A_n is invertible for $n \geq N$ and $M = \sup_{n \geq N} \|A_n^{-1}\| < \infty$. Then for $n \geq N$,

$$A = A_n + A - A_n = A_n(I + A_n^{-1}(A - A_n)). \tag{8.1}$$

Since $\|A_n^{-1}(A - A_n)\| \leq M\|A - A_n\| \to 0$, it follows from Theorem 8.1 that there exists $N_1 \geq N$ such that $I + A_n^{-1}(A - A_n)$ is invertible. Since A_n is invertible for $n \geq N_1$, we have from (8.1) that A is invertible. Hence $\|A_n^{-1} - A^{-1}\| \to 0$ by Corollary 8.2.

If we define the distance $d(A, B)$ between the operators A and B to be $\|A - B\|$, then Corollary 8.2 shows that the set 0 of invertible operators in $\mathcal{L}(\mathcal{H})$ is an open set in the sense that if A is in 0, then there exists an $r > 0$ such that $d(A, B) < r$ implies $B \in 0$.

□

2.9 Infinite Systems of Linear Equations

In this section we shall solve certain infinite systems of linear equations

$$\sum_{j=1}^{\infty} a_{ij}x_j = y_i, i = 1, 2, \ldots, \tag{9.1}$$

where $\{y_j\} \in \ell_2$ is given and $\{x_j\}$ is the desired solution in ℓ_2.

This system is important because any equation $Ax = y$, where A is in $\mathcal{L}(\mathcal{H})$ and $\mathcal{H}$ is a separable Hilbert space, can be written in the form (9.1). Indeed, let $\{\varphi_j\}$ be an orthonormal basis for $\mathcal{H}$. Then from matrix equation (4.4) in Section 4, we see that (9.1) holds when

$$x_j = \langle x, \varphi_j \rangle, \ y_j = \langle y, \varphi_j \rangle \text{ and } a_{ij} = \langle A\varphi_j, \varphi_i \rangle.$$

A natural approach to the problem of finding a solution to an infinite system of linear equations is to approximate the solution by solutions to finite sections of the system. This method does not always work (see Example 2 in Section 17). However, we do have the following result.

Theorem 9.1 *Given the matrix* $(a_{ij})_{i,j=1}^\infty$, *where* $\sum_{i,j=1}^\infty |a_{ij}|^2 < 1$, *the system of equations*

$$x_i - \sum_{j=1}^\infty a_{ij}x_j = y_i, \quad i = 1, 2, \dots \tag{9.2}$$

has a unique solution $\xi = (\xi_1, \xi_2, \dots) \in \ell_2$ *for every* $y = (y_1, y_2, \dots) \in \ell_2$. *The truncated system of equations*

$$x_i - \sum_{j=1}^n a_{ij}x_j = y_i, \quad 1 \le i \le n, \tag{9.3}$$

has a unique solution $(x_1^{(n)}, \dots, x_n^{(n)})$ *and* $z_n = (x_1^{(n)}, \dots, x_n^{(n)}, 0, 0, \dots)$ *converges in* ℓ_2 *to* ξ *as* $n \to \infty$.

Proof: Define $A \in \mathcal{L}(\ell_2)$ by $A(\alpha_1, \alpha_2, \dots) = (\beta_1, \beta_2, \dots)$, where

$$(a_{ij}) \begin{pmatrix} \alpha_1 \\ \alpha_2 \\ \vdots \end{pmatrix} = \begin{pmatrix} \beta_1 \\ \beta_2 \\ \vdots \end{pmatrix}.$$

By Section 2, Example 3, $\|A\|^2 \le \sum_{i,j=1}^\infty |a_{ij}|^2 < 1$. Hence we have from Theorem 8.1 that $I - A$ is invertible, from which it follows that (9.2) has a unique solution in ℓ_2. To approximate this solution, let $A_n \in \mathcal{L}(\ell_2), n = 1, 2, \dots$, be defined by $A_n(\alpha_1, \alpha_2, \dots) = (\beta_1, \beta_2, \dots)$, where

$$\begin{pmatrix} \begin{matrix} a_{11} & \dots & a_{1n} \\ \vdots & & \vdots \\ a_{n1} & \dots & a_{nn} \end{matrix} & 0 \\ & \\ 0 & \end{pmatrix} \begin{pmatrix} \alpha_1 \\ \alpha_2 \\ \vdots \end{pmatrix} = \begin{pmatrix} \beta_1 \\ \beta_2 \\ \vdots \end{pmatrix}$$

Then $\|A_n\|^2 \le \sum_{i,j=1}^n |a_{ij}|^2 < 1$ and

$$\|A - A_n\|^2 \le \sum_{i=n+1}^\infty \sum_{j=1}^\infty |a_{ij}|^2 + \sum_{j=n+1}^\infty \sum_{i=1}^\infty |a_{ij}|^2 \to 0 \text{ as } n \to \infty.$$

Hence $I - A_n$ is invertible and

$$\|(I - A)^{-1} - (I - A_n)^{-1}\| \to 0 \tag{9.4}$$

by Corollary 8.3. Taking $(x_1^{(n)}, x_2^{(n)}, \ldots) = (I - A_n)^{-1} (y_1, y_2, \ldots)$, it follows from the definition of A_n that $(x_1^{(n)}, \ldots, x_n^{(n)})$ is the unique solution to (9.3) and $x_j^{(n)} = y_j$, for $j > n$. Since $\sum_{j=1}^{\infty} |y_j|^2 < \infty$, (9.4) implies that

$$\lim_{n \to \infty} (x_1^{(n)}, \ldots, x_n^{(n)}, 0, 0, \ldots) = \lim_{n \to \infty} (x_1^{(n)}, \ldots, x_n^{(n)}, y_{n+1}, y_{n+2}, \ldots)$$
$$= (I - A)^{-1}(y_1, y_2, \ldots). \tag{9.5}$$

In fact, (9.4) shows that as $(y_1, y_2, \ldots)$ ranges over the 1-ball of ℓ_2, the corresponding sequences in (9.5) converge uniformly. □

2.10 Integral Equations of the Second Kind

Given $k \in L_2([a, b] \times [a, b])$, we know from Section 2, Example 5 that

$$(Kf)(t) = \int_a^b k(t, s) f(s) \, ds$$

is a bounded linear operator from $L_2([a, b])$ into $L_2([a, b])$ with $\|K\| \le \|k\|$. Suppose $\|k\| < 1$. By Theorem 8.1, $I - K$ is invertible. Thus for each $g \in L_2([a, b])$, there exists a unique $f \in L_2([a, b])$ such that

$$f(t) - \int_a^b k(t, s) f(s) \, ds = g(t). \tag{10.1}$$

Here " $=$ " means equal almost everywhere.
  Equation (10.1) is called an integral equation of the second kind.
  We shall show that $(I - K)^{-1} = I + \widetilde{K}$, where $\widetilde{K}$ is an integral operator. By Theorem 8.1,

$$(I - K)^{-1} g = \sum_{n=0}^{\infty} K^n g. \tag{10.2}$$

From the definition of K and Fubini's theorem (Appendix 2),

$$
\begin{aligned}
(K^2 g)(t) &= \int_a^b k(t, x)(Kg)(x)dx \\
&= \int_a^b k(t, x) \left\{ \int_a^b k(x, s)g(s)ds \right\} dx \\
&= \int_a^b g(s) \left\{ \int_a^b k(t, x)k(x, s)dx \right\} ds \\
&= \int_a^b k_2(t, s)g(s)\, ds,
\end{aligned}
$$

where

$$
k_2(t, s) = \int_a^b k(t, x)k(x, s)dx.
$$

The function k_2 is in $L_2([a, b] \times [a, b])$; for, by Schwarz's inequality,

$$
|k_2(t, s)|^2 \le \left\{ \int_a^b |k(t, x)|^2 dx \right\} \left\{ \int_a^b |k(x, s)|^2 dx \right\}.
$$

Therefore,

$$
\int_a^b |k_2(t, s)|^2 ds \le \left\{ \int_a^b |k(t, x)|^2 dx \right\} \left\{ \int_a^b \int_a^b |k(x, s)|^2 dx\, ds \right\}
$$

and

$$
\int_a^b \int_a^b |k_2(t, s)|^2 ds\, dt
$$

$$
\le \left\{ \int_a^b \int_a^b |k(t, x)|^2 dx\, dt \right\} \left\{ \int_a^b \int_a^b |k(x, s)|^2 dx\, ds \right\}.
$$

Proceeding in this manner, an induction argument establishes that

$$
(K^n g)(t) = \int_a^b k_n(t, s)g(s)\, ds, \quad n = 1, 2, \ldots, \tag{10.3}
$$

and $\|k_n\| \le \|k\|^n$, where

$$
k_1(t, s) = k(t, s) \text{ and } k_n(t, s) = \int_a^b k(t, x)k_{n-1}(x, s)\, dx, n > 1, \tag{10.4}
$$

is in $L_2([a, b] \times [a, b])$. Since

$$
\sum_{n=1}^\infty \|k_n\| \le \sum_{n=1}^\infty \|k\|^n < \infty,
$$

$\sum_{n=1}^{\infty} k_n = \widetilde{k}$ for some $\widetilde{k} \in L_2([a, b] \times [a, b])$. Let $\widetilde{K}$ be the integral operator with kernel function $\widetilde{k}$. It follows from (10.3) and Schwarz's inequality that

$$\left\| \sum_{n=1}^{p} K^n g - \widetilde{K} g \right\| \leq \left\| \sum_{n=1}^{p} K_n - \widetilde{K} \right\| \|g\| \to 0$$

as $p \to \infty$. Hence

$$(I - K)^{-1} g = \sum_{n=0}^{\infty} K^n g = (I + \widetilde{K})g,$$

which shows that the solution (a. e.) to (10.1) is given by

$$f(t) = g(t) + \int_a^b \widetilde{k}(t, s)g(s)\, ds. \tag{10.5}$$

If, in addition,

$$c = \sup_t \int_a^b |k(t, s)|^2 ds < \infty,$$

then

$$((I - K)^{-1} g)(t) = \sum_{n=0}^{\infty} (K^n g)(t) \quad \text{a.e.}$$

The series converges absolutely and uniformly on $[a, b]$. Indeed, for any $h \in L_2([a, b])$, we get from Schwarz's inequality,

$$|(Kh)(t)| \leq \|h\| \left(\int_a^b |k(t, s)|^2\, ds \right)^{1/2} \leq \sqrt{c}\|h\|.$$

Replacing h by $K^{n-1} g$ gives

$$\|(K^n g)(t)\| \leq \sqrt{c}\|K^{n-1} g\| \leq \sqrt{c}\|K\|^{n-1}\|g\|, \quad n = 1, 2, \ldots.$$

Hence

$$\sum_{n=1}^{\infty} |(K^n g)(t)| \leq \sqrt{c}\|g\| \sum_{n=1}^{\infty} \|K\|^{n-1} < \infty, \quad t \in [a, b].$$

Example: To solve the integral equation

$$f(t) - \lambda \int_0^1 e^{(t-s)} f(s)\, ds = g(t) \in L_2([0, 1]), \tag{10.6}$$

let $k_1(t, s) = \lambda e^{t-s}$. Then from (10.4) in the discussion above,

$$k_2(t, s) = \lambda^2 \int_0^1 e^{t-x} e^{x-s}\, dx = \lambda^2 e^{t-s}.$$

In general, $k_n(t, s) = \lambda^n e^{t-s}$, and from (10.5),

$$f(t) = g(t) + \sum_{n=1}^{\infty} \lambda^n \int_0^1 e^{t-s} g(s) \, ds$$

$$= g(t) + \frac{\lambda}{1 - \lambda} \int_0^1 e^{t-s} g(s) \, ds \qquad (10.7)$$

is the solution to the integral equation for $|\lambda| < 1$.

Even though the series converges only for $|\lambda| < 1$, a straighforward computation verifies that for all $\lambda \neq 1$, the expression after the second equality in (10.7) is still a solution to (10.6).

Another way to find the solution to an integral equation of the second kind is to make use of the results in Section 9. Namely, suppose $\|k\| < 1$. Let $\{\varphi_j\}$ be an orthonormal basis for $L_2([a, b])$. Then for $a_{ij} = \langle K\varphi_j, \varphi_i \rangle$, we have from Example 3, Section 5, that $\sum_{i,j=1}^{\infty} |a_{ij}|^2 = \|k\|^2 < 1$. Now by Theorem 9.1, the finite system of equations

$$x_i - \sum_{j=1}^{n} a_{ij} x_j = \langle g, \varphi_i \rangle, \quad 1 \leq i \leq n,$$

has a unique solution

$$\begin{pmatrix} x_1^{(n)} \\ \vdots \\ x_n^{(n)} \end{pmatrix} = B_n \begin{pmatrix} \langle g, \varphi_1 \rangle \\ \vdots \\ \langle g, \varphi_n \rangle \end{pmatrix},$$

where B_n is the inverse of the matrix $(\delta_{jk} - a_{jk})_{j,k=1}^n$.

The sequence of vectors $(x_1^{(n)}, \ldots, x_n^{(n)}, 0, 0, \ldots)$ converges in ℓ_2 to some vector, say, $(x_1, x_2, \ldots)$. The solution to the integral equation is $f = \sum_{j=1}^{\infty} x_j \varphi_j$ (convergence in $L_2([a, b])$).

2.11 Adjoint Operators

In this section we discuss the infinite dimensional counterpart to the adjoint of a matrix, namely the adjoint of a bounded linear operator.

Frequently, knowledge about an operator can be gained from information about its adjoint, which may be simpler to deal with.

Suppose $A \in \mathcal{L}(\mathcal{H}_1, \mathcal{H}_2)$. For each $y \in \mathcal{H}_2$, the functional $f_y(x) = \langle Ax, y \rangle$ is a bounded linear functional. Hence the Riesz representation theorem guarantees the existence of a *unique* $y^* \in \mathcal{H}_1$ such that for all $x \in \mathcal{H}_1$,

$$\langle Ax, y \rangle = f_y(x) = \langle x, y^* \rangle.$$

This gives rise to an operator $A^* : \mathcal{H}_2 \to \mathcal{H}_1$ defined by $A^* y = y^*$. Thus for all $x \in \mathcal{H}_1$,

$$\langle Ax, y \rangle = \langle x, y^* \rangle = \langle x, A^* y \rangle.$$

The operator A^* is called the *adjoint* of A.

$$A^* \text{ is in } \mathcal{L}(\mathcal{H}_2, \mathcal{H}_1) \text{ and } \|A\| = \|A^*\|.$$

Indeed, given u and $v \in \mathcal{H}_2$ and $\alpha, \beta \in \mathbb{C}$,

$$\langle Ax, \alpha u + \beta V \rangle = \bar{\alpha} \langle Ax, u \rangle + \bar{\beta} \langle Ax, v \rangle$$
$$= \langle x, \alpha A^* u + \beta A^* v \rangle$$

for all $x \in \mathcal{H}_1$. Thus, by definition, $A^*(\alpha u + \beta v) = \alpha A^* u + \beta A^* v$. Also, we have from (1.1) in Section 1, that

$$\|A\| = \sup_{\|x\|=\|y\|=1} |\langle Ax, y \rangle| = \sup_{\|x\|=\|y\|=1} |\langle x, A^* y \rangle| = \|A^*\|.$$

Examples 11.1 1. We have $0^* = 0$ and $I^* = I$.

2. Let S_r be the forward shift on ℓ_2. Given $y = (y_1, y_2, \ldots)$ and $x = (x_1, x_2, \ldots)$ in ℓ_2,

$$\langle S_r x, y \rangle = \langle (0, x_1, x_2, \ldots), (y_1, y_2, \ldots) \rangle$$
$$= x_1 \bar{y}_2 + x_2 \bar{y}_3 + \cdots = \langle x, y^* \rangle.$$

where $y^* = (y_2, y_3, \ldots)$. Thus S_r^* is the backward shift operator.

3. Let S_ℓ be the backward shift operator on ℓ_2. For $x = (x_1, x_2, \ldots)$ and $y = (y_1, y_2, \ldots)$ in ℓ_2,

$$\langle S_\ell x, y \rangle = x_2 \bar{y}_1 + x_3 \bar{y}_2 + \cdots = \langle x, y^* \rangle,$$

where $y^* = (0, y_1, y_2, \ldots)$. Thus S_ℓ^* is the forward shift operator.

In a similar way one shows that the adjoint of the bilateral forward shift V on $\ell_2(\mathbb{Z})$ is equal to the bilateral backward shift on $\ell_2(\mathbb{Z})$. Thus in this case $V^* = V^{-1}$.

4. Let $K \in \mathcal{L}(\mathcal{H}_1, \mathcal{H}_2)$ be the operator of finite rank defined by $Kx = \sum_{j=1}^n \langle x, u_j \rangle v_j, u_j \in \mathcal{H}_1, v_j \in \mathcal{H}_2$. Then for all $x \in \mathcal{H}_1$ and $y \in \mathcal{H}_2$,

$$\langle Kx, y \rangle = \sum_{j=1}^n \langle x, u_j \rangle \langle v_j, y \rangle = \left\langle x, \sum_{j=1}^n \langle y, v_j \rangle u_j \right\rangle.$$

Hence

$$K^* y = \sum_{j=1}^n \langle y, v_j \rangle u_j.$$

5. Let $a(t)$ be a bounded complex valued Lebesgue measurable function on $[a, b]$, and let A: $L_2([a, b]) \rightarrow L_2([a, b])$ be the bounded linear operator defined by

$$(Af)(t) = a(t)f(t).$$

For all f, g in $L_2([a, b])$,

$$\langle Af, g \rangle = \int_a^b a(t)f(t)\bar{g}(t)dt = \langle f, \bar{a}g \rangle.$$

Thus $(A^*g)(t) = \bar{a}(t)g(t)$.

6. Let $\mathcal{H}$ be a Hilbert space with an orthonormal basis $\varphi_1, \varphi_2, \ldots$. For $A \in \mathcal{L}(\mathcal{H})$, the matrix (a_{ij}) corresponding to A and $\varphi_1, \varphi_2, \ldots$ is $(a_{ij}) = (\langle A\varphi_j, \varphi_i \rangle)$ (cf. Section 4). Since

$$\langle A^*\varphi_j, \varphi_i \rangle = \langle \varphi_j, A\varphi_i \rangle = \bar{a}_{ji},$$

The matrix corresponding to A^* and $\varphi_1, \varphi_2, \ldots$ is the *conjugate transpose* $(\bar{a}_{ji})$ of the matrix (a_{ij}).

The next example is important enough to be designated as a theorem.

Theorem 11.2 *Let $K : L_2([a, b]) \rightarrow L_2([a, b])$ be the bounded linear operator defined by*

$$(Kf)(t) = \int_a^b k(t, s)f(s)ds,$$

where k is in $L_2([a, b] \times [a, b])$. For all $g \in L_2([a, b])$,

$$(K^*g)(t) = \int_a^b \overline{k(s, t)}g(s)ds.$$

Proof: By Fubini's theorem,

$$\langle Kf, g \rangle = \int_a^b \left(\int_a^b k(t, s)f(s)ds \right) \bar{g}(t)dt$$

$$= \int_a^b f(s) \left(\overline{\int_a^b \overline{k(t, s)}g(t)dt} \right) ds = \langle f, g^* \rangle,$$

where

$$g^*(s) = \int_a^b \overline{k(t, s)}g(t) \, dt.$$

$\square$

Theorem 11.3 *If A and B are in* $\mathcal{L}(\mathcal{H}_1, \mathcal{H}_2)$, *then*

(i) $(A + B)^* = A^* + B^*$;

(ii) $(\alpha A)^* = \bar{\alpha} A^*, \alpha \in \mathbb{C}$;

(iii) $A^{**} = A$;

(iv) *If D is in* $\mathcal{L}(\mathcal{H}_2, \mathcal{H}_3)$, *where* $\mathcal{H}_3$ *is a Hilbert space, then* $(DA)^* = A^*D^*$.

Proof: (i) Given $x \in \mathcal{H}_1$ and $y \in \mathcal{H}_2$,

$$\langle (A + B)x, y \rangle = \langle Ax, y \rangle + \langle Bx, y \rangle = \langle x, A^*y + B^*y \rangle.$$

Therefore, by definition, $(A + B)^*y = A^*y + B^*y$.

(ii) $\langle \alpha Ax, y \rangle = \alpha \langle x, A^*y \rangle = \langle x, \bar{\alpha} A^*y \rangle$.

Hence $(\alpha A)^*y = \bar{\alpha} A^*y$.

(iii) $\langle A^*y, x \rangle = \overline{\langle x, A^*y \rangle} = \langle y, Ax \rangle$.

Thus, by the definition of the adjoint of A^*, $A^{**}x = Ax$.

(iv) For $v \in \mathcal{H}_3$,

$$\langle DAx, v \rangle = \langle Ax, D^*v \rangle = \langle x, A^*D^*v \rangle.$$

Hence $(DA)^*v = A^*D^*v$. ☐

The next theorem gives some useful relationships between the kernels and ranges of an operator and its adjoint.

Theorem 11.4 *For* $A \in \mathcal{L}(\mathcal{H}_1, \mathcal{H}_2)$, *the following dual properties hold.*

(i) $\text{Ker}A = \Im A^{*\perp}$.

(ii) $\text{Ker}A^* = \Im A^{\perp}$.

(iii) $\overline{\Im A} = \text{Ker}A^{*\perp}$.

(iv) $\overline{\Im A^*} = \text{Ker}A^{\perp}$.

Proof: (i) Given $x \in \text{Ker } A$ and $y \in \mathcal{H}_2$,
$$0 = \langle Ax, y \rangle = \langle x, A^*y \rangle.$$

Thus $\text{Ker}A \subset \Im A^{*\perp}$. On the other hand, if $v \in \Im A^{*\perp}$, then for all $y \in \mathcal{H}_2$,

$$0 = \langle v, A^*y \rangle = \langle Av, y \rangle.$$

In particular, $0 = \langle Av, Av \rangle$. Hence $v \in \text{Ker}A$ and therefore $\text{Im}A^{*\perp} \subset \text{Ker}A$.

(ii) Applying (i) to A^* and $A^{**} = A$, we get

$$\operatorname{Ker} A^* = \operatorname{Im} A^{**\perp} = \operatorname{Im} A^\perp.$$

(iii) Corollary I.8.3, the continuity of the inner product and (ii) imply

$$\overline{\Im A} = \overline{(\Im A)}^{\perp\perp} = (\Im A)^{\perp\perp} = \operatorname{Ker} A^{*\perp}.$$

(iv) Apply (iii) to A^* and $A^{**} = A$.

$\square$

2.12 Self Adjoint Operators

The requirement that an operator be self adjoint is very restrictive. Nevertheless, a wide variety of problems in mathematics and physics give rise to self adjoint operators. Some of these operators are studied in subsequent chapters.

Definition: An operator $A \in \mathcal{L}(\mathcal{H})$ is called *self adjoint* if $A^* = A$.
As a trivial consequence of Theorem 11.4(iv) we have the following result.

Theorem 12.1 *If $A \in \mathcal{L}(\mathcal{H})$ is self adjoint, then* $\operatorname{Ker} A^\perp = \overline{\Im A}$. *Thus* $\mathcal{H} = \operatorname{Ker} A \oplus \overline{\Im A}$.

Examples 12.2 1. Let $K \in \mathcal{L}(\mathcal{H})$ be the operator of finite rank given by $Kx = \sum_{j=1}^{n} \langle x, u_j \rangle v_j$. By 11.1, Example 4, K is self adjoint if and only if for all $x \in \mathcal{H}$,

$$\sum_{j=1}^{n} \langle x, u_j \rangle v_j = \sum_{j=1}^{n} \langle x, v_j \rangle u_j.$$

2. Let $(Af)(t) = a(t)f(t)$ be the operator defined in 11.1, Example 5. Then A is self adjoint if and only if $a(t) = \overline{a(t)}$.

3. Let K be the integral operator with kernel function k defined in Theorem 11.2. If $\overline{k(t, s)} = k(s, t)$ a.e., then A is self adjoint. The converse also holds.

4. If $A \in \mathcal{L}(\mathcal{H})$, then A^*A is self adjoint since

$$(A^*A)^* = A^*A^{**} = A^*A.$$

Theorem 12.3 *An operator $A \in \mathcal{L}(\mathcal{H})$ is self adjoint if and only if $\langle Ax, x \rangle$ is real for all $x \in \mathcal{H}$.*

Proof: If $A = A^*$, then for $x \in \mathcal{H}$,

$$\langle Ax, x \rangle = \langle x, Ax \rangle = \overline{\langle Ax, x \rangle}.$$

Thus $\langle Ax, x \rangle$ is real

Suppose $\langle Au, u \rangle$ is real for all $u \in \mathcal{H}$. Then for all $x, y \in \mathcal{H}$ and $\lambda \in \mathbb{C}$,

$$\langle A(x + \lambda y), (x + \lambda y) \rangle = \langle x + \lambda y, A(x + \lambda y) \rangle$$

which, together with the assumption, implies that

$$\overline{\lambda} \langle Ax, y \rangle + \lambda \langle Ay, x \rangle = \overline{\lambda} \langle x, Ay \rangle + \lambda \langle y, Ax \rangle.$$

Hence

$$\Im \lambda \langle Ay, x \rangle = \Im \lambda \langle y, Ax \rangle.$$

Taking $\lambda = 1$ and $\lambda = i$, it follows that $\langle Ay, x \rangle = \langle y, Ax \rangle$. Thus $A = A^*$. $\square$

2.13 Orthogonal Projections

There is a special class of self adjoint operators, called orthogonal projections, which are building blocks for the theory of self adjoint operators, in the sense that every bounded linear self adjoint operator is the limit, in norm, of a sequence of linear combinations of orthogonal projections. This result is proved in Chapter IV, but only for a certain class of self adjoint operators.

Definition: Given a closed subspace M of $\mathcal{H}$, an operator P defined on $\mathcal{H}$ is called the *orthogonal projection* onto M if

$$P(m + n) = m, \text{ for all } m \in M \text{ and } n \in M^{\perp}.$$

It is easy to see that P is linear on $\mathcal{H}$, $\Im P = M$, $\operatorname{Ker} P = M^{\perp}$ and $Pm = m$ for all $m \in M$. Clearly, $I - P$ is an orthogonal projection onto $M^{\perp}$ with kernel M.

The projection P is called orthogonal since $\operatorname{Ker} P$ is orthogonal to $\Im P$.

If $M \neq 0$, then $\|P\| = 1$; for given $x = u + v$, where $u \in M$ and $v \in M^{\perp}$, we have by the Pythagorean theorem,

$$\|Px\|^2 = \|u\|^2 \le \|u\|^2 + \|v\|^2 = \|u + v\|^2 = \|x\|^2.$$

Thus $\|P\| \le 1$. But for $m \neq 0$ in M, $\|P\|\|m\| \ge \|Pm\| = \|m\|$. Hence $\|P\| \ge 1$.

From Theorems I.4.3 and I.8.1 we know that if M is a closed subspace of $\mathcal{H}$, then for each $y \in \mathcal{H}$, there exists a unique $w \in M$ such that $(y - w) \perp M$ or, equivalently, $\|y - w\| = d(y, M)$. If we define $Py = w$, then P is the orthogonal projection of y onto M since $y = w + y - w$, $w \in M$, $y - w \in M^{\perp}$.

Theorem 13.1 *An operator $P \in \mathcal{L}(\mathcal{H})$ is an orthogonal projection if and only if $P^2 = P$ and P is self adjoint.*

Proof: Suppose P is the orthogonal projection onto a closed subspace M of $\mathcal{H}$. Given $x = m + n, m \in M$ and $n \in M^\perp$,

$$P(Px) = Pm = m = Px.$$

Thus $P^2 = P$. Now for $y = m_1 + n_1, m_1 \in M, n_1 \in M^\perp, \langle Px, y \rangle = \langle m, m_1 \rangle = \langle m + n, m_1 \rangle = \langle x, Py \rangle$. Hence $P = P^*$.

Assume $P^2 = P$ and $P^* = P$. Let $M = \Im P$. Since $M = \text{Ker}(I - P)$. M is closed and by Theorem 11.4,

$$M^\perp = \Im P^\perp = \text{Ker} P.$$

Therefore, given $u \in M$ and $v \in M^\perp$,

$$P(u + v) = Pu + Pv = Pu = u$$

since $u \in M = \Im P$. Hence P is the orthogonal projection onto M. $\qquad \square$

2.14 Two Fundamental Theorems

The following two theorems, which play a major role in functional analysis, will be used often throughout the remaining chapters of the text. The proof of Theorem 14.1 which is given in the more general setting of Banach spaces, appears in Theorem XII.4.4.

Theorem 14.1 *Suppose $A \in L(\mathcal{H}_1, \mathcal{H}_2)$ has the properties that $\text{Ker} A = \{0\}$ and $\text{Im} A = \mathcal{H}_2$. Then its inverse A^{-1} is bounded.*

Corollary 14.2 *An operator $A \in L(\mathcal{H}_1, \mathcal{H}_2)$ is one-one and has a closed range if and only if there exists an $m > 0$ such that*

$$\|Ax\| \geq m\|x\| \text{ for all } x \in H_1. \tag{14.1}$$

Proof: Assume inequality (14.1). Clearly, A is one-one. Suppose $Ax_n \to y$. Then the sequence $\{Ax_n\}$ is Cauchy and therefore the sequence $\{x_n\}$ is Cauchy by (14.1). Hence $x = \lim_{n \to \infty} x_n$ exists since $\mathcal{H}_1$ is complete. Thus $Ax_n \to Ax = y$, which shows that $\Im A$ is closed.

Conversely, suppose that A is one-one and $\Im A$ is closed. Let A_1 be the operator A considered as a map from $\mathcal{H}_1$ onto the Hilbert space $\Im A$. Then A_1^{-1} is bounded by Theorem 14.1. Hence

$$\|x\| = \|A_1^{-1}Ax\| \leq \|A_1^{-1}\|\|Ax\|$$

and inequality (14.1) follows. $\qquad \square$

The proof of the next theorem is elementary. It differs from the usual proofs which depend on the closed graph theorem or the Baire category theorem.

Theorem 14.3 *Suppose $\{A_n\}$ is a sequence in $\mathcal{L}(\mathcal{H}_1, \mathcal{H}_2)$ with the property that $\sup_n \|A_n x\| < \infty$ for each $x \in \mathcal{H}_1$. Then $\sup_n \|A_n\| < \infty$.*

Proof: Assume $\sup_n \|A_n\| = \infty$. Let $\{\alpha_k\}$ be a sequence of positive numbers which we shall choose later. There exists operators $A_{n1}, A_{n2}, \ldots$ and vectors $x_1, x_2, \ldots$ in $\mathcal{H}_1$ such that

$$\|A_{n_k}\| > \alpha_k, \ \|x_k\| = \frac{1}{4^k} \tag{14.2}$$

and

$$\|A_{n_k} x_k\| > \frac{2}{3} \|A_{n_k}\| \|x_k\| = \frac{2}{3 \cdot 4^k} \|A_{n_k}\|. \tag{14.3}$$

Let $x = \sum_{j=1}^{\infty} x_j$. Note that the series converges since $\mathcal{H}_1$ is complete and $\sum_{j=1}^{\infty} \|x_j\| = \sum_{j=1}^{\infty} 4^{-j} < \infty$. Since

$$A_{n_k} x = A_{n_k} \alpha_k + A_{n_k} \left(\sum_{j=1}^{k-1} x_j \right) + A_{n_k} \left(\sum_{j=k+1}^{\infty} x_j \right), k \geq 1,$$

it follows from (14.2) and (14.3) that for $k > 1$

$$\|A_{n_k} x\| \geq \|A_{n_k} x_k\| - C_{k-1} - \|A_{n_k}\| \frac{1}{3} 4^{-k}$$

$$\geq \frac{1}{3} 4^{-k} \|A_{n_k}\| - C_{k-1} > \frac{\alpha_k}{3} 4^{-k} - C_{k-1},$$

where

$$C_{k-1} = \sup_n \left\| A_n \left(\sum_{j=1}^{k-1} x_j \right) \right\| < \infty.$$

Now choose $\alpha_k > 3 \cdot 4^k (C_{k-1} + k)$, $k = 1, 2, \ldots$. Then $\sup_n \|A_n x\| = \infty$, which is a contradiction. $\qquad \square$

Corollary 14.4 *Let $\{A_n\}$ and $\{B_n\}$ be sequences in $\mathcal{L}(\mathcal{H}_1, \mathcal{H}_2)$ and $\mathcal{L}(\mathcal{H}_2, \mathcal{H}_3)$, respectively, where $\mathcal{H}_i$ is a Hilbert space, $i = 1, 2, 3$. Suppose $Ax = \lim_{n \to \infty} A_n x$ and $By = \lim_{n \to \infty} B_n y$ exist for each $x \in H_1$ and $y \in H_2$. Then A and B are bounded linear operators and*

$$B_n A_n x \to BA x \text{ for each } x \in \mathcal{H}_1.$$

Proof: Since the sequence $\{B_n y\}$ converges for each $y \in \mathcal{H}_2$, $\sup_n \|B_n y\| < \infty$. Hence $M = \sup_n \|B_n\| < \infty$ by Theorem 14.3. Thus $\|By\| = \lim_{n \to \infty} \|B_n y\| \leq$

$M\|y\|$. Clearly B is linear. Hence $B \in L(\mathcal{H}_2, \mathcal{H}_3)$. Similarly $A \in L(\mathcal{H}_1, \mathcal{H}_2)$. Therefore

$$\|B_n A_n x - BAx\| \le \|B_n A_n x - B_n Ax\| + \|B_n Ax - BAx\|$$
$$\le M\|A_n x - Ax\| + \|B_n Ax - BAx\| \to 0.$$

$\square$

2.15 Projections and One-Sided Invertibility of Operators

Definition: An operator P acting on a Hilbert space $\mathcal{H}$ is called a *projection* if P is a bounded linear operator on $\mathcal{H}$ and $P^2 = P$.
   If P is a projection on $\mathcal{H}$, then $I - P$ is also a projection on H since $(I - P)^2 = I - 2P + P^2 = I - P$.

Examples: Each operator P defined on ℓ^2 in $a) - c)$ is a projection.

(a) $P(\alpha_1, \alpha_2, \ldots) = (0, 0, \ldots 0, \alpha_n, \alpha_{n+1}, \ldots)$

(b) $P(\alpha_1, \alpha_2, \ldots) = (\alpha_1, 0, \alpha_3, 0, \alpha_5, \ldots)$

(c) $P(\alpha_1, \alpha_2, \ldots) = (\underbrace{\alpha_1, \alpha_1, \ldots, \alpha_1}_{n}, \alpha_{n+1}, \alpha_{n+2}, \ldots)$.

(d) Let $\mathcal{H} = L_2([0, 1])$ and let E be a measurable subset of $[0, 1]$. Define P on $\mathcal{H}$ by

$$(Pf)(t) = f(t)C_E(t),$$

where $C_E(t)$ is the characteristic function of E. Clearly, P is a projection on $\mathcal{H}$.

Theorem 15.1 *Let P be a projection on a Hilbert space $\mathcal{H}$. Then*

(i) $\Im P = \mathrm{Ker}\,(I - P)$.

(ii) $\Im P$ *is closed.*

(iii) *Every vector $v \in \mathcal{H}$ can be uniquely written in the form $v = x + y$, where $Px = 0$ and $Py = y$.*

Proof: (i) Since $(I - P)Px = Px - Px = 0$, we have $\Im P \subseteq \mathrm{Ker}\,(I - P)$. If $y \in \mathrm{Ker}\,(I - P)$, then $y - Py = 0$. Thus $y = Py \in \Im P$. Equality (i) follows

(ii) Statement ii) is an immediate consequence of (i)

(iii) Take $x = (I - P)v$ and $y = Pv$. Then x and y have the properties stated above. To prove the uniqueness of the representation, suppose $v = x_1 + y_1$, where $Px_1 = 0$ and $Py_1 = y_1$. Then $x - x_1 = y_1 - y$ and $0 = P(x - x_1) = P(y_1 - y) = y_1 - y$.

Hence $y = y_1$ and $x = x_1$. $\square$

Recall that we have already met projections in Section 13.

Definition: A Hilbert space $\mathcal{H}$ is said to be the *direct sum* of subspaces M and N, written $\mathcal{H} = M \oplus N$, if every vector $v \in \mathcal{H}$ has a *unique* representation of the form $v = x + y$, where $x \in M$ and $y \in N$.

The subspace M is said to be *complemented* in $\mathcal{H}$ if there exists a *closed* subspace N of $\mathcal{H}$ such that $\mathcal{H} = M \oplus N$. Note that the representation of $v = x + y, x \in M, y \in N$ is unique if and only if $M \cap N = \{0\}$.

We have seen in Theorem 15.1 that if P is a projection on $\mathcal{H}$, then

$$H = \operatorname{Ker} P \oplus \Im P. \tag{15.1}$$

Examples: (i) Let P be the projection in a). Then

$$\Im P = \{(\alpha_k) \in \ell_2 | \alpha_k = 0, 0 \le k < n\},$$
$$\operatorname{Ker} P = \{(\beta_k), \beta_k = 0, \ k \ge n\}$$

(ii) Let P be the projection in b). Then

$$\Im P = \{(\alpha_k) \in \ell_2 | \alpha_{2k} = 0, k = 1, 2, \ldots\},$$
$$\operatorname{Ker} P = \{(\beta_k) \in \ell_2 | \beta_{2k+1} = 0, k = 0, 1, \ldots\}$$

(iii) Let P be the projection in c). Then

$$\Im P = \{(\alpha_k) | \alpha_k = \alpha_1, 1 \le k \le n\},$$
$$\operatorname{Ker} P = \{(\beta_k) | \beta_k = 0, k = 1 \text{ and } k > n\}$$

(iv) Let P be the projection in (d). Then

$$\Im P = \{f \in L_2([0, 1]) | f(t) = 0, t \notin E\},$$
$$\operatorname{Ker} P = \{g \in L_2([0, 1]) | g(t) = 0, t \in E\}$$

The projections P in examples (i), (ii) and (iv) are orthogonal projections since $\Im P$ is orthogonal to $\operatorname{Ker} P$. This is not the case for the projection in example (iii) since $\langle (\alpha_k), (\beta_k) \rangle = \sum_{k=1}^{m} \alpha_k \bar{\beta}_k$ need not be zero for $(\alpha_k) \in \Im P$ and $(\beta_k) \in \operatorname{Ker} P$.

Let $\mathcal{H}_1$ and $\mathcal{H}_2$ be Hilbert spaces. An operator $A \in \mathcal{L}(\mathcal{H}_1, \mathcal{H}_2)$ is said to be *left invertible* if there exists an operator $L \in \mathcal{L}(\mathcal{H}_2, \mathcal{H}_1)$ such that $LA = I$. The operator L is called a *left inverse* of A.

Examples: 1. Let A be the forward shift operator on ℓ_2, i.e., $A(\alpha_1, \alpha_2, \ldots) = (0, \alpha_1, \alpha_2, \ldots)$. Then A is left invertible and every left inverse L has the form

$$L(\alpha_1, \alpha_2, \ldots) = \alpha_1 \{\beta_k\}_{k=1}^{\infty} + (\alpha_2, \alpha_3, \ldots) = \{\alpha_1 \beta_k + \alpha_{k+1}\}_{k=1}^{\infty}, \tag{15.2}$$

where $\{\beta_k\}$ is an arbitrary vector in ℓ_2. Indeed, Let $Le_1 = \{\beta_k\}_{k=1}^{\infty}$. Since $Le_k = LAe_{k-1} = e_{k-1}, k > 1$, equality (15.2) follows.

2. Let A_1 be the bounded linear operator defined on ℓ_2 by

$$A_1(\alpha_1, \alpha_2, \ldots) = \left(0, \alpha_1, \alpha_2 - \frac{1}{2}\alpha_1, \alpha_3 - \frac{1}{2}\alpha_2, \ldots\right).$$

Then A_1 is left invertible and every left inverse L_1 of A_1 is of the form

$$L_1((\alpha_k)) = (r_k), \text{ where } r_k = \alpha_1\beta_k + \alpha_{k+1} + \sum_{j=1}^{k} \frac{1}{2^{k+1-j}}\alpha_j, \qquad (15.3)$$

and (β_k) is an arbitrary vector in ℓ_2. This can be shown as follows: First we note that $A_1 = S - \frac{1}{2}S^2 = S(I - \frac{1}{2}S)$, where S is a forward shift operator on ℓ_2. If L_1 is a left inverse of A, then

$$I = L_1 A = L_1\left(I - \frac{1}{2}S\right)S. \qquad (15.4)$$

Since S is a forward shift, we have from (15.4) and the above example that $L_1 \times (I - \frac{1}{2}S) = L$, where L is given in (15.2). Now $\|\frac{1}{2}S\| = \frac{1}{2}$. Hence $I - \frac{1}{2}S$ is invertible, $L_1 = L(I - \frac{1}{2}S)^{-1}$, and a simple computation shows that

$$\left(I - \frac{1}{2}S\right)^{-1}((\alpha_k)) = (\eta_k), \text{ where}$$

$$\eta_1 = \alpha_1, \eta_k = \alpha_k + \sum_{j=1}^{k-1} \frac{1}{2^{k-j}}\alpha_j, k > 1. \qquad (15.5)$$

Thus from (15.2) and (15.5) we get

$$L_1((\alpha_k)) = L\left(I - \frac{1}{2}S\right)^{-1}((\alpha_k)) = L((\eta_k))$$

$$= \left(\alpha_1\beta_k + \alpha_{k+1} + \sum_{j=1}^{k} \frac{1}{2^{k-j+1}}\alpha_j\right).$$

3. Let $\mathcal{H} = L_2([0, 1])$. For $0 < \alpha < 1$, define the operators R_α and $R_{1/\alpha}$ on $\mathcal{H}$ by

$$(R_\alpha f)(t) = f(\alpha t), \quad 0 \le t \le 1, \qquad (15.6)$$

and

$$(R_{1/\alpha} f)(t) = \begin{cases} f(\frac{1}{\alpha}t), & 0 \le t \le \alpha, \\ 0, & \alpha < t \le 1. \end{cases} \qquad (15.7)$$

Clearly R_α is linear and

$$\|R_\alpha f\|^2 = \int_0^1 |f(\alpha t)|^2 dt = \frac{1}{\alpha} \int_0^\alpha |f(0)|^2 ds \le \frac{1}{\alpha} \|f\|^2. \qquad (15.8)$$

Thus $\|R_\alpha\| \le 1/\sqrt{\alpha}$. On the other hand, define

$$\varphi(t) = \begin{cases} \frac{1}{\sqrt{\alpha}}, & 0 \le t \le \alpha, \\ 0, & \alpha < t \le 1. \end{cases}$$

Then $\|\varphi\| = 1$, and from (15.8) we have

$$\|R_\alpha\|^2 \ge \|R_\alpha \varphi\|^2 = \frac{1}{\alpha} \|\varphi\|^2 = \frac{1}{\alpha}.$$

Hence $\|R_\alpha\| \ge 1/\sqrt{\alpha}$. We have shown that

$$\|R_\alpha\| = \frac{1}{\sqrt{\alpha}}. \qquad (15.9)$$

Similarly

$$\|R_{\frac{1}{\alpha}} \varphi\|^2 = \int_0^\alpha \left| \varphi \left(\frac{1}{\alpha} t \right) \right|^2 dt = \alpha \int_0^1 |\varphi(s)|^2 ds = \alpha \|\varphi\|.$$

Hence

$$\|R_{\frac{1}{\alpha}}\| = \sqrt{\alpha}. \qquad (15.10)$$

Now

$$(R_\alpha R_{\frac{1}{\alpha}} \varphi)(t) = (R_{\frac{1}{\alpha}} \varphi)(\alpha t) = \varphi(t), 0 \le t \le 1, \qquad (15.11)$$

which shows that $R_{\frac{1}{\alpha}}$ is a right inverse of R_α.

If follows from (15.6) and (15.11) that

$$\text{Ker } R_\alpha = \{\varphi \in \mathcal{H} | \varphi = 0 \text{ on } [0, \alpha]\}.$$

and

$$\Im R_\alpha = \mathcal{H}.$$

Thus R_α has a right inverse but is not invertible. Also, from (15.7) and (15.11) we have this

$$\text{Ker } R_{\frac{1}{\alpha}} = \{0\}$$

and

$$\Im R_{\frac{1}{\alpha}} = \{g \in \mathcal{H} | g = 0 \text{ on } [\alpha, 1]\}. \qquad (15.12)$$

Indeed, given such a g, define $\varphi(t) = g(\alpha t), 0 \le t \le 1$. Then $R_{1/\alpha} \varphi = g$.

Finally, the adjoint $R_\alpha^* = \frac{1}{\alpha} R_{1/\alpha}$. Indeed

$$\langle R_\alpha f, g \rangle = \int_0^1 f(\alpha t) \, \overline{g(t)} \, dt = \frac{1}{\alpha} \int_0^\alpha f(s) \, \overline{g\left(\frac{1}{\alpha} s\right)} \, ds$$

$$= \left\langle f, \frac{1}{\alpha} R_{1/\alpha} g \right\rangle,$$

where the last equality follows from (15.7).

Theorem 15.2 *Let operator $A \in \mathcal{L}(\mathcal{H}_1, \mathcal{H}_2)$ have a left inverse L. Then $\Im A$ is closed and complemented in $\mathcal{H}$ by Ker L. The operator AL is a projection onto $\Im A$ and Ker AL = Ker L.*

Proof: Suppose A has a left inverse L. Clearly, Ker $A \subseteq$ Ker $LA =$ Ker $I = \{0\}$. The bounded operator AL is a projection since $(AL)^2 = A(LA)L = AL$. From $\Im AL \subseteq \Im A$ and $\Im A = \Im(AL)A \subseteq \Im AL$, we have $\Im AL = \Im A$. Also, Ker $AL \subseteq$ Ker $LAL =$ Ker L and Ker $L \subseteq$ Ker AL. Thus Ker $AL =$ Ker L and

$$\mathcal{H} = \Im A \oplus \text{Ker } L. \qquad (15.13)$$

$\square$

Definition: Let M and N be subspaces (not necessarily closed) and let $\mathcal{H} = M \oplus N$. The *codimension* of M, written codim M, is defined to be dim N (finite or infinite).

The codim M is independent of the subspace N. Indeed, suppose $\mathcal{H} = M \oplus Z$. Define $\varphi : N \to Z$ as follows:

For each $v \in N$, there exists a unique $u \in M$ and a unique $z \in Z$ such that $v = u + z$. Let $\varphi(v) = z$. The linear operator φ is a $1 - 1$ map from N onto Z. Hence dim $Z =$ dim N.

Theorem 15.3 *Let the operator $A \in \mathcal{L}(\mathcal{H}_1, \mathcal{H}_2)$ have a left inverse $L \in \mathcal{L}(\mathcal{H}_2, \mathcal{H}_1)$. If B is an operator in $\mathcal{L}(\mathcal{H}_1, \mathcal{H}_2)$ and $\|A - B\| < \|L\|^{-1}$, then B has a left inverse L_1 given by*

$$L_1 = L(I - (A - B)L)^{-1} = L \left(\sum_{k=0}^\infty [(A - B)L]^k \right). \qquad (15.14)$$

Moreover,

$$\text{codim } \Im B = \text{codim } \Im A. \qquad (15.15)$$

Proof: Since $\|A - B\| < \|L\|^{-1}$, the operator $I - (A - B)L$ is invertible and $(I - (A - B)L)^{-1} = \sum_{k=0}^\infty [(A - B)L]^k$ by Theorem 8.1. Hence

$$L_1 B = L(I - (A - B)L)^{-1} B = L(I - (A - B)L)^{-1}[I - (A - B)L]A$$

$$= LA = I.$$

From the identity $B = [I - (A - B)L]A$ and the invertibility of $I - (A - B)L$, equality (15.15) is valid. Indeed, for $C = (I - (A - B)L)$ it follows from the invertibility of C and equality (15.13) that

$$\mathcal{H}_2 = C\mathcal{H}_2 = \Im CA \oplus C \mathrm{Ker}\, L = \Im B \oplus C\, \mathrm{Ker}\, L.$$

Hence

$$\mathrm{codim}\, \Im B = \dim C\, \mathrm{Ker}\, L = \dim \mathrm{Ker}\, L = \mathrm{codim}\, \Im A$$

□

If an operator $A \in L(\mathcal{H}_1, \mathcal{H}_2)$ has a left inverse L but A is not invertible (e.g., A is a forward shift operator on ℓ_2), then A has infinitely many left inverses; namely, for each $\lambda \in \mathbb{C}$, $L + \lambda(I - AL)$ is a left inverse of A. Indeed,

$$[L + \lambda(I - AL)]A = LA + \lambda A - \lambda ALA = LA = I.$$

Definition: An operator $A \in \mathcal{L}(\mathcal{H}_1, \mathcal{H}_2)$ is said to be *right invertible* if there exists an operator $R \in L(\mathcal{H}_2, \mathcal{H}_1)$ such that $AR = I$. The operator R is called a *right inverse* of A.

Example: Let A be the backward shift operator on ℓ_2, i.e., $A(\alpha_1, \alpha_2, \ldots) = (\alpha_2, \alpha_3, \ldots)$ Then A is right invertible and every right inverse R of A has the form

$$R(c_1, c_2, \ldots) = \sum_{k=1}^{\infty} c_k R e_k = \left(\sum_{k=1}^{\infty} c_k \gamma_k, c_1, c_2, \ldots \right), \qquad (15.16)$$

where (γ_k) is in ℓ_2 and $\{e_k\}$ is the standard basis in ℓ_2. To see this, let $Re_k = (\beta_1, \beta_2, \ldots)$ now

$$e_k = A R e_k = (\beta_2, \beta_3, \ldots).$$

Thus β_1 is arbitrary, $\beta_{k+1} = 1$, and $\beta_j = 0$ when $j \neq 1$ and $j \neq k + 1$. Hence $Re_k = \gamma_k e_1 + e_{k+1}$ for $k = 1, 2, \ldots$, and (15.16) follows. To see that (γ_k) is in ℓ_2 let f be the bounded linear functional on ℓ_2 defined by $f((\alpha_k)) = \alpha_1$. Since fR is a bounded linear functional on ℓ_2, there exists, by the Riesz representation theorem, a unique vector (z_k) in ℓ_2 such that for all $(c_k) \in \ell_2$ we have

$$\sum_{k=1}^{\infty} c_k \gamma_k = f R((c_k)) = \langle (c_k), (z_k) \rangle = \sum_{k=1}^{\infty} c_k \bar{z}_k.$$

Hence $(\gamma_k) = (\bar{z}_k) \in \ell_2$.

If the operator A has a right inverse R, then the equation $Ax = y$ has a solution $x = Ry$.

If R is a right inverse of A, then R^* is a left inverse of A^*.

Theorem 15.4 *Let $A \in \mathcal{L}(\mathcal{H}_1, \mathcal{H}_2)$ have a right inverse R. Then $\Im A = \mathcal{H}_2$ and the operator $I - RA$ is a projection from $\mathcal{H}_1$ onto Ker A with kernel $\Im R$. Thus $\Im R$ is closed and*

$$\mathcal{H}_1 = \Im R \oplus \text{Ker } A. \tag{15.17}$$

Proof: Clearly $\mathcal{H}_2 = \Im AR \subseteq \Im A \subseteq \mathcal{H}_2$. The bounded operator RA is a projection since $(RA)^2 = R(AR)A = RA$. From Im $RA \subseteq$ Im R and Im $R =$ Im $RAR \subseteq$ Im RA, we have Im $RA =$ Im R. Also, Ker $A \subseteq$ Ker RA and Ker $RA \subseteq$ Ker $ARA =$ Ker A. Thus Im $(I - RA) =$ Ker $RA =$ Ker A and Ker $(I - RA) = \Im RA = \Im R$. □

Theorem 15.5 *Let the operator $A \in \mathcal{L}(\mathcal{H}_1, \mathcal{H}_2)$ have a right inverse $R \in \mathcal{L}(\mathcal{H}_2, \mathcal{H}_1)$. If B is an operator in $\mathcal{L}(\mathcal{H}_1, \mathcal{H}_2)$ and*

$$\|A - B\| < \|R\|^{-1},$$

then B has a right inverse $R_1 \in \mathcal{L}(\mathcal{H}_2, \mathcal{H}_1)$ given by

$$R_1 = (I - R(A - B))^{-1} R = \sum_{k=0}^{\infty} [R(A - B)]^k R.$$

Moreover,

$$\dim \text{Ker } A = \dim \text{Ker } B.$$

Proof: Since $\|R(A - B)\| < 1$, the operator $I - R(A - B)$ is invertible and $(I - R(A - B))^{-1} = \sum_{k=0}^{\infty} [R(A - B)]^k$ by Theorem 8.1. Hence

$$BR_1 = B(I - R(A - B))^{-1} R = A[I - R(A - B)] \left[(I - R(A - B))^{-1} R \right]$$
$$= AR = I.$$

Let $C = (I - R(A - B))$. From the identity $B = A[I - R(A - B)] = AC$ and the invertibilty of C, we have

$$\dim \text{Ker } B = \dim \text{Ker } AC = \dim C^{-1} \text{Ker } A = \dim \text{Ker } A.$$

□

If an operator $A \in \mathcal{L}(H_1, H_2)$ has a right inverse R but A is not invertible (e.g., A is a backward shift operator on ℓ_2), then A has infinitely many right inverses, namely for each $\lambda \in \mathbb{C}$, $R + \lambda(I - RA)$ is a right inverse of A. Indeed,

$$A[R + \lambda(I - RA)] = AR + \lambda A - \lambda ARA = AR = I.$$

2.16 Compact Operators

A very important class of bounded linear operators which arise in the study of integral equations is the class of compact operators. This section presents some properties and examples of compact operators which are used in subsequent chapters.

Definition: An operator $K \in \mathcal{L}(\mathcal{H}_1, \mathcal{H}_2)$ is *compact* if for each sequence $\{x_n\}$ in $\mathcal{H}_1$, $\|x_n\| = 1$, the sequence $\{Kx_n\}$ has a subsequence which converges in $\mathcal{H}_2$.

(a) If $K \in \mathcal{L}(\mathcal{H}_1, \mathcal{H}_2)$ is of finite rank, then K is compact. For suppose $\{x_n\} \subset H_1$, $\|x_n\| = 1$. Then $\{Kx_n\}$ is a bounded sequence in the finite dimensional space $\Im K$. Since $\Im K$ is linearly isometric to $\mathbb{C}^k$ for some k, it follows that $\{Kx_n\}$ has a convergent subsequence.

(b) If $\dim \mathcal{H}_1 < \infty$, then any linear map K from $\mathcal{H}_1$ into $\mathcal{H}_2$ is bounded by Example 1 in Section 2. Since $\Im K = KH_1$ is finite dimensional, K is compact by (a).

(c) The identity operator on an infinite dimensional Hilbert space is not compact; for if $\varphi_1, \varphi_2, \ldots$ is an infinite orthonormal set in $\mathcal{H}$, then

$$\|I\varphi_n - I\varphi_m\| = \sqrt{2},$$

which implies that $\{I\varphi_n\}$ has no convergent subsequence even though $\|\varphi_n\| = 1$.

If $K \in \mathcal{L}(\mathcal{H}_1, \mathcal{H}_2)$ is compact and $\{z_n\}$ is a bounded sequence in $\mathcal{H}_1$, then $\{Kz_n\}$ has a convergent subsequence.

Theorem 16.1 *Suppose K and L are compact operators in $\mathcal{L}(\mathcal{H}_1, \mathcal{H}_2)$. Then*

(i) *$K + L$ is compact.*

(ii) *If $A \in \mathcal{L}(\mathcal{H}_3, \mathcal{H}_1)$ and $B \in \mathcal{L}(\mathcal{H}_2, \mathcal{H}_3)$, where $\mathcal{H}_3$ is a Hilbert space, then KA and BK are compact.*

Proof: (i) Given $\{x_n\} \subset \mathcal{H}_1$, $\|x_n\| = 1$, the sequence $\{Kx_n\}$ has a convergent subsequence $\{Kx_{n''}\}$. Since L is compact, $\{Lx_{n'}\}$ has a convergent subsequence $\{Lx_{n''}\}$. Thus $\{(K + L)x_{n''}\}$ converges.

(ii) Given $\{z_n\} \subset \mathcal{H}_3$, $\|z_n\| = 1$, the sequence $\{Az_n\}$ is bounded. Therefore it follows from the compactness of K that $\{KAz_n\}$ has a convergent subsequence. Hence KA is compact.

If $\{x_n\} \subset \mathcal{H}_1$, $\|x_n\| = 1$, there is a subsequence $\{Kx_{n'}\}$ of $\{Kx_n\}$ which converges. Therefore, by the continuity of B, $\{BKx_{n'}\}$ converges. Thus BK is compact. $\qquad\square$

Theorem 16.2 *An operator $A \in \mathcal{L}(\mathcal{H}_1, \mathcal{H}_2)$ is compact if and only if its adjoint A^* is compact.*

Proof: Suppose A is compact. Let $\{x_n\} \subset \mathcal{H}_2$, $\|x_n\| = 1$. Since AA^* is compact by Theorem 16.1, there exists a subsequence $\{x_{n'}\}$ of $\{x_n\}$ such that $\{AA^*x_{n'}\}$ converges. Thus

$$\|A^*x_{n'} - A^*x_{m'}\|^2 = \langle AA^*(x_{n'} - x_{m'}), x_{n'} - x_{m'} \rangle$$
$$\leq \|AA^*(x_{n'} - x_{m'})\| \|x_{n'} - x_{m'}\|$$
$$\leq 2\|AA^*x_{n'} - AA^*x_{m'}\| \to 0$$

as n' and $m' \to \infty$.

The convergence of $\{A^*x_{n'}\}$ is ensured by the completeness of $\mathcal{H}_1$. Therefore A^* is compact.

If A^* is compact, then by the result we just proved, $A^{**} = A$ is compact. $\square$

In order to exhibit some important examples of compact operators, we use the following result.

Theorem 16.3 *Suppose $\{K_n\}$ is a sequence of compact operators in $\mathcal{L}(\mathcal{H}_1, \mathcal{H}_2)$ and $\|K_n - K\| \to 0$, where K is in $\mathcal{L}(\mathcal{H}_1, \mathcal{H}_2)$. Then K is compact.*

Proof: The proof employs a diagonalization procedure as follows.

Let $\{x_n\}$ be a sequence in $\mathcal{H}_1$, $\|x_n\| = 1$. Since K_1 is compact, there exists a subsequence $\{x_{1n}\}$ of $\{x_n\}$ such that $\{K_1x_{1n}\}$ converges. Since K_2 is compact, there exists a subsequence $\{x_{2n}\}$ of $\{x_{1n}\}$ such that $\{K_2x_{2n}\}$ converges. Continuing in this manner, we obtain for each integer $j \geq 2$, a subsequence $\{x_{jn}\}_{n=1}^{\infty}$ of $\{x_{(j-1)n}\}_{n=1}^{\infty}$ such that $\{K_jx_{jn}\}_{n=1}^{\infty}$ converges. We now show that the "diagonal" sequence $\{Kx_{nn}\}$ converges, which proves that K is compact.

Given $\varepsilon > 0$, there exists, by hypothesis, an integer p such that

$$\|K - K_p\| < \varepsilon/2. \tag{16.1}$$

Now $\{K_px_{nn}\}$ converges since $n \geq p$ implies that $\{K_px_{nn}\}$ is a subsequence of the convergent sequence $\{K_px_{pn}\}$. By (16.1),

$$\|Kx_{nn} - Kx_{mm}\| \leq \|Kx_{nn} - K_px_{nn}\| + \|K_px_{nn} - K_px_{mm}\|$$
$$+ \|K_px_{mm} - Kx_{mm}\|$$
$$\leq 2\|K_p - K\| + \|K_px_{nn} - K_px_{mm}\|$$
$$< \varepsilon + \|K_px_{nn} - K_px_{mm}\| \to \varepsilon$$

as $n, m \to \infty$. Thus it follows that $\{Kx_{nn}\}$ is a Cauchy sequence which must converge since $\mathcal{H}_2$ is complete. $\square$

Examples: 1. Let $\{\lambda_k\}$ be a sequence in $\mathbb{C}$ which converges to zero. Define $K \in \mathcal{L}(\ell_2)$ by

$$K(\alpha_1, \alpha_2, \ldots) = (\lambda_1 \alpha_1, \lambda_2 \alpha_2, \ldots).$$

For each positive integer n, let $K_n \in \mathcal{L}(\ell_2)$ be the operator defined by

$$K_n(\alpha_1, \alpha_2, \ldots) = (\lambda_1 \alpha_1, \ldots, \lambda_n \alpha_n, 0, 0, \ldots).$$

K_n is of finite rank and therefore is compact. Since

$$\|K_n - K\| \leq \sup_{k \geq n} |\lambda_k| \to 0, \quad \text{as } n \to \infty,$$

K is also compact by Theorem 16.3.

2. Given the infinite matrix $(a_{ij})_{i,j=1}^{\infty}$, where $\sum_{i,j=1}^{\infty} |a_{ij}|^2 < \infty$, let $A \in \mathcal{L}(\ell_2)$ be the operator corresponding to this matrix as defined in Section 2, Example 3. To show that A is compact, let $A_n \in \mathcal{L}(\ell_2), n = 1, 2, \ldots$, be the operator corresponding to the matrix $(a_{ij}^{(n)})_{i,j=1}^{\infty}$, where $a_{ij}^{(n)} = a_{ij}, 1 \leq i, j \leq n$, and $a_{ij}^{(n)} = 0$ otherwise. Since A_n is of finite rank, it is compact. Moreover,

$$\|A - A_n\| \to 0.$$

Hence A is compact.

3. Let $K \colon L_2([a, b]) \to L_2([a, b])$ be the integral operator

$$(Kf)(t) = \int_a^b k(t, s) f(s)\,ds,$$

where k is in $L_2([a, b] \times [a, b])$. By Section 2, Example 5, K is a bounded linear operator and

$$\|K\| \leq \left(\int_a^b \int_a^b |k(t, s)|^2\,ds\,dt \right)^{1/2} = \|k\|. \tag{16.2}$$

We construct a sequence of operators of finite rank which converges, in norm, to K as follows:

Let $\varphi_1, \varphi_2, \ldots$ be an orthonormal basis for $L_2([a, b])$. Then $\Phi_{ij}(t, s) = \varphi_i(t) \times \bar{\varphi}_j(s), i, j = 1, 2, \ldots$ is an orthonormal basis for $L_2([a, b] \times [a, b])$ by Theorem I.14.1.

Thus $k = \sum_{i,j=1}^{\infty} \langle k, \Phi_{ij} \rangle \Phi_{ij}$. Define

$$k_n(t, s) = \sum_{i,j=1}^{n} \langle k, \Phi_{ij} \rangle \Phi_{ij}(t, s).$$

Then

$$\|k - k_n\| \to 0. \tag{16.3}$$

Let K_n be the integral operator defined on $L_2([a, b])$ by

$$(K_n f)(t) = \int_a^b k_n(t, s) f(s)\,ds.$$

Now K_n is a bounded linear operator of finite rank since $\Im K_n \subset \mathrm{sp}\{\varphi_1, \ldots, \varphi_n\}$. By (16.3) and (16.2) applied to $K - K_n$.

$$\|K - K_n\| \le \|k - k_n\| \to 0.$$

Hence K is compact.

4. Given a complex valued function $a(t)$ which is continuous on $[a, b]$, let $A : L_2([a, b]) \to L_2([a, b])$ be the bounded linear operator given by

$$(Af)(t) = a(t)f(t).$$

If $a(t_0) \ne 0$ for some $t_0 \in [a, b]$, then A is not compact. Indeed, it follows from the continuity of $a(t)$ that $|a(t)| \ge \frac{|a(t_0)|}{2} > 0$ for all t in some compact interval J containing t_0. Let $\{\widetilde{\varphi}_n\}$ be an orthonormal basis for $L_2(J)$. Define φ_n to be $\widetilde{\varphi}_n$ on J and zero elsewhere on $[a, b]$. Then $\|\varphi_n\| = 1$ and for $n \ne m$,

$$\|A\varphi_n - A\varphi_m\|^2 = \int_a^b |a(t)|^2 |\varphi_n(t) - \varphi_m(t)|^2 dt$$

$$\ge \frac{|a(t_0)|^2}{4} \int_J |\widetilde{\varphi}_n(t) - \widetilde{\varphi}_m(t)|^2 dt = \frac{|a(t_0)|^2}{2}.$$

Hence $\{A\varphi_n\}$ does not have a convergent subsequence.

We conclude this section with a compactness result for band operators. This requires some preparations.

An infinite matrix $(a_{jk})_{j,k=0}^\infty$ is called a *band matrix* if all its nonzero entries are in a finite number of diagonals parallel to the main diagonal, that is, there exists an integer N such that

$$a_{jk} = 0 \text{ if } |j - k| > N. \tag{16.4}$$

If this number N is equal to 1, then the matrix is called *tridiagonal*.

An operator A on ℓ_2 induced by a band matrix is called a *band operator*. In other words, an operator A on ℓ_2 is a band operator if and only if the matrix of A with respect to the standard orthogonal basis $e_0, e_1, e_2, \ldots$ of ℓ_2 is a band matrix.

Theorem 16.4 *In order that a band matrix $(a_{jk})_{j,k=0}^\infty$ induce a bounded linear operator A on ℓ_2 it is necessary and sufficient that*

$$\sup_{j,k=1,2,\ldots} |a_{jk}| < \infty, \tag{16.5}$$

and in this case

$$\|A\| \le \sum_{\nu=-N}^N \sup_{j-k=\nu} |a_{jk}|, \tag{16.6}$$

where N is the integer in (16.4). Furthermore, the band operator A is compact if and only if

$$\lim_{j,k\to\infty} a_{jk} = 0. \tag{16.7}$$

Proof: If the operator A induced by the matrix $(a_{jk})_{j,k=0}^{\infty}$ is bounded, then

$$|a_{jk}| = |\langle Ae_k, e_j \rangle| \leq \|A\|$$

for each j and k. Hence (16.5) holds. To prove the reverse implication, assume (16.5) holds. For $\nu \geq 0$ let D_ν and $D_{-\nu}$ be the operators on ℓ_2 defined by

$$D_\nu(x_0, x_1, x_2, \ldots) = (\underbrace{0, \ldots 0}_{\nu}, a_{\nu 0}x_0, a_{\nu 1}x_1, \ldots),$$

$$D_{-\nu}(x_0, x_1, x_2, \ldots) = (a_{0\nu}x_\nu, a_{1\nu}x_{\nu+1}, a_{2\nu}x_{\nu+2}, \ldots).$$

Condition (16.5) implies that the operator D_n is bounded for each integer n and

$$\|D_n\| = \sup_{j-k=n} |a_{jk}|. \tag{16.8}$$

Let $A = \sum_{n=-N}^{N} D_n$. Then A is a bounded linear operator on ℓ_2, and the matrix of A with respect to the standard basis of ℓ_2 is precisely the band matrix $(a_{jk})_{j,k=0}^{\infty}$. Thus (16.5) implies that the matrix $(a_{jk})_{j,k=0}^{\infty}$ induces a bounded linear operator on ℓ_2. Futhermore, $\|A\| \leq \sum_{n=-N}^{N} \|D_n\|$. By combining this with (16.8) we obtain (16.6). Notice that (16.6) implies that

$$\|A\| \leq (2N+1) \sup_{j,k=1,2,\ldots} |a_{jk}|. \tag{16.9}$$

Now, let condition (16.7) hold. Let P_n be the orthogonal projection on ℓ_2 defined by

$$P_n(x_0, x_1, x_2, \ldots) = (x_0, \ldots, x_n, 0, 0, \ldots),$$

and set $Q_n = I - P_n$. Relative to the standard orthogonal basis of ℓ_2 the matrix of the operator $Q_n A Q_n$ has the following block form

$$Q_n A Q_n = \begin{bmatrix} 0 & 0 \\ 0 & A_n \end{bmatrix},$$

where A_n is the band matrix $(a_{jk})_{j,k=n+1}^{\infty}$. Notice that

$$A - Q_n A Q_n = P_n A P_n + P_n A Q_n + Q_n A P_n,$$

and hence $A - Q_n A Q_n$ is finite rank for each n. From (16.9) it follows that

$$\|Q_n A Q_n\| \leq (2N+1) \sup_{j,k=n+1,n+2,\ldots} |a_{jk}|,$$

and hence (16.7) implies that $\|Q_n A Q_n\| \to 0$ for $n \to \infty$. Thus in the operator norm, A is the limit of a sequence of finite rank operators, and hence A is compact.

Conversely, let A be compact. Then from Lemma 17.8 (which is independent of the results in this section) it follows that $\| Q_n A Q_n \| \to 0$ for $n \to \infty$. For $j, k \geq n + 1$ we have

$$|a_{jk}| = |\langle A e_k, e_j \rangle| = |\langle Q_n A Q_n e_k, e_j \rangle|$$
$$\leq \| Q_n A Q_n \|.$$

Since $\| Q_n A Q_n \| \to 0$ for $n \to \infty$ we see that (16.7) holds. $\qquad\square$

2.17 The Projection Method for Inversion of Linear Operators

A scheme for approximating the solution to an infinite system of linear equations was given in Theorem 9.1. The proof of the theorem relied on the fact that the operator corresponding to the coefficient matrix was the limit, in norm, of a sequence of operators of finite rank.

If the restrictions on the coefficient matrix are weakened so that the sequence of these operators does not necessarily converge in norm, then other arguments may be used.

Let us begin with the infinite system

$$\begin{aligned} c_{11}x_1 + c_{12}x_2 + \cdots &= y_1 \\ c_{21}x_1 + c_{22}x_2 + \cdots &= y_2 \\ \vdots \qquad \vdots \qquad\quad &\ \ \vdots \end{aligned} \tag{i}$$

where $(y_1, y_2, \ldots)$ is in ℓ_2. As in Theorem 9.1, the natural approach to approximating a solution to (i) is the following:

For all n sufficiently large, find a solution $(x_1^{(1)}, \ldots, x_n^{(n)})$ to the finite system

$$\begin{aligned} c_{11}x_1 + \cdots + c_{1n}x_n &= y_1 \\ c_{21}x_1 + \cdots + c_{2n}x_n &= y_2 \\ \vdots \qquad \vdots \qquad\ \ \vdots \qquad & \\ c_{n1}x_1 + \cdots + c_{nn}x_n &= y_n. \end{aligned} \tag{ii}$$

Then, hopefully, $v_n = (x_1^{(n)}, \ldots, x_n^{(n)}, 0, 0, \ldots)$ converges in ℓ_2 to a solution of (i).

To put this scheme within the framework of operator theory, define the operator C on ℓ_2 by

$$C(\alpha_1, \alpha_1, \ldots) = (\beta_1, \beta_2, \ldots),$$

where

$$\beta_j = \sum_{k=1}^{\infty} c_{jk}\alpha_k, \quad j = 1, 2, \ldots.$$

We shall assume that $\sum_{j=1}^{\infty} \sum_{k=1}^{\infty} |c_{jk}|^2 < \infty$ so that C is in $\mathcal{L}(\ell_2)$ (Example 3 in Section 2). Define $P_n \in \mathcal{L}(\ell_2)$ by

$$P_n(\alpha_1, \alpha_2, \ldots) = (\alpha_1, \alpha_2, \ldots, \alpha_n, 0, 0, \ldots),$$

we see that (ii) is equivalent to

$$P_n C P_n(x_1, x_2, \ldots) = P_n y.$$

Note that

$$P_n^2 = P_n, \ \|P_n\| = 1 \text{ and } P_n x \to x \text{ for every } x \in \ell_2.$$

The scheme prescribes that $v_n = (x_1^{(n)}, \ldots, x_n^{(n)}, 0, 0, \ldots)$ be found so that

$$P_n C P_n v_n = P_n y$$

and $\{v_n\}$ converges in ℓ_2 to a solution of (i).

We now consider a more general approximation scheme.

Definition: Let $\{P_n\}$ be a bounded sequence of projections in $\mathcal{L}(\mathcal{H})$ with the property that $P_n x \to x$ for all $x \in \mathcal{H}$. Given an invertible operator $A \in \mathcal{L}(\mathcal{H})$, the *projection method* for A seeks to approximate a solution to the equation $Ax = y$ by a sequence $\{x_n\}$ of solutions to the equations

$$P_n A P_n x = P_n y, \quad n = 1, 2, \ldots \tag{17.1}$$

This projection method for A is said to *converge* if there exists an integer N such that for each $y \in \mathcal{H}$ and $n \geq N$, there exists a *unique* solution x_n to equation (17.1) and, in addition, the sequence $\{x_n\}$ converges to $A^{-1} y$.

We denote by $\Pi(P_n)$ the set of invertible operators A for which the above projection method for A converges.

Unless a specific sequence of projections is used, we shall always assume that we have the general sequence $\{P_n\}$ with the properties above.

Since $P_n x \to x$ for each $x \in \mathcal{H}$, $\sup_n \|P_n x\| < \infty$ for each $x \in \mathcal{H}$. Hence

$$\sup_n \|P_n\| < \infty \tag{17.2}$$

by Theorem 14.3. Also, since P_n is a projection, we know from Theorem 15.1 that $\Im P_n$ is closed and therefore is a Hilbert space.

Theorem 17.1 *Let $A \in \mathcal{L}(H)$ be invertible. Then A is in $\Pi(P_n)$ if and only if there exists an integer N such that for $n \geq N$, the restriction of the operatro $P_n A P_n$ to $\Im P_n$ has a bounded inverse, denoted by $(P_n A P_n)^{-1}$, and*

$$\sup_{n \geq N} \|(P_n A P_n)^{-1}\| < \infty. \tag{17.3}$$

Proof: Suppose there exists an integer N such that for $n \geq N$ and $y \in \mathcal{H}$, the equation $P_n A P_n x = P_n y$ has a unique solution $x_n \in \Im P_n$. In addition, assume (17.3) holds. By (17.2) and (17.3),

$$M = \sup_{n \geq N} \{\|(P_n A P_n)^{-1}\| + \|P_n\|\} < \infty.$$

Then for $x = A^{-1}y$,

$$\|x_n - P_n x\| = \|(P_n A P_n)^{-1} P_n A P_n (x_n - P_n x)\| \leq M \|P_n A P_n x_n - P_n A P_n x\|$$
$$= M \|P_n y - P_n A P_n x\| \leq M^2 \|y - A P_n x\| \to M^2 \|y - Ax\| = 0.$$

Hence

$$x_n = x_n - P_n x + P_n x \to x = A^{-1} y,$$

which shows that $A \in \Pi(P_n)$. Conversely, assume $A \in \Pi(P_n)$. Then there exists an integer N such that for $n \geq N$, the operator $P_n A P_n$ restricted to $\Im P_n$ is a one-one map from the Hilbert space $\Im P_n$ onto $\Im P_n$. Hence $(P_n A P_n)^{-1}$ is a bounded linear operator on $\Im P_n$ by Theorem 14.1. Since $x_n = (P_n A P_n)^{-1} P_n y$ is the unique solution to equation (17.1) for $n \geq N$ and since $A \in \Pi(P_n)$, we have that

$$(P_n A P_n)^{-1} P_n y \to A^{-1} y \text{ for each } y \in \mathcal{H}.$$

Hence

$$m = \sup_{n \geq N} \|(P_n A P_n)^{-1} P_n\| < \infty$$

by Theorem 14.3. Thus

$$\|(P_n A P_n)^{-1} P_n y\| = \|(P_n A P_n)^{-1} P_n P_n y\| \leq m \|P_n y\|$$

which shows that

$$\sup_{n \geq N} \|(P_n A P_n)^{-1}\| \leq m.$$

$\square$

Corollary 17.2 *Let* $\{P_n\}$ *be a sequence of projections on a Hilbert space* $\mathcal{H}$. *Suppose* $\|P_n\| = 1$, $n = 1, 2, \ldots$ *and* $P_n x \to x$ *for every* $x \in \mathcal{H}$. *If* $A \in \mathcal{L}(\mathcal{H})$ *with* $\|I - A\| < 1$, *then* $A \in \Pi(P_n)$.

Proof: For any $x \in \mathcal{H}$,

$$P_n A P_n x = (I - P_n(I - A) P_n) P_n x. \tag{17.4}$$

Let $B = I - A$. Since $\|P_n B P_n\| \leq \|B\| < 1$, both $A = I - B$ and $I - P_n B P_n$ are invertible. It is easy to see that $(I - P_n B P_n)^{-1}$ maps $\Im P_n$ onto $\Im P_n$. Thus it follows from (17.4) that $P_n A P_n$ is invertible on $\Im P_n$ and

$$(P_n A P_n)^{-1} P_n y = (I - P_n(I - A) P_n)^{-1} P_n y. \tag{17.5}$$

By Theorem 8.1,

$$\|(I - P_n(I - A)P_n)^{-1}\| \leq \frac{1}{1 - \|P_n(I - A)P_n\|}$$
$$\leq (1 - \|I - A\|)^{-1}. \tag{17.6}$$

The corollary now follows from (17.5), (17.6) and Theorem 17.1.

Let P_n be the orthogonal projection of ℓ_2 onto the first n coordinates, i.e.,

$$P_n(x_0, x_1, x_2, \ldots) = (x_0, \ldots, x_{n-1}, 0, 0, \ldots).$$

For any $x = (x_0, x_1, x_2, \ldots) \in \ell_2$ we have that

$$\|x - P_n x\|^2 = \sum_{j \geq n} |x_j|^2 \to 0 \ (n \to \infty),$$

and hence $P_n x \to x$ for each $x \in \ell_2$. Now, let A be invertible on ℓ_2. We say that the *finite section method converges* for A if the projections method for A relative to the projections $\{P_n\}_{n=0}^{\infty}$ converges, that is, $A \in \Pi\{P_n\}$.

In this setting identify $\Im P_n$ with $\mathbb{C}^n$, and we refer to $P_n A P_n$ as the nth *section* of A.

Similarly, for an invertible operator A on $\ell_2(\mathbb{Z})$ we say that the *finite section method converges* for A if $A \in \Pi\{P_n\}$, where now P_n is the orthogonal projection on $\ell_2(\mathbb{Z})$ given by

$$P_n(x_j)_{j \in \mathbb{Z}} = (\ldots, 0, 0, x_{-n}, \ldots, x_n, 0, 0, \ldots),$$

with $n = 0, 1, 2, \ldots$ notice that

$$\|P_n x - x\|^2 = \sum_{|j| > n} |x_j|^2 \to 0 \ (n \to \infty),$$

and thus $P_n x \to x$ for each $x \in \ell_2(\mathbb{Z})$. $\qquad\square$

Examples: 1. Let A be the operator on $\ell_2(\mathbb{Z})$ given by the following two diagonal doubly infinite matrix

$$A = \begin{pmatrix} \ddots & & & \\ & 1 & 0 & 0 \\ & -c & \boxed{1} & 0 \\ & 0 & -c & 1 \\ & & & \ddots \end{pmatrix}.$$

Here $|c| < 1$, and, as usual, the box indicates the entry in the zero-zero position. We shall show that the finite section method converges for A.

Notice that $A = I - cV$, where V is the forward shift on $\ell_2(\mathbb{Z})$. Since $\|V\| = 1$, we have $\|cV\| < 1$, and hence A is invertible with

$$A^{-1}x = \sum_{j=0}^{\infty} c^j V^j \; x \in \ell_2(\mathbb{Z}).$$

It follows that the matrix of A^{-1} relative to the standard basis of $\ell_2(\mathbb{Z})$ is given by

$$A^{-1} = \begin{pmatrix} \ddots & & \ddots & & & & \\ & 1 & 0 & 0 & & 0 & 0 \\ \ddots & c & 1 & 0 & & 0 & 0 \\ & c^2 & c & \boxed{1} & & 0 & 0 \\ & c^3 & c^2 & c & & 1 & 0 \\ & & & & & & \ddots \\ & c^4 & c^3 & c^2 & & c & 1 \\ & & & & & & & \ddots \end{pmatrix}.$$

Next, consider the truncation $A_n = P_n A P_n |\Im P_n$, where P_n is as in the paragraph preceding this example. Identifying $\Im P_n$ with $\mathbb{C}^{2n+1}$ we see that the matrix of A_n is given by

$$A_n = \begin{pmatrix} 1 & 0 & 0 & \cdots & 0 \\ -c & 1 & 0 & \cdots & 0 \\ 0 & -c & 1 & & 0 \\ \vdots & & & \ddots & \\ 0 & 0 & 0 & & 1 \end{pmatrix}.$$

A straight forward calculation shows that A_n is invertible and

$$A_n^{-1} = \begin{pmatrix} 1 & 0 & 0 & \cdots & 0 \\ c & 1 & 0 & \cdots & 0 \\ c^2 & c & 1 & & 0 \\ \vdots & \vdots & \vdots & \ddots & \\ c^{2n+1} & c^{2n} & c^{2n-1} & \cdots & 1 \end{pmatrix}.$$

By comparing the matrices for A^{-1} and A_n^{-1} it is clear that

$$A_n^{-1} = P_n A^{-1} |\Im P_n,$$

and hence for the operator A the finite section method converges.

The above conclusion can also be derived directly from Corollory 17.2. Indeed, $I - A = cV$, and thus $\|I - A\| < 1$.

2. In this example we consider the operator B on $\ell_2(\mathbb{Z})$ given by

$$B = \begin{pmatrix} \ddots & & & \\ -c & 0 & 0 & \\ 1 & \boxed{-c} & 0 & \\ 0 & 1 & -c & \\ & & & \ddots \end{pmatrix}.$$

Again the matrix is two diagonal and $|c| < 1$. Notice that B differs from A in the previous example only in the fact that the order of the two diagonals with nonzero entries is interchanged.

Recall that V is invertible, and that V^{-1} is the backward shift on $\ell_2(\mathbb{Z})$. Thus

$$B = -cI + V = (I - cV^{-1})V.$$

Since $\|cV^{-1}\| = |c| < 1$, we conclude that B is invertible, and

$$B^{-1}x = V^{-1}(I - cV^{-1})^{-1}x = \sum_{j=0}^{\infty} c^j V^{-j-1}x,$$

for each $x \in \ell_2(\mathbb{Z})$. Thus

$$B^{-1} = \begin{pmatrix} \ddots & & & & & \\ 0 & 1 & c & c^2 & c^3 & \\ 0 & 0 & 1 & c & c^2 & \\ 0 & 0 & \boxed{0} & 1 & c & \\ 0 & 0 & 0 & 0 & 1 & \\ 0 & 0 & 0 & 0 & 0 & \\ & & & & & \ddots \end{pmatrix}.$$

Let us now consider the truncated operator $B_n = P_n B P_n$, where P_n is as in previous example.

Identifying $\Im P_n$ with $\mathbb{C}^{2n+1}$ we see that

$$B_n = \begin{pmatrix} -c & 0 & 0 & \cdots & 0 \\ 1 & -c & 0 & \cdots & 0 \\ 0 & 1 & -c & \cdots & 0 \\ \vdots & & & & \\ 0 & 0 & 0 & & -c \end{pmatrix}.$$

A straightforward calculation shows that

$$B_n^{-1} = -\frac{1}{c} \begin{pmatrix} 1 & 0 & 0 & \cdots & 0 \\ 1/c & 1 & 0 & \cdots & 0 \\ 1/c^2 & 1/c & 1 & & 0 \\ \vdots & \vdots & \vdots & & \vdots \\ 1/c^{2n+1} & 1/c^{2n} & 1/c^{2n-1} & \cdots & 1 \end{pmatrix}.$$

By comparing the matrix of B^{-1} with those for B_0^{-1}, B_1^{-1}, B_2^{-1} ... we see that for B the finite section method does not converge. This conclusion also follows from Theorem 17.1 because in this case

$$\|(P_n B P_n)^{-1}\| = \|B_n^{-1}\| \geq \left(\frac{1}{c}\right)^{2n+1} \to \infty \quad (n \to \infty).$$

Let us remark that in this example the operator B is invertible and the equation

$$P_n A P_n x = P_n y$$

has a unique solution $x_n \in \Im P_n$ for each $y \in \ell_2(\mathbb{Z})$, but the sequence $x_0, x_1, x_2, \ldots$ need not converge. For instance, if $y = (\delta_{j0})_{j \in \mathbb{Z}}$, then

$$x_n = -\frac{1}{c}\left(1, \frac{1}{c}, \ldots, \frac{1}{c^{2n+1}}\right), \quad n = 0, 1, 2, \ldots,$$

and this sequence does not converge.

A further analysis of the different behavior of the finite section approximation in these two examples and the reason behind it will be presented in the last section of Chapter III.

Lemma 17.3 *Let $\mathcal{H}$ be a Hilbert space. Given $A \in \mathcal{L}(\mathcal{H})$, suppose there exists a number $c > 0$ such that*

$$|\langle Ax, x\rangle| \geq c\|x\|^2 \tag{17.7}$$

for all $x \in \mathcal{H}$. Then A is invertible and $\|A^{-1}\| \leq \frac{1}{c}$.

Proof: Since

$$\|Ax\|\|x\| \geq |\langle Ax, x\rangle| \geq c\|x\|^2$$

we have

$$\|Ax\| \geq c\|x\|. \tag{17.8}$$

Thus A is injective and its range $\Im A$ is closed by Corollary 14.2. We show that $\Im A = \mathcal{H}$. Let P be the orthogonal projection onto $\Im A$. Given $y \in \mathcal{H}$, we know that $y - Py \in \operatorname{Ker} P = \Im P^{\perp} = \Im A^{\perp}$. Hence

$$0 = |\langle A(y - Py), y - Py\rangle| \geq c\|y - Py\|^2.$$

Therefore, $y = Py \in \Im A$. The inequality $\|A^{-1}\| \leq \frac{1}{c}$ follows from inequality (17.8). $\qquad\square$

Theorem 17.4 *Let $\mathcal{H}$ be a Hilbert space. Given $A \in \mathcal{L}(\mathcal{H})$, suppose there exists a number $c > 0$ such that*

$$|\langle Ax, x\rangle| \geq c\|x\|^2$$

for all $x \in \mathcal{H}$. Let $\{P_n\}$ be a sequence of orthogonal projections on $\mathcal{H}$ with the property that $P_n x \to x$ for all $x \in \mathcal{H}$. Then $A \in \Pi(P_n)$.

Proof: Since P_n is a self adjoint projection,

$$|\langle P_n AP_n x, P_n x\rangle| = |\langle AP_n x, P_n x\rangle| \geq c\|P_n x\|^2, \ x \in \mathcal{H}.$$

It follows from Lemma 17.3, that $P_n AP_n$, restricted to $(\Im P_n)$, is invertible and $\|(P_n AP_n)^{-1}\| \leq \frac{1}{c}, \ m = 1, 2, \ldots$

Hence $A \in \Pi(P_n)$ by Theorem 17.1. $\qquad\square$

Corollary 17.5 *Given $A \in \mathcal{L}(\mathcal{H})$, suppose there is a number $c > 0$ such that*

$$Re\langle Ax, x\rangle \geq c\|x\|^2.$$

Let $\{P_n\}$ be a sequence of orthogonal projections on $\mathcal{H}$ with the property that $P_n x \to x$ for all $x \in \mathcal{H}$. Then $A \in \Pi(P_n)$.

Next we show that the projection method is stable under certain perturbations.

Theorem 17.6 *Suppose $A \in \Pi(P_n)$. There exists a $\gamma > 0$ such that if $B \in L(\mathcal{H})$ and $\|B\| < \gamma$, then $A + B \in \Pi(P_n)$.*

Proof: There exists an integer N such that for $n \geq N$, the operator $P_n AP_n$, restricted to $\Im P_n$, is invertible and

$$M = \sup_{n \geq N} (\|(P_n AP_n)^{-1}\| + \|P_n\|) < \infty \tag{17.9}$$

now for $n \geq N$,

$$P_n(A + B)P_n = P_n AP_n (I + (P_n AP_n)^{-1} P_n B) P_n \tag{17.10}$$

and

$$\|(P_n AP_n)^{-1} P_n B\| < \frac{1}{2} \text{ if } \|B\| < \frac{1}{2M^2}, \ n \geq N. \tag{17.11}$$

Hence $I + (P_n AP_n)^{-1} P_n B$ is invertible and

$$\|(I + (P_n AP_n)^{-1} P_n B)^{-1}\| \leq \frac{1}{1 - \|(P_n AP_n) P_n B\|} \leq 2, \ n \geq N. \tag{17.12}$$

by Theorem 8.1 and inequality (17.11).

Since $(I + (P_n AP_n)^{-1} P_n B)^{-1}$ maps $\Im P_n$ onto $\Im P_n$, it follows from (17.9), (17.10) and (17.12) that for $n \geq N$,

$$(P_n(A + B)P_n)^{-1} = (I + (P_n AP_n)^{-1} P_n B)^{-1}(P_n AP_n)^{-1} \tag{17.13}$$

and

$$\sup_{n \geq N} \|(P_n(A + B)P_n)^{-1}\| \leq 2M. \tag{17.14}$$

Hence for $\|B\| < \min(\frac{1}{2M^2}, \frac{1}{\|A^{-1}\|})$,

$$A + B = A(I + A^{-1}B)$$

is invertible and $\sup_{n \geq N} \|(P_n(A + B)P_n)^{-1}\| < \infty$. The theorem now follows from Theorem 17.1. $\qquad\square$

Theorem 17.7 *Suppose $A \in \Pi(P_n)$ and suppose K is a compact operator in $\mathcal{L}(\mathcal{H})$ with the property that $A + K$ is invertible. Then $A + K \in \Pi(P_n)$.*

First we prove the following Lemma.

Lemma 17.8 *Suppose $\{T_n\}$ is a sequence of operators in $\mathcal{L}(\mathcal{H}_2)$ such that $T_n y \to Ty$ for all $y \in \mathcal{H}_2$. If K is a compact operator in $\mathcal{L}(\mathcal{H}_1, \mathcal{H}_2)$, then*

$$\|T_n K - TK\| \to 0. \tag{17.15}$$

Proof: It follows from Theorem 14.3 that $M = \sup_n \|T_n\| < \infty$. Suppose (17.15) does not hold. Then there exists an $\varepsilon > 0$, and a subsequence $\{n'\}$ of $\{n\}$, vectors $\{x_{n'}\}$ such that $\|x_{n'}\| = 1$ and

$$\|T_n K x_{n'} - TK x_{n'}\| \geq \varepsilon. \tag{17.16}$$

Since K is compact, there exists a subsequence $\{x_{n''}\}$ of $\{x_{n'}\}$ such that $K x_{n''} \to y \in \mathcal{H}_2$. Hence

$$
\begin{aligned}
\|T_{n''} K x_{n''} - TK x_{n''}\| &\leq \|T_{n''} K x_{n''} - T_{n''} y\| + \|T_{n''} y - Ty\| \\
&\quad + \|Ty - TK x_{n''}\| \\
&\leq M \|K x_{n''} - y\| + \|T_{n''} y - Ty\| + \|T\| \|y - K x_{n''}\| \to 0
\end{aligned}
$$

which contradicts (17.16).

Proof of Theorem 17.7 Since the projection method for A converges there exists an interger N such that for $n \geq N$, the operator $P_n A P_n$, restricted to $\Im P_n$, is invertible and

$$\sup_{n \geq N} (\|(P_n A P_n)^{-1}\| + \|P_n\|) < \infty \tag{17.17}$$

now for $m \geq N$,

$$P_n(A + K)P_n = P_n A P_n (I + (P_n A P_n)^{-1} P_n K) P_n \tag{17.18}$$

moreover

$$\|(P_n A P_n)^{-1} P_n K - A^{-1} K\| \to 0. \tag{17.19}$$

Indeed, from the definition of convergence of the projection method for A, we know that

$$(P_n A P_n)^{-1} P_n K x \to A^{-1} K x, \quad x \in \mathcal{H}.$$

Hence (17.19) follows from Lemma 17.8. Since $I + A^{-1}K = A^{-1}(A + K)$ is invertible, we have from (17.19) and Corollary I.8.3 that $I + (P_n A P_n)^{-1} P_n K$ is invertible for $n \geq N_1 \geq N$ and

$$\sup_{m \geq N_1} \|(I + (P_n A P_n)^{-1} P_n K)^{-1}\| < \infty. \tag{17.20}$$

Since $(I + P_n A P_n)^{-1}$ maps $\Im P_n$ onto $\Im P_n$, it follows from (17.17), (17.18) and (17.20) that

$$(P_n(A + K)P_n)^{-1} = (I + (P_n A P_n)^{-1} P_n K)^{-1} (P_n A P_n)^{-1}$$

and

$$\sup_{n \geq N_1} \|(P_n(A + K)P_n)^{-1}\| < \infty.$$

The theorem is now a consequence of Theorem 17.1.

In our theorem on the convergence of the projection method we always assume that the operator A is invertible. However, if we only assume

$$\sup_n \|(P_n A P_n)^{-1}\| < \infty,$$

then A is one-one and has a closed range. Indeed suppose $M = \sup\|(P_n A P_n)^{-1}\|$. Then for all $x \in \mathcal{H}$,

$$\|P_n A P_n x\| \geq \frac{1}{M}\|P_n x\|.$$

If follows from Corollary 14.4 that $P_n A P_n x \to A x$.

Hence

$$\|Ax\| = \lim_{n \to \infty} \|P_n A P_n x\| \geq \frac{1}{M} \lim_{n \to \infty} \|P_n x\| = \frac{1}{M}\|x\|.$$

Thus

$$\|Ax\| \geq \frac{1}{M}\|x\| \text{ for all } x \in \mathcal{H}.$$

By Corollary 14.2, A is one-one and has a closed range. $\qquad\square$

2.18 The Modified Projection Method

Let $\{P_n\}$ be a sequence of projections in $\mathcal{L}(\mathcal{H})$ with the property that $P_n x \to x$ for all $x \in \mathcal{H}$. In the ordinary projection method for $\{P_n\}$, the solution x of an equation $Ax = y$, when A is an invertible operator on $\mathcal{H}$, in approximated by solutions $x_n \in \Im P_n$ of the equations

$$P_n A x_n = P_n y, \quad n = 1, 2, \ldots. \tag{18.1}$$

In some cases it is more convenient to approximate the original equation by the equations

$$F_n x_n = P_n y, \quad n = 1, 2, \ldots, \tag{18.2}$$

where $F_n \in \mathcal{L}(\Im P_n)$ for each n and

$$\lim_{n \to \infty} \|F_n P_n x - A x\| = 0, \quad x \in \mathcal{H}. \tag{18.3}$$

We refer to this way of inverting A as the *modified projection method*. One of the reasons for replacing (18.1) by (18.2) is that it is sometimes easier to invert the operators F_n than the operators $P_n A|\Im P_n$.

The next result is the analogue of Theorem 17.1 for the modified projection method

Theorem 18.1 *Let $\{P_n\}$ be a sequence of projections in $\mathcal{L}(\mathcal{H})$ such that $P_n x \to x$ for each $x \in \mathcal{H}$, and let A be an invertible operator on $\mathcal{H}$. Also, let $F_n \in \mathcal{L}(\Im P_n)$, $n = 1, 2, \ldots$, be a sequence of operators satisfying (18.3), and assume that for $n \geq N$ the operator F_n is invertible. Then*

$$\lim_{n\to\infty} \|F_n^{-1} P_n y - A^{-1} y\| = 0, \quad y \in \mathcal{H}, \tag{18.4}$$

if and only if

$$\sup_{n\geq N} \|F_n^{-1}\| < \infty. \tag{18.5}$$

Proof: The necessity of (18.4) is an immediate corollary of Theorem 14.3. To prove the sufficiency, take $n \geq N$. Then F_n is invertible, and for each $y \in \mathcal{H}$ we have

$$\|F_n^{-1} P_n y - A^{-1} y\| \leq \|F_n^{-1} P_n y - P_n A^{-1} y\| + \|(I - P_n)A^{-1} y\|. \tag{18.6}$$

The second term in the right hand side converges to zero. The first can be evaluated as follows:

$$\|F_n^{-1} P_n y - P_n A^{-1} y\| \leq \|F_n^{-1}\| \, \|P_n y - F_n P_n A^{-1} y\|$$
$$\leq \|F_n^{-1}\| (\|P_n y - y\| + \|(F_n P_n - A)A^{-1} y\|).$$

Recall that $P_n y \to y$. Using (18.3) with $x = A^{-1} y$ and (18.4) we see that the preceding inequality implies that the first term in the right hand side of (18.6) converges to zero when $n \to \infty$. Thus (18.4) is proved. $\square$

Example: Let the operator A on ℓ_2 be defined by the infinite tridiagonal matrix

$$A = \begin{pmatrix} 6 & -3 & 0 & 0 & \cdots \\ -2 & 7 & -3 & 0 & \\ 0 & -2 & 7 & -3 & \\ 0 & 0 & -2 & 7 & \\ \vdots & & & & \ddots \end{pmatrix}.$$

We shall use the modified projection method to show that A is invertible and to find its inverse.

For each n we let P_n be the usual orthogonal projection onto the first $n + 1$ coordinates:

$$P_n(x_0, x_1, x_2, \ldots) = (x_0, \ldots, x_n, 0, 0, \ldots).$$

For F_n, $n = 0, 1, 2, \ldots$, we choose the operators

$$F_n = \begin{pmatrix} 6 & -3 & 0 & 0 & \ldots & 0 & 0 \\ -2 & 7 & -3 & 0 & \ldots & 0 & 0 \\ 0 & -2 & 7 & -3 & & 0 & 0 \\ \vdots & \vdots & \vdots & \vdots & & \vdots & \vdots \\ 0 & 0 & 0 & 0 & & 7 & -3 \\ 0 & 0 & 0 & 0 & & -2 & 6 \end{pmatrix}$$

Here we identify $\Im P_n$ with $\mathbb{C}^{n+1}$. Notice that F_n differs from $P_n A P_n$ only in the entry in the right lower corner. It follows that

$$\lim_{n \to \infty} \| F_n P_n x - P_n A P_n x \| = 0, \quad x \in \mathcal{H},$$

and hence (18.3) holds. A direct computation shows that

$$F_n^{-1} = \frac{1}{5} \begin{pmatrix} 1 & 2^{-1} & 2^{-2} & \ldots & 2^{-n} \\ 3^{-1} & 1 & 2^{-1} & \ldots & 2^{-n+1} \\ 3^{-2} & 3^{-1} & 1 & \ldots & 2^{-n+2} \\ \vdots & \vdots & \vdots & & \vdots \\ 3^{-n} & 3^{-n+1} & 3^{-n+2} & \ldots & 1 \end{pmatrix}$$

By letting n go to infinity it is natural to expect that the inverse of A is equal to the operator R given by

$$R = \frac{1}{5} \begin{pmatrix} 1 & 2^{-1} & 2^{-2} & \ldots \\ 3^{-1} & 1 & 2^{-1} & \ldots \\ 3^{-2} & 3^{-1} & 1 & \\ \vdots & \vdots & & \ddots \end{pmatrix} \tag{18.7}$$

Notice that R is a bounded linear operator on ℓ_2. In fact (cf., the proof of Theorem II.16.4) we have

$$\| R \| \leq \frac{1}{5} \left(\left(1 - \frac{1}{2} \right)^{-1} + \frac{1}{3} \left(1 - \frac{1}{3} \right)^{-1} \right) = \frac{1}{2} \tag{18.8}$$

By a direct computation one sees that $RA = AR = I$, and hence A is invertible and $A^{-1} = R$.

Since $F_n^{-1} = P_n R | \Im P_n$, we have $\| F_n^{-1} \| \leq \| R \|$, and hence

$$\sup_{n \geq 0} \| F_n^{-1} \| < \infty.$$

Thus we can apply Theorem 8.1 to show that for this example the modified projection method converges, that is, for each $y \in \mathcal{H}$

$$\lim_{n \to \infty} \| F_n^{-1} P_n y - A^{-1} y \| = 0.$$

2.19 Invariant Subspaces

Sometimes properties of an operator $A \in \mathcal{L}(\mathcal{H})$ can be determined rather easily by considering simpler operators which are restrictions of A to certain subspaces. We now present some elementary facts about invariant subspaces.

Definition: Let A be in $\mathcal{L}(\mathcal{H})$. A subspace M of $\mathcal{H}$ is called A-*invariant* if $AM \subset M$.

It follows readily from the continuity of A that if M is A-invariant, then so is $\overline{M}$.

Examples: 1. Let $\{\varphi_1, \varphi_2, \ldots\}$ be an orthonormal basis for $\mathcal{H}$. Suppose that the matrix corresponding to $A \in \mathcal{L}(\mathcal{H})$ and $\{\varphi_n\}$ (cf. Section 4) is upper triangular, i.e.,

$$a_{ij} = \langle A\varphi_j, \varphi_i \rangle = 0, \quad i > j.$$

2. If, in Example 1, (a_{ij}) is lower triangular, i.e., $a_{ij} = 0$ if $i < j$, then for each n, $\overline{\text{sp}}\{\varphi_{n+1}, \varphi_{n+2}, \ldots\}$ is A-invariant.

3. Given $k \in L_2([a, b] \times [a, b])$, define K on $L_2([a, b])$ by

$$(Kf)(t) = \int_a^t k(t, s) f(s) \, ds.$$

For each $t \in [a, b]$, the subspace

$$M_t = \{f \in L_2([a, b]) : f = 0 \text{ a.e. on } [a, t]\}$$

is invariant under K.

Theorem 19.1 *If a subspace M is A-invariant, then $M^\perp$ is A^*-invariant. In particular, if A is self adjoint, then $M^\perp$ is A-invariant.*

Proof: Suppose $v \in M^\perp$. For any $u \in M$, Au is also in M. Hence

$$0 = \langle Au, v \rangle = \langle u, A^*v \rangle.$$

Therefore, $A^*v \in M^\perp$.   □

The following theorem gives a connection between invariant subspaces and orthogonal projections.

Theorem 19.2 *A closed subspace $M \subset \mathcal{H}$ is A-invariant if and only if $AP = PAP$, where P is the orthogonal projection onto M.*

Proof: If $AM \subset M$, then for each $u \in \mathcal{H}$, $APu \in AM \subset M$. Therefore, $PAPu = APu$. Conversely, if $PAP = AP$, then for $v \in M$,

$$Av = APv = PAPv \in M.$$

$\square$

Given $A \in \mathcal{L}(\mathcal{H})$ and given a closed subspace $M \subset \mathcal{H}$, let P be the orthogonal projection onto M. Then $Q = I - P$ is the orthogonal projection onto $M^\perp$ and $x = Px + Qx$. If we identify with each $x \in \mathcal{H}$ the column vector $\begin{pmatrix} Px \\ Qx \end{pmatrix}$, then the operator A can be represented as a block matrix

$$\begin{pmatrix} A_{11} & A_{12} \\ A_{21} & A_{22} \end{pmatrix}.$$

To be specific, suppose $Ax = y$. Since

$$A = PAP + QAP + PAQ + QAQ$$

and $x = Px + Qx$, $y = Py + Qy$, we get

$$\begin{pmatrix} PAP & PAQ \\ QAP & QAQ \end{pmatrix} \begin{pmatrix} Px \\ Qx \end{pmatrix} = \begin{pmatrix} Py \\ Qy \end{pmatrix}.$$

It is easy to see that M is A-invariant if and only if $A_{21} = QAP = 0$, i.e., the block matrix is upper triangular. Similarly, $M^\perp$ is A-invariant if and only if $A_{12} = PAQ = 0$, i.e., the block matrix is lower triangular. These remarks give alternate proofs of Theorems 19.1 and 19.2.

A problem which is still unsolved after years of effort by numerous mathematicians is whether every bounded linear operator on a Hilbert space has a nontrivial closed invariant subspace. It is known, for example, that every compact linear operator on a Hilbert space (more generally on a Banach space) has a nontrivial closed invariant subsapce. For an exposition of the invariant subspace problem see [PS].

2.20 The Spectrum of an Operator

For a linear operator on a finite dimensional Hilbert space $\mathcal{H}$ it is well known from linear algebra that the equation $\lambda x - Ax = y$ has a unique solution for every $y \in \mathcal{H}$ if and only if $\det(\lambda \delta_{ij} - a_{ij}) \neq 0$, where δ_{ij} is the Kronecker delta and (a_{ij}) is the matrix corresponding to A and some orthonormal basis of $\mathcal{H}$. Therefore, in this case $\lambda I - A$ is invertible for all but a finite number of λ.

If $\mathcal{H}$ is infinite dimensional, then the set $\sigma(A)$ of those λ for which $\lambda I - A$ is not invertible in more complex. In this section we touch upon some properties of $\sigma(A)$.

Definition: Given $A \in \mathcal{L}(\mathcal{H})$, a point $\lambda \in \mathbb{C}$ is called a *regular point* of A if $\lambda I - A$ is invertible. The set $\rho(A)$ of regular points is called the *resolvent set* of A. The *spectrum* $\sigma(A)$ of A is the complement of $\rho(A)$.

Theorem 20.1 *The resolvent set of $A \in \mathcal{L}(\mathcal{H})$ is an open set containing $\{\lambda \mid |\lambda| > \|A\|\}$. Hence $\sigma(A)$ is a closed bounded set contained in $\{\lambda \mid |\lambda| \leq \|A\|\}$. Furthermore, for $\lambda \in \partial\sigma(A)$, the boundary of $\sigma(A)$, the operator $\lambda I - A$ is neither left nor right invertible.*

Proof: If $|\lambda| > \|A\|$, then $\lambda I - A = \lambda(I - \frac{1}{\lambda}A)$ is invertible since $\|\frac{1}{\lambda}A\| < 1$. Thus $\lambda \in \rho(A)$. Suppose $\lambda_0 \in \rho(A)$. Since $\lambda_0 I - A$ is invertible, we have by Corollary 8.2 that $\lambda I - A$ is invertible if $|\lambda - \lambda_0| < \|(\lambda_0 - A)^{-1}\|^{-1}$. Hence $\rho(A)$ is an open set.

Next, take $\lambda \in \partial\sigma(A)$, and assume $\lambda I - A$ is left invertible. Since $\sigma(A)$ is closed. λ belongs to $\sigma(A)$, and hence $\lambda I - A$ is not two-sided invertible. It follows that codim $\Im(\lambda I - A) \neq 0$. By Theorem 15.3 we have

$$\mathrm{codim}\Im(\lambda' I - A) = \mathrm{codim}\Im(\lambda I - A)$$

whenever $|\lambda - \lambda'|$ is sufficiently small. But each open neighborhood of λ contains points of resolvent set of A. In particular, each open neighborhood of λ contains points λ' such that codim $\Im(\lambda' I - A) = 0$. We reached a contradiction. Thus $\lambda I - A$ is not left invertible. In a similar way, using Theorem 15.5, one proves that $\lambda I - A$ cannot be right invertible. $\qquad\square$

Notice that an operator $A \in \mathcal{L}(\mathcal{H})$ is invertible if and only if its adjoint A^* is invertible, and in that case

$$(A^*)^{-1} = (A^{-1})^*.$$

Indeed, if $AB = BA = I$, then by Theorem 11.3(iv) we have $B^*A^* = A^*B^* = I$. Hence A^* is invertible, and the above identity holds. Since $(A^*)^* = A$, the invertibility of A^* also implies that of A.

From the above remark it follows that $\lambda I - A$ is invertible if and only if $\bar{\lambda}I - A^*$ is invertible, and hence $\lambda \in \sigma(A)$ if and only if $\bar{\lambda} \in \sigma(A^*)$. We conclude that

$$\sigma(A^*) = \{\bar{\lambda} \mid \lambda \in \sigma(A)\}$$

Examples: 1. Let K be the finite rank operator on the Hilbert space $\mathcal{H}$ given by

$$Kx = \sum_{j=1}^{n} \langle x, \varphi_j \rangle \, \psi_j.$$

If $\mathcal{H}$ is infinite dimensional, then the spectrum

$$\sigma(K) = \{\lambda \neq 0 \mid \det(\lambda \delta_{ij} - \langle \psi_j, \varphi_i \rangle)_{i,j=1}^n = 0\} \cup \{0\}.$$

This follows directly from Theorem 7.1.

2. Let $\varphi_1, \varphi_x, \ldots$ be an orthonormal basis of the Hilbert space $\mathcal{H}$, and let $\{\lambda_1, \lambda_2, \ldots\}$ be a bounded sequence of complex numbers. For $x \in \mathcal{H}$ define

$$Ax = \sum_{k=1}^{\infty} \lambda_K \langle x, \varphi_k \rangle \varphi_k.$$

From Example 2 in Section 2 we know that A is a bounded linear operator on $\mathcal{H}$. Notice that the matrix of A corresponding to the basis $\varphi_1, \varphi_2, \ldots$ is an infinite diagonal matrix with diagonal entries $\lambda_1, \lambda_2, \ldots$, i.e.,

$$A = \begin{pmatrix} \lambda_1 & 0 & 0 & \cdots \\ 0 & \lambda_2 & 0 & \\ 0 & 0 & \lambda_3 & \\ \vdots & & & \ddots \end{pmatrix} = \text{diag}(\lambda_1, \lambda_2, \lambda_3, \ldots).$$

We claim that the spectrum of A is equal to the closure of the set $\{\lambda_j \mid j = 1, 2, \ldots\}$. In other words,

$$\sigma(A) = \left\{ \lambda \in \mathbb{C} \mid \inf_j |\lambda - \lambda_j| = 0 \right\}.$$

To prove this result, it is sufficient to show that $\lambda \in \rho(A)$ if and only if

$$\inf_j |\lambda - \lambda_j| > 0. \tag{20.1}$$

Assume condition (20.1) holds. Then, according to Example 2 in Section 2, the formula

$$Bx = \sum_{j=1}^{\infty} \frac{1}{\lambda - \lambda_j} \langle x, \varphi_j \rangle \varphi_j \ (x \in \mathcal{H})$$

defines a bounded linear opeator on $\mathcal{H}$. In fact,

$$\|B\| = \sup_j \left| \frac{1}{\lambda - \lambda_j} \right| = \left\{ \inf_j |\lambda - \lambda_j| \right\}^{-1} < \infty.$$

It is straightforward to check that $AB = BA = I$, and thus $B = (\lambda I - A)^{-1}$, and $\lambda \in \rho(A)$.

Conversely, let $\lambda \in \rho(A)$, and assume (20.1) is not satisfied. Then there is a subsequence $\{\lambda_{k'}\}$ of $\{\lambda_k\}$ such that $\lambda_{k'} \to \lambda$. But then

$$\|(\lambda I - A)\varphi_{k'}\| = |(\lambda - \lambda_{k'})\varphi_{k'}| = |\lambda - \lambda_{k'}| \to 0.$$

But this is impossible, because

$$1 = \|\varphi_{k'}\| = \|(\lambda I - A)^{-1}(\lambda I - A)\varphi_{k'}\| \le \|(\lambda I - A)^{-1}\|\|(\lambda I - A)\varphi_{k'}\|,$$

and thus $\|(\lambda I - A)\varphi_{k'}\| \ge \|(\lambda I - A)^{-1}\|^{-1}$. Thus $\lambda \in \rho(A)$ implies (20.1).

3. Let A be the backward shift on ℓ_2, i.e.,

$$A(\alpha_1, \alpha_2, \ldots) = (\alpha_2, \alpha_3, \ldots).$$

We shall show that $\sigma(A) = \{\lambda \in \mathbb{C} \mid |\lambda| \le 1\}$. If $|\lambda| > 1$, then $\lambda I - A = \lambda(I - \lambda^{-1}A)$ is invertible since $\|\lambda^{-1}A\| = |\lambda^{-1}| < 1$. If $|\lambda| < 1$, then $\lambda \in \sigma(A)$ because $x = (1, \lambda, \lambda^2, \ldots) \in \operatorname{Ker}(\lambda I - A)$. Finally, since $\sigma(A)$ is a closed set, it follows that $\sigma(A)$ is the closed unit disc in $\mathbb{C}$.

The spectrum of the forward shift S on ℓ_2 is also equal to the closed unit disc. To see this note that $S = A^*$, and use that

$$\sigma(A^*) = \{\bar{\lambda} \mid \lambda \in \alpha(A)\}.$$

Since $|\lambda| \le 1$ if and only if $|\bar{\lambda}| \le 1$, we get that $\sigma(S) = \sigma(A^*)$ is the closed unit disc.

4. Let $a(\cdot)$ be a continuous complex valued function defined on the closed bounded interval $[c, d]$. Define a bounded linear operator A on $\mathcal{H} = L_2([c, d])$ by

$$(Af)(t) = a(t)f(t), \quad c \le t \le d.$$

Since $((\lambda I - A)f)(t) = (\lambda - a(t))f(t)$,

We have from Example 2 in Section 7 that $\lambda I - A$ is invertible if and only if $\lambda - a(t) \ne 0$ for all $t \in [c, d]$,

$$((\lambda I - A)^{-1}g)(t) = \frac{1}{\lambda - a(t)}\, g(t),$$

and

$$\|(\lambda I - A)^{-1}\| = \max_{t \in [c,d]} |\lambda - a(t)|^{-1}.$$

Hence $\sigma(A) = \{a(t) \mid t \in [c, d]\}$.

5. Let $\mathcal{H} = L_2([0, 1])$. For $0 < \alpha < 1$ let R_α and $R_{\frac{1}{\alpha}}$ be the operators introduced in Section 15, Example 3. As we have proved there, the operator $R_{\frac{1}{\alpha}}$ is a right inverse of R_α and R_α is not left invertible. Also, $\|R_\alpha\| = \frac{1}{\sqrt{\alpha}}$ and $\|R_{\frac{1}{\alpha}}\| = \sqrt{\alpha}$. We show that

$$\sigma(R_\alpha) = \left\{\lambda \mid |\lambda| \le \frac{1}{\sqrt{\alpha}}\right\}. \tag{20.2}$$

If $|\lambda| < \frac{1}{\sqrt{\alpha}}$, then $\|\lambda R_{\frac{1}{\alpha}}\| < 1$ and therefore $I - \lambda R_{\frac{1}{\alpha}}$ is invertible. Since R_α is not invertible and

$$\lambda I - R_\alpha = R_\alpha(\lambda R_{\frac{1}{\alpha}} - I),$$

it follows that $\lambda I - R_\alpha$ is not invertible, i.e., $\lambda \in \sigma(R_\alpha)$. If $|\lambda| > \frac{1}{\sqrt{\alpha}} = \| R_\alpha \|$, then $\lambda \in \rho(R_\alpha)$ by Theorem 19.1. Since $\sigma(R_\alpha)$ is closed and $\{\lambda \mid |\lambda| < \frac{1}{\sqrt{\alpha}}\} \subseteq \sigma(R_\alpha)$, equality (20.2) follows.

We shall also determine the spectrum of the operator R_α on $L_2([0, \infty])$; see Example 6 below. For this purpose we need the following results.

Theorem 20.2 *Given $A \in \mathcal{L}(\mathcal{H})$, where $\mathcal{H}$ is a Hilbert space, let $\rho_1(A)$ be an open, connected subset of $\rho(A)$. Let $\mathcal{H}_0$ be a closed A-invariant subspace of $\mathcal{H}$. Define $A_0 = A/\mathcal{H}_0$, the restriction of A to $\mathcal{H}_0$. Then either*

$$\text{(i) } \rho_1(A) \subset \rho(A_0) \text{ or (ii)} \rho_1(A) \subset \sigma(A_0).$$

If $\rho_1(A) \subset \sigma(A_0)$, then for all $\lambda \in \rho_1(A)$, the operator $\lambda I - A_0$ is strictly left invertible and $0 < \mathrm{codim}(\lambda I - A_0)$ is constant on $\rho_1(A)$.

Proof: Obviously,

$$\rho_1(A) = (\rho_1(A) \cap \rho(A_0)) \cup (\rho_1(A) \cap \sigma(A_0)). \tag{20.3}$$

The set $\rho_1(A) \cap \rho(A_0)$ is open. Also, $\rho_1(A) \cap \sigma(A_0)$ is open. To see this, suppose $\lambda \in \rho_1(A) \cap \sigma(A_0)$. Let P be the orthogonal projection onto $\mathcal{H}_0$. Then $(\lambda I - A)^{-1}/\mathcal{H}_0$ its a left inverse for $\lambda I - A_0$ on $\mathcal{H}_0$ since

$$P(\lambda - A)^{-1}(\lambda I - A_0)x_0 = P(\lambda - A)^{-1}(\lambda - A)x_0 = Px_0 = x_0.$$

Since $\lambda \in \sigma(A_0)$, $\lambda I - A_0$ is strictly left invertible. Hence it follows from Theorem 15.3 that λ is an interior point of $\rho_1(A) \cap \sigma(A)$. Therefore $\rho_1(A) \cap \sigma(A)$ is an open set. Since $\rho_1(A)$ is an open connected set, it cannot be expressed as the union of two non-empty mutually disjoint open sets. Thus from (20.3) we have that either $\rho_1(A) \cap \rho(A_0) = \emptyset$ or $P_1(A) \cap \sigma(A_0) = \emptyset$. Hence either (i) or (ii) holds.

If $\rho_1(A) \subseteq \sigma(A_0)$, then by Theorem 15.3, the function $\varphi(\lambda) = \mathrm{codim}(\lambda I - A_0)$ is continuous on $\rho_1(A)$. Since $\rho_1(A)$ is connected, $\varphi(\lambda)$ is constant (and greater than zero). □

Corollary 20.3 *If $\rho(A)$ is connected, $\sigma(A_0) \subset \sigma(A)$.*

Proof: The open set $\rho(A)$ is not contained in $\sigma(A)$ since $\rho(A)$ is unbounded. Hence $\rho(A) \subset \rho(A_0)$ by Theorem 19.2. Thus $\sigma(A_0) \subset \sigma(A)$. □

Corollary 20.4 *If λ is in the boundary of $\sigma(A_0)$, i.e.,*

$$\lambda \in \sigma(A_0) \cap \overline{\rho(A_0)}, \text{ then } \lambda \in \sigma(A).$$

Proof: Suppose $\lambda \in \rho(A)$. Then there exists an open disc D such that $\lambda \in D \subset \rho(A)$. Since $\lambda \in \overline{\rho(A_0)}$, $D \cap \rho(A_0) \neq \emptyset$. Also, $\lambda \in D \cap \sigma(A_0)$. This contradicts Theorem 19.2 as D is an open connected subset of $\rho(A)$.

An operator $U \in \mathcal{L}(\mathcal{H})$ is called an *isometry* if $\|Ux\| = \|x\|$ for each $x \in \mathcal{H}$. If, in addition, $\Im U = \mathcal{H}$, the operator U is said to be *unitary*. Notice that a unitary operator U is invertible, and both U and U^{-1} have norm 1. This implies that for a unitary operator U the spectrum belongs to the unit circle, i.e.,

$$\sigma(U) \subset \{\lambda \in \mathbb{C} |\ |\lambda| = 1\}. \tag{20.4}$$

Indeed, for $|\lambda| > 1$ we have $\lambda \in \rho(U)$ by Theorem 19.1, and for $|\lambda| < 1$ we have $\|\lambda U^{-1}\| < 1$, and thus $\lambda I - U = -U(I - \lambda U^{-1})$ is again invertible. In general, the inclusion in (20.4) can be strict. For instance, if $U = I$, where I is the identity operator on $\mathcal{H}$, then $\sigma(U) = \{1\}$. $\qquad\square$

Examples: 6. Let $\mathcal{H} = L_2([0, \infty])$. For $0 < \alpha < 1$, define R_α on $\mathcal{H}$ by

$$(R_\alpha \varphi)(t) = \varphi(\alpha t).$$

The operator $\sqrt{\alpha} R_\alpha$ is an isometry since

$$\|R_\alpha \varphi\|^2 = \int_0^\infty |\varphi(\alpha t)|^2 \, dt = \frac{1}{\alpha} \int_0^\infty |\varphi(S)|^2 \, ds = \frac{1}{\alpha} \|\varphi\|^2. \tag{20.5}$$

also $\Im(\sqrt{\alpha} R_\alpha \varphi) = \mathcal{H}$, for given $\psi \in \mathcal{H}$, take $\varphi(t) = \psi(\frac{1}{\alpha} t), 0 \leq t < \infty$. Then $(R_\alpha \varphi)(t) = R_\alpha \psi(\frac{1}{\alpha} t) = \psi(t)$. Hence $\sqrt{\alpha} R_\alpha$ is unitary and by (20.4) we have $\sigma(\sqrt{\alpha} R_\alpha) \subset \{\lambda | |\lambda| = 1\}$ which implies

$$\sigma(R_\alpha) \subset \left\{\lambda |\ |\lambda| = \frac{1}{\sqrt{\alpha}}\right\}. \tag{20.6}$$

By making use of Corollary 20.4 we shall show that

$$\sigma(R_\alpha) = \left\{\lambda |\ |\lambda| = \frac{1}{\sqrt{\alpha}}\right\}. \tag{20.7}$$

Let $\mathcal{H}_0 = \{\varphi \in \mathcal{H} |\ \varphi(t) = 0,\ 0 \leq t \leq 1\}$. Clearly $\mathcal{H}_0$ is a closed R_α-invariant subspace of $\mathcal{H}$. Define the linear operators R_α^0 and $R_{\frac{1}{\alpha}}^0$ on $\mathcal{H}_0$ as follows:

$$R_\alpha^0 = R_\alpha | \mathcal{H}_0$$

and

$$(R_{\frac{1}{\alpha}}^0 \varphi)(t) = \begin{cases} 0, & 0 \leq t \leq 1, \\ \varphi(\frac{1}{\alpha} t), & t > 1. \end{cases}$$

Then

$$\|R^0_{\frac{1}{\alpha}}\varphi\|^2 = \int_1^\infty \left|\varphi\left(\frac{1}{\alpha}t\right)\right|^2 dt = \alpha \int_{\frac{1}{\alpha}}^\infty |\varphi(s)|^2 ds \leq \alpha \|\varphi\|^2. \tag{20.8}$$

Hence $\|R^0_{\frac{1}{\alpha}}\| \leq \sqrt{\alpha}$. Infact,

$$\|R^0_{\frac{1}{\alpha}}\| = \sqrt{\alpha} \tag{20.9}$$

Indeed, define

$$\varphi(t) = \begin{cases} 1, & t \in [\frac{1}{\alpha}, \frac{2}{\alpha}] \\ 0, & t \notin [\frac{1}{\alpha}, \frac{2}{\alpha}] \end{cases}.$$

Then $\varphi \in \mathcal{H}_0$ and from (20.8) we have

$$\|R^0_{\frac{1}{\alpha}}\varphi\|^2 = \alpha \int_{\frac{1}{\alpha}}^{\frac{2}{\alpha}} 1 ds = 1$$

and

$$\|\varphi\|^2 = \int_{\frac{1}{\alpha}}^{\frac{2}{\alpha}} 1 dt = \frac{1}{\alpha}.$$

Hence

$$\|R^0_{\frac{1}{\alpha}}\|\frac{1}{\sqrt{\alpha}} = \|R^0_{\frac{1}{\alpha}}\|\|\varphi\| \geq \|R^0_{\frac{1}{\alpha}}\varphi\| = 1$$

or $\|R^0_{\frac{1}{\alpha}}\| \geq \sqrt{\alpha}$, Thus (20.9) holds. Also

$$R^0_{\frac{1}{\alpha}} R^0_\alpha \varphi = \varphi, \quad \varphi \in \mathcal{H}_0, \tag{20.10}$$

since

$$(R^0_{\frac{1}{\alpha}} R^0_\alpha \varphi)(t) = \begin{cases} 0, & 0 \leq t \leq 1 \\ R^0_\alpha(\varphi(\frac{1}{\alpha}t)) = \varphi(t), & t > 1. \end{cases}$$

The operator R^0_α is not invertible since Im $R^0_\alpha \neq \mathcal{H}_0$. To see this, define

$$\psi(t) = \begin{cases} 1, & t \in [1, \frac{1}{\alpha}], \\ 0, & t \notin [1, \frac{1}{\alpha}]. \end{cases}$$

Clearly $\psi \in \mathcal{H}_0$ and for all $\varphi \in \mathcal{H}_0$,

$$(R^0_\alpha \varphi)(t) = \varphi(\alpha t) = 0, \quad 1 < t \leq \frac{1}{\alpha}.$$

Thus $\psi \notin \Im R^0_\alpha$. Next we show that

$$\sigma(R^0_\alpha) = \left\{\lambda \mid |\lambda| \leq \frac{1}{\sqrt{\alpha}}\right\}. \tag{20.11}$$

From (20.5) we have $\|R_\alpha^0\| = \frac{1}{\sqrt{\alpha}}$. Hence if $|\lambda| > \frac{1}{\sqrt{\alpha}}$, then $\lambda \in \rho(R_\alpha^0)$. For $|\lambda| < \frac{1}{\sqrt{\alpha}}$, $\|\lambda R_{\frac{1}{\alpha}}^0\| < \frac{1}{\sqrt{\alpha}} \cdot \sqrt{\alpha} = 1$ by (20.9). Thus $I - \lambda R_{\frac{1}{\alpha}}^0$ is invertible. Equality (20.10) shows that

$$\lambda I - R_\alpha^0 = (\lambda R_{\frac{1}{\alpha}}^0 - I) R_\alpha^0 \qquad (20.12)$$

Since $(\lambda R_{\frac{1}{\alpha}}^0 - I)$ is invertible and R_α^0 is not, it follows from (20.12) that $\lambda I - R_\alpha^0$ is not invertible, i.e., $\lambda \in \sigma(R_\alpha^0)$, $|\lambda| < \frac{1}{\sqrt{\alpha}}$. To summarize, we have shown that

$$\left\{ \lambda \,|\, |\lambda| < \frac{1}{\sqrt{\alpha}} \right\} \subset \sigma(R_\alpha^0) \subset \left\{ \lambda \,|\, |\lambda| \leq \frac{1}{\alpha} \right\}.$$

Since $\sigma(R_\alpha^0)$ is closed, equality (20.11) follows. Suppose $|\lambda| = \frac{1}{\sqrt{\alpha}}$. Then by (20.11), λ is in the boundary of $\sigma(R_\alpha^0)$, hence we may conclude from Corollary 20.4 that $\lambda \in \sigma(R_\alpha)$. Equality (20.7) is now a consequence of (20.6).

7. Let $\mathcal{H} = L_2([-\pi, \pi])$, and let

$$a(t) = a_{-1}e^{-it} + a_0 + a_1 e^{it}, \quad -\pi \leq t \leq \pi,$$

where a_{-1}, a_0 and a_1 are real numbers. On $\mathcal{H}$ we consider the operator A of multiplications by a, that is

$$(Af)(t) = a(t)f(t), \quad -\pi \leq t \leq \pi.$$

Example 4 shows that the spectrum $\sigma(A)$ consists of those $\lambda \in \mathbb{C}$ for which

$$a_{-1}e^{-it} + a_0 - \lambda + a_1 e^{it} = 0 \qquad (20.13)$$

for some t. Taking the real and imaginary parts in (20.13) and using $e^{it} = \cos t + i \sin t$ gives

$$\mathrm{Re}\lambda = a_0 + (a_{-1} + a_1) \cos t, \qquad (20.14)$$

$$\Im\lambda = (a_1 - a_{-1}) \sin t. \qquad (20.15)$$

To describe $\sigma(A)$ a bit better we consider two cases.

Case 1: First assume that $|a_1| \neq |a_{-1}|$. In this case

$$\frac{a_0 - \mathrm{Re}\,\lambda}{a_{-1} + a_1} = \cos t, \qquad \frac{\Im\lambda}{a_1 - a_{-1}} = \sin t. \qquad (20.16)$$

Hence

$$\left(\frac{a_0 - \mathrm{Re}\lambda}{a_{-1} + a_1}\right)^2 + \left(\frac{\Im\lambda}{a_1 - a_{-1}}\right)^2 = 1. \qquad (20.17)$$

Thus λ lies on the ellipse given by (20.17). Conversely, if λ lies on the above ellipse, then there exists a $t \in [-\pi, \pi]$ such that (20.16) holds. Hence for this choice of t, equation (20.13) is valid. So, if $|a_1| \neq |a_{-1}|$, then $\sigma(A)$ is precisely the ellipse given by (20.17).

Case 2: Assume $a_1 = a_{-1}$. Then $\Im\lambda = 0$, and from (20.14) and (20.15) we get $\lambda = a_0 + 2a_1 \cos t$. Thus $\sigma(A)$ is a real line segment, namely

$$\sigma(A) = \{r \in \mathbb{R} \mid a_0 - 2|a_1| \leq r \leq a_0 + 2|a_1|\}.$$

Case 3: Finally, assume $a_1 = -a_{-1}$. Then we see from (20.14) and (20.15) that

$$\Re\lambda = a_0, \qquad \Im\lambda = 2a_1 \sin t.$$

Thus

$$\sigma(A) = \{\lambda = a_0 + ir \mid -2|a_1| \leq r \leq 2|a_1|\}.$$

We conclude with a remark about the operator A. Recall (see Section I.13), that the functions

$$\varphi_n(t) = \frac{1}{\sqrt{2\pi}} e^{int}, \quad n \in \mathbb{Z}, \tag{20.18}$$

form an orthonormal basis of $\mathcal{H} = L_2([-\pi, \pi])$.

The matrix of A with respect to this basis in the doubly infinite tridiagonal matrix

$$
\begin{pmatrix}
\ddots & & & & & \\
 & a_0 & a_1 & 0 & 0 & 0 \\
 & a_{-1} & a_0 & a_1 & 0 & 0 \\
 & 0 & a_{-1} & \boxed{a_0} & a_1 & 0 \\
 & 0 & 0 & a_{-1} & a_0 & a_1 \\
 & 0 & 0 & 0 & a_{-1} & a_0 \\
 & & & & & & \ddots
\end{pmatrix}
$$

Here the box indicates the entry in the $(0, 0)$-position. Operators defined by matrices of the above type are examples of Laurent operators. The latter operators will be studied in the first section of the next chapter.

From Example 7 above we derive the following result.

Corollary 20.5 *The spectrum of the forward shift V on $\ell_2(\mathbb{Z})$ is equal to the unit circle.*

Proof: Let $\mathcal{F}$ be the map which assigns to each f in $L_2([-\pi, \pi])$ the sequence of Fourier coefficients of f relative to the orthonormal basis (20.18). Then $\mathcal{F}$ is an invertible operator from $L_2([-\pi, \pi])$ onto $\ell_2(\mathbb{Z})$. Put $A = \mathcal{F}^{-1}V\mathcal{F}$. Then $\sigma(V) = \sigma(A)$. But A is the operator considered in Example 7 with $a_1 = 1$ and $a_0 = a_{-1} = 0$. The spectrum of A is given by (20.17), which in this case is the unit circle. $\qquad\square$

Exercises II

1. Let $(\omega_j)_{j=1}^{\infty}$ be a sequence of complex numbers. Define an operator D_ω on ℓ_2 by
$$D_\omega \xi = (\omega_1 \xi_1, \omega_2 \xi_2, \omega_3 \xi_3, \ldots).$$

   (a) Prove that D_ω is bounded if and only if $(\omega_j)_{j=1}^{\infty}$ is bounded. In this case $\|D_\omega\| = \sup_j |\omega_j|$.

   (b) Prove that $\inf_j |\omega_j| \|\xi\| \le \|D_\omega \xi\|$.

   (c) Compute D_ω^K for any $k \in \mathbb{N}$.

   (d) Prove that D_ω is invertible if and only if $\inf_j |\omega_j| > 0$. Give a formula for D_ω^{-1}.

2. Let D_ω be as in problem 1 and let $\inf_j |\omega_j| > 0$ and $\sup|w_j| < \infty$. Which of the following equalities or inequalities hold for any ω?

   (a) $\|D_\omega\| = \frac{1}{\|D_\omega^{-1}\|}$

   (b) $\|D_\omega\| \ge \frac{1}{\|D_\omega^{-1}\|}$

   (c) $\|D_\omega\| \le \frac{1}{\|D_\omega^{-1}\|}$

   (d) $\|D_\omega\| < \frac{1}{\|D_\omega^{-1}\|}$

   (e) $\|D_\omega\| > \frac{1}{\|D_\omega^{-1}\|}$.

3. Let ω be such that $\sup_j |\omega_j| < \infty$, find $\overline{\Im D_\omega}$ and $\operatorname{Ker} D_\omega$.

4. Let A be an operator on ℓ_2 given by the matrix $(a_{jk})_{j,k=1}^{\infty}$ with

$$\sum_{k=-\infty}^{\infty} \sup_j |a_{jj-k}| < \infty.$$

   Prove that A is bounded and that

$$\|A\| \le \sum_{k=-\infty}^{\infty} \sup_j |a_{jj-k}|.$$

5. Let $\mathcal{H}_1$ and $\mathcal{H}_2$ be Hilbert spaces. Define $\mathcal{H} = \mathcal{H}_1 \oplus \mathcal{H}_2$ to be the Hilbert space consisting of all pairs $\{(u_1, u_2) : u_1 \in \mathcal{H}_1, u_2 \in H_2\}$ with

$$(u_1, u_2) + (v_1, v_2) = (u_1 + v_1, u_2 + v_2); \quad \alpha(u_1, u_2) = (\alpha u_1, \alpha u_2),$$

   and an inner product

$$\langle (u_1, u_2), (v_1, v_2) \rangle = \langle u_1, v_1 \rangle_{\mathcal{H}_1} + \langle u_2, v_2 \rangle_{\mathcal{H}_2}.$$

$\mathcal{H}$ is called the *direct sum* of $\mathcal{H}_1$, $\mathcal{H}_2$. Given $A_1 \in \mathcal{L}(\mathcal{H}_1)$ and $A_2 \in \mathcal{L}(\mathcal{H}_2)$, define A on $\mathcal{H}$ by the matrix

$$A = \begin{pmatrix} A_1 & 0 \\ 0 & A_2 \end{pmatrix}$$

i.e., $A(u_1, u_2) = (A_1 u_1, A_2 u_2)$. Prove that A is in $\mathcal{L}(\mathcal{H})$ and $\|A\| = \max(\|A_1\|, \|A_2\|)$.

6. For each $n \in \mathbb{N}$, let U_n be an operator on ℓ_2 given by $U_n \xi = (\xi_n, \xi_{n-1}, \ldots, \xi_1, 0, 0, \ldots)$. Find $\Im U_n$, Ker U_n, $\|U_n\|$ and a matrix representation for U_n with respect to the standard basis.

7. Let $a(t)$ be a continuous complex valued function on $[a, b]$. Define A: $L_2[a, b] \to L_2[a, b]$ by $(Af)(t) = a(t)f(t)$. Find Ker A and $\Im A$.

8. Given the operator A in problem 7, prove that $\Im A$ is closed if and only if $a(t) \neq 0$ for $t \in [a, b]$ or $a(t)$ is identically zero on $[a, b]$.

9. Does the statement of problem 8 remain true if $a(t)$ is allowed to be discontinuous?

10. Let D_ω be an operator on ℓ_2 as defined in problem 1. Prove that $\Im D_\omega$ is closed if and only if

$$\inf_{\omega_j \neq 0} |\omega_j| > 0.$$

11. Let K be an operator of finite rank on a Hilbert space $\mathcal{H}$. For $\varphi \in \mathcal{H}$,

$$K\varphi = \sum_{i=1}^{n} \langle \varphi, \varphi_i \rangle \psi_i.$$

Suppose $\psi_i \in \text{span}\{\varphi_1, \ldots, \varphi_n\}^{\perp}$ for $i = 1, \ldots, n$. Prove that $I + \alpha K$ is invertible for any α and find its inverse.

12. Which of the following operators

$$K \colon L_2[a, b] \to L_2[a, b]$$

have finite rank and which do not?

(a) $(Kf)(t) = \sum_{j=1}^{n} \varphi_j(t) \int_a^b \psi_j(s) f(s) ds.$
(b) $(Kf)(t) = \sum_{j=1}^{n} \varphi_j(t) \int_a^t \psi_j(s) f(s) ds.$

13. For the following operators K of finite rank, find the orthonormal basis for which K has the diagonal matrix representation

$$K = \begin{pmatrix} \alpha_1 & & & & & \\ & \alpha_2 & & & 0 & \\ & & \ddots & & & \\ & & & \alpha_n & & \\ & 0 & & & 0 & \\ & & & & & \ddots \end{pmatrix}.$$

(a) $K\xi = (3\xi_1 + \xi_2, \xi_1 + 3\xi_2, 0, 0, \ldots)$, $K\colon \ell_2 \to \ell_2$.

(b) $K = (\sum_{j=1}^{\infty} \xi_j (\frac{1}{\sqrt{2}})^{j-1} + \xi_1, \frac{1}{\sqrt{2}}\xi_1, \frac{1}{\sqrt{4}}\xi_1, \ldots)$, $K\colon \ell_2 \to \ell_2$.

(c) $(K\varphi)(t) = t \int_{-\pi}^{\pi} \varphi(x) \cos x \, dx + \cos t \int_{-\pi}^{\pi} x\varphi(x) \, dx$, $K\colon L_2[-\pi, \pi]$
$\to L_2[-\pi, \pi]$.

14. For the operators from problem 13 find K^{1980}.

15. Given vectors $\varphi_1, \ldots, \varphi_n, \psi_1, \ldots, \psi_n$ in a Hilbert space $\mathcal{H}$, let $N = \text{span}$ $\{\varphi_1, \ldots, \varphi_n, \psi_1, \ldots, \psi_n\}$. Define $K \in \mathcal{L}(\mathcal{H})$ by $Kv = \sum_{j=1}^{n} \langle v, \varphi_j \rangle \times$ ψ_j. Prove that $KN \subset N$ and that $KN^{\perp} = 0$.

16. Given a separable Hilbert space $\mathcal{H}$, Let $K \in \mathcal{L}(\mathcal{H})$ be an operator of finite rank. Prove that there exists an orthonormal basis in $\mathcal{H}$ such that with respect to this basis, K has the matrix representation of the form $(a_{ij})_{i,j=1}^{\infty}$, where $a_{ij} = 0$ if $i > n$ or $j > n$.

17. Let $\mathcal{H}$ be a Hilbert space and let $K \in L(\mathcal{H})$ be an operator of finite rank given by

$$Kv = \sum_{j=1}^{n} \langle v, \varphi_j \rangle \psi_j, \quad \varphi_j, \psi_j \in \mathcal{H}, j = 1, \ldots, n.$$

Which of the following statements are true and which are not?

(a) rank $K = \dim sp\{\varphi_1, \ldots, \varphi_n\}$.

(b) rank $K = \dim sp\{\psi_1, \ldots, \psi_n\}$.

(c) rank $K = \min[\dim sp\{\varphi_1, \ldots, \varphi_n\}, \dim sp\{\psi_1, \ldots, \psi_n\}]$.

(d) rank $K = \max[\dim sp\{\varphi_1, \ldots, \varphi_n\}, \dim sp\{\psi_1, \ldots, \psi_n\}]$.

(e) rank $K \leq \min[\dim sp\{\varphi_1, \ldots, \varphi_n\}, \dim sp\{\psi_1, \ldots, \psi_n\}]$.

18. Let $\mathcal{H}$ be a Hilbert space and let $K_1, K_2 \in \mathcal{L}(\mathcal{H})$ be two operators of finite rank. Prove that rank $(K_1 + K_2) \leq$ rank $K_1 +$ rank K_2

19. Let $\mathcal{H}$ be a Hilbert space and let $B, C.D \in \mathcal{L}(\mathcal{H})$. On $\mathcal{H}^{(3)} = \mathcal{H} \oplus \mathcal{H} \oplus \mathcal{H}$ define A by the matrix

$$A = \begin{pmatrix} 0 & D & B \\ 0 & 0 & C \\ 0 & 0 & 0 \end{pmatrix}.$$

Prove

(a) $A \in \mathcal{L}(\mathcal{H}^{(3)})$

(b) $A^3 = 0$

(c) $I - \alpha A$ is invertible for any $\alpha \in \mathbb{C}$ and $I + \alpha A + \alpha^2 A$ is its inverse.

20. Let $\mu = (\mu_k)_{k=1}$ be a given sequence of complex numbers with $\sup_k |\mu_k| < 1$. Prove that the following two systems have a unique solution for any right side in ℓ_2. Find the solution for $\eta_k = \delta_{1k}$ and $\mu_k = \frac{1}{2^{k-1}}$.

(a) $\xi_k - \mu_k \xi_{k+1} = \eta_k, \quad k = 1, 2, \ldots$

(b) $\xi_k - \mu_k \xi_{k-1} = \eta_k, \quad k = 2, 3, \ldots, \xi_1 = \eta_1$.

21. Let $\mu = (\mu_k)_{k=1}^{\infty}$ and $\nu = (\nu_k)_{k=1}^{\infty}$ be two sequences of complex numbers with $\sup_k |\mu_k| + \sup_k |\nu_k| < 1$. Prove that the following system of equations has a unique solution in ℓ_2 for any $\{\eta_k\}$ in ℓ_2.

$$\xi_1 + \mu_1 \xi_2 = \eta_1$$
$$\nu_{k-1}\xi_{k-1} + \xi_k + \mu_k \xi_{k+1} = \eta_k, \quad k = 2, 3, \ldots.$$

22. Find the inverse of $I - K$, where K is of the form

(a) $K\xi = (\frac{1}{2}\xi_1 + \xi_2, \frac{1}{4}\xi_1, \frac{1}{8}\xi_1, \ldots), \quad K: \ell_2 \to \ell_2$.

(b) $(Kf)(t) = [\int_{-\pi/2}^{\pi/2} f(t)\cos t\, dt]\frac{4}{\pi}\cos t + [\int_{-\pi/2}^{\pi/2} f(t) t\, dt]\sin t$
$K: L_2[-\frac{\pi}{2}, \frac{\pi}{2}] \to L_2[-\frac{\pi}{2}, \frac{\pi}{2}]$.

23. Let $K: L_2[-\pi, \pi] \to L_2[-\pi, \pi]$ be given by

$$(K\varphi)(t) = \int_{-\pi}^{\pi} k(t-s)\varphi(s)\, ds.$$

Find the matrix of K with respect to the basis $\{e^{int}\}_{n\in\mathbb{Z}}$ in each of the following cases:

(a) $k(t) = |t|$

(a) $k(t) = \sin t$

24. Solve the integral equation

$$\varphi(t) = \sin t + \int_0^t \varphi(s)\, ds.$$

25. Let $K: L_2[0, \infty) \to L_2[0, \infty)$ be given by $(K\varphi)(t) = \int_0^{\infty} k(t+s)\varphi(s)ds$ where $k(t)$ is continuous and $\int_0^{\infty} |k(t)|\, dt < \infty$. Show that K is compact,

26. Find the adjoint of the operator K given in problem 22.

27. When is the operator D_ω given in problem 1 self adjoint?

28. Which of the following operators are self adjoint and which are not?

    (a) The operator $K: L_2[-\pi, \pi] \rightarrow L_2[-\pi, \pi]$ defined by $(K\varphi)(t) = \int_{-\pi}^{\pi} e^{i(t-s)}\varphi(s) \, ds$.

    (b) The operator $K: L_2[-\pi, \pi] \rightarrow L_2[-\pi, \pi]$ defined by $(K\varphi)(t) = \int_{-\pi}^{\pi} \cos(t - s)\varphi(s) \, ds$.

    (c) The operator $K_2: L_2[0, 1] \rightarrow L_2[0, 1]$ defined by $(K\varphi)(t) = \int_0^t \varphi(s) \, ds$.

29. Let A be in $\mathcal{L}(\mathcal{H})$, where $\mathcal{H}$ is a Hilbert space. Define on the direct sum $\mathcal{H}^{(2)} = \mathcal{H} \oplus \mathcal{H}$ (cf. exercise 7) the operator B by

$$B = \begin{pmatrix} 0 & iA \\ -iA^* & 0 \end{pmatrix}.$$

    Prove that $\|A\| = \|B\|$ and that B is self adjoint.

30. Suppose $A \in \mathcal{L}(\mathcal{H})$ and dim Im $A = 1$. When is A self adjoint?

31. Let $\mathcal{X}$ and φ be given vectors in a Hilbert space $\mathcal{H}$. When does there exist a self adjoint operator A in $\mathcal{L}(\mathcal{H})$ such that $A\mathcal{X} = \varphi$? When is A of rank 1?

32. Give a necessary and sufficient condition for the product of two self adjoint operators to be self adjoint.

33. Let $A_{jk} \in \mathcal{L}(\mathcal{H})$, $j, k = 1, \ldots, n$. Define on the direct sum $\mathcal{H}^{(n)} = \underbrace{\mathcal{H} \oplus \mathcal{H} \oplus \ldots \oplus \mathcal{H}}_{n}$ an operator A by

$$A = \begin{pmatrix} A_{11} & \cdots & A_{1n} \\ \vdots & & \vdots \\ A_{n1} & \cdots & A_{nn} \end{pmatrix}.$$

    Give necessary and sufficient conditions for A to be self adjoint.

34. Let $A: \ell_2 \rightarrow \ell_2$ be given by the matrix

$$A = \begin{pmatrix} a_0 & & & \\ a_1 & a_0 & & 0 \\ a_2 & a_1 & a_0 & \\ & a_2 & a_1 & \ddots \\ \vdots & & \ddots & \ddots \end{pmatrix}$$

    with $\sum_{j=0}^{\infty} |a_j| < \infty$.

    (a) Prove that A is bounded and that $\|A\| \leq \sum_{j=0}^{\infty} |a_j|$.

    (b) Find the matrix representation of A^*.

35. Let $\omega = (\lambda_1, \lambda_2, \ldots)$, where $\lambda_j > 0, j = 1, 2, \ldots$ be such that $\sup_j \lambda_j < \infty$ and $\inf_j \lambda_j > 0$. Define $\ell_2(\omega)$ as in Exercise I-15. For $\xi = (\xi_1, \xi_2, \xi_3, \ldots) \in \ell_2(\omega)$, define

$$S_r^\lambda(\xi_1, \xi_2, \ldots) = (0, \xi_1, \xi_2, \ldots).$$

Prove that $(S_r^\lambda)^*$ is given by

$$(S_r^\lambda)^*(\eta_1, \eta_2, \ldots) = \left(\frac{\lambda_2}{\lambda_1}\eta_2, \frac{\lambda_3}{\lambda_2}\eta_3, \ldots\right).$$

36. Describe all selfadjoint operators of finite rank for which $K^{1980} = 0$.

37. Define an operator U on ℓ_2 by $U\xi = (\xi_n, \xi_{n-1}, \ldots, \xi_1, \xi_{n+1})$. Prove that
    (a) $\|U\xi\| = \|\xi\|$ for all $\xi \in \ell_2$.
    (b) $U^{-1} = U = U^*$ and $U^2 = I$.
    (c) if $\alpha \neq \pm 1, \alpha \in \mathbb{C}$, then $I - \alpha U$ is invertible and $(I - \alpha U)^{-1} = \frac{1}{1-\alpha^2}(I + \alpha U)$.
    (d) Give a matrix representation for U.

38. Define the operators $S_r^{(\tau)}$ and $S_\ell^{(\tau)}$ on $L_2[0, 1]$ by

$$(S_r^{(\tau)}\varphi)(t) = \begin{cases} \varphi(t - \tau) & \text{for} \quad 1 \geq t \geq \tau \\ 0 & \text{for} \quad 0 \leq t < \tau \end{cases}$$

$$(S_\ell^{(\tau)}\varphi)(t) = \begin{cases} \varphi(t + \tau) & \text{for} \quad 0 \leq t \leq 1 - \tau \\ 0 & \text{for} \quad 1 - \tau < t \leq 1. \end{cases}$$

Prove that
    (a) $[S_r^{(\tau)}]^* = S_\ell^{(\tau)}$
    (b) $S_r^{(\tau_1)} S_r^{(\tau_2)} = S_r^{(\tau_1+\tau_2)}$
    (c) $S_\ell^{(\tau_1)} S_\ell^{(\tau_2)} = S_\ell^{(\tau_1+\tau_2)}$
    (d) $S_\ell^{(\tau_2)} S_\ell^{(\tau_1)} = \begin{cases} P_{\tau_2} S_\ell^{(\tau_1-\tau_2)} & \text{for} \quad \tau_1 \geq \tau_2 \\ P_{\tau_2} S_r^{(\tau_2-\tau_1)} & \text{for} \quad \tau_2 \geq \tau_1 \end{cases}$
    where P_{τ_2} is the projection onto $\{\varphi \in L_2[a, b] \mid \varphi(t) = 0, 0 \leq t \leq \tau_2\}$.
    (e) $\|S_r^{(\tau)}\| = \|S_\ell^{(\tau)}\| = 1|$ for $\tau \in [0, 1]$.

39. Let $\mathcal{H}$ be a Hilbert space. Prove that for any $A_1, A_2 \in L(\mathcal{H})$,
    (a) $\{\Im A_1 + \Im A_2\}^\perp = \operatorname{Ker} A_1^* \cap \operatorname{Ker} A_2^*$
    (b) $\overline{\Im A_1} \cap \overline{\Im A_2} = (\operatorname{Ker} A_1^* + \operatorname{Ker} A_2^*)^\perp$.

40. Give a formula for the orthogonal projection P onto $\operatorname{sp}\{\varphi_1, \varphi_2, \varphi_3\}$, where
    (a) $\varphi_1 = (1, 0, 0, 1, 0, \ldots), \varphi_2 = (1, 0, 1, 0, \ldots), \varphi_3 = (1, 1, 0, \ldots)$ are in ℓ_2.
    (b) $\varphi_1(t) = \cos t, \varphi_2(t) = e^t, \varphi_3(t) = t$ are in $L_2[-\pi, \pi]$.

41. Let L_0 be the subspace of all odd functions in $L_2[-\pi, \pi]$ and let L_E be the subspace of all even functions. Denote by P_0 and P_E the orthogonal projections onto L_0 and L_E, respectively. Give a formula for P_0 and P_E.

42. Let $N_1 = \{(\xi_1, \xi_1, \xi_1, \xi_2, \xi_2, \xi_2, \ldots)\}$,

$$N_2 = \{(\xi_1, \varepsilon\xi_1, \varepsilon^2\xi_1, \xi_2, \varepsilon\xi_2, \varepsilon^2\xi_2, \ldots)\},$$
$$N_3 = \{(\xi_1, \varepsilon^2\xi_1, \varepsilon\xi_1, \xi_2, \varepsilon^2\xi_2, \varepsilon\xi_2, \ldots)\},$$

be subspaces in ℓ_2, where $\varepsilon = -\frac{1}{2} + \frac{1}{2}\sqrt{3}i$. Given an expression for each of the orthogonal projectors P_j onto the subspace N_j, $j = 1, 2, 3$.

43. Find the orthogonal projections onto the following subspaces of ℓ_2:

   (a) $\overline{\mathrm{sp}}\{(1, 2, 0, \ldots), (0, 1, 2, 0, \ldots), (0, 0, 1, 2, 0, \ldots), \ldots\}$
   (b) $\overline{\mathrm{sp}}(1, -5, 6, 0, \ldots), (0, 1, -5, 6, 0, \ldots), \ldots\}$.

44. Find the orthogonal projection onto the intersection of the following pair of subspaces in ℓ_2:

$$\overline{\mathrm{sp}}\{(1, 2, 0, \ldots), \quad (0, 1, 2, 0, \ldots), \ldots\},$$
$$\overline{\mathrm{sp}}\{(1, 4, 0, \ldots), \quad (0, 1, 4, 0, \ldots), \ldots\}.$$

45. Given $A_{jk} \in \mathcal{L}(\mathcal{H})$, $j, k = 1, 2$, define on $\mathcal{H}^{(2)} = \mathcal{H} \oplus \mathcal{H}$ an operator A by

$$A = \begin{pmatrix} A_{11} & A_{12} \\ A_{21} & A_{12} \end{pmatrix}.$$

   Prove that A is compact if and only if each A_{jk} is compact.

46. Suppose $A, B \in \mathcal{L}(\mathcal{H})$ and AB is compact. Which statements must be true?

   (a) Both A and B are compact.
   (b) At least A or B is compact.

47. Which of the following statements about compact operators on a Hilbert space are true?

   (a) There exists a compact operator with a closed image.
   (b) The image of any compact operator is closed.
   (c) The image of any compact operator is not closed.
   (d) There exists a compact operator with a nonclosed image.
   (e) There exist a compact operator with a finite dimensional kernel.
   (f) The kernel of any compact operator is finite dimensional.

48. Let $(a_j)_{j=1}^\infty$ be a sequence of complex numbers with $\sum_{j=1}^\infty |a_j| < \infty$. Define an operator on ℓ_2 by the matrix

$$A = \begin{pmatrix} a_1 & a_2 & a_3 & \cdots \\ a_2 & a_3 & \cdots & \\ a_3 & \cdots & & \\ \cdots & & & \end{pmatrix}.$$

Prove that A is compact.

49. Let $A \in \mathcal{L}(\ell_2)$ be given by the matrix

$$A = \begin{pmatrix} \omega_1 & & & \\ & \omega_2 & & 0 \\ & & \omega_3 & \\ 0 & & & \ddots \end{pmatrix}$$

i.e., $A = D_\omega$ as in exercise 1. Show that for any $k_1 \le k_2 \le \cdots \le k_m$ in $\mathbb{N}$, the subspace $\overline{\text{sp}}\{e_{k_1}, e_{k_2}, \ldots, e_{k_m}\}$ is A-invariant.

50. Let A be the operator in problem 49. Suppose $\omega_i \in \mathbb{R}, i = 1, 2, \ldots$ and $\omega_i \ne \omega_j$ for $i \ne j$. Prove that all the closed A-invariant subspaces of A have the form described in problem 49.

51. Let $A \in \mathcal{L}(\mathcal{H})$ be invertible and let M be an A-invariant subspace. Prove or disprove the following statements.

(a) $M^\perp$ is A^{-1} invariant.

(b) M is A^{-1} invariant if M is finite dimensional.

(c) M is A^{-1} invariant. Hint: Define A on $\ell_2(\mathbb{Z})$ by $Ae_j = e_{2j-1}, 1 \le j$, $Ae_{-2j} = e_{2j}, 0 \le j, Ae_{-2j+1} = e_{-j}, 1 \le j$.

52. Given $A \in \mathcal{L}(\mathcal{H})$, let L be a closed A invariant subspace. Denote by P_L the orthogonal projection onto L. Prove or disprove the following statements.

(a) $P_L A = A P_L$

(b) $(I - P_L)A = A(I - P_L)$

(c) $P_L A P_L = A P_L$

(d) $P_L A P_L = P_L A$

(e) $(I - P_L)A(I - P_L) = A(I - P_L)$

(f) $(I - P_L)A(I - P_L) = (I - P_L)A$.

53. What can one say about the A-invariance of L and $L^\perp$ for each of the properties (a)–(f) in problem 52?

54. Given $A \in \mathcal{L}(\mathcal{H})$, let M be a closed A-invariant subspace. Denote by A_M the restriction of A to M. Show that $(A_M)^* = P_M A^*|_M$, where $A^*|_M$ is the restriction of A^* to M and P_M is the orthogonal projection onto M.

55. Find an invertible operator A and an A-invariant subspace M such that $M^\perp$ is not A-invariant but is A^{-1}-invariant. Hint: exercise 51.

56. Let M be a closed subspace of a Hilbert space $\mathcal{H}$. With respect to the decomposition $\mathcal{H} = M \oplus M^\perp$, let $A_1, A_2, A_3 \in \mathcal{L}(\mathcal{H})$ be given by the matrices

$$A_1 = \begin{pmatrix} A_{11} & A_{12} \\ 0 & A_{22} \end{pmatrix}, A_2 = \begin{pmatrix} B_{11} & 0 \\ B_{21} & B_{22} \end{pmatrix}, A_3 = \begin{pmatrix} C_{11} & 0 \\ 0 & C_{22} \end{pmatrix}.$$

List all obvious invariant subspaces for these operators.

57. Let $\mathcal{H}_1$, $\mathcal{H}_2$ and $\mathcal{H}_3$ be mutually orthogonal subspaces of a Hilbert space $\mathcal{H}$ such that $\mathcal{H} = \mathcal{H}_1 \oplus \mathcal{H}_2 \oplus \mathcal{H}_3$. Let A, B, C and $D \in \mathcal{L}(\mathcal{H})$ be given by the matrices

$$A = \begin{pmatrix} A_{11} & A_{12} & A_{13} \\ 0 & A_{22} & A_{23} \\ 0 & 0 & A_{33} \end{pmatrix}, \quad B = \begin{pmatrix} B_{11} & 0 & 0 \\ B_{21} & B_{22} & 0 \\ B_{31} & B_{32} & B_{33} \end{pmatrix},$$

$$C = \begin{pmatrix} C_{11} & C_{12} & 0 \\ 0 & C_{22} & 0 \\ 0 & 0 & C_{33} \end{pmatrix} \quad D = \begin{pmatrix} D_{11} & 0 & 0 \\ 0 & D_{22} & 0 \\ 0 & D_{32} & D_{33} \end{pmatrix}.$$

List all the obvious invariant subspaces for A, B, C and D.

58. List some A-invariant subspaces different from $\overline{\mathrm{sp}}\{e_k, e_{k+1}, \ldots\}$, $k = 1, 2, \ldots$, where the operator $A: \ell_2 \to \ell_2$ is given by the matrix

$$A = \begin{pmatrix} 1 & & & & & 0 \\ 2 & 1 & & & \\ & 2 & 1 & & \\ & & & \ddots & \\ & & & 2 & \\ 0 & & & & \ddots \end{pmatrix}.$$

59. List some A-invariant subspaces different from $\overline{\mathrm{sp}}\{e_1, \ldots, e_k\}, k = 1, 2, \ldots,$ where the operator $A: \ell_2 \to \ell_2$ is given by the matrix

$$A = \begin{pmatrix} 1 & 2 & & & 0 \\ & 1 & 2 & & \\ & & 1 & 2 & \\ 0 & & & \ddots & \ddots \end{pmatrix}.$$

60. Given $A \in \mathcal{L}(\mathcal{H})$, suppose L is A-invariant. Prove that

   (a) L is A^{-1}-invariant if and only if the restriction of A to L is invertible on L (assuming A is invertible).

   (b) The closed subspace L is A^*-invariant if and only if $L^\perp$ is A-invariant.

61. Let $\mathcal{H}$ be a Hilbert space. Construct a bounded linear operator A on $\mathcal{H}$ such that

   (a) for a given pair of vectors x and y in $\mathcal{H}$ we have $Ax = y$.
   (b) for two given systems of vectors $x_1, \ldots, x_n$ and $y_1, \ldots, y_n$ in $\mathcal{H}$ we have

   $$Ax_j = y_j, \quad j = 1, \ldots, n.$$

62. Let M and N be closed subspaces of a separable Hilbert space $\mathcal{H}$. Under what conditions on M and N does there exist an operator $A \in \mathcal{L}(\mathcal{H})$ such that $AM = N$?

63. Do the two previous exercises under the additional constraint that A is self adjoint.

64. Do Exercises 61 and 62 under the additional constraint that A is invertible.

65. Do Exercises 61 and 62 under the additional constraint that A is right (left) invertible but not two-sided invertible.

66. Let N be a given closed subspace of a separable Hilbert space $\mathcal{H}$. When does there exist a right (left) invertible operator $A \in \mathcal{L}(\mathcal{H})$ such that its kernel (image) is N?

67. Let M and N be closed subspaces of a separable Hilbert space $\mathcal{H}$. Does there exist an operator $A \in \mathcal{L}(\mathcal{H})$ such that

   $$M = \operatorname{Im} A, \quad N = \operatorname{Ker} A?$$

   When can A be chosen to be a projection?

68. Let M and N be as in the previous exercise. Does there exist an operator $A \in \mathcal{L}(\mathcal{H})$ such that

   $$M = \operatorname{Im} A^*, \quad N = \operatorname{Ker} A^*?$$

69. For $i = 1, 2$ let M_i and N_i be closed subspaces of the separable Hilbert space $\mathcal{H}_i$ such that $\mathcal{H}_i = M_i \oplus N_i$. Find (if possible) an operator $A \in \mathcal{L}(\mathcal{H}_1, \mathcal{H}_2)$ with the property that

   $$N_1 = \operatorname{Ker} A, \qquad N_2 = \operatorname{Ker} A^*,$$
   $$M_1 = \operatorname{Im} A^*, \qquad M_2 = \operatorname{Im} A.$$

70. Let n be a positive integer. If for $A \in \mathcal{L}(\mathcal{H})$ the operator $I - A^n$ is one-sided invertible, then $I - A$ is invertible from the same side. Prove or disprove this statement.

71. Let $\mathcal{H} = \mathcal{H}_1 \oplus \mathcal{H}_2$, where $\mathcal{H}_1$ and $\mathcal{H}_2$ are closed subspaces of the Hilbert space $\mathcal{H}$, and let $T \in \mathcal{L}(\mathcal{H})$ have the following operator matrix representation

$$T = \begin{pmatrix} A & C \\ 0 & B \end{pmatrix}$$

relative to the given decomposition of $\mathcal{H}$. Assume T is invertible. Determine which of the following statements are true:

   (a) A is invertible,
   (b) A is left invertible,
   (c) A is right invertible,
   (d) B is invertible,
   (e) B is right invertible,
   (f) B is left invertible.

72. As in the previous exercise, let $\mathcal{H} = \mathcal{H}_1 \oplus \mathcal{H}_2$ and $T \in \mathcal{L}(\mathcal{H})$ be given by

$$T = \begin{pmatrix} A & C \\ 0 & B \end{pmatrix}.$$

Let A and B both be left (right) invertible. Is the operator T left (right) invertible?

73. Do Exercise 71 with the operator matrix representation of T being replaced by

$$T = \begin{pmatrix} A & 0 \\ C & B \end{pmatrix}.$$

74. Do Exercise 72 with the operator matrix representation of T being replaced by

$$T = \begin{pmatrix} A & 0 \\ C & B \end{pmatrix}.$$

75. (a) Given an example of a product of two orthogonal projections that is not an orthogonal projection.

   (b) When is the product of two orthogonal projections again an orthogonal projection?

76. Let $\omega = (1, a, a^2, \ldots)$, where $0 < a \in \mathbb{R}$, and consider the corresponding Hilbert spaces $\ell_2(\omega)$; see Exercise I-15. Define S_f to be the forward shift on $\ell_2(\omega)$ and S_b the backward shift.

   (a) Prove that S_f and S_b are bounded linear operators on $\ell_2(\omega)$ and compute their norms.

   (b) Find the spectra of S_f and S_b.

77. Let ω be as in the previous exercise, and fix $0 \neq q \in \mathbb{C}$. Let $S_{f,q}$ and $S_{b,q}$ be the weighted shift operators on $\ell_2(\omega)$ given by

$$S_{f,q}(x_0, x_1, x_2, \ldots) = (0, qx_0, q^2x_1, \ldots),$$
$$S_{b,q}(x_0, x_1, x_2, \ldots) = (x_1, qx_2, q^2x_3, \ldots).$$

Solve the problems (a) and (b) in the previous exercise with $S_{f,q}$ and $S_{b,q}$ in place of S_f and S_b, respectively.

78. Do Exercise 76 with the sequence $\omega = (1, a, a^2, \ldots)$ being replaced by $\omega = (\omega_0, \omega_1, \omega_2 \ldots)$, where $\omega_j > 0$ for $j = 0, 1, 2, \ldots$.

79. Do Exercise 77 with $\omega = (\omega_0, \omega_1, \omega_2, \ldots)$ and $\omega_j > 0$ for $j = 0, 1, 2, \ldots$.

80. Do Exercise 77 with $\omega = (\omega_0, \omega_1, \omega_2, \ldots)$ as in the previous exercise and with $S_{f,Q}$ and $S_{b,Q}$ being given by

$$S_{f,Q}(x_0, x_1, x_2, \ldots) = (0, q_1x_0, q_2x_1, \ldots),$$
$$S_{b,Q}(x_0, x_1, x_2, \ldots) = (q_0x_1, q_1x_2, q_2x_3, \ldots).$$

Here $0 \neq q_j \in \mathbb{C}$ for $j = 0, 1, 2, \ldots$.

81. Let $S_{f,Q}$ and $S_{b,Q}$ be the operators on $\ell_2(\omega)$ defined in the previous exercise. Find the matrix representations of these operators relative to the standard basis of $\ell_2(\omega)$.

82. Let $\omega = (\omega_0, \omega_1, \omega_2, \ldots)$ with $\omega_j > 0$ for each j, and let U be the operator from $\ell_2(\omega)$ into ℓ_2 defined by

$$U(x_0, x_1, x_2, \ldots) = (\sqrt{\omega_0}x_0, \sqrt{\omega_1}x_1, \sqrt{\omega_2}x_2, \ldots).$$

(a) Show that U is invertible, and determine U^{-1} and U^*.

(b) Let $S_{f,Q}$ and $S_{b,Q}$ be the operators on $\ell_2(\omega)$ defined in Exercise 80. What kind of operators are the operators $US_{f,Q}U^{-1}$ and $US_{b,Q}U^{-1}$?

83. Let $\mathcal{H} = \mathcal{H}_1 \oplus \mathcal{H}_2 \oplus \mathcal{H}_3$ be the direct sum of the Hilbert spaces $\mathcal{H}_1$, $\mathcal{H}_2$ and $\mathcal{H}_3$ (cf., Exercise 5). Relative to this decompostion of $\mathcal{H}$ let the operator $A \in \mathcal{L}(\mathcal{H})$ be given by the lower triangular operator matrix

$$A = \begin{pmatrix} A_{11} & 0 & 0 \\ A_{21} & A_{22} & 0 \\ A_{31} & A_{32} & A_{33} \end{pmatrix}.$$

What can we conclude about the operators A_{11}, A_{22}, A_{33} when the operator A is invertible? Illustrate the answers with examples.

84. Do the previous exercise for the case when the operator A is given by

$$A = \begin{pmatrix} A_{11} & A_{12} & A_{13} \\ 0 & A_{22} & A_{23} \\ 0 & 0 & A_{33} \end{pmatrix}.$$

85. Fix $t > 0$. Let W be the operator on $L_2(0, \infty)$ defined by

$$(Wf)(x) = \begin{cases} f(x-t) & \text{for} \quad x > t, \\ 0 & \text{for} \quad 0 < x \leq t. \end{cases}$$

(a) Show that W is an isometry and W is not unitary.
(b) Prove that its adjoint is given by

$$(W^*f)(x) = f(x+t), \quad x > 0.$$

(c) Show that the spectrum of W is equal to the closed unit disc

86. Fix $t > 0$. Let U be the operator on $L_2(-\infty, \infty)$ defined by

$$(Uf)(x) = f(x+t), \quad x \in \mathbb{R}$$

(a) Prove that U is unitary and determine U^{-1}.
(b) Show that $\alpha(U) = \mathbb{T}$.

87. Let $A \in \mathcal{L}(\ell_2)$, and let P_n be the orthogonal projection onto span $\{e_0, \ldots, e_n\}$, where $e_0, e_1, \ldots$ is the standard basis of ℓ_2. Show that the operator A is a band operator if and only if for some nonnegative integer m the following equalities hold

(i) $P_k A(I - P_{m+k+1}) = 0$, $\quad k = 0, 1, 2, \ldots$,
(ii) $(I - P_{m+k+1}) A P_k = 0$, $\quad k = 0, 1, 2, \ldots$.

Give an interpretation of these equalities in terms of the spaces Im P_n and Ker P_n, $n = 0, 1, 2, \ldots$.

88. Let L_0 be the linear span of the vectors $e_0, e_1, \ldots$ in the standard basis of ℓ_2. If A on ℓ_2 is a band operator, show that

$$AL_0 \subset L_0, \quad A^*L_0 \subset L_0. \qquad (*)$$

Is the converse implication also true?

89. Let A be an operator on ℓ_2 satisfying $(*)$ in the previous exercise. Describe the form of the matrix of A relative to the standard basis of ℓ_2.

90. Let L_0 be as in Exercise 88, and let $A \in \mathcal{L}(\ell_2)$. Describe in terms of the matrix of A relative to the standard basis of ℓ_2 the property that

$$AL_0 \subset L_0.$$

91. Do the previous exercise with A^* in place of A.

92. Is the inverse of a lower triangular invertible band operator always given by a lower triangular matrix?

93. Let T be the operator on ℓ_2 given by the matrix

$$T = \begin{pmatrix} 1 & a & a^2 & \cdots \\ b & 1 & a & \cdots \\ b^2 & b & 1 & \\ \vdots & \vdots & & \ddots \end{pmatrix}$$

where a and b are complex number with $|a| < 1$ and $|b| < 1$.

(a) Prove that indeed T is a bounded operator and find an evaluation of the norm of T.

(b) Use the finite section method to find the conditions of invertibility of T and to invert T.

94. Do the previous exercise with $\ell_2(\mathbb{Z})$ in place of ℓ_2 and with

$$T = \begin{pmatrix} \ddots & & & & & & \\ & 1 & a & a^2 & a^3 & a^4 & \\ & b & 1 & a & a^2 & a^3 & \\ \cdots & b^2 & b & \boxed{1} & a & a^2 & \cdots \\ & b^3 & b^2 & b & 1 & a & \\ & b^4 & b^3 & b^2 & b & 1 & \\ & & & & & & \ddots \end{pmatrix}$$

95. Consider on $\ell_2(\mathbb{Z})$ the operator T given by the two-diagonal lower triangular matrix

$$T = \begin{pmatrix} \ddots & & & \\ \ddots & b & 0 & 0 \\ & a & \boxed{b} & 0 \\ & 0 & a & b \\ & & & \ddots \end{pmatrix}$$

Use the finite section method to find that T is invertible and to determine its inverse for the following cases:

(i) $a = 1$ and $b = -4$;

(ii) $a = -4$ and $b = 1$;

(iii) $a, b \in \mathbb{C}$ and $|a| < |b|$;

(iv) $a, b \in \mathbb{C}$ and $|b| < |a|$.

96. Solve the problems from the previous exercise in an alternative way by writing

$$T = bI + aV.$$

where V is the forward shift on $\ell_2(\mathbb{Z})$.

97. Do Exercise 95 with the matrix of T replaced by its transpose.

98. Use a modified finite section method to invert the operator T on $\ell_2(\mathbb{Z})$ given by the following tridiagonal matrix

$$T = \begin{pmatrix} \ddots & \ddots & & & \\ \ddots & 7 & -3 & 0 & \\ & -2 & \boxed{7} & -3 & \\ & 0 & -2 & 7 & \ddots \\ & & & \ddots & \ddots \end{pmatrix}$$

99. Use the finite section method to find the conditions of invertibility of $T \in \ell_2$ and to construct its inverse when T is given by the following tridiagonal matrix

$$T = \begin{pmatrix} 1 + |\alpha|^2 & \bar{\alpha} & 0 & 0 & \cdots \\ \alpha & 1 + |\alpha|^2 & \bar{\alpha} & 0 & \cdots \\ 0 & \alpha & 1 + |\alpha|^2 & \bar{\alpha} & \\ 0 & 0 & \alpha & 1 + |\alpha|^2 & \\ \vdots & \vdots & & & \ddots \end{pmatrix}$$

100. Do the previous exercise with ℓ_2 replaced by $\ell_2(\mathbb{Z})$ and with the matrix replaced by the corresponding doubly infinite tridiagonal analog.

101. Let $A \in \mathcal{L}(\mathcal{H}_1, \mathcal{H}_2)$ and $B \in \mathcal{L}(\mathcal{H}_2, \mathcal{H}_1)$. Assume that the product AB is invertible. Which of the following statements is true?

   (a) A is invertible.
   (b) B is invertible.
   (c) A and B are both invertible.
   (d) A right invertible and B is left invertible.
   (e) A is left invertible and B is right invertible.

102. If AB is invertible, does it follow that BA is invertible?

103. Find the spectrum $\sigma(A)$ and the inverse $(\lambda I - A)^{-1}$ for λ in the resolvent set of A, when $A \in \mathcal{L}(\mathcal{H})$ is one of the following operators:

   (a) $Ax = \langle x, \varphi \rangle \psi$ where $\langle \varphi, \psi \rangle = 0$;
   (b) $Ax = \sum_{j=1}^{n} \langle x, \varphi_j \rangle \psi_j$ where $\langle \varphi_j, \psi_k \rangle = 0$ for $j, k = 1, \ldots, n$;
   (c) A is given by

$$Ax = \sum_{j=1}^{\infty} \frac{1}{2^j} \langle x, \varphi_j \rangle \psi_j,$$

   where $\langle \varphi_j, \psi_k \rangle = 0$ and $\|\varphi_j\| = \|\psi_k\| = 1$ for $j, k = 1, 2, 3, \ldots$.

104. Find the connection between the spectra of the operators A_1 and A_2 when

    (a) $A_1 = \begin{pmatrix} 0 & B_1 \\ B_2 & 0 \end{pmatrix}$, $\quad A_2 = B_1 B_2$;

    (b) $A = \begin{pmatrix} 0 & 0 & B_1 \\ 0 & B_2 & 0 \\ B_3 & 0 & 0 \end{pmatrix}$, $\quad A_2 = B_1 B_2 B_3$.

    (c) What can you say additionally if the operators B_j are compact?

105. Find $X \in \mathcal{L}(\mathcal{H})$ such that X is a solution of the operator equation

$$X - AXB = Y,$$

    where A and B are bounded linear operators on $\mathcal{H}$ of norm strictly less than one.

106. Let A and B be as in the previous exercise, and let F be the linear operator on $\mathcal{L}(\mathcal{H})$ defined by

$$F(X) = X - AXB.$$

    (a) Prove that F is bounded and give an evaluation of its norm.
    (b) Show that F is invertible and determine its inverse.
    (c) Determine the spectrum of F.

Chapter III
Laurent and Toeplitz Operators

This chapter deals with operators on $\ell_2(\mathbb{Z})$ and ℓ_2 with the property that the matrix relative to the standard basis in these spaces has a special structure, namely the elements on diagonals parallel to the main diagonal are the same, i.e., the matrix entries a_{jk} depend on the difference $j - k$ only. On $\ell_2(\mathbb{Z})$ these operators are called Laurent operators (and in that case the matrix is doubly infinite); on ℓ_2 they are called Toeplitz operators. These operators form important classes of operators and they appear in many applications. They also have remarkable properties. For instance, there are different methods to invert explicitly these operators, and to compute their spectra. This chapter reviews these results starting from the simplest class.

3.1 Laurent Operators

A *Laurent operator* A is a bounded linear operator on $\ell_2(\mathbb{Z})$ with the property that the matrix of A with respect to the standard orthonormal basis $\{e_j\}_{j=-\infty}^{\infty}$ of $\ell_2(\mathbb{Z})$ is of the form

$$
\begin{bmatrix}
\ddots & \ddots & & & \\
\ddots & a_0 & a_{-1} & a_{-2} & \\
 & a_1 & \boxed{a_0} & a_{-1} & \\
 & & a_1 & \ddots & \\
 & a_2 & a_1 & a_0 & \ddots \\
 & & & \ddots & \ddots
\end{bmatrix} \tag{1.1}
$$

Here $\boxed{a_0}$ denotes the entry a_0 located in the zero row zero column position. In other words, a bounded linear operator A on $\ell_2(\mathbb{Z})$ is a Laurent operator if and only if $\langle Ae_k, e_j \rangle$ depends on the difference $j - k$ only.

Proposition 1.1 *A bounded linear operator A on $\ell_2(\mathbb{Z})$ is a Laurent operator if and only if A commutes with the bilateral shift on $\ell_2(\mathbb{Z})$.*

Proof: Let $\{e_j\}_{j=-\infty}^{\infty}$ be the standard orthonormal basis of $\ell_2(\mathbb{Z})$, and put $a_{jk} = \langle Ae_k, e_j \rangle$. Recall that the bilateral shift V on $\ell_2(\mathbb{Z})$ is given by

$$
V(\ldots, \xi_{-1}, \boxed{\xi_0}, \xi_1, \ldots) = (\ldots, \xi_{-2}, \boxed{\xi_{-1}}, \xi_0, \ldots).
$$

Thus $Ve_j = e_{j+1}$ for each $j \in \mathbb{Z}$. Using $V^* = V^{-1}$, we have

$$\langle VAe_k, e_j \rangle = \langle Ae_k, V^{-1}e_j \rangle = \langle Ae_k, e_{j-1} \rangle = a_{j-1,k}.$$

Also, $\langle AVe_k, e_j \rangle = \langle Ae_{k+1}, e_j \rangle = a_{j,k+1}$. Next, observe that $VA = AV$ if and only if

$$\langle VAe_k, e_j \rangle = \langle AVe_k, e_j \rangle, \quad j, k \in \mathbb{Z}.$$

Thus A and V commute if and only if $a_{j-1,k} = a_{j,k+1}$ for each $j, k \in \mathbb{Z}$, that is, if and only if A is a Laurent operator. $\square$

In a somewhat different form we have already met Laurent operators in Example 2 of Section II.4. Indeed, let a be a bounded complex valued Lebesgue measurable function on $[-\pi, \pi]$, and let M be the corresponding operator of multiplication by a on $L_2([-\pi, \pi])$, that is,

$$(Mf)(t) = a(t)f(t), \quad f \in L_2([-\pi, \pi]).$$

Example 4 in Section II.2 shows that M is a bounded linear operator on $L_2([-\pi, \pi])$ and the matrix of M with respect to the orthonormal basis $\varphi_n(t) = \frac{1}{\sqrt{2\pi}}e^{int}$, $n = 0, \pm 1, \pm 2, \ldots$, is given by (1.1), where

$$a_n = \frac{1}{\sqrt{2\pi}} \int_{-\pi}^{\pi} a(t)e^{-int} \, dt, \quad n \in \mathbb{Z}. \tag{1.2}$$

It follows that the operator $A = \mathcal{F}M\mathcal{F}^{-1}$, where $\mathcal{F}$ is the Fourier transform (see the last paragraph of Section I.18) on $L_2([-\pi, \pi])$, is a bounded linear operator on $\ell_2(\mathbb{Z})$ and the matrix of A with respect standard basis of $\ell_2(\mathbb{Z})$ is also given by (1.1). Thus A is a Laurent operator. In this case we say that A is the Laurent operator *defined by* the function a, and we refer to a as the *defining function* of A. Often the defining function a will be of the form $a(t) = \omega(e^{it})$, where ω is defined on the unit circle. In that case we refer to ω as the *symbol* of A.

The procedure described in the previous section yields all Laurent operators on $\ell_2(\mathbb{Z})$. In other words, given a Laurent operator A on $\ell_2(\mathbb{Z})$, we then find a bounded complex valued Lebesgue measurable function on $[-\pi, \pi]$ such that A is the Laurent operator defined by a. More precisely, if A is the Laurent operator given by the infinite matrix (1.1), then one can find a bounded complex valued Lebesgue measurable function a on $[-\pi, \pi]$ such that (1.2) holds. For this result, which we will not need in this chapter, we refer the reader to Section XXIII.2 of [GGK2].

Let A be the Laurent operator on $\ell_2(\mathbb{Z})$ defined by the continuous function a, and let M be the corresponding operator of multiplication on $L_2([-\pi, \pi])$. Since the Fourier transform preserves the norm, $\|A\| = \|M\|$, and we can use Example 4 in Section II.2 to show that

$$\|A\| = \max_{-\pi \le t \le \pi} |a(t)|. \tag{1.3}$$

The relation $A = \mathcal{F}M\mathcal{F}^{-1}$ also yields the following inversion theorem.

Theorem 1.2 *Let A be the Laurent operator defined by the continuous function a. Then A is invertible if and only if $a(t) \neq 0$ for each $-\pi \le t \le \pi$, and in this case, A^{-1} is the Laurent operator defined by the function $1/a$, that is*

$$A^{-1} = \begin{bmatrix} \ddots & & & & \\ & b_0 & b_{-1} & b_{-2} & \\ & b_1 & b_0 & b_{-1} & \\ & b_2 & b_1 & b_0 & \\ & & & & \ddots \end{bmatrix},$$

where

$$b_n = \frac{1}{\sqrt{2\pi}} \int_{-\pi}^{\pi} \frac{1}{a(t)} e^{-int}\, dt, \quad n \in \mathbb{Z}.$$

Proof: Let M be the operator of multiplication by a on $L_2([-\pi, \pi])$. From Example 2 in Section II.7 we know that M is invertible if and only if $a(t) \neq 0$ for $-\pi \le t \le \pi]$, and in that case M^{-1} is the operator of multiplication by $b = 1/a$. Since $A = \mathcal{F}M\mathcal{F}^{-1}$, the operator A is invertible if and only if M is invertible , and then $A^{-1} = \mathcal{F}M^{-1}\mathcal{F}^{-1}$. By combining these results the theorem follows. $\square$

Corollary 1.3 *The spectrum of the Laurent operator A defined by the continuous functions consists of all points on the curve parametrized by a, that is*

$$\sigma(A) = \{a(t) \mid -\pi \le t \le \pi\}. \tag{1.4}$$

Proof: Take $\lambda \in \mathbb{C}$. Notice that $\lambda I - A$ is again a Laurent operator. In fact, $\lambda I - A$ is the Laurent operator defined by the continuous function $t \mapsto \lambda - a(t)$. But then we can use the previous theorem to show that $\lambda I - A$ is not invertible if and only if $\lambda - a(t) = 0$ for some $t \in [-\pi, \pi]$. It follows that $\lambda \in \sigma(A)$ if and only if $\lambda = a(t)$ for some $t \in [-\pi, \pi]$, which proves (1.4). $\square$

The set $\mathcal{A}$ of all Laurent operators with a continuous defining function is closed under addition and multiplication. Moreover multiplication in $\mathcal{A}$ is commutative. These facts follow immediately from the fact that $\mathcal{F}\mathcal{A}\mathcal{F}^{-1}$ consists of all operators of multiplication by a continuous function. Thus $\mathcal{A}$ has the same algebraic structure as the ring of all complex valued continuous functions on $[-\pi, \pi]$.

Examples: 1. Let A be the Laurent operator on $\ell_2(\mathbb{Z})$ given by the following tridiagonal matrix representation:

$$A = \begin{bmatrix} \ddots & \ddots & & & & \\ \ddots & 7/5 & -3/5 & 0 & 0 & 0 \\ & -2/5 & 7/5 & -3/5 & 0 & 0 \\ \cdots & 0 & -2/5 & \boxed{7/5} & -3/5 & 0 & \cdots \\ & 0 & 0 & -2/5 & 7/5 & -3/5 \\ & & & & & \\ & 0 & 0 & 0 & -2/5 & 7/5 & \ddots \\ & & & & & \ddots & \ddots \end{bmatrix}$$

The function

$$a(t) = -\frac{2}{5}e^{it} + \frac{7}{5} - \frac{3}{5}e^{-it}$$

is the defining function of A. It follows that $a(t) = \omega(e^{it})$, where

$$\omega(\lambda) = -\frac{2}{5}\lambda + \frac{7}{5} - \frac{3}{5}\lambda^{-1} = \frac{1}{5}\left(\frac{1}{\lambda} - 2\right)(\lambda - 3).$$

Here $a(t) \neq 0$ for each t, and we can apply Theorem 1.2 to show that A is invertible. To compute A^{-1}, notice that

$$\frac{1}{\omega(\lambda)} = \frac{\frac{1}{2}\lambda^{-1}}{1 - \frac{1}{2}\lambda^{-1}} + \frac{1}{1 - \frac{1}{3}\lambda}, \qquad |\lambda| = 1,$$

and hence

$$b(t) = \frac{1}{a(t)} = \sum_{j=1}^{\infty} \frac{1}{2^j}e^{-ijt} + \sum_{j=0}^{\infty} \frac{1}{3^j}e^{ijt}.$$

Thus the inverse operator A^{-1} has the form

$$A^{-1} = \begin{bmatrix} \ddots & \ddots & & & & \\ \ddots & 1 & 1/2 & 1/2^2 & 1/2^3 & 1/2^4 \\ & 1/3 & 1 & 1/2 & 1/2^2 & 1/2^3 \\ \cdots & 1/3^2 & 1/3 & \boxed{1} & 1/2 & 1/2^2 & \cdots \\ & 1/3^3 & 1/3^2 & 1/3 & 1 & 1/2 \\ & & & & & \\ & 1/3^4 & 1/3^3 & 1/3^2 & 1/3 & 1 & \ddots \\ & & & & & \ddots & \ddots \end{bmatrix}.$$

Since $e^{it} = \cos t + i\sin t$, we see that the defining function a is also given by

$$a(t) = \left(\frac{7}{5} - \cos t\right) + \frac{1}{5}i\sin t. \tag{1.5}$$

Thus $\cos t = \frac{7}{5} - \Re a(t)$ and $\sin t = 5\Im a(t)$. Using Corollary 1.3 we see that the spectrum of A is precisely given by the ellipse

$$\left(\Re \lambda - \frac{7}{5}\right)^2 + 25(\Im \lambda)^2 = 1.$$

2. Next we consider the Laurent operator

$$A = \begin{bmatrix} \ddots & \ddots & & & \\ \ddots & 1 & 1/2 & 1/2^2 & \\ & \frac{1}{2} & \boxed{1} & 1/2 & \\ & 1/2^2 & 1/2 & 1 & \ddots \\ & & & \ddots & \ddots \end{bmatrix}.$$

The defining function is given by

$$a(t) = \sum_{j=1}^{\infty} \frac{1}{2^j} e^{-ijt} + \sum_{j=0}^{\infty} \frac{1}{2^j} e^{ijt}.$$

Thus $a(t) = \omega(e^{it})$ with ω being given by

$$\omega(\lambda) = \frac{1}{2}\lambda^{-1}\left(1 - \frac{1}{2}\lambda^{-1}\right)^{-1} + \left(1 - \frac{1}{2}\lambda\right)^{-1} = \left(\frac{5}{3} - \frac{2}{3}\lambda^{-1} - \frac{2}{3}\lambda\right)^{-1}.$$

Since $\omega(\lambda) \neq 0$ for $|\lambda| = 1$, we see that $a(t) \neq 0$ for each t, and hence the operator A is invertible by Theorem 1.2, and the defining function b of A^{-1} is given by

$$b(t) = \frac{1}{a(t)} = -\frac{2}{3}e^{-it} + \frac{5}{3} - \frac{2}{3}e^{it}.$$

We conclude that A^{-1} has the following tridiagonal matrix representation:

$$A^{-1} = \begin{bmatrix} \ddots & \ddots & & & & \\ \ddots & 5/3 & -2/3 & 0 & 0 & 0 \\ & -2/3 & 5/3 & -2/3 & 0 & 0 \\ \cdots & 0 & -2/3 & \boxed{5/3} & -2/3 & 0 & \cdots \\ & 0 & 0 & -2/3 & 5/3 & -2/3 \\ & 0 & 0 & 0 & -2/3 & 5/3 & \ddots \\ & & & & & \ddots & \ddots \end{bmatrix}$$

Notice that $\Re b(t) = -\frac{4}{3}\cos t + \frac{5}{3}$ and $\Im b(t) = 0$. Thus, by Corollary 1.3,

$$\sigma(A^{-1}) = \{r \in \Re \mid 1/3 \le r \le 3\}.$$

Since $\sigma(A^{-1}) = \{\lambda^{-1} \mid \lambda \in \sigma(A)\}$, we obtain $\sigma(A) = \{r \in \mathbb{R} \mid 1/3 \le r \le 3\}$. The fact that $\sigma(A)$ is real also follows from the fact that A is selfadjoint.

3. The final example is the tridiagonal Laurent operator

$$
A = \frac{10}{9} \begin{bmatrix}
\ddots & \ddots & & & & \\
\ddots & 8/9 & 1/3 & 1/3 & 0 & 0 \\
& -1/3 & 8/9 & 1/3 & 0 & 0 \\
\cdots & 0 & -1/3 & \boxed{8/9} & 1/3 & 0 & \cdots \\
& 0 & 0 & -1/3 & 8/9 & 1/3 \\
& & & & & \\
& 0 & 0 & & 0-1/3 & 8/9 & \ddots \\
& & & & & \ddots & \ddots
\end{bmatrix}
$$

of which the defining function a is given by $a(t) = \omega(e^{it})$, where

$$\omega(\lambda) = \frac{10}{9}\left(-\frac{1}{3}\lambda + \frac{8}{9} + \frac{1}{3}\lambda^{-1}\right) = \frac{10}{9}\left(1 + \frac{1}{3}\lambda^{-1}\right)\left(1 - \frac{1}{3}\lambda\right).$$

We see that $a(t) \ne 0$ for each t, and

$$b(t) = \frac{1}{a(t)} = \frac{1}{1 + \frac{1}{3}e^{-it}} + \frac{\frac{1}{3}e^{it}}{1 - \frac{1}{3}e^{it}} = \sum_{j=0}^{\infty}\left(-\frac{1}{3}\right)^j e^{-ijt} + \sum_{j=1}^{\infty}\left(\frac{1}{3}\right)^j e^{ijt}.$$

By Theorem 1.2 the operator A is invertible, and its inverse is given by

$$
A^{-1} = \begin{bmatrix}
\ddots & \ddots & & & & \\
\ddots & 1 & -1/3 & 1/3^2 & -1/3^3 & 1/3^4 \\
& 1/3 & 1 & -1/3 & 1/3^2 & -1/3^3 \\
\cdots & 1/3^2 & 1/3 & \boxed{1} & -1/3 & 1/3^2 & \cdots \\
& 1/3^3 & 1/3^2 & 1/3 & 1 & -1/3 \\
& & & & & \\
& 1/3^4 & 1/3^3 & 1/3^2 & 1/3 & 1 & \ddots \\
& & & & & \ddots & \ddots
\end{bmatrix}.
$$

Notice that $\Re a(t) = \frac{80}{81}$ and $\Im a(t) = -\frac{20}{27}\sin t$. Thus by Corollary 1.3, we have

$$\sigma(A) = \left\{z = \frac{80}{81} + ir \,\middle|\, -\frac{20}{27} \le r \le \frac{20}{27}\right\}.$$

3.2 Toeplitz Operators

Let T be a bounded linear operator on ℓ_2. In the sequel the formula

$$T = \begin{bmatrix} a_{11} & a_{12} & a_{13} & \cdots \\ a_{21} & a_{22} & a_{23} & \cdots \\ a_{31} & a_{32} & a_{33} & \\ \vdots & \vdots & & \ddots \end{bmatrix} \tag{2.1}$$

means that the matrix in the right hand side of (2.1) is the matrix corresponding to T and the standard orthonormal basis $e_1, e_2, \ldots$ in ℓ_2. We say that T is a *Toeplitz operator* if

$$T = \begin{bmatrix} a_0 & a_{-1} & a_{-2} & \cdots \\ a_1 & a_0 & a_{-1} & \cdots \\ a_2 & a_1 & a_0 & \\ \vdots & \vdots & & \ddots \end{bmatrix}. \tag{2.2}$$

In this case we refer to the right hand side of (2.2) as the standard matrix representation of T. Recall that $e_j = (0, \ldots, 0, 1, 0, \ldots)$, where the one appears in the j-th entry. Thus T is a Toeplitz operator if and only if $\langle Te_k, e_j \rangle$ depends only on the difference $j - k$. This remark yields the following theorem.

Theorem 2.1 *An operator T on ℓ_2 is a Toeplitz operator if and only if $T = S^*TS$, where S is the forward shift on ℓ_2.*

Proof: Assume T is given by (2.1). By definition, $Se_n = e_{n+1}$ for $n = 1, 2 \ldots$. Thus

$$\langle S^*TSe_k, e_j \rangle = \langle TSe_k, Se_j \rangle = \langle Te_{k+1}, e_{j+1} \rangle = a_{k+1,j+1}.$$

Therefore, $T = S^*TS$ if and only if $a_{kj} = a_{k+1,j+1}$ for all $j, k = 1, 2, \ldots$. Thus $T = S^*TS$ if and only if T is Toeplitz. $\square$

With each continuous complex valued function on the unit circle $\mathbb{T}$ in $\mathbb{C}$ we can associate a Toeplitz operator on ℓ_2. Indeed, let ω be such a function, and put $a(t) = \omega(e^{it})$. Notice that a is continuous on $[-\pi, \pi]$ and $a(-\pi) = a(\pi)$. Now let A be the Laurent operator on $\ell_2(\mathbb{Z})$ defined by a, i.e.,

$$A = \begin{bmatrix} \ddots & \ddots & & & \\ \ddots & a_0 & a_{-1} & a_{-2} & \\ & a_1 & \boxed{a_0} & a_{-1} & \\ & a_2 & a_1 & a_0 & \ddots \\ & & & \ddots & \ddots \end{bmatrix}, \tag{2.3}$$

where

$$a_n = \frac{1}{\sqrt{2\pi}} \int_{-\pi}^{\pi} \omega(e^{it}) e^{-int} \, dt, \quad n \in \mathbb{Z}. \tag{2.4}$$

In the sequel we identify the space ℓ_2 with the closed subspce of $\ell_2(\mathbb{Z})$ consisting of all sequences $(\xi_j)_{j \in \mathbb{Z}}$ such that $\xi_j = 0$ for $j \leq -1$. Put

$$T = P_{\ell_2} A | \ell_2, \tag{2.5}$$

where P_{ℓ_2} is the orthogonal projection of $\ell_2(\mathbb{Z})$ onto ℓ_2. Then T is a bounded linear operator on ℓ_2, and from (2.3) we see that (2.2) holds with a_n given by (2.4). Thus T is a Toeplitz operator. We shall refer to T as the Toeplitz operator with continuous *symbol* ω.

Theorem 2.2 *The norm of a Toeplitz operator T with continuous symbol ω is given by*

$$\|T\| = \max_{|\lambda|=1} |\omega(\lambda)|.$$

Proof: We continue to use the notation introduced in the paragraph preceding the present theorem. From (2.5) and formula (1.3) in the previous section we see that

$$\|T\| \leq \|A\| = \max_{-\pi \leq t \leq \pi} |a(t)| = \max_{|\lambda|=1} |\omega(\lambda)|.$$

So it suffices to prove that $\|T\| \geq \|A\|$. To do this we first make some preparations.

For $N = 0, 1, 2, \ldots$, let $\mathcal{H}_N$ be the closed subspace of $\ell_2(\mathbb{Z})$ consisting of all sequences $(\xi_j)_{j \in \mathbb{Z}}$ such that $\xi_j = 0$ for $j \leq -N - 1$. Thus $\mathcal{H}_0 = \ell_2$ and $\mathcal{H}_0 \subset \mathcal{H}_1 \subset \mathcal{H}_2 \subset \cdots$. Denote by $P_{\mathcal{H}_N}$ the orthogonal projection of $\ell_2(\mathbb{Z})$ onto $\mathcal{H}_N$. Notice that for each $x = (\xi_j)_{j \in \mathbb{Z}}$ in $\ell_2(\mathbb{Z})$ we have

$$\|x - P_{\mathcal{H}_N} x\|^2 = \sum_{j=-\infty}^{-N-1} |\xi_j|^2 \to 0 \quad (N \to \infty). \tag{2.6}$$

Also, if $\ldots, e_{-1}, e_0, e_1, \ldots$ is the standard basis of $\ell_2(\mathbb{Z})$, then $e_{-N}, e_{-N+1}, e_{-N+2}, \ldots$ is an orthonormal basis of $\mathcal{H}_N$, and from (2.5) if follows that with respect this basis the matrix of $P_{\mathcal{H}_N} A | \mathcal{H}_N$ is given by

$$\begin{bmatrix} a_0 & a_{-1} & a_{-2} & \cdots \\ a_1 & a_0 & a_{-1} & \cdots \\ a_2 & a_1 & a_0 & \\ \vdots & \vdots & & \ddots \end{bmatrix}.$$

It follows that $\|P_{\mathcal{H}_N} A | \mathcal{H}_N\| = \|T\|$ for each N.

Now, fix $\varepsilon > 0$, and choose $x \in \ell_2(\mathbb{Z})$ such that $\|x\| = 1$ and $\|Ax\| > \|A\| - \varepsilon$. By (2.6), we have $P_{\mathcal{H}_N} x \to x$ and $A P_{\mathcal{H}_N} x \to Ax$ for $N \to \infty$. Thus we can find a positive integer N' such that $x' = P_{\mathcal{H}_{N'}} x$ has the following properties:

$$0 \neq \|x'\| < 1 + \varepsilon, \quad \|Ax'\| > \|A\| - \varepsilon. \tag{2.7}$$

According to (2.6), we also have $P_{\mathcal{H}_N} Ax' \to Ax'$ for $N \to \infty$. Thus we can choose $N'' \geq N'$ such that $\| P_{\mathcal{H}_{N''}} Ax' \| > \|A\| - \varepsilon$. Since $x' \in \mathcal{H}_{N'}$ and $N'' \geq N'$, we have $x' \in \mathcal{H}_{N''}$, and thus

$$\| P_{\mathcal{H}_{N''}} A | \mathcal{H}_{N''} \| \geq \frac{\| P_{\mathcal{H}_{N''}} Ax' \|}{\|x'\|} > \frac{\|A\| - \varepsilon}{1 + \varepsilon}.$$

Therefore, $\|T\| > (1 + \varepsilon)^{-1}(\|A\| - \varepsilon)$. This holds for each $\varepsilon > 0$. Hence $\|T\| \geq \|A\|$. □

The construction given in the paragraph preceding Theorem 2.2 can be carried out for larger classes of functions than continuous ones. Also Theorem 2.2 holds in a more general setting. See, for instance, Section XXIII.3 in [GGK2].

3.3 Band Toeplitz Operators

In this section we study the invertibility of a Toeplitz operator that has the additional property that in its standard matrix representation all diagonals are zero with the exception of a finite number. In that case all the non-zero entries in the standard matrix representation are located in a band, and for that reason we refer to such an operator as a band Toeplitz operator (cf., the last part of Section II.16). Notice that a band Toeplitz operator has a continuous symbol of the form

$$\omega(\lambda) = \sum_{k=-p}^{q} \lambda^k a_k. \tag{3.1}$$

Thus the symbol of a band Toeplitz operator is a trigonometric polynomial.

The function ω in (3.1) can be represented in the form

$$\omega(\lambda) = c\lambda^{-r}(\lambda - \alpha_1)\dots(\lambda - \alpha_\ell)(\lambda - \beta_1)\dots(\lambda - \beta_m), \tag{3.2}$$

where c is a non-zero constant, r is a nonnegative integer, $\alpha_1, \dots, \alpha_\ell$ are complex numbers inside the open unit disk, and $\beta_1, \dots, \beta_m$ are complex numbers on the unit circle or outside the closed unit disk. The representation (3.2) of ω is not unique, but the number $\kappa = \ell - r$ is. In fact, κ is just equal to the number of zeros (multiplicities taken into account) of ω in the open unit disk minus the order of the pole of ω at zero. If $\omega(\lambda) \neq 0$ for each $|\lambda| = 1$, i.e., if $|\beta_j| > 1$ for each j, then κ is equal to the number of times the oriented curve $t \mapsto \omega(e^{it})$ circles around zero when t runs from $-\pi$ to π, and in that case we call κ the *winding number* of ω relative to zero. In other words

$$\kappa = \frac{1}{2\pi} [\arg \omega(e^{it})]_{t=-\pi}^{\pi}.$$

We shall prove the following theorems.

Theorem 3.1 *Let T be the band Toeplitz operator with symbol ω given by (3.2). Then T is two-sided invertible if and only if $\omega(e^{it}) \neq 0$ for $-\pi \leq t \leq \pi$ and $\kappa = l - r = 0$. In that case*

$$T^{-1} = \frac{1}{c} \Pi_{j=1}^{m} (S - \beta_j I)^{-1} \Pi_{i=1}^{\ell} (I - \alpha_i S^*)^{-1}, \tag{3.3}$$

where S is the forward shift on ℓ_2.

Theorem 3.2 *Let T be the band Toeplitz operator with symbol ω given by (3.1), and put $\kappa = \ell - r$. Then T is left or right invertible if and only if $\omega(e^{it}) \neq 0$ for $-\pi \leq t \leq \pi$. Furthermore, if this condition is satisfied, we have*

(i) *T is left invertible if and only if $\kappa \geq 0$, and in that case codim Im $T = \kappa$, and a left inverse of T is given by*

$$T^{(-1)} = \frac{1}{c} \Pi_{j=1}^{m} (S - \beta_j I)^{-1} (S^*)^{\kappa} \Pi_{i=1}^{\ell} (I - \alpha_i S^*)^{-1}; \tag{3.4}$$

(ii) *T is right invertible if and only if $\kappa \leq 0$, and in that case dim Ker $T = -\kappa$ and a right inverse of T is given by*

$$T^{(-1)} = \frac{1}{c} \Pi_{j=1}^{m} (S - \beta_j I)^{-1} S^{-\kappa} \Pi_{i=1}^{\ell} (I - \alpha_i S^*)^{-1}. \tag{3.5}$$

Here S is the forward shift on ℓ_2.

The proofs of Theorems 3.1 and 3.2 are considerably more involved than the inversion theorem for Laurent operators. There are two reasons for this. The first is the fact that in general, in contrast to the Laurent operators in Section 3.1, Toeplitz operators do not commute. Secondly, also unlike Laurent operators, the product of two Toeplitz operators does not have to be a Toeplitz operator. These phenomena can already be illustrated in the forward shift S and the backward shift S^*. Both S and S^* are Toeplitz operators, $S^* S$ is the identity operator on ℓ_2, and

$$SS^* = \begin{bmatrix} 0 & 0 & 0 & 0 & \cdots \\ 0 & 1 & 0 & 0 & \cdots \\ 0 & 0 & 1 & 0 & \\ 0 & 0 & 0 & 1 & \\ \vdots & \vdots & & & \ddots \end{bmatrix}$$

Thus $SS^* \neq S^* S$ (and thus the Toeplitz operators S and S^* do not commute), and the product SS^* is not Toeplitz (but a projection).

Notice that for $n \geq 0$ and $m \geq 0$ we have

$$(S^*)^m S^n = \begin{cases} (S^*)^{m-n}, & m \geq n, \\ S^{n-m}, & n \geq m \end{cases} \tag{3.6}$$

These identities will turn out to be very useful in the proofs.

In the proofs of Theorems 3.1 and 3.2 we will also need the following lemma.

Lemma 3.3 *For* $|\alpha| < 1$ *and* $|\beta| > 1$ *the operators* $I - \alpha S^*$ *and* $S - \beta I$ *are invertible, and the respective inverses are given by*

$$(I - \alpha S^*)^{-1} = \begin{bmatrix} 1 & \alpha & \alpha^2 & \alpha^3 & \cdots \\ 0 & 1 & \alpha & \alpha^2 & \cdots \\ 0 & 0 & 1 & \alpha & \\ 0 & 0 & 0 & 1 & \\ & & & & \\ \vdots & \vdots & & & \ddots \end{bmatrix},$$

$$(S - \beta I)^{-1} = -\beta \begin{bmatrix} 1 & 0 & 0 & 0 & \cdots \\ \beta^{-1} & 1 & 0 & 0 & \cdots \\ \beta^{-2} & \beta^{-1} & 1 & 0 & \\ \beta^{-3} & \beta^{-2} & \beta^{-1} & 1 & \\ & & & & \\ \vdots & \vdots & \vdots & & \ddots \end{bmatrix}. \tag{3.7}$$

For $|\alpha| = |\beta| = 1$ *the operators* $I - \alpha S^*$ *and* $S - \beta I$ *are neither left nor right invertible.*

Proof: For $|\alpha| < 1$ we have $\|\alpha S^*\| < 1$, and hence, by Theorem II.8.1, the operator $I - \alpha S^*$ is invertible and its inverse is given by the first part of (3.7). Since $S - \beta I = -\beta(I - \beta^{-1}S)$, a similar argument shows that $S - \beta I$ is invertible, and that its inverse is given by the second part of (3.7). From the results in Section II.20 we know that the spectra of S and S^* are both equal to the closed unit disc. Thus the perturbation results Theorems II.15.3 and II.15.5 imply the operators $I - \alpha S^*$ and $S - \beta I$ are neither left nor right invertible when $|\alpha| = |\beta| = 1$. $\qquad\square$

Now, let R be a band Toeplitz operator, and let its symbol ω be given by (3.1). Since we allow a_{-p} and a_q to be zero, we may assume that p and q are positive integers, and in that case (3.1) means that

$$R = a_{-p}S^{*p} + \cdots + a_{-1}S^* + a_0 I + a_1 S + \cdots + a_q S^q,$$

where S is the forward shift on ℓ_2. The fact that $S^* S = I$ implies that for each pair of complex a and b the operator $R(a + bS)$ is again a band Toeplitz operator. In fact, the symbol of $R(a + bS)$ is given by $\omega(\lambda)(a + \lambda b)$. Similarly, for each pair of complex numbers c and d the operator $(c + dS^*)R$ is a Toeplitz operator and its symbol is the function $(c + d\lambda^{-1})\omega(\lambda)$. As we remarked above (see (3.6)) in these product formulas the order of the factors is important. In general, $(a + bS)R$ and $R(c + dS^*)$ are not Toeplitz operators. We are now ready to prove Theorems 3.1 and 3.2.

Proof of Theorem 3.2 We split the proof in two parts. In the first part we assume that $\omega(e^{it}) \neq 0$ for $-\pi \leq t \leq \pi$ and prove statements (i) and (ii). In the second part we prove that $\omega(e^{it}) \neq 0$ for $-\pi \leq t \leq \pi$ is necessary for left or right invertibility.

Part 1. Assume $\omega(e^{it}) \neq 0$ for $-\pi \leq t \leq \pi$. Here ω is given by (3.2). It follows that in (3.2) the numbers $\beta_1, \ldots, \beta_m$ are outside the closed unit disk, i.e., $|\beta_j| > 1$ for $j = 1, \ldots, m$.

Assume $\kappa = \ell - r \geq 0$. Then from the results proved in the paragraph preceding the present proof we see that

$$T = c\Pi_{i=1}^{\ell}(I - \alpha_i S^*)S^\kappa \Pi_{j=1}^{m}(S - \beta_j I). \tag{3.8}$$

Indeed, $S^{\ell-r}$ is the band Toeplitz operator with symbol $\lambda^{\ell-r}$, and thus by repeatedly applying the above mentioned multiplication rules, we see that the right hand side of (3.8) is the band Toeplitz operator with symbol

$$\tilde{\omega}(\lambda) = c\Pi_{i=1}^{\ell}(I - \alpha_i \lambda^{-1})\lambda^{\ell-r}\Pi_{i=1}^{m}(\lambda - \beta_i).$$

Since $\tilde{\omega} = \omega$, it follows that (3.8) holds.

The fact that $|\alpha_i| < 1$ and $|\beta_j| > 1$ for each i and j implies (see Lemma 3.3) that the factors $I - \alpha_i S^*$ and $S - \beta_j I$ appearing in the right hand side of (3.8) are invertible operators, and hence the operators

$$\Pi_{i=1}^{\ell}(I - \alpha_i S^*), \quad \Pi_{j=1}^{m}(S - \beta_j I) \tag{3.9}$$

are invertible operators. Also, the operator $T^{(-1)}$ in (3.4) is well-defined, and, using $(S^*)^\kappa S^\kappa = I$, we see that $T^{(-1)}T = I$. Thus T is left invertible and $T^{(-1)}$ is a left inverse of T. From (3.8) and the invertibility of the operators in (3.9) we obtain that

$$\text{codim Im } T = \text{codim Im } S^\kappa = \kappa,$$

which completes the proof of statement (i).

Next, assume $\kappa \leq 0$, and hence $\ell \leq r$. Then, by repeated application of the product rules proved in the paragraph preceding the present proof, we have

$$T = c\Pi_{i=1}^{\ell}(I - \alpha_i S^*)(S^*)^{-\kappa}\Pi_{j=1}^{m}(S - \beta_j I). \tag{3.10}$$

Since the operators $I - \alpha_i S^*$ and $S - \beta_j I$ are invertible for each i and j, the operator $T^{(-1)}$ in (3.5) is well-defined. From $(S^*)^{-\kappa}S^{-\kappa} = I$ we conclude that $TT^{(-1)} = I$, and hence $T^{(-1)}$ is a right inverse of T. Furthermore, using the invertibility of the operators in (3.9) we see that

$$\text{codim Ker } T = \dim \text{Ker } (S^*)^{-\kappa} = -\kappa,$$

which completes the proof of (ii).

Part 2. In this part T is left or right invertible. We want to show that $\omega(e^{it}) \neq 0$ for $-\pi \leq t \leq \pi$. The proof is by contradiction. So we assume that ω has a zero on the unit circle $|\lambda| = 1$, that is, in the representation (3.2) we have $|\beta_k| = 1$ for at least one k. Choose such a β_k and put

$$\omega_1(\lambda) = \frac{\omega(\lambda)}{\lambda - \beta_k}, \quad \omega_2(\lambda) = \lambda\frac{\omega(\lambda)}{\lambda - \beta_k}.$$

Then ω_1 and ω_2 are functions of the same type as ω. Let T_1 and T_2 be the band Toeplitz operators with symbols ω_1 and ω_2, respectively. Since

$$\omega(\lambda) = \omega_1(\lambda)(\lambda - \beta_k), \quad \omega(\lambda) = (1 - \beta_k \lambda^{-1})\omega_2(\lambda),$$

the product rules from the paragraph preceding the present proof yield

$$T = T_1(S - \beta_k I), \quad T = (I - \beta_k S^*)T_2. \tag{3.11}$$

By Lemma 3.3 the operator $S - \beta_k I$ is not left invertible, because $|\beta_k| = 1$. Hence the first equality in (3.11) shows that T cannot be left invertible. It follows that T must be right invertible, but then the second equality in (3.11) implies that $I - \beta_k S^*$ is right invertible. Thus $I - \bar{\beta}_k S$ is left invertible. However the latter is impossible, again by Lemma 3.3 and the fact that $|\beta_k| = 1$.

Proof of Theorem 3.1 By Theorem 3.2 (i) and (ii) the operator T is both left and right invertible if and only if $\omega(e^{it}) \neq 0$ for $-\pi \leq t \leq \pi$ and $\kappa = \ell - r = 0$. Moreover in that case, formula (3.4) with $\kappa = 0$ yields the formula for the inverse in (3.3). $\square$

Corollary 3.4 *Let T be the band Toeplitz operator with symbol ω. Then the spectrum $\sigma(T)$ of T is given by*

$$\sigma(T) = \mathbb{C} \setminus \Omega, \tag{3.12}$$

where Ω is the set of all points λ in $\mathbb{C}$ such that $\lambda \neq \omega(e^{it})$ for all $-\pi \leq t \leq \pi$ and the winding number of $\omega(\cdot) - \lambda$ relative to zero is equal to zero. In particular, the spectral radius of T is equal to the norm of T.

Proof: Notice that for each λ in $\mathbb{C}$ the operator $T - \lambda I$ is the band Toeplitz operator with symbol $\omega(\cdot) - \lambda$. Thus Theorem 3.1 shows that $\lambda \in \sigma(T)$ if and only if $\lambda \in \Omega$, which proves (3.12).

Since $\omega(e^{it}) \in \sigma(T)$ for each $-\pi \leq t \leq \pi$, the spectral radius $r(T) = \sup\{|\lambda| \mid \lambda \in \sigma(T)\}$ is larger than or equal to $\max_{-\pi \leq t \leq \pi} |\omega(e^{it})|$. On the other hand, the latter quantity is equal to $\|T\|$, by Theorem 2.2. Since $\|T\| \geq r(T)$, we conclude that $r(T) = \|T\|$. $\square$

Example: In this example we consider the Toeplitz operator T on ℓ_2 defined by the tridiagonal matrix:

$$T = \begin{bmatrix} 7/5 & -3/5 & 0 & \cdots \\ -2/5 & 7/5 & -3/5 & \\ 0 & -2/5 & 7/5 & \\ \vdots & & & \ddots \end{bmatrix}$$

The corresponding Laurent operator has been considered in Example 1 of the first section. The operator T is bounded, and its symbol ω is given by

$$\omega(\lambda) = -\frac{2}{5}\lambda + \frac{7}{5} - \frac{3}{5}\lambda^{-1} = -\frac{2}{5}\lambda^{-1}\left(\lambda - \frac{1}{2}\right)(\lambda - 3). \tag{3.13}$$

The aim is to study the invertibility properties of $zI - T$ for each z in $\mathbb{C}$, and to compute an inverse (one-sided or two-sided) whenever it exists. We split the text into five parts.

Part 1. Notice that $\omega(\lambda) \neq 0$ for each $|\lambda| = 1$. Moreover, for this ω the numbers ℓ and r in (3.2) are given by $\ell = 1$ and $r = 1$. Thus $\kappa = 0$. By Theorem 3.1 the operator T is invertible, and

$$T^{-1} = -\frac{5}{2}(S - 3I)^{-1}\left(I - \frac{1}{2}S^*\right)^{-1}. \tag{3.14}$$

By formula (3.7) in Lemma 3.3 we have

$$\left(I - \frac{1}{2}S^*\right)^{-1} = \sum_{j=0}^{\infty} \frac{1}{2^j}(S^*)^j, \tag{3.15}$$

and

$$-(S - 3I)^{-1} = (3I - S)^{-1} = \frac{1}{3}\sum_{j=0}^{\infty}\frac{1}{3^j}S^j, \tag{3.16}$$

and the formula for T^{-1} can be rewritten in matrix form as follows

$$T^{-1} = \frac{5}{6}\begin{bmatrix} 1 & 0 & 0 & \cdots \\ 1/3 & 1 & 0 & \\ 1/3^2 & 1/3 & 1 & \ddots \\ \vdots & & \ddots & \ddots \end{bmatrix}\begin{bmatrix} 1 & 1/2 & 1/2^2 & \cdots \\ 0 & 1 & 1/2 & \ddots \\ 0 & 0 & 1 & \\ \vdots & \ddots & & \ddots \end{bmatrix}$$

$$= \frac{5}{6}\begin{bmatrix} 1 & 1/2 & 1/2^2 & \cdots \\ 1/3 & 7/6 & 7/12 & \cdots \\ 1/9 & 7/18 & 43/36 & \cdots \\ \vdots & \vdots & & \ddots \end{bmatrix}.$$

Let us examine the matrix entries $t_{jk}^{\times}$ of T^{-1} in more detail. From (3.14)–(3.16) we see that

$$t_{jk}^{\times} = \frac{5}{6}\sum_{\nu=0}^{\min(j,k)}\left(\frac{1}{3}\right)^{j-\nu}\left(\frac{1}{2}\right)^{k-\nu}.$$

An easy calculation shows that

$$t_{jk}^{\times} = \frac{1}{3^{j+1}2^{k+1}}(6^{j+1} - 1) \quad \text{for } j \leq k,$$

and

$$t_{jk}^{\times} = \frac{1}{3^{j+1}2^{k+1}}(6^{k+1} - 1) \quad \text{for } j \geq k.$$

From these formulas it follows that $t_{jk}^{\times}$ can be represented in the form

$$t_{jk}^{\times} = g_{j-k} + f_{jk}, \qquad j, k = 0, 1, \dots,$$

where

$$g_{j-k} = \begin{cases} 3^{k-j}, & \text{for } k \le j, \\ 2^{j-k}, & \text{for } k \ge j, \end{cases}$$

and

$$f_{jk} = \frac{1}{3^{j+1}2^{k+1}}, \qquad j, k = 0, 1, \dots.$$

Notice that the operator

$$G = (g_{j-k})_{j, k=0}^{\infty}$$

is a Toeplitz operator and the operator

$$F = (f_{jk})_{j, k=0}^{\infty}$$

is an operator of rank one.

Now let us compare this with Example 1 in the first section. There we considered the Laurent operator A with the same symbol ω as for T, and we showed that A is invertible with its inverse being given by

$$A^{-1} = (a_{j-k}^{\times})_{j, k=0}^{\infty},$$

where

$$a_j^{\times} = \begin{cases} 3^{-j}, & \text{for } j \ge 0, \\ 2^j, & \text{for } j \le 0. \end{cases}$$

This means that $G = PA^{-1}P|\operatorname{Im}P$, where P is the projection defined by the equality

$$P(\dots \xi_{-1}, \xi_0, \xi_1 \dots) = (\dots 0, \xi_0, \xi_1, \dots).$$

We proved that

$$T^{-1} = PA^{-1}P|\operatorname{Im}P + F.$$

The latter equality can be derived for other Toeplitz operators T too and can be used for the computation of T^{-1}.

Part 2. Let us determine the spectrum $\sigma(T)$ of the Toeplitz operator T with symbol ω given by (3.13). As we have seen in Example 1 of the first section the curve $t \mapsto \omega(e^{it})$ is precisely equal to the ellipse

$$\left(\Re\lambda - \frac{7}{5}\right)^2 + 25(\Im\lambda)^2 = 1. \tag{3.17}$$

Thus we can apply Corollary 3.4 to show that $\sigma(T)$ consists of all points $\lambda \in \mathbb{C}$ that lie on or are inside the ellipse (3.17), that is,

$$\sigma(T) = \left\{\lambda \in \mathbb{C} \mid \left(\Re\lambda - \frac{7}{5}\right)^2 + 25(\Im\lambda)^2 \le 1\right\}.$$

On the other hand, as we have seen in Example 1 of the first section, the spectrum of the Laurent operator A with symbol ω given by (3.13) is precisely the ellipse (3.17). Thus in this case $\sigma(A)$ is just the boundary $\partial\sigma(T)$ of the spectrum of T.

Part 3. We continue with an analysis of $z_0 I - T$ for the case when z_0 is a point strictly inside the ellipse (3.17). From the result in the previous part we know that $z_0 I - T$ is not two-sided invertible. We claim that $z_0 I - T$ is right invertible with dim Ker $(z_0 I - T) = 1$. To see this, notice that $\omega_0(\lambda) = \omega(\lambda) - z_0$ is the symbol of the Toeplitz operator $T_0 = T - z_0 I$, and $\lambda \in \sigma(T_0)$ if and only if $\lambda - z_0 \in \sigma(T)$. Now, apply Theorem 3.2 to T_0. Since z_0 is inside the ellipse (3.17), we have $\omega(e^{it}) - z_0 \neq 0$ for each t, and hence $\omega_0(e^{it}) \neq 0$ for each t. Furthermore, the winding number of the curve $t \mapsto \omega_0(e^{it})$ with respect to zero is precisely equal to the winding number of the curve $t \mapsto \omega(e^{it})$ with respect to z_0. Since (cf., formula (5) in the first section)

$$\omega(e^{it}) = \frac{7}{5} - \cos t + \frac{1}{5}i \sin t,$$

the orientation on the curve $t \mapsto \omega(e^{it})$ is clockwise, and hence the winding number is -1. Thus we can apply Theorem 3.2 to show that $T_0 = T - z_0 I$ is right invertible and dim Ker $T_0 = 1$.

Let us specify further the above for the case when $z_0 = 7/5$ (the center of the ellipse (3.17)). Then the symbol $\omega_0(\lambda)$ is given by

$$\omega_0(\lambda) = -\frac{2}{3}\lambda - \frac{3}{5}\lambda^{-1} = -\frac{2}{5}\lambda^{-1}\left(\lambda - i\sqrt{\frac{3}{2}}\right)\left(\lambda + i\sqrt{\frac{3}{2}}\right).$$

Notice that the roots $\pm i\sqrt{3/2}$ are outside the unit circle. Thus in this case the number $\kappa = \ell - r$ corresponding to (3.2) is equal to -1 (which we already know from the previous paragraph). Furthermore, by Theorem 3.2 (ii), the operator $T - \frac{7}{5}I$ is right invertible and a right inverse is given by

$$\left(T - \frac{7}{5}I\right)^{(-1)} = -\frac{5}{2}\left(S - i\sqrt{\frac{3}{2}}\right)^{-1}\left(S + i\sqrt{\frac{3}{2}}\right)^{-1}S.$$

More generally, let z_0 be an arbitrary point strictly inside the ellipse (3.17). Then

$$\omega_0(\lambda) = -\frac{2}{5}\lambda^{-1}\left[\lambda^2 - \frac{5}{2}\left(\frac{7}{5} - z_0\right) + \frac{3}{2}\right],$$

and the roots a_1, a_2 of the quadratic polynomial $\lambda^2 - \frac{5}{2}(\frac{7}{5} - z_0) + \frac{3}{2}$ are outside the unit circle. To see this, recall that $\omega_0(e^{it}) \neq 0$ for each t and each choice of z_0 strictly inside the ellipse (3.17). Next, notice that the roots a_1 and a_2 depend continuously on the point z_0, and for $z_0 = 7/5$ they are outside the unit circle. Thus, if for $i = 1$ or $i = 2$ the root a_i would not be outside the unit circle, then the root a_i has to cross the unit circle, which contradicts the fact that $\omega_0(e^{it}) \neq 0$ for each t. Thus

$$\omega_0(\lambda) = -\frac{2}{5}\lambda^{-1}(\lambda - a_1)(\lambda - a_2),$$

with $|a_1| > 1$, $|a_2| > 1$, and by Theorem 3.2 (ii) the operator $T - z_0 I$ is right invertible and a right inverse is given by

$$(T - z_0 I)^{(-1)} = -\frac{5}{2}(S - a_1 I)^{-1}(S - a_2 I)^{-1}S.$$

By using the second part of formula (3.7) in Lemma 3.3 we obtain that

$$(z_0 I - T)^{(-1)} = -\frac{5}{2}\frac{1}{a_1 a_2}\sum_{n=0}^{\infty}\left(\sum_{j+k=n}\left(\frac{1}{a_1}\right)^j\left(\frac{1}{a_2}\right)^k\right)S^{n+1}.$$

Thus $(z_0 I - T)^{(-1)}$ is a strictly lower triangular Toeplitz operator, namely

$$(z_0 I - T)^{(-1)} = -\frac{5}{2}\frac{1}{a_1 a_2}\begin{bmatrix} 0 & 0 & 0 & 0 & \cdots \\ b_1 & 0 & 0 & 0 & \cdots \\ b_2 & b_1 & 0 & 0 \\ b_3 & b_3 & b_1 & 0 \\ \vdots & \vdots & & \ddots \end{bmatrix},$$

where

$$b_n = \begin{cases} \frac{a_2 a_1}{a_2 - a_1}(a_1^{-n-1} - a_2^{-n-1}), & a_1 \neq a_2, \\ (n+1)a_1^{-n}, & a_1 = a_2. \end{cases}$$

Part 4. Next let z_0 be a point strictly outside the ellipse (3.17). We already know from Part 2 that in that case $z_0 \notin \sigma(T)$. Thus $z_0 I - T$ is invertible. To compute its inverse, we consider again the corresponding symbol

$$\omega_0(\lambda) = -\frac{2}{5}\lambda^{-1}\left[\lambda^2 - \frac{5}{2}\left(\frac{7}{5} - z_0\right) + \frac{3}{2}\right].$$

In this case one root, a_1 say, of the quadratic polynomial

$$\lambda^2 - \frac{5}{2}\left(\frac{7}{5} - z_0\right) + \frac{3}{2}$$

is inside the open unit disc, and one root, a_2 say, is outside the unit circle. For $z_0 = 0$ we know this from Example 1. Indeed, for $z_0 = 0$ we have $a_1 = 1/2$ and $a_2 = 3$. For an arbitrary z_0 it follows by using a continuity argument similar to the one used in the third paragraph of the previous part. Thus Theorem 3.1 yields that

$$(z_0 I - T)^{-1} = \frac{5}{2}(S - a_2 I)^{-1}(I - a_1 S^*)^{-1}.$$

Using the formulas appearing in (3.7) we see that

$$(z_0 I - T)^{-1} = -\frac{5}{2a_2} \begin{bmatrix} 1 & 0 & 0 & \cdots \\ a_2^{-1} & 1 & 0 & \\ a_2^{-2} & a_2^{-1} & 1 & \\ \vdots & & & \ddots \end{bmatrix} \begin{bmatrix} 1 & a_1 & a_1^2 & \cdots \\ 0 & 1 & a_1 & \\ 0 & 0 & 1 & \\ \vdots & & & \ddots \end{bmatrix}.$$

This product we can compute in the same way as we compute T^{-1} at the end of Part 1. We have

$$(z_0 I - T)^{-1} = (g_{j-k})_{j,k=0}^{\infty} + (f_{jk})_{j,k=0}^{\infty}$$

where

$$g_j = \begin{cases} a_2^{-j} & \text{for} \quad j = -1, -2, \ldots \\ a_1^{j} & \text{for} \quad j = 0, 1, 2, \ldots . \end{cases}$$

and

$$f_{jk} = \frac{a_1^{k+1}}{a_2^{j+1}}, \qquad j, k = 0, 1, 2, \ldots .$$

Part 5. Finally, let z_0 be a point on the ellipse (3.17). As we have already seen in Part 2, this implies that $z_0 \in \partial\sigma(T)$. We claim that in this case $z_0 I - T$ is neither left nor right invertible. Indeed, let $z_0 I - T$ be left or right invertible. Since z_0 belongs to the spectrum of T, the operator $z_0 I - T$ is not two-sided invertible. Thus by the stability results of Section II.15 there is an open neighborhood of z_0 such that for each point z_0' in this neighborhood the operator $z_0' I - T$ is one-sided invertible but not two-sided invertible. Since $z_0 \in \partial\sigma(T)$, this is impossible.

3.4 Toeplitz Operators with Continuous Symbols

In this section we study invertibility, one-sided or two-sided, of Toeplitz operators with a continuous symbol ω. To state the main theorem we need the notion of a winding number.

Let ω be a continuous function on the unit circle $\mathbb{T}$, and assume that $\omega(e^{it}) \neq 0$ for $-\pi \leq t \leq \pi$. Then the closed curve parametrized by $t \mapsto \omega(e^{it})$, where $-\pi \leq t \leq \pi$, does not pass through zero. Again we define the *winding number* κ of ω relative to zero to be the number of times the oriented curve $t \mapsto \omega(e^{it})$ circles around 0 when t runs from $-\pi \leq t \leq \pi$. In other words, the winding number is equal to $1/2\pi$ times the total variation of the argument function $\arg(\omega(e^{it}))$ when the variable t varies over $-\pi \leq t \leq \pi$. In the sequel we shall need the following fact (which is known as Rouché's theorem): if ω and $\tilde{\omega}$ are both continuous functions on the unit circle $\mathbb{T}$ and

$$|\omega(e^{it}) - \tilde{\omega}(e^{it})| < |\tilde{\omega}(e^{it})|, \quad -\pi \leq t \leq \pi, \tag{4.1}$$

then the two curves $t \mapsto \omega(e^{it})$ and $t \mapsto \tilde{\omega}(e^{it})$, with $-\pi \leq t \leq \pi$, do not pass through zero and the corresponding winding numbers are equal.

The following theorem is the main result of this section.

Theorem 4.1 *Let T be the Toeplitz operator with continuous symbol ω. Then T is left or right invertible if and only if $\omega(e^{it}) \neq 0$ for $-\pi \leq t \leq \pi$. Furthermore, if this condition is satisfied, then*

   (i) *T is left invertible if and only if $\kappa \geq 0$, and in that case*

$$\text{codim Im } T = \kappa, \tag{4.2}$$

  (ii) *T is right invertible if and only if $\kappa \leq 0$, and in that case*

$$\dim \text{Ker } T = -\kappa. \tag{4.3}$$

Here κ is the winding number of ω relative to zero. In particular, T is two-sided invertible if and only if $\omega(e^{it}) \neq 0$ for $-\pi \leq t \leq \pi$ and the winding number of ω relative to zero is equal to zero.

Proof: We split the proof into two parts. In the first part we assume that ω has no zero on the unit circle and prove (i) and (ii). The second part concerns the necessity of the above condition.

Part 1. Assume $\omega(e^{it}) \neq 0$ for $-\pi \leq t \leq \pi$. We prove (i) and (ii) by an approximation argument using Theorem 3.2. By the second Weierstrass approximation theorem (see Section I.13) there exists a sequence $\omega_1(e^{it}), \omega_2(e^{it}), \ldots$ of trigonometric polynomials such that

$$\max_{-\pi \leq t \leq \pi} |\omega(e^{it}) - \omega_n(e^{it})| \to 0 \quad (n \to \infty). \tag{4.4}$$

From (4.4) it follows that for n sufficiently large $\omega_n(\lambda) \neq 0$ for $|\lambda| = 1$ and $\omega_n(\lambda)^{-1}\omega(\lambda) \to 1$ uniformly on $|\lambda| = 1$ when $n \to \infty$. Thus we can find a trigonometric polynomial $\tilde{\omega}$ such that $\tilde{\omega}(e^{it}) \neq 0$ for $-\pi \leq t \leq \pi$ and

$$|\tilde{\omega}(e^{it})^{-1}\omega(e^{it}) - 1| < 1, \quad -\pi \leq t \leq \pi. \tag{4.5}$$

The function $\tilde{\omega}$ can be written in the form

$$\tilde{\omega}(\lambda) = c\Pi_{i=1}^{\ell}(1 - \alpha_i\lambda^{-1})\lambda^{\ell-r}\Pi_{j=1}^{m}(\lambda - \beta_j), \tag{4.6}$$

where c is a non-zero constant, r is a non-negative integer, $\alpha_1, \ldots, \alpha_\ell$ are complex numbers inside the open unit disk, and $\beta_1, \ldots, \beta_m$ are complex numbers outside the closed unit disk. Using (4.5) and the remark related to the inequality (4.1) we see that the winding numbers of ω and $\tilde{\omega}$ relative to zero are equal. Thus the number $\ell - r$ in (4.6) is equal to the winding number κ of ω relative to zero.

Let C be the Toeplitz operator with continuous symbol equal to $\gamma(\lambda) = \tilde{\omega}(\lambda)^{-1} \times \omega(\lambda) - 1$. From (4.5) and Theorem 2.2 we see that $\|C\| < 1$, and the Toeplitz operator $I + C$ is invertible. Next put

$$\tilde{T}_- = \Pi_{i=1}^\ell (I - \alpha_i S^*)^{-1}, \quad \tilde{T}_+ = \Pi_{j=1}^m (S - \beta_j I). \tag{4.7}$$

These operators are also invertible. This follows from Lemma 3.3 and the fact that $|\alpha_i| < 1$ and $|\beta_j| > 1$ for each i and j. We claim that

$$T = c\tilde{T}_-(I + C)S^\kappa \tilde{T}_+ \quad (\kappa \geq 0), \tag{4.8}$$

or

$$T = c\tilde{T}_-(S^*)^{-\kappa}(I + C)\tilde{T}_+ \quad (\kappa \leq 0). \tag{4.9}$$

Indeed, if $\kappa \geq 0$, then $\kappa = \ell - r$ and the product rules appearing in the paragraph before the proof of Theorem 3.2 imply that the right hand side of (4.8) is the Toeplitz operator of which the symbol $\omega^\#$ is given by

$$\omega^\#(\lambda) = c\Pi_{i=1}^\ell(1 - \alpha_i\lambda^{-1})(\tilde{\omega}(\lambda)^{-1}\omega(\lambda))\lambda^{\ell-r}\Pi_{j=1}^m(\lambda - \beta_j).$$

Since $\tilde{\omega}$ is given by (4.6) and all scalar functions commute, we see that $\omega^\# = \omega$, and hence (4.8) holds. The identity (4.9) is proved in a similar way.

From (4.8) and the invertibility of the operators $\tilde{T}_-$, $\tilde{T}_+$ and $I + C$ we conclude that T is left invertible if $\kappa \geq 0$. In fact, in that case a left inverse $T^{(-1)}$ of T is given by

$$T^{(-1)} = \frac{1}{c}\tilde{T}_+^{-1}(S^*)^\kappa(I + C)^{-1}\tilde{T}_-^{-1}. \tag{4.10}$$

Moreover,

$$\text{codim Im } T = \text{codim Im } S^\kappa = \kappa.$$

Similarly, (4.9) yields that T is right invertible for $\kappa \leq 0$, a right inverse of $T^{(-1)}$ of T is given by

$$T^{(-1)} = \frac{1}{c}\tilde{T}_+^{-1}(I + C)^{-1}S^{-\kappa}\tilde{T}_-^{-1}, \tag{4.11}$$

and

$$\dim \text{Ker } T = \dim \text{Ker } (S^*)^{-\kappa} = -\kappa.$$

This completes the proof of (i) and (ii).

Part 2. In this part T is left or right invertible, and we prove that $\omega(e^{it}) \neq 0$ for $-\pi \leq t \leq \pi$. The proof is by contradiction. We assume that $\omega(\lambda_0) = 0$ for some $|\lambda_0| = 1$. Since T is left or right invertible, the perturbation results of Section I.15 show that there exists $\varepsilon > 0$ such that the operator $\hat{T}$ on ℓ_2 is left or right invertible whenever

$$\|\hat{T} - T\| < \varepsilon. \tag{4.12}$$

By the second Weierstrass approximation theorem (see Section I.13) there exists a trigonometric polynomial $\tilde{\omega}(e^{it})$ such that

$$|\tilde{\omega}(e^{it}) - \omega(e^{it})| < \frac{1}{2}\varepsilon, \quad -\pi \leq t \leq \pi. \tag{4.13}$$

Now let $\hat{T}$ be the Toeplitz operator with symbol $\hat{\omega}(\lambda) = \tilde{\omega}(\lambda) - \tilde{\omega}(\lambda_0)$. Since $\omega(\lambda_0) = 0$, we see from (4.13) that $|\tilde{\omega}(\lambda_0)| < \frac{1}{2}\varepsilon$. Hence, again using (4.13), we have

$$|\hat{\omega}(e^{it}) - \omega(e^{it})| < \varepsilon, \quad -\pi \leq t \leq \pi.$$

But then we can use Theorem 2.2 to show that $\hat{T}$ satisfies (4.12), and hence $\hat{T}$ is left or right invertible. However, $\hat{\omega}(\lambda_0) = 0$ by construction, and thus Theorem 3.2 tells us that $\hat{T}$ cannot be left or right invertible. So we reached a contradiction. We conclude that $\omega(e^{it}) \neq 0$ for $-\pi \leq t \leq \pi$. □

Corollary 4.2 *Let T be the Toeplitz operator with continuous symbol ω. Then the spectrum of T is given by*

$$\sigma(T) = \mathbb{C} \setminus \Omega, \tag{4.14}$$

where Ω is the set of all points λ in $\mathbb{C}$ such that $\lambda \neq \omega(e^{it})$ for all $-\pi \leq t \leq \pi$ and the winding number of $\omega(\cdot) - \lambda$ relative to zero is equal to zero. In particular, the spectral radius of T is equal to the norm of T.

The proof of the above corollary is the same as that of Corollary 3.4, except that the reference to Theorem 3.1 has to be replaced by a reference to the final part of Theorem 4.1. We omit further details. □

We conclude this section with the case when the symbol of ω of the Toeplitz T is a rational function. In that case the process of inverting T can be carried out without the approximation step described in the first part of the proof of Theorem 4.1. Indeed, let

$$\omega(\lambda) = \frac{p_1(\lambda)}{p_2(\lambda)}, \quad |\lambda| = 1,$$

where p_1 and p_2 are polynomials which have no common zeros. Furthermore, we assume that p_1 and p_2 have no zeros on $\mathbb{T}$, which means that we assume that ω is continuous on $\mathbb{T}$ and $\omega(e^{it}) \neq 0$ for each t.

Because of our assumptions on p_1 and p_2 we can write

$$p_1(\lambda) = c_1 \prod_{j=1}^{k^+} (\lambda - t_j^+) \prod_{j=1}^{k^-} (\lambda - t_j^-),$$

$$p_2(\lambda) = c_2 \prod_{j=1}^{\ell^+} (\lambda - \tau_j^+) \prod_{j=1}^{\ell^-} (\lambda - \tau_j^-),$$

where t_j^+ and τ_j^+ are inside the unit circle and the points t_j^- and τ_j^- are outside $\mathbb{T}$. It follows that ω can be represented in the form

$$\omega(\lambda) = \omega_-(\lambda)\lambda^\kappa \omega_+(\lambda), \qquad \lambda \in \mathbb{T}, \tag{4.15}$$

where $\kappa = k^+ - \ell^+$, and the factors ω_- and ω_+ are rational functions which have their zeros and poles inside $\mathbb{T}$ and outside $\mathbb{T}$, respectively. In fact,

$$\omega_-(\lambda) = \frac{c_1 \prod_{j=1}^{k^+}(1 - t_j^+ \lambda^{-1})}{c_2 \prod_{j=1}^{\ell^+}(1 - \tau_j^+ \lambda^{-1})}, \qquad \omega_+(\lambda) = \frac{\prod_{j=1}^{k^-}(\lambda - t_j^-)}{\prod_{j=1}^{\ell^-}(\lambda - \tau_j^-)}.$$

Notice that $\kappa = k^+ - \ell^+$ is precisely the winding number of ω relative to zero. Thus T will be invertible if and only if $\omega = \omega_-\omega_+$.

Let T_- and T_+ be the Toeplitz operators with symbols ω_- and ω_+, respectively. Using the multiplication rules appearing in the paragraphs before the proof of Theorem 3.2 we see that

$$T_- = \frac{c_1}{c_2} \prod_{j=1}^{k^+}(I - t_j^+ S^*) \prod_{j=1}^{\ell^+}(I - \tau_j^+ S^*)^{-1},$$

$$T_+ = \prod_{j=1}^{k^-}(S - t_j^- I) \prod_{j=1}^{\ell^-}(S - \tau_j^- I)^{-1},$$

and

$$T = T_- S^{(\kappa)} T_+, \tag{4.16}$$

where

$$S^{(n)} = \begin{cases} S^n & \text{for } n = 0, 1, 2, \ldots, \\ (S^*)^{-n} & \text{for } n = -1, -2, \ldots. \end{cases}$$

Notice that because of commutativity the order of the factors in the formulas for T_- and T_+ is not important. However in (4.16) the order is essential.

From Lemma 3.3 and the location of the points $t_j^\pm$ and $\tau_j^\pm$ it follows that the Toeplitz operators T_- and T_+ are invertible and their inverses are given by

$$T_-^{-1} = \frac{c_2}{c_1} \prod_{j=1}^{\ell^+}(I - \tau_j^+ S^*) \prod_{j=1}^{k^+}(I - t_j^+ S^*)^{-1},$$

$$T_+^{-1} = \prod_{j=1}^{k^-}(S - \tau_j^- I) \prod_{j=1}^{k^-}(S - t_j^- I)^{-1}.$$

In other words, T_-^{-1} and T_+^{-1} are the Toeplitz operators with symbols $1/\omega_-$ and $1/\omega_+$, respectively. Furthermore, we see that T_-^{-1} and T_+^{-1} are of the form

$$T_-^{-1} = \begin{bmatrix} \gamma_0^- & \gamma_{-1}^- & \gamma_{-2}^- & \cdots \\ 0 & \gamma_0^- & \gamma_{-1}^- & \\ 0 & 0 & \gamma_0^- & \\ \vdots & & & \ddots \end{bmatrix}, \tag{4.17}$$

$$T_+^{-1} = \begin{bmatrix} \gamma_0^+ & 0 & 0 & \cdots \\ \gamma_1^+ & \gamma_0^+ & 0 & \\ \gamma_2^+ & \gamma_1^+ & \gamma_0^+ & \\ \vdots & & & \ddots \end{bmatrix}, \tag{4.18}$$

where

$$\omega_-(\lambda) = \sum_{j=-\infty}^{0} \gamma_j^- \lambda^j, \qquad |\lambda| \geq 1, \tag{4.19}$$

$$\omega_+(\lambda) = \sum_{j=0}^{\infty} \gamma_j^+ \lambda^j, \qquad |\lambda| \leq 1. \tag{4.20}$$

The first conclusion is that the Toeplitz operator with the (rational) symbol (4.15) is invertible if and only if $\kappa = 0$, and in that case its inverse is given by

$$T^{-1} = (\gamma_{jk})_{j,k=0}^{\infty},$$

where

$$\gamma_{jk} = \sum_{r=0}^{\min(j,k)} \gamma_{j-r}^+ \gamma_{r-k}^-$$

with γ_j^- and γ_j^+ being given by (4.19) and (4.20), respectively.

Furthermore, if $\kappa < 0$, then the operator T is right invertible (but not left) and a right inverse is given by

$$T^{(-1)} = T_+^{-1} S^{-\kappa} T_-^{-1}. \tag{4.21}$$

If $\kappa > 0$, then the operator T is left invertible (but not right), and a left inverse is given by

$$T^{(-1)} = T_+^{-1} (S^*)^\kappa T_-^{-1}. \tag{4.22}$$

The formulas (4.21) and (4.22) are the analogs of formulas (3.4) and (3.5) in Theorem 3.2. As in the case $\kappa = 0$, formulas (4.17) and (4.18) can be used to obtain the matrix entries in the matrix representations of the operators $T^{(-1)}$ in (4.21) and (4.22).

Example: Consider the Toeplitz operator on ℓ_2 given by

$$T = \begin{bmatrix} 1 & 1/2 & 1/4 & \cdots \\ 1/3 & 1 & 1/2 & \\ 1/3^2 & 1/3 & 1 & \\ \vdots & & & \ddots \end{bmatrix}.$$

We met the corresponding Laurent operator as the inverse of the operator A in Example 1 of Section 1. The symbol ω of T is easy to compute:

$$\omega(\lambda) = \sum_{j=0}^{\infty} \left(\frac{1}{2}\right)^j \lambda^{-j} + \sum_{k=1}^{\infty} \left(\frac{1}{3}\right)^k \lambda^k = \frac{1}{1 - \lambda^{-1/2}} + \frac{\lambda/3}{1 - \lambda/3}$$

$$= \frac{5}{6} \left(1 - \frac{1}{2}\lambda^{-1}\right)^{-1} \left(1 - \frac{1}{3}\lambda\right)^{-1}.$$

Thus the symbol is a rational function, and we have the representation (4.15) with $\kappa = 0$ and

$$\omega_-(\lambda) = \frac{5}{6} \left(1 - \frac{1}{2}\lambda^{-1}\right)^{-1}, \qquad \omega_+(\lambda) = \left(1 - \frac{1}{3}\lambda\right)^{-1}.$$

We conclude that T is invertible and

$$T^{-1} = \frac{6}{5} \left(I - \frac{1}{3}S\right) \left(I - \frac{1}{2}S^*\right).$$

Hence T^{-1} is the tridiagonal operator

$$T^{-1} = \begin{bmatrix} 6/5 & -3/5 & 0 & \cdots \\ -2/5 & 7/5 & -3/5 & \\ 0 & -2/5 & 7/5 & \\ \vdots & & & \ddots \end{bmatrix}.$$

Notice that $T^{-1} = G - F$, where

$$G = \begin{bmatrix} 7/5 & -3/5 & 0 & \cdots \\ -2/5 & 7/5 & -3/5 & \\ 0 & -2/5 & 7/5 & \\ \vdots & & & \ddots \end{bmatrix}, \quad F = \begin{bmatrix} 1 & 0 & 0 & \cdots \\ 0 & 0 & 0 & \\ 0 & 0 & 0 & \\ \vdots & & & \ddots \end{bmatrix}.$$

Thus T^{-1} is a rank one perturbation of a tridiagonal Toeplitz operator G. Moreover the Laurent operator corresponding to G is the operator A in Example 1 of Section 1.

3.5 Finite Section Method

Let T be a Toeplitz operator on ℓ_2 with continuous symbol ω. Thus

$$T = \begin{bmatrix} a_0 & a_{-1} & a_{-2} & \cdots \\ a_1 & a_0 & a_{-1} & \cdots \\ a_2 & a_1 & a_0 & \\ \vdots & \vdots & & \ddots \end{bmatrix}, \tag{5.1}$$

where

$$a_n = \frac{1}{2\pi} \int_{-\pi}^{\pi} \omega(e^{it}) e^{-int} \, dt, \quad n \in \mathbb{Z}. \tag{5.2}$$

Given a non-negative integer N, the *N-th section* T_N of T is the linear operator on $\mathbb{C}^N$ given by

$$T_N = \begin{bmatrix} a_0 & a_{-1} & \cdots & a_{-N} \\ a_1 & a_0 & \cdots & a_{-N+1} \\ \vdots & \vdots & \ddots & \vdots \\ a_N & a_{N-1} & \cdots & a_0 \end{bmatrix}. \tag{5.3}$$

Here we identify a linear operator on $\mathbb{C}^n$ with its matrix with respect to the standard basis of $\mathbb{C}^n$. Assume T is invertible. In this section we study the problem to get the solution x of the equation $Tx = y$, where $y = (y_0, y_1, y_2, \ldots)$ in ℓ_2, as a limit of solutions of the truncated equation

$$T_N \begin{bmatrix} x_0(N) \\ \vdots \\ x_N(N) \end{bmatrix} = \begin{bmatrix} y_0 \\ \vdots \\ y_N \end{bmatrix}. \tag{5.4}$$

More precisely, we say that the *finite section method for T converges* if T_N is invertible for N sufficiently large, $N \geq N_0$ say, and for each $y = (y_0, y_1, y_2, \ldots)$ in

ℓ_2 the vector $x(N) = (x_0(N), \ldots, x_N(N), 0, 0, \ldots)$, where $(x_0(N), \ldots, x_N(N))$ is the solution of (5.4) for $N \geq N_0$, converges in the norm of ℓ_2 to the solution x of $Tx = y$.

In the sequel we identify the space $\mathbb{C}^N$ with the subspace $\mathcal{H}_N$ of ℓ_2 consisting of all $x = (x_j)_{j=0}^{\infty}$ in ℓ_2 such that $x_j = 0$ for $j > N$, and we let P_N be the orthogonal projection of ℓ_2 onto $\mathcal{H}_N = \mathbb{C}^N$. It then follows that the finite section method for T converges if and only if the projection method relative to $\{P_N\}_{N=0}^{\infty}$ is applicable to T, that is, using the notation of Section II.17, the operator T belongs to $\Pi\{P_N\}$.

The following theorem is the main result of this section.

Theorem 5.1 *Let T be a Toeplitz operator with a continuous symbol ω. If T is invertible, then the finite section method for T converges.*

Before we prove the theorem it will be convenient to make some preparations. The first is a general proposition and the second a lemma.

Proposition 5.2 *Let A be an invertible operator on ℓ_2, and assume that with respect to the standard orthonormal basis of ℓ_2 the matrices of A and A^{-1} are both lower triangular or both upper triangular. Then for each N the N-th section A_N of A is invertible and*

$$(A_N)^{-1} P_N = P_N A^{-1} \quad \text{(lower triangular case)} \tag{5.5}$$

$$(A_N)^{-1} P_N = A^{-1} P_N \quad \text{(upper triangular case)} \tag{5.6}$$

where P_N is the orthogonal projection of ℓ_2 onto the space spanned by the first $N + 1$ vectors in the standard basis. Furthermore

$$\lim_{N \to \infty} (A_N)^{-1} P_N y = A^{-1} y, \quad y \in \ell_2. \tag{5.7}$$

Proof: First consider the case when the standard matrices of A and A^{-1} are both lower triangular. So

$$
A = \begin{bmatrix} a_{00} & & & \\ a_{10} & a_{11} & & \\ a_{20} & a_{21} & a_{22} & \\ \vdots & & & \ddots \end{bmatrix}, \quad
A^{-1} = \begin{bmatrix} b_{00} & & & \\ b_{10} & b_{11} & & \\ b_{20} & b_{21} & b_{22} & \\ \vdots & & & \ddots \end{bmatrix},
$$

with zero elements in the strictly upper triangular part. Then the N-th sections of A and A^{-1} are given by

$$
A_N = \begin{bmatrix} a_{00} & & & \\ a_{10} & a_{00} & & \\ \vdots & \vdots & \ddots & \\ a_{N0} & a_{N1} & & a_{NN} \end{bmatrix}, \quad
(A^{-1})_N = \begin{bmatrix} b_{00} & & & \\ b_{10} & b_{11} & & \\ \vdots & \vdots & \ddots & \\ b_{N0} & b_{N1} & & b_{NN} \end{bmatrix}.
$$

Since $AA^{-1} = A^{-1}A$ is the identity on ℓ_2, it follows that A_N is invertible and $(A_N)^{-1} = (A^{-1})_N$. Now, take $y \in \ell_2$. Then

$$(A_N)^{-1} P_N y = (A^{-1})_N P_N y = P_N A^{-1} y,$$

which proves (5.5). Moreover, (5.7) follows by using that $\|A^{-1}y - P_N A^{-1}y\| \to 0$ if $N \to \infty$. This completes the proof for the case when the standard matrices of A and A^{-1} are both lower triangular. The upper triangular case is proved in a similar way.   □

Lemma 5.3 *Let T be an arbitrary Toeplitz operator, and let $R = \Pi_{j=1}^{m}(S - \beta_j I)$, where S is the forward shift on ℓ_2 and $\beta_1, \ldots, \beta_m$ are complex numbers. Then*

$$\operatorname{rank}(RT - TR) < \infty. \tag{5.8}$$

Proof: First notice that $ST - TS$ has rank at most one. Indeed, let $e_0, e_1, e_2, \ldots$ be the standard orthonormal basis of ℓ_2, and let

$$T = \begin{bmatrix} a_0 & a_{-1} & a_{-2} & \cdots \\ a_1 & a_0 & a_{-1} & \cdots \\ a_2 & a_1 & a_0 & \\ \vdots & \vdots & & \ddots \end{bmatrix}.$$

Then for each $n = 0, 1, 2, \ldots$ we have

$$(ST - TS)e_n = S \begin{bmatrix} a_{-n} \\ a_{-n+1} \\ a_{-n+2} \\ \vdots \end{bmatrix} - T e_{n+1} = -a_{-n-1}e_0.$$

Thus rank $(ST - TS) \leq 1$. Next observe that

$$S^p T - T S^p = \sum_{j=1}^{p} S^{p-j}(ST - TS)S^{j-1},$$

and thus

$$\operatorname{rank}(S^p T - T S^p) \leq p, \quad p = 1, 2, \ldots. \tag{5.9}$$

Now notice that R is a polynomial in S of degree m, and hence (5.9) yields that the rank of $RT - TR$ is at most mp.   □

Proof of Theorem 5.1: Since T is assumed to be invertible, we know from Theorem 4.1 that $\omega(e^{it}) \neq 0$ for $-\pi \leq t \leq \pi$ and the winding number κ of ω relative to zero is equal to zero. But then we can use the first part of the proof of Theorem 4.1 to conclude that T can be represented in the form

$$T = c\tilde{T}_-(I + C)\tilde{T}_+. \tag{5.10}$$

Here c is a non-zero constant, C is a Toeplitz operator with continuous symbol γ satisfying $|\gamma(e^{it})| < 1$ for $-\pi \le t \le \pi$, and

$$\tilde{T}_- = \Pi_{i=1}^{\ell}(I - \alpha_i S^*), \quad \tilde{T}_+ = \Pi_{j=1}^{m}(S - \beta_j I), \tag{5.11}$$

where S is the forward shift on ℓ_2, and $|\alpha_i| < 1$, $|\beta_j| > 1$ for each i and j. As we have seen before, the fact that $|\alpha_i| < 1$ and $|\beta_j| > 1$ for each i and j implies that the operators $\tilde{T}_-$ and $\tilde{T}_+$ are invertible (by Lemma 3.3). In fact,

$$\tilde{T}_-^{-1} = \Pi_{i=1}^{\ell}(I - \alpha_i S^*)^{-1}, \quad \tilde{T}_+^{-1} = \Pi_{j=1}^{m}(S - \beta_j I)^{-1}. \tag{5.12}$$

From (5.11) and (5.12) we also see that with respect to the standard basis of ℓ_2 the matrices of $\tilde{T}_-$ and $\tilde{T}_-^{-1}$ are upper triangular, and those of $\tilde{T}_+$ and $\tilde{T}_+^{-1}$ are lower triangular.

Consider the operator

$$F = c\tilde{T}_+(I + C)\tilde{T}_-. \tag{5.13}$$

By comparing (5.10) and (5.13) we see that F is obtained from T by interchanging the factors $\tilde{T}_-$ and $\tilde{T}_+$. According to Lemma 5.3 the operators $\tilde{T}_+C - C\tilde{T}_+$ and $\tilde{T}_+\tilde{T}_- - \tilde{T}_-\tilde{T}_+$ are of finite rank. It follows that $T - c\tilde{T}_+\tilde{T}_-(I + C)$ is an operator of finite rank. Also, by taking adjoints in Lemma 5.3 we see that $\tilde{T}_-C - C\tilde{T}_-$ is of finite rank. So we have proved that the operator $T - F$ is of finite rank.

Notice that the three factors in the right hand side of (5.13) are invertible, and hence F is invertible. But then, by Theorem II.17.7, the result of the previous paragraph shows that $T \in \Pi\{P_N\}$ if and only if $F \in \Pi\{P_N\}$. Moreover, since $\|C\| < 1$ we also know (see Theorem II.17.6) that $I + C \in \Pi\{P_N\}$.

Let $\tilde{T}_{+,N}$ and $\tilde{T}_{-,N}$ be the N-th sections of $\tilde{T}_+$ and $\tilde{T}_-$, respectively, and let F_N and C_N be those of F and C, respectively. Since $\tilde{T}_+$ is upper triangular and $\tilde{T}_-$ lower triangular with respect to the standard basis of ℓ_2, we have

$$P_N\tilde{T}_+ = \tilde{T}_{+,N}P_N, \quad P_N\tilde{T}_{-,N} = \tilde{T}_-P_N,$$

and thus

$$F_N = c\tilde{T}_{+,N}(I + C_N)\tilde{T}_{-,N}. \tag{5.14}$$

From the triangular properties of the operators in (5.11) and (5.13) it follows that we can apply Proposition 5.2 to show that $\tilde{T}_{+,N}$ and $\tilde{T}_{-,N}$ are invertible. Also, $\|C_N\| \le \|C\| < 1$, and therefore $I + C_N$ is invertible. Using (5.14) we see that for each N the N-th section F_N is invertible, and

$$F_N^{-1} = \frac{1}{c}(\tilde{T}_{-,N})^{-1}(I + C_N)^{-1}(\tilde{T}_{+,N})^{-1}. \tag{5.15}$$

Next, take $y \in \ell_2$. Formula (5.5) applied to $A = \tilde{T}_+$ yields

$$(\tilde{T}_{+,N})^{-1}P_N y = P_N\tilde{T}_+^{-1}y.$$

Now use that $I + C \in \Pi\{P_N\}$. It follows that

$$x_N := (I + C_N)^{-1}(\tilde{T}_{+,N})^{-1}P_N y = (I + C_N)^{-1}P_N \tilde{T}_+^{-1} y \to (I + C)^{-1}\tilde{T}_+^{-1} y$$

for $N \to \infty$. Put $x = (I + C)^{-1}\tilde{T}_+^{-1}y$. Since $\tilde{T}_-$ and $\tilde{T}_-^{-1}$ are upper triangular, Proposition 5.2 applied to $A = \tilde{T}_-$ yields

$$(\tilde{T}_{-,N})^{-1}x_N = \tilde{T}_-^{-1}x_N \to \tilde{T}_-^{-1}x \quad (N \to \infty).$$

We conclude that

$$F_N^{-1}P_N y = \frac{1}{c}\tilde{T}_-^{-1}(I + C)^{-1}\tilde{T}_+^{-1}y = F^{-1}y \quad (N \to \infty),$$

and hence $F \in \Pi\{P_N\}$. $\square$

3.6   The Finite Section Method for Laurent Operators

In this section we consider the finite section method for Laurent operators. Let L be an invertible Laurent operator on $\ell_2(\mathbb{Z})$. Recall (see Section II.17) that the finite section method is said to converge for L if $L \in \Pi\{Q_n\}$, where Q_n is the orthogonal projection of $\ell_2(\mathbb{Z})$ onto span $\{e_{-n}, \ldots, e_n\}$. Here $\ldots, e_{-1}, e_0, e_1, \ldots$ is the standard orthonormal basis of $\ell_2(\mathbb{Z})$.

In general, in contrast to Toeplitz operators, from the invertibility of L it does not follow that the finite section method converges for L. For instance, assume that $L = V$, where V is the bilateral forward shift on $\ell_2(\mathbb{Z})$. Then L is invertible, but

$$L_n = Q_n V Q_n = \begin{bmatrix} 0 & 0 & 0 & \ldots & 0 & 0 \\ 1 & 0 & 0 & \ldots & 0 & 0 \\ 0 & 1 & 0 & \ldots & 0 & 0 \\ \vdots & & & & & \vdots \\ 0 & 0 & 0 & \ldots & 1 & 0 \end{bmatrix}.$$

Hence there is no n for which L_n is invertible, and therefore $L = V \notin \Pi\{Q_n\}$. To get the convergence of the finite section method for an invertible Laurent operator an additional condition is required (see the next theorem) which in the Toeplitz case follows from invertibility.

Theorem 6.1 *Let L be an invertible Laurent operator on $\ell_2(\mathbb{Z})$ defined by the function $a(t) = \omega(e^{it})$, where ω is continuous on the unit circle $\mathbb{T}$ (satisfying $\omega(e^{it}) \neq 0$ for each $-\pi \leq t \leq \pi$). In order that the finite section method converge for L it is necessary and sufficient that the winding number κ of ω relative to zero is equal to zero.*

Proof: The sufficiency follows from Theorems 4.1 and 5.1. Indeed, assume the winding number κ of ω relative to zero is equal to zero. Let T be the Toeplitz operator on ℓ_2 with symbol ω. Since $\kappa = 0$, Theorem 4.1 implies that T is invertible. But then we can use Theorem 5.1 to show that the finite section method converges for T. Thus for n sufficiently large, $n \geq n_\circ$ say, the operator $P_n T P_n$ is invertible on Im P, and

$$\sup_{n \geq n_\circ} \|(P_n T P_n)^{-1}|\text{Im } P_n\| < \infty.$$

Here P_n is the orthogonal projection onto the space spanned by the first n vectors in the standard orthonormal basis of ℓ_2. Notice that we may identify both Im P_{2n+1} and Im Q_n with $\mathbb{C}^{2n+1}$, and in that case we have $P_{2n+1} T P_{2n+1} = Q_n L Q_n$. We conclude that $Q_n L Q_n$ is invertible on Im Q_n for $n \geq n_\circ$ and $\sup_{n \geq n_\circ} \|(Q_n L Q_n)^{-1}\text{Im } Q_n\| < \infty$. By Theorem II.17.1 this implies that the finite section method converges for L.

For the proof of the necessity of the winding number condition we refer to the proof of Theorem XVI.5.2 where the necessity is established in a somewhat different setting. The proof given there carries over to the case considered here. $\quad\square$

To understand Theorem 6.1 better, consider the Laurent operator

$$L = \begin{bmatrix} \ddots & \ddots & & & \\ \ddots & a_0 & a_{-1} & a_{-2} & \\ & a_1 & \boxed{a_0} & a_{-1} & \\ & a_2 & a_1 & a_0 & \ddots \\ & & & \ddots & \ddots \end{bmatrix}$$

Let ω be the symbol of L which we take to be continuous on $\mathbb{T}$. Assume L is invertible, i.e., $\omega(e^{it}) \neq 0$ for each $-\pi \leq t \leq \pi$. Let κ be the corresponding winding number. Notice $L_1 = V^{-\kappa} L$ is again a Laurent operator. In fact,

$$L_1 = \begin{bmatrix} \ddots & \ddots & & & \\ \ddots & a_\kappa & a_{\kappa-1} & a_{\kappa-2} & \\ & a_{\kappa+1} & \boxed{a_\kappa} & a_{\kappa-1} & \\ & a_{\kappa+2} & a_{\kappa+1} & a_\kappa & \ddots \\ & & & \ddots & \ddots \end{bmatrix}$$

The difference between the matrices for L and L_1 is only in the location of the entry in the zero-zero position.

The symbol of L_1 is equal to $\omega_1(\lambda) = \lambda^{-\kappa}\omega(\lambda)$. We have $\omega_1(e^{it}) \neq 0$ for each t and the corresponding winding number is equal to zero. Thus L_1 is invertible, and Theorem 6.1 tells us that the finite section method converges for L_1. This means that the matrices

$$Q_n L_1 Q_n | \text{Im } Q_n = (a_{\kappa+j-r})^n_{j,r=-n}$$

are invertible for n large enough and

$$\lim_{n\to\infty} \|(Q_n L_1 Q_n)^{-1} Q_n y - L_1^{-1} y\| = 0.$$

Since $L_1 = V^{-\kappa}L$, the latter limit can be rewritten in the form

$$\lim_{n\to\infty} \|(V^\kappa Q_n V^{-\kappa} L Q_n)^{-1} V^\kappa Q_n V^{-\kappa} y - Ly\| = 0.$$

Notice that

$$V^\kappa Q_n V^{-\kappa} = Q_{n,\kappa},$$

where $Q_{n,\kappa}$ is the orthogonal projection of $\ell_2(\mathbb{Z})$ onto span $\{e_{-n+\kappa}, \ldots, e_{n+\kappa}\}$ with $\ldots, e_{-1}, e_0, e_1, \ldots$ being the standard orthonormal basis of $\ell_2(\mathbb{Z})$. We arrive at the conclusion that for the operator L the projection method with respect to two sequences of projections $\{Q_{n,\kappa}, Q_n\}_{n\in\mathbb{Z}}$ is convergent. In other words to have the finite section method converging for an invertible Laurent operator L one has to take the sections around the κ-th diagonal, where κ is the winding number of the symbol of L.

As we have seen in Section II. 18 it is sometimes convenient to a consider a modified finite section method. This is also true for Laurent operators.

Example: Consider the Laurent operator L with symbol $\omega(\lambda) = -2\lambda + 7 - 3\lambda^{-1}$. Since

$$\omega(\lambda) = 6\left(1 - \frac{1}{3}\lambda\right)\left(1 - \frac{1}{2}\lambda^{-1}\right),$$

we see that $\omega(e^{it}) \neq 0$ for each t and relative to zero the winding number of the corresponding curve is equal to zero. It follows that the finite section method converges for L.

To find the inverse of L it is convenient to use a modified finite section method. Notice that the n-th section of L is the following $(2n+1) \times (2n+1)$ matrix

$$L_n = \begin{bmatrix} 7 & -3 & 0 & 0 & \ldots & 0 & 0 \\ -2 & 7 & -3 & 0 & \ldots & 0 & 0 \\ 0 & -2 & 7 & -3 & & 0 & 0 \\ 0 & 0 & -2 & 7 & & 0 & 0 \\ \vdots & & & & \ddots & & \\ 0 & 0 & 0 & 0 & & 7 & -3 \\ 0 & 0 & 0 & 0 & & -2 & 7 \end{bmatrix}.$$

Now replace the entries in the left upper corner and in the right lower corner by 6, and let F_n be the resulting matrix. As we have seen in Section II.18 the inverse of F_n is given by

$$
F_n^{-1} = \frac{1}{5} = \begin{bmatrix}
1 & 2^{-1} & 2^{-2} & \cdots & 2^{-2n} \\
3^{-1} & 1 & 2^{-1} & \cdots & 2^{-2n+1} \\
3^{-2} & 3^{-1} & 1 & \cdots & 2^{-2n+2} \\
\vdots & \vdots & \vdots & \ddots & \vdots \\
3^{-2n} & 3^{-2n+1} & 3^{-2n+2} & \cdots & 1
\end{bmatrix}.
$$

Hence one expects the inverse of L to be given by

$$
L^{-1} = \frac{1}{5} = \begin{bmatrix}
\ddots & & & & & & \\
& 1 & 2^{-1} & 2^{-2} & 2^{-3} & 2^{-4} & \\
& 3^{-1} & 1 & 2^{-1} & 2^{-2} & 2^{-3} & \\
\cdots & 3^{-2} & 3^{-1} & \boxed{1} & 2^{-1} & 2^{-2} & \cdots \\
& 3^{-3} & 3^{-2} & 3^{-1} & 1 & 2^{-1} & \\
& 3^{-4} & 3^{-3} & 3^{-2} & 3^{-1} & 1 & \\
& & & & & & \ddots
\end{bmatrix}.
$$

It is straightforward to check that indeed this is the correct inverse of L.

The sources for the material covered in this chapter can be found in [Kre], Section 3.13, [GF], Chapters I-III; see also [BS] and [GGK2], Part VI. In these monographs one can also find further developments of the contents of this chapter. For a partial extension of the theory to Banach spaces, see Chapter XVI. Finally we would like to note that there is some difference in terminology between the above mentioned sources and Chapters III and XVI of the present book.

Exercises III

1. Let A be the Laurent operator on $\ell_2(\mathbb{Z})$ defined by

$$
a(t) = -4e^{-it} + 17 - 4e^{it}.
$$

   (a) Prove that the operator A is *strictly positive*, that is, for some $\varepsilon > 0$

   $$
   \langle Ax, x \rangle \geq \varepsilon \|x\|^2 \qquad (x \in \ell_2(\mathbb{Z})).
   $$

   (b) Show that A is invertible, and find the matrix of A^{-1} with respect to the standard basis of $\ell_2(\mathbb{Z})$.

   (c) Find the spectra of the operators A and A^{-1}.

2. Let $|\alpha| \neq 1$. Do the previous exercise for the Laurent operator A on $\ell_2(\mathbb{Z})$ defined by

$$
a(t) = \alpha e^{-it} + 1 + |\alpha|^2 + \bar{\alpha} e^{it}.
$$

3. Consider on $\ell_2(\mathbb{Z})$ the Laurent operator

$$
A = \begin{bmatrix}
\ddots & & & & & & & \\
& 1 & a & a^2 & a^3 & a^4 & & \\
& b & 1 & a & a^2 & a^3 & & \\
\cdots & b^2 & b & \boxed{1} & a & a^2 & \cdots & , \\
& b^3 & b^2 & b & 1 & a & & \\
& b^4 & b^3 & b^2 & b & 1 & & \\
& & & & & & \ddots &
\end{bmatrix}
$$

where $|a| < 1$ and $|b| < 1$.

(a) Compute the symbol of A.

(b) Find the inverse of A when it exists.

(c) Determine the spectrum of A.

4. Let $|a| \neq 1$, and let A be the Laurent operator on $\ell_2(\mathbb{Z})$ with symbol

$$
\omega(\lambda) = \frac{1 - |a|^2}{1 + |a|^2 - a\lambda - \overline{a}\lambda^{-1}}, \qquad \lambda \in \mathbb{T}.
$$

(a) Prove that A is invertible, and determine its inverse.

(b) Find the spectra of A and A^{-1}.

5. Can it happen that the inverse of a lower triangular Laurent operator is upper triangular?

6. Prove or disprove the statement: a left invertible Laurent operator is always two-sided invertible.

7. Fix $t \in \mathbb{R}$, and let U be the operator on $L_2(-\infty, \infty)$ given by

$$
(Uf)(x) = f(x + t), \qquad x \in \mathbb{R}.
$$

Recall (see Exercise 86 to Chapter II) that U is unitary and $\sigma(U) = \mathbb{T}$.

(a) Prove that the operator

$$
A = -2U + 7I - 3U^{-1}.
$$

is invertible and determine its inverse.

(b) Given $g \in L_2(-\infty, \infty)$ solve in $L_2(-\infty, \infty)$ the functional equation

$$
-2f(x + t) + 7f(x) - 3f(x - t) = g(x), \qquad x \in \mathbb{R}.
$$

(c) Determine the spectrum of the operator A defined in (a).

8. Fix $\alpha > 0$. Let R_α be the operator on $L_2(0, \infty)$ defined by

$$(R_\alpha f)(t) = f(\alpha t), \qquad t \geq 0.$$

(a) Given $g \in L_2(0, \infty)$ solve in $L_2(0, \infty)$ the functional equation

$$\sum_{j=1}^{\infty} b^j f(\alpha^{-j} t) + f(t) + \sum_{j=1}^{\infty} a^j f(\alpha^j t) = g(t), \quad t \geq 0.$$

Here $|a| < \sqrt{a}$ and $|b| < \sqrt{\alpha^{-1}}$.

(b) Determine the spectrum of the operator A given by

$$(Af)(t) = \sum_{j=1}^{\infty} b^j f(\alpha^{-j} t) + \sum_{j=1}^{\infty} a^j f(\alpha^j t), \quad t \geq 0,$$

where $|a| < \sqrt{\alpha}$ and $|b| < \sqrt{\alpha^{-1}}$.

Hint: Use the results of Example 6 in Section II.20.

9. Let T be the Toeplitz operator on ℓ_2 given by the following tridiagonal matrix representation:

$$T = \begin{bmatrix} 17 & -4 & 0 & 0 & \cdots \\ -4 & 17 & -4 & 0 & \cdots \\ 0 & -4 & 17 & -4 & \\ 0 & 0 & -4 & 17 & \\ \vdots & \vdots & & & \ddots \end{bmatrix}.$$

(a) Prove that the operator T is strictly positive, that is, for some $\varepsilon > 0$

$$\langle Tx, x \rangle \geq \varepsilon \|x\|^2 \qquad (x \in \ell_2).$$

(b) Show that T is invertible, and find the matrix of T^{-1} with respect to the standard basis of ℓ_2.

(c) Find the spectra of T and T^{-1}.

10. Let $|\alpha| \neq 1$. Do the previous exercise for the Toeplitz operator T on ℓ_2 with symbol

$$\alpha(\lambda) = \alpha \lambda^{-1} + 1 + |\alpha|^2 + \overline{\alpha}\lambda, \qquad \lambda \in \mathbb{T}.$$

11. Consider on ℓ_2 the Toeplitz operator T given by

$$T = \begin{bmatrix} 1 & a & a^2 & \cdots \\ b & 1 & a & \\ b^2 & b & 1 & \\ \vdots & & & \ddots \end{bmatrix}.$$

where $|a| < 1$ and $|b| < 1$.

(a) Compute the symbol of T.

(b) Find the inverse of T when it exists.

(c) Determine the spectrum of T.

12. Let $|a| \neq 1$, and let T be the Toeplitz operator on ℓ_2 with symbol

$$\omega(\lambda) = \frac{1 - |a|^2}{1 + |a|^2 - a\lambda - \bar{a}\lambda^{-1}}, \qquad \lambda \in \mathbb{T}.$$

(a) Find the matrix of T with respect to the standard basis of ℓ_2.

(b) Prove that T is invertible, and determine its inverse.

(c) Find the spectra of T and T^{-1}.

13. Let T be the Toeplitz operator on ℓ_2 with the following tridiagonal matrix representation:

$$T = \begin{bmatrix} 0 & a & 0 & 0 & \cdots \\ b & 0 & a & 0 & \cdots \\ 0 & b & 0 & a & \\ 0 & 0 & b & 0 & \ddots \\ \vdots & \vdots & & \ddots & \ddots \end{bmatrix}.$$

(a) For which values of a and b in $\mathbb{C}$ is the operator T left, right or two-sided invertible? For these values determine the one- or two-sided inverse of T.

(b) Solve the equation $Tx = y$, where $y = (1, q, q^2, \ldots)$ with $|q| < 1$.

14. Fix $t > 0$. Let W be the operator on $L_2(0, \infty)$ defined by

$$(Wf)(x) = \begin{cases} f(x - t), & \text{for} \quad x > t, \\ 0, & \text{for} \quad 0 < x \leq t. \end{cases}$$

Recall (see Exercise 85 to Chapter II) that W is an isometry and $\sigma(W)$ is the closed unit disc.

(a) Determine Ker W^* and Im W.

(b) Prove that the operator

$$T = -2W + 7I - 3W^*$$

is invertible and determine the inverse.

(c) Given $g \in L_2(0, \infty)$ solve in $L_2(0, \infty)$ the functional equation

$$-2f(x + t) + 7f(x) - 3f(x - t) = g(x), \qquad x > 0.$$

Here we follow the convention that $f(x - t) = 0$ for $0 < x \leq t$.

(d) Find the spectrum of the operator T in (b).

15. Fix $0 < \alpha < 1$. Let M_α be the operator $L_2([0, 1])$ defined by

$$(M_\alpha f)(t) = \begin{cases} \frac{1}{\sqrt{\alpha}} f(\frac{1}{\alpha}t), & 0 \le t \le \alpha, \\ 0, & \alpha < t \le 1. \end{cases}$$

(a) Prove that M_α is an isometry, and show that

$$(M_\alpha^* f)(t) = \sqrt{\alpha} f(\alpha t), \qquad 0 \le t \le 1.$$

(b) Determine Im M_α and Ker M_α^*.

(c) Find the spectra of the operators M_α and M_α^*.

16. Fix $0 < \alpha < 1$. On $L_2([0, 1])$ consider the operator

$$T = -4M_\alpha + 17I - 4M_\alpha^*,$$

where M_α is defined in the previous exercise. For the operator T solve the problems posed in (a), (b) and (c) of Exercise 9.

17. Fix $0 < \alpha < 1$. On $L_2([0, 1])$ consider the operator

$$T = aM_\alpha + bM_\alpha^*,$$

where M_α is defined in Exercise 15.

(a) For which values of a and b in $\mathbb{C}$ is the operator T left, right or two-sided invertible? For these values determine the one- or two-sided inverse of T.

(b) Find the spectrum of T.

18. Fix $0 < \alpha < 1$. Find the spectrum of T, where T is the operator on $L_2([0, 1])$ given by

$$T = \sum_{j=1}^{\infty} a^j M_\alpha + I + \sum_{j=1}^{\infty} b^j (M_\alpha^*)^j.$$

Here M_α is as in Exercise 15, and $|a| < 1$, $|b| < 1$.

19. Let U be an isometry on a Hilbert space H, and U is not unitary. Generalize the results of Section 3 to operators of the form

$$T = \sum_{j=0}^{p} c_j U_j + \sum_{j=1}^{q} d_j (U^*)^j.$$

In particular, for the operator T prove the analogs of Theorems 3.1 and 3.2, and determine its spectrum.

Chapter IV
Spectral Theory of Compact Self Adjoint Operators

One of the fundamental results in linear algebra is the spectral theorem which states that if $\mathcal{H}$ is a finite dimensional Hilbert space and $A \in \mathcal{L}(\mathcal{H})$ is self adjoint, then there exists an orthonormal basis $\varphi_1, \ldots, \varphi_n$ for $\mathcal{H}$ and real numbers $\lambda_1, \ldots, \lambda_n$ such that

$$A\varphi_i = \lambda_i \varphi_i, \quad 1 \le i \le n.$$

The matrix $(a_{ij}) = (\langle A\varphi_j, \varphi_i \rangle)$ corresponding to A and $\varphi_1, \ldots, \varphi_n$ is the diagonal matrix

$$\begin{pmatrix} \lambda_1 & & 0 \\ & \ddots & \\ 0 & & \lambda_n \end{pmatrix}$$

A natural question is whether this spectral theorem can be generalized to the case where A is self adjoint and $\mathcal{H}$ is infinite dimensional. That is to say, is there an orthonormal basis $\varphi_1, \varphi_2, \ldots$ for $\mathcal{H}$ and numbers $\lambda_1, \lambda_2, \ldots$ such that

$$A\varphi_i = \lambda_i \varphi_i, \quad 1 \le i?$$

This means that the matrix corresponding to A and $\varphi_1, \varphi_2, \ldots$ is an infinite diagonal matrix.

In this chapter it is shown that the spectral theorem admits an important generalization to compact self adjoint operators.

Let us first consider an example which indicates the possibility of a generalization.

4.1 Example of an Infinite Dimensional Generalization

Let h be a continuous complex valued function of period 2π. The operator K defined on $L_2([-\pi, \pi])$ by

$$(Kf)(t) = \int_{-\pi}^{\pi} h(t-s) f(s) \, ds$$

is a bounded linear operator with range in $L_2([-\pi, \pi])$. Taking $\varphi_n(t) = \frac{1}{\sqrt{2\pi}}e^{int}$, $n = 0, \pm 1, \ldots$ as the orthonormal basis for $L_2([-\pi, \pi])$, it follows from the periodicity of h and φ_n that

$$(K\varphi_n)(t) = \int_{-\pi}^{\pi} h(t-s)\varphi_n(s)\,ds = \int_{t-\pi}^{t+\pi} h(s)\varphi_n(t-s)\,ds$$

$$= \frac{1}{\sqrt{2\pi}}e^{int}\int_{-\pi}^{\pi} h(s)e^{-ins}\,ds = \lambda_n\varphi_n(t),$$

where

$$\lambda_n = \int_{-\pi}^{\pi} h(s)e^{-ins}\,ds, \quad n = 0, \pm 1, \ldots.$$

The matrix corresponding to the operator K and $\{\varphi_n\}_{n=-\infty}^{\infty}$, is the doubly infinite diagonal matrix

$$\begin{pmatrix} \ddots & & & & & \\ & \lambda_{-2} & & & & \\ & & \lambda_{-1} & & 0 & \\ & 0 & & \lambda_0 & & \\ & & & & \lambda_1 & \\ & & & & & \lambda_2 \\ & & & & & & \ddots \end{pmatrix}$$

4.2 The Problem of Existence of Eigenvalues and Eigenvectors

The examples described above show that the spectral representation theory starts with the problem of the existence of eigenvalues and eigenvectors.

Definition: A complex number λ is called an *eigenvalue* of $A \in \mathcal{L}(\mathcal{H})$ if there exists a $\varphi \neq 0$ in $\mathcal{H}$ such that $A\varphi = \lambda\varphi$. The vector φ is called an *eigenvector* of A corresponding to the eigenvalue λ.

We shall see later the significance of eigenvalues and eigenvectors which appear in various problems in mathematical analysis and mechanics.

Example: Let K be the integral operator defined on $\mathcal{H} = L_2([0.1])$ with kernel function ik, where

$$k(t, s) = \begin{cases} 1, & s \leq t \\ -1, & s > t. \end{cases}$$

Example 5 in II.2 shows that K is in $\mathcal{L}(\mathcal{H})$. To find the eigenvalues and eigenvectors of K, let us suppose that $\varphi \neq 0$ and

$$(K\varphi)(t) = i\int_0^t \varphi(s)\,ds - i\int_t^1 \varphi(s)\,ds = \lambda\varphi(t) \text{ a.e.} \tag{2.1}$$

Differentiating each side of (2.1) yields

$$2i\varphi(t) = \lambda\varphi'(t) \text{ a.e.} \tag{2.2}$$

Thus $\lambda = 0$ is not an eigenvalue of K. If $\lambda \neq 0$, it follows from (2.2) that

$$\varphi(t) = ce^{\frac{2i}{\lambda}t} \text{ a.e., } c \neq 0. \tag{2.3}$$

By identifying functions which are equal a.e., we may assume (2.3) holds for all t. Now (2.1) implies that

$$0 = \varphi(0) + \varphi(1) = c(l + e^{\frac{2i}{\lambda}}), \ c \neq 0.$$

Hence

$$\frac{2}{\lambda} = (2k+1)\pi, \ k = 0, \pm 1, \ldots.$$

By reversing our steps, it follows that

$$\lambda_k = \frac{2}{(2k+1)\pi}, \ k = 0, \pm 1, \ldots$$

are the eigenvalues of K and $e^{i(2k+1)\pi t}$ are eigenvectors corresponding to λ_k.

Every linear operator on a finite dimensional Hilbert space over $\mathbb{C}$ has an eigenvalue. However, even a self adjoint operator on an infinite dimensional Hilbert space need not have an eigenvalue.

For example, let $A : L_2([a, b]) \to L_2([a, b])$ be the operator defined by

$$(Af)(t) = tf(t).$$

Now A is a bounded linear self adjoint operator by II.12.2 – Example 2. However, A has no eigenvalue; for if $A\varphi = \lambda\varphi$, then

$$(t - \lambda)\varphi(t) = 0 \text{ a.e}$$

Thus $\varphi(t) = 0$ a.e., which means that $\varphi = 0$ when considered as a vector in $L_2([a, b])$.

The following results are used throughout this chapter.

2.1 (a) *Any eigenvalue of a self adjoint operator is real.* For if A is self adjoint and $Ax = \lambda x, x \neq 0$, then

$$\lambda\|x\|^2 = \langle Ax, x\rangle = \langle x, Ax\rangle = \bar{\lambda}\|x\|^2,$$

whence $\lambda = \bar{\lambda}$.

(b) *Eigenvectors corresponding to distinct eigenvalues of a self adjoint operator are orthogonal.* Indeed, if A is self adjoint and

$$Ax = \lambda x, \; Ay = \mu y, \; y \neq 0,$$

then μ is real and

$$\lambda \langle x, y \rangle = \langle Ax, y \rangle = \langle x, Ay \rangle = \mu \langle x, y \rangle.$$

Hence $\lambda \neq \mu$ implies $\langle x, y \rangle = 0$.

(c) *If λ is an eigenvalue of $A \in \mathcal{L}(\mathcal{H})$, then $|\lambda| \leq \|A\|$.* For if $Ax = \lambda x$, then

$$\|A\| \|x\| \geq \|Ax\| = |\lambda| \|x\|.$$

4.3 Eigenvalues and Eigenvectors of Operators of Finite Rank

Let $K \in \mathcal{L}(\mathcal{H})$ be an operator of finite rank, say

$$Kx = \sum_{j=1}^{n} \langle x, \varphi_j \rangle \psi_j.$$

The eigenvalues and eigenvectors of K are determined as follows.
Suppose $\lambda \neq 0$ is an eigenvalue of K with eigenvector x. Then

$$\lambda x - \sum_{j=1}^{n} \langle x, \varphi_j \rangle \psi_j = 0 \tag{3.1}$$

and

$$\lambda \langle x, \varphi_k \rangle - \sum_{j=1}^{n} \langle x, \varphi_j \rangle \langle \psi_j, \varphi_k \rangle = 0, \; 1 \leq k \leq n. \tag{3.2}$$

Now $\langle x, \varphi_j \rangle \neq 0$ for some j, otherwise $x = 0$ by (3.1). Thus $\{\langle x, \varphi_j \rangle\}_{j=1}^{n}$ is a non trivial solution to the system of equations

$$\lambda c_k - \sum_{j=1}^{n} \langle \psi_j, \varphi_k \rangle c_j = 0, \; 1 \leq k \leq n. \tag{3.3}$$

Therefore, by Cramer's rule,

$$\det \left(\lambda \delta_{jk} - \langle \psi_j, \varphi_k \rangle \right) = 0. \tag{3.4}$$

Conversely, if (3.4) holds, then there exist $c_1, \ldots, c_n$, not all zero, which satisfy (3.3). Guided by (3.1), we take $x = \frac{1}{\lambda} \sum_{j=1}^{n} c_j \psi_j$ and get $Kx = \lambda x$.

To summarize, we have shown that $\lambda \neq 0$ is an eigenvalue of K if and only if

$$\det (\lambda \delta_{kj} - \langle \psi_j, \varphi_k \rangle) = 0,$$

and x is an eigenvector corresponding to λ if and only if it is a non-zero vector of the form. $\frac{1}{\lambda} \sum_{j=1}^{n} \alpha_j \psi_j$, where

$$\begin{pmatrix} \langle \psi_1, \varphi_1 \rangle - \lambda & \langle \psi_2, \varphi_1 \rangle & \cdots & \langle \psi_n, \varphi_1 \rangle \\ \langle \psi_1, \varphi_2 \rangle & \langle \psi_2, \varphi_2 \rangle - \lambda & \cdots & \langle \psi_n, \varphi_2 \rangle \\ \vdots & \vdots & & \vdots \\ \langle \psi_1, \varphi_n \rangle & \langle \psi_2, \varphi_n \rangle & \cdots & \langle \psi_n, \varphi_n \rangle - \lambda \end{pmatrix} \begin{pmatrix} \alpha_1 \\ \alpha_2 \\ \vdots \\ \alpha_n \end{pmatrix} = \begin{pmatrix} 0 \\ 0 \\ \vdots \\ 0 \end{pmatrix}$$

If $\{\psi_1, \psi_2, \ldots, \psi_n\}$ is linearly independent, then it is clear from the definition of K that zero is an eigenvalue of K with eigenvector x if and only if $x \neq 0$ and $x \perp \varphi_j$, $1 \leq j \leq n$.

The above results can also be obtained by applying Theorem II.7.1 to $I - \frac{1}{\lambda} K$.

4.4 Existence of Eigenvalues

In this section it is shown that every compact self adjoint operator has an eigenvalue.

Let us start with the following theorem.

Theorem 4.1 *If $A \in \mathcal{L}(\mathcal{H})$ is self adjoint, then*

$$\|A\| = \sup_{\|x\|=1} |\langle Ax, x \rangle|.$$

Proof: Let m = $\sup_{\|x\|=1} |\langle Ax, x \rangle|$. Then for $\|x\| = 1$,

$$|\langle Ax, x \rangle| \leq \|Ax\| \leq \|A\|.$$

Hence m $\leq \|A\|$. To prove that $m \geq \|A\|$, let x and y be arbitrary vectors in $\mathcal{H}$. Then

$$\langle A(x \pm y), x \pm y \rangle = \langle Ax, x \rangle \pm 2 \Re \langle Ax, y \rangle + \langle Ay, y \rangle.$$

Therefore,

$$4 \Re \langle Ax, y \rangle = \langle A(x + y), x + y \rangle - \langle A(x - y), x - y \rangle.$$

Combining this with the definition of m and the parallelogram law (Theorem I.3.1), we get

$$4 \Re \langle Ax, y \rangle \leq m(\|x + y\|^2 + \|x - y\|^2) = 2m(\|x\|^2 + \|y\|^2). \tag{4.1}$$

Now $\langle Ax, y \rangle = |\langle Ax, y \rangle| \, e^{i\theta}$ for some real number θ. Substituting $e^{-i\theta} x$ for x in (4.1) yields

$$|\langle Ax, y \rangle| \leq \frac{m}{2} \, (\|x\|^2 + \|y\|^2). \tag{4.2}$$

Suppose $Ax \neq 0$. Then taking $y = \frac{\|x\|}{\|Ax\|} Ax$ in (4.2), we get

$$\|Ax\| \, \|x\| \leq m \|x\|^2.$$

Hence $\|Ax\| \leq m \|x\|$ for all $x \in \mathcal{H}$ and $\|A\| \leq m$. $\qquad\square$

The following corollary is an immediate consequence of Theorems II.12.3 and 4.1.

Corollary 4.2 *If $A \in \mathcal{L}(\mathcal{H})$ and $\langle Ax, x \rangle = 0$ for all $x \in \mathcal{H}$, then $A = 0$.*

Now that we know that the least upper bound of the set $\{|\langle Ax, x \rangle| : \|x\| = 1\}$ is $\|A\|$ (A self adjoint), the next problem is to determine if $\|A\|$ is attained, i.e., $\|A\| = |\langle Ax_0, x_0 \rangle|$ for some $x_0, \|x_0\| = 1$. The next theorem shows that if this is the case, then at least one of the numbers $\|A\|$ or $-\|A\|$ is an eigenvalue of A. Thus if A is a self adjoint operator which does not have an eigenvalue, as in the example in Section 2, then

$$\|A\| > |\langle Ax, x \rangle| \text{ for all } x, \|x\| = 1.$$

Theorem 4.3 *Suppose A is self adjoint. Let*

$$\lambda = \inf_{\|x\|=1} \langle Ax, x \rangle.$$

If there exists an $x_0 \in \mathcal{H}$ such that $\|x_0\| = 1$ and

$$\lambda = \langle Ax_0, x_0 \rangle,$$

then λ is an eigenvalue of A with corresponding eigenvector x_0.
    Let

$$\mu = \sup_{\|x\|=1} \langle Ax, x \rangle.$$

If there exists an $x_1 \in \mathcal{H}$ such that $\|x_1\| = 1$ and

$$\mu = \langle Ax_1, x_1 \rangle,$$

then μ is an eigenvalue of A with corresponding eigenvector x_1.

Proof: For every $\alpha \in \mathbb{C}$ and every $v \in \mathcal{H}$, it follows from the definition of λ that

$$\langle A(x_0 + \alpha v), x_0 + \alpha v \rangle \geq \lambda \langle x_0 + \alpha v, x_0 + \alpha v \rangle.$$

Expanding the inner products and setting $\lambda = \langle Ax_0, x_0 \rangle$, we obtain the inequality

$$2 \, \Re \, \alpha \langle v, (A - \lambda)x_0 \rangle + |\alpha|^2 \, \langle (A - \lambda)v, v \rangle \geq 0. \tag{4.3}$$

Taking $\alpha = \overline{r \langle v, (A - \lambda) x_0 \rangle}$, where r is an arbitrary real number, it is easy to see that the inequality can only hold for all α if

$$\langle v, (A - \lambda)x_0 \rangle = 0.$$

Since v is arbitrary, $Ax_0 - \lambda x_0 = 0$.

The second conclusion of the theorem follows from the above result applied to the self adjoint operator $-A$. $\square$

When are we guaranteed that $\langle Ax, x \rangle$ has at least a largest or a smallest value as x ranges over the 1-sphere of $\mathcal{H}$? The next corollary supplies an answer to this question.

Theorem 4.4 *If $A \in \mathcal{L}(\mathcal{H})$ is compact and self adjoint, $\mathcal{H} \neq \{0\}$, then at least one of the numbers $\|A\|$ or $-\|A\|$ is an eigenvalue of A.*

Proof: The theorem is trivial if $A = 0$. Assume $A \neq 0$. It follows from Theorem 4.1 that there exists a sequence $\{x_n\}$ in $\mathcal{H}$, $\|x_n\| = 1$, and a real number λ such that $|\lambda| = \|A\| \neq 0$ and $\langle Ax_n, x_n \rangle \to \lambda$. To prove that λ is an eigenvalue of A, we first note that

$$0 \leq \|Ax_n - \lambda x_n\|^2 = \|Ax_n\|^2 - 2\lambda \langle Ax_n, x_n \rangle + \lambda^2$$
$$\leq 2\lambda^2 - 2\lambda \langle Ax_n, x_n \rangle \to 0.$$

Thus

$$Ax_n - \lambda x_n \to 0. \tag{4.4}$$

Since A is compact, there exists a subsequence $\{Ax_{n'}\}$ of $\{Ax_n\}$ which converges to some $y \in \mathcal{H}$. Consequently, (4.3) implies that $x_{n'} \to \frac{1}{\lambda} y$, and by the continuity of A,

$$y = \lim_{n \to \infty} Ax_{n'} = \frac{1}{\lambda} Ay.$$

Hence $Ay = \lambda y$ and $y \neq 0$ since $\|y\| = \lim_{n \to \infty} \|\lambda x_{n'}\| = |\lambda| = \|A\|$. Thus λ is an eigenvalue of A. $\square$

Corollary 4.5 *If $A \in \mathcal{L}(\mathcal{H})$ is compact and self adjoint, then $\max_{\|x\|=1} |\langle Ax, x \rangle|$ exists and equals $\|A\|$.*

Proof: By Theorem 4.4, there exists a λ which is an eigenvalue of A with $|\lambda| = \|A\|$. Let φ be an eigenvector corresponding to λ with $\|\varphi\| = 1$. Then for $\|x\| = 1$,

$$|\langle A\varphi, \varphi \rangle| = |\lambda| = \|A\| \geq |\langle Ax, x \rangle|.$$ $\square$

A compact operator need not have an eigenvalue if it is not self adjoint.

Examples: Let K: $L_2([0, 1]) \rightarrow L_2([0, 1])$ be defined by

$$(Kf)(t) = \int_0^t f(s) \, ds.$$

Since

$$(Kf)(t) = \int_0^1 k(t, s) \, f(s) ds,$$

Where $k(t, s) = 1$ if $0 \leq s \leq t$ and zero otherwise, it follows from II.16, Example 3, that K is compact. However, K has no eigenvalue. For if $K\varphi = \lambda\varphi$, then

$$\int_0^t \varphi(s) \, ds = \lambda\varphi(t) \text{ a.e.} \qquad (4.5)$$

Since $L_2[0, 1]$ consists of equivalence classes of square integrable functions which are equal almost everywhere, we may redefine φ so that (4.1) holds for all t. Hence

$$\varphi(t) = \lambda\varphi'(t), \quad \varphi(0) = 0,$$

from which it follows that $\varphi = 0$.

4.5 Spectral Theorem

We shall now prove the spectral theorem for compact self adjoint operators. The proof depends on successive applications of Theorem 4.4.

Theorem 5.1 *Suppose A is a compact self adjoint operator on $\mathcal{H}$. There exist an orthonormal system $\varphi_1, \varphi_2, \ldots$ of eigenvectors of A and corresponding eigenvalues $\lambda_1, \lambda_2, \ldots$ such that for all $x \in \mathcal{H}$,*

$$Ax = \sum_k \lambda_k \langle x, \varphi_k \rangle \varphi_k.$$

If $\{\lambda_k\}$ is an infinite sequence, then it converges to zero.

It is clear from the theorem that if $\{\varphi_k\}$ is an orthonormal basis for $\mathcal{H}$, then the matrix corresponding to A and $\{\varphi_k\}$ is the diagonal matrix

$$\begin{pmatrix} \lambda_1 & & \\ & \lambda_2 & \\ & & \ddots \end{pmatrix}.$$

Proof: Let $\mathcal{H}_1 = \mathcal{H}$ and $A_1 = A$. By Theorem 4.4, there exists an eigenvalue λ_1 of A_1 and a corresponding eigenvector φ_1 such that $\|\varphi_1\| = 1$ and $|\lambda_1| = \|A_1\|$. Now $\mathcal{H}_2 = \{\varphi_1\}^\perp$ is a closed subspace of $\mathcal{H}_1$ and $A\mathcal{H}_2 \subset \mathcal{H}_2$ by

Theorem II.18.1. Let A_2 be the restriction of A to $\mathcal{H}_2$. Then A_2 is a compact self adjoint operator in $\mathcal{L}(\mathcal{H}_2)$. If $A_2 \neq 0$, there exists an eigenvalue λ_2 of A_2 and a corresponding eigenvector φ_2 such that $\|\varphi_2\| = 1$ and

$$|\lambda_2| = \|A_2\| \leq \|A\| = |\lambda_1|.$$

Clearly, $\{\varphi_1, \varphi_2\}$ is orthonormal. Now $\mathcal{H}_3 = \{\varphi_1, \varphi_2\}^\perp$ is a closed subspace of $\mathcal{H}$, $\mathcal{H}_3 \subset \mathcal{H}_2$ and $A\mathcal{H}_3 \subset \mathcal{H}_3$. Letting A_3 be the restriction of A to $\mathcal{H}_3$, we have that A_3 is a compact self adjoint operator in $\mathcal{L}(\mathcal{H}_3)$. Continuing in this manner, the process either stops when $A_n = 0$ or else we get a sequence $\{\lambda_n\}$ of eigenvalues of A and a corresponding orthonormal set $\{\varphi_1, \varphi_2, \ldots\}$ or eigenvectors such that

$$|\lambda_{n+1}| = \|A_{n+1}\| \leq \|A_n\| = |\lambda_n|, \quad n = 1, 2, \ldots.$$

If $\{\lambda_n\}$ is an infinite sequence, then $\lambda_n \to 0$.

Assume this is not the case. Since $|\lambda_n| \geq |\lambda_{n+1}|$, there exists an $\varepsilon > 0$ such that $|\lambda_n| \geq \varepsilon$ for all n. Hence for $n \neq m$,

$$\|A\varphi_n - A\varphi_m\|^2 = \|\lambda_n \varphi_n - \lambda_m \varphi_m\|^2 = \lambda_n^2 + \lambda_m^2 > \varepsilon^2.$$

But this is impossible since $\{A\varphi_n\}$ has a convergent sub-sequence due to the compactness of A.

We are now ready to prove the representation of A as asserted in the theorem. Let $x \in \mathcal{H}$ be given.

Case 1: $A_n = 0$ for some n.

Since $x_n = x - \sum_{k=1}^n \langle x, \varphi_k \rangle \varphi_k$ is orthogonal to φ_1, $1 \leq i \leq n$, the vector x_n is in $\mathcal{H}_n$. Hence

$$0 = A_n x_n = Ax - \sum_{k=1}^n \lambda_k \langle x, \varphi_k \rangle \varphi_k.$$

Case 2: $A_n \neq 0$ for all n.

From what we have seen in case 1,

$$\left\| Ax - \sum_{k=1}^n \lambda_k \langle x, \varphi_k \rangle \varphi_k \right\| = \|A_n x_n\| \leq \|A_n\| \, \|x_n\|$$

$$= |\lambda_n| \, \|x_n\| \leq |\lambda_n| \, \|x\| \to 0,$$

which means that

$$Ax = \sum_{k=1}^\infty \lambda_k \langle x, \varphi_k \rangle \varphi_k.$$

$\square$

4.6 Basic Systems of Eigenvalues and Eigenvectors

For convenience we introduce the following definition.

Definition: An orthonormal system $\varphi_1, \varphi_2, \ldots$ of eigenvectors of $A \in \mathcal{L}(\mathcal{H})$ with corresponding *non-zero* eigenvalues $\lambda_1, \lambda_2, \ldots$ is called a *basic system* of eigenvectors and eigenvalues of A if for all $x \in \mathcal{H}$.

$$Ax = \sum_k \lambda_k \langle x, \varphi_k \rangle \varphi_k. \tag{6.1}$$

The existence of a basic system of eigenvectors and eigenvalues of any non-zero compact self adjoint operator is ensured by the spectral theorem.

The following observations (a)–(d) pertain to a compact self adjoint operator A which has a basic system of eigenvectors $\{\varphi_n\}$ and eigenvalues $\{\lambda_n\}$.

6.1. (a) *The orthonormal system $\{\varphi_n\}$ is an orthonormal basis for $\overline{\mathrm{Im}\,A} = \mathrm{Ker}\,A^\perp$, and for each $x \in \mathcal{H}$,*

$$x = P_0 x + \sum_n \langle x, \varphi_n \rangle \varphi_n, \tag{6.2}$$

where P_0 is the orthogonal projection onto $\mathrm{Ker}\,A$.

Indeed, since $\varphi_n = \frac{1}{\lambda_n} A\varphi_n \in \mathrm{Im}\,A$, formula (6.1) in the definition implies that $\mathrm{sp}\{\varphi_n\}$ is dense in $\mathrm{Im}\,A$. It now follows from Theorem II.11.4 that

$$\overline{\mathrm{sp}}\,\{\varphi_n\} = \overline{\mathrm{Im}\,A} = \mathrm{Ker}\,A^\perp.$$

Thus $\{\varphi_n\}$ is an orthonormal basis for $\mathrm{Ker}\,A^\perp$. Hence, given $x \in \mathcal{H}$, there exists a $v \in \mathrm{Ker}\,A^\perp$ such that

$$x = P_0 x + v = P_0 x + \sum_k \langle v, \varphi_k \rangle \varphi_k,$$

which implies (6.2) since $\langle v, \varphi_k \rangle = \langle x, \varphi_k \rangle$.

If A_0 is the restriction of A to $\overline{\mathrm{Im}}\,A$, then the matrix corresponding to A_0 and $\{\varphi_n\}$ is the diagonal matrix $(\delta_{ij} \lambda_i)$.

(b) $\mathcal{H}$ *has an orthonormal basis consisting of a basic system of eigenvectors of A if and only if* $\mathrm{Ker}\,A = (0)$.

The statement follows from (6.2).

If $\mathrm{Ker}\,A \neq (0)$ is separable, let $\{\psi_k\}$ be an orthonormal basis for $\mathrm{Ker}\,A$. Then $\mathcal{H}$ has an orthonormal basis consisting of eigenvectors of A, namely $\{\psi_k\} \cup \{\varphi_n\}$.

(c) *If $\lambda \neq 0$ is an eigenvalue of A, then $\lambda = \lambda_k$ for some k.*

Otherwise any eigenvector v with eigenvalue λ would be orthogonal to each φ_k by 2.1 (b). Hence

$$\lambda v = Av = \sum_k \lambda_k \langle v, \varphi_k \rangle \varphi_k = 0,$$

which is impossible since $\lambda \neq 0$ and $v \neq 0$.

It is shown in Theorem 8.1 that $\lambda I - A$ is invertible if $\lambda \neq 0$ and $\lambda \neq \lambda_k$ for all k. In addition, a formula for $(\lambda I - A)^{-1}$ is given.

(d) *Each λ_j is repeated in the sequence $\{\lambda_n\}$ exactly p_j times, where $p_j =$* dim Ker$(\lambda_j I - A)$.

To see this, let us recall from the spectral theorem that $\lambda_n \to 0$ if $\{\lambda_n\}$ is an infinite sequence. Consequently, λ_j appears only a finite number of times in the sequence $\{\lambda_n\}$. Suppose $\lambda_j = \lambda_{n_i}, i = 1, 2, \ldots, p$. Then

$$\text{Ker}(\lambda_j - A) = \text{sp}\{\varphi_{n_i}\}_{i=1}^p.$$

For if this is not the case, then there exists a $v \neq 0$ in Ker$(\lambda_j - A)$ such that $v \perp \varphi_{n_i}, 1 \leq i \leq p$. But for $k \neq n_i, 1 \leq i \leq p$, v is also orthogonal to φ_k since $\lambda_j \neq \lambda_k$. Hence

$$\lambda_j v = Av = \sum_k \lambda_k \langle v, \varphi_k \rangle \varphi_k = 0$$

which is impossible since $\lambda_j \neq 0$ and $v \neq 0$.

The dimension of Ker$(\lambda_j - A)$ is called the *multiplicity* of λ_j. In view of the above results, every basic system of eigenvectors and eigenvalues of a compact self adjoint operator A can be obtained as follows.

Suppose $\{\mu_1, \mu_2, \ldots\}$ is the set of non-zero eigenvalues of A. Choose any orthonormal basis B_j for Ker$(\mu_j I - A)$. Then B_j is a finite set by (d). Let $\{\varphi_n\}$ be a sequence consisting of all the vectors in $\cup_j B_j$. Since eigenvectors corresponding to distinct eigenvalues of A are orthogonal, it follows that $\{\varphi_n\}$ is an orthonormal set of eigenvectors of A. Let λ_n be the eigenvalue of A corresponding to φ_n. Then $\{\varphi_n\}, \{\lambda_n\}$ is a basic system of eigenvectors and eigenvalues of A.

The converse of the spectral theorem also holds.

Theorem 6.2 *Suppose $\{\varphi_k\}$ is an orthonormal system in $\mathcal{H}$ and $\{\lambda_k\}$ is a sequence of real numbers which is either a finite sequence or converges to zero. The linear operator A defined on $\mathcal{H}$ by*

$$Ax = \sum_k \lambda_k \langle x, \varphi_k \rangle \varphi_k$$

is compact and self adjoint.

Proof: A is in $\mathcal{L}(\mathcal{H})$ by II. 2 Example 2. If $\{\lambda_k\}$ is a finite sequence, then A is of finite rank and is therefore compact. Assume $\lambda_n \rightarrow 0$. Let $A_n \in \mathcal{L}(\mathcal{H})$ be the compact operator defined by

$$A_n x = \sum_{k=1}^{n} \lambda_k \langle x, \varphi_k \rangle \varphi_k.$$

Since

$$\|A - A_n\|^2 = \sup_{\|x\|=1} \left\| \sum_{k=n+1}^{\infty} \lambda_k \langle x, \varphi_k \rangle \varphi_k \right\|^2 \leq \sup_{k>n} |\lambda_k|^2 \rightarrow 0$$

as $n \rightarrow \infty$, A is compact. For x and y in $\mathcal{H}$,

$$\langle Ax, y \rangle = \sum_{k} \lambda_k \langle x, \varphi_k \rangle \langle \varphi_k, y \rangle = \langle x, Ay \rangle.$$

Thus A is self adjoint. $\square$

4.7 Second Form of the Spectral Theorem

The construction above of a basic system of eigenvectors and eigenvalues sheds light on how to approximate a compact self adjoint operator by linear combinations of orthogonal projections. This is done in the following theorem.

Theorem 7.1 *Given a compact self adjoint operator* $A \in \mathcal{L}(\mathcal{H})$, *let* $\{\mu_j\}$ *be the set of non-zero eigenvalues of* A *and let* P_j *be the orthogonal projection onto* $\mathrm{Ker}(\mu_j I - A)$. *Then*

(i) $P_j P_k = 0, j \neq k$;

(ii) $A = \sum_j \mu_j P_j$,

   where convergence of the series is with respect to the norm on $\mathcal{L}(\mathcal{H})$.

(iii) *For each* $x \in \mathcal{H}$,

$$x = P_0 x + \sum_{j} P_j x,$$

   where P_0 *is the orthogonal projection onto* $\mathrm{Ker}\, A$.

Proof: Let $\{\varphi_n\}$, $\{\lambda_n\}$ be a basic system of eigenvectors and eigenvalues of A. For each k, a subset of $\{\varphi_n\}$ is an orthonormal basis for $\text{Ker}(\mu_k - A)$, say φ_{n_i}, $1 \le i \le p$. Then $P_k x = \sum_{i=1}^{p} \langle x, \varphi_{n_i} \rangle \varphi_{n_i}$, and since each λ_n is a μ_k, it follows from (6.2) that

$$x = P_0 x + \sum_k P_k x \text{ and } Ax = \sum_k \mu_k P_k x.$$

Furthermore, $P_j P_k = 0$, $j \ne k$, since $\text{Ker}(\mu_j - A) \perp \text{Ker}(\mu_k - A)$. Finally, if $\{\lambda_j\}$ is an infinite sequence, then

$$\left\| A - \sum_{k=1}^{n} \mu_k P_k \right\|^2 = \sup_{\|x\|=1} \left\| Ax - \sum_{k=1}^{n} \mu_k P_k x \right\|^2$$

$$\le \sup_{\|x\|=1} \sum_{j \ge n} \lambda_j^2 |\langle x, \varphi_j \rangle|^2 \le \sup_{j \ge n} \lambda_j^2 \to 0$$

as $n \to \infty$. $\qquad\qquad\square$

4.8 Formula for the Inverse Operator

The following theorem gives sufficient conditions for the equation $\lambda x - Kx = y$ to have a unique solution for every $y \in \mathcal{H}$.

Theorem 8.1 *Let K be a compact self adjoint operator in $\mathcal{L}(\mathcal{H})$ with a basic system of eigenvectors $\{\varphi_n\}$ and eigenvalues $\{\lambda_n\}$. If $\lambda \ne 0$ and $\lambda \ne \lambda_n$, $n \ge 1$, then $\lambda I - K$ is invertible and*

$$(\lambda I - K)^{-1} y = \frac{1}{\lambda} \left[y + \sum_k \frac{\lambda_k}{\lambda - \lambda_k} \langle y, \varphi_k \rangle \varphi_k \right].$$

Proof: Suppose $(\lambda I - K)x = y$. Then

$$\lambda x = y + Kx = y + \sum_k \lambda_k \langle x, \varphi_k \rangle \varphi_k, \qquad (8.1)$$

which implies that

$$\lambda \langle x, \varphi_j \rangle = \langle y, \varphi_j \rangle + \lambda_j \langle x, \varphi_j \rangle$$

or

$$(\lambda - \lambda_j) \langle x, \varphi_j \rangle = \langle y, \varphi_j \rangle, \qquad 1 \le j. \qquad (8.2)$$

Thus, from (8.1) and (8.2),

$$x = \frac{1}{\lambda} \left[y + \sum_k \frac{\lambda_k}{\lambda - \lambda_k} \langle y, \varphi_k \rangle \varphi_k \right]. \qquad (8.3)$$

Conversely, if x is defined by (8.3), (the series converges since the sequence $\{\frac{\lambda_k}{\lambda-\lambda_k}\langle y, \varphi_k\rangle\}$ is in ℓ_2) then a straightforward computation verifies that $(\lambda I - K)x = y$. We have shown that for each $y \in \mathcal{H}$, there exists a unique $x \in \mathcal{H}$ such that $(\lambda I - K)x = y$ and this x is given by (8.3). Hence

$$(\lambda I - K)^{-1}y = \frac{1}{\lambda}(I + K_1)y, \tag{8.4}$$

where $K_1 y = \sum_k \frac{\lambda_k}{\lambda-\lambda_k}\langle y, \varphi_k\rangle\varphi_k$.

Now K_1 is in $\mathcal{L}(\mathcal{H})$. In fact, Theorem 6.2 shows that K_1 is compact. Hence (8.4) implies that $\lambda I - K$ is invertible. $\qquad\qquad\square$

We shall show in Section VIII.1, equation (1.14) that

$$\|(\lambda I - K)^{-1}\| = \sup\left\{\frac{1}{|\lambda - \mu|} : \mu \text{ an eigenvalue of } K\right\}.$$

If $\mathcal{H}$ is infinite dimensional and $K \in \mathcal{L}(\mathcal{H})$ is compact, then K is not invertible, otherwise, $I = K^{-1}K$ would be compact.

The following theorem characterizes Im K when K is compact and self adjoint.

Theorem 8.2 *Let $K \in \mathcal{L}(\mathcal{H})$ be a compact self adjoint operator with a basic system of eigenvectors $\{\varphi_n\}$ and eigenvalues $\{\lambda_n\}$. Given $y \in \mathcal{H}$, the equation $Kx = y$ has a solution if and only if*

(i) $y \perp \text{Ker } K$

    and

(ii) $\sum_n \frac{1}{\lambda_n^2}|\langle y, \varphi_n\rangle|^2 < \infty.$

    Every solution is of the form

$$x = u + \sum_n \frac{1}{\lambda_n}\langle y, \varphi_n\rangle\varphi_n, \, u \in \text{Ker} K.$$

Proof: Suppose (i) and (ii) hold. Then $x_0 = \sum_n \frac{1}{\lambda_n}\langle y, \varphi_n\rangle$ is in $\mathcal{H}$ and it follows from (2) in Section 6 that

$$Kx_0 = \sum_n \frac{1}{\lambda_n}\langle y, \varphi_n\rangle K\varphi_n = \sum_n \langle y, \varphi_n\rangle\varphi_n = y.$$

If also $Kx = y$, then $u = x - x_0 \in \text{Ker } K$ and

$$x = u + \sum_n \frac{1}{\lambda_n}\langle y, \varphi_n\rangle\varphi_n.$$

Conversely, suppose $Kx = y$. Then $y \in \text{Im } K \subset \text{Ker } K^\perp$ and

$$\sum_n \langle y, \varphi_n \rangle \varphi_n = y = Kx = \sum_n \lambda_n \langle x, \varphi_n \rangle \varphi_n.$$

Hence $\langle y, \varphi_n \rangle = \lambda_n \langle x, \varphi_n \rangle$ and

$$\sum_n \frac{1}{\lambda_n^2} |\langle y, \varphi_n \rangle|^2 = \sum_n |\langle x, \varphi_n \rangle|^2 \le \|x\|^2.$$

$\square$

Suppose that in Theorem 8.1, $\lambda = \lambda_n \ne 0$ for some n. By an argument similar to the one given in the proof of Theorem 8.2, the following result can be shown.
   The equation $\lambda x - Kx = y$ *has a solution if and only if* $y \in \text{Ker}(\lambda_n - K)^\perp$. *In this case the general solution to the equation is*

$$x = \frac{1}{\lambda} y + \frac{1}{\lambda} \sum_{\lambda_k \ne \lambda} \frac{\lambda_k}{\lambda - \lambda_k} \langle y, \varphi_k \rangle \varphi_k + z,$$

where $z \in \text{Ker}(\lambda_n - K)$.

4.9 Minimum-Maximum Properties of Eigenvalues

Definition: An operator $A \in \mathcal{L}(\mathcal{H})$ is called *non-negative or positive if* $\langle Ax, x \rangle \ge 0$ for all $x \in \mathcal{H}$.

A compact self adjoint operator A is non-negative if and only if its eigenvalues are non-negative. For suppose $\{\varphi_n\}$, $\{\lambda_n\}$ is a basic system of eigenvectors and eigenvalues of A. If A is non-negative, then $\lambda_k = \langle A\varphi_k, \varphi_k \rangle \ge 0$. On the other hand, if each λ_k is non-negative, then it follows from 6.1(a) that

$$\langle Ax, x \rangle = \left\langle \sum_k \lambda_k \langle x, \varphi_k \rangle \varphi_k, P_0 x + \sum_k \langle x, \varphi_k \rangle \varphi_k \right\rangle$$

$$= \sum_k \lambda_k |\langle x, \varphi_k \rangle|^2 \ge 0.$$

Let A be compact and non-negative. In the proof of the spectral theorem, a basic system $\{\varphi_n\}$, $\{\lambda_n\}$ of eigenvectors and eigenvalues of A was obtained by taking

$$\lambda_1 = \max_{\|x\|=1} \langle Ax, x \rangle = \langle A\varphi_1, \varphi_1 \rangle$$

$$\lambda_2 = \max_{\substack{\|x\|=1 \\ x \perp \text{sp}\{\varphi_1\}}} \langle Ax, x \rangle = \langle A\varphi_2, \varphi_2 \rangle$$

$$\vdots \qquad \vdots$$

$$\lambda_n = \max_{\substack{\|x\|=1 \\ x \perp \text{sp}\{\varphi_1,\dots,\varphi_{n-1}\}}} \langle Ax, x \rangle = \langle A\varphi_n, \varphi_n \rangle \qquad (9.1)$$

The following theorem shows that it is unnecessary to find the eigenvectors $\varphi_1, \ldots, \varphi_{n-1}$ in order to determine λ_n as in (9.1). The result has numerous applications-especially to numerical methods for estimating eigenvalues.

Theorem 9.1 *Let $A \in \mathcal{L}(\mathcal{H})$ be compact and non-negative and let $\lambda_1 \geq \lambda_2 \geq \ldots$ be the basic system of eigenvalues of A. Then for each positive integer n,*

$$\lambda_n = \min_{\substack{M \\ \dim M = n-1}} \max_{\substack{\|x\|=1 \\ x \perp M}} \langle Ax, x \rangle.$$

Proof: We note that $\max\{\langle Ax, x \rangle : \|x\| = 1, \ x \perp M\}$ is attained. This can be seen from Corollary 4.5 applied to the restriction of PA to $M^\perp$, where P is the orthogonal projection onto $M^\perp$. For $n = 1$, the only subspace of dimension zero is (0). Thus the formula for λ_1 reduces to $\lambda_1 = \max_{\|x\|=1} \langle Ax, x \rangle$, which we already know.

Let $\{\varphi_n\}$ be a basic system of eigenvectors of A corresponding to $\{\lambda_n\}$. Given any subspace M of dimension $n - 1$, Lemma I.16.1 implies that there exists an $x_0 \in \mathrm{sp}\{\varphi_1, \ldots, \varphi_n\}$ such that $x_0 \perp M$ and $\|x_0\| = 1$. Suppose $x_0 = \sum_{k=1}^n \alpha_k \varphi_k$. Since $\lambda_k \geq \lambda_n$, $1 \leq k \leq n$,

$$\max_{\substack{\|x\|=1 \\ x \perp M}} \langle Ax, x \rangle \geq \langle Ax_0, x_0 \rangle = \left\langle \sum_{k=1}^n \lambda_k \alpha_k \varphi_k, \sum_{k=1}^n \alpha_k \varphi_k \right\rangle$$

$$= \sum_{k=1}^n \lambda_k |\alpha_k|^2 \geq \lambda_n \|x_0\|^2 = \lambda_n. \tag{9.2}$$

But

$$\lambda_n = \max_{\substack{\|x\|=1 \\ x \perp \mathrm{sp}\{\varphi_1, \ldots, \varphi_{n-1}\}}} \langle Ax, x \rangle. \tag{9.3}$$

Since M is an arbitrary subspace of $\dim n - 1$, the theorem follows from (9.2) and (9.3). □

Now for some simple applications of the min-max theorem. Suppose A and B are compact non-negative operators in $\mathcal{L}(\mathcal{H})$. Let $\varphi_1, \varphi_2, \ldots$ and $\lambda_1(A) \geq \lambda_2(A) \geq \ldots$ be a basic system of eigenvectors and eigenvalues of A. Let $\psi_1, \psi_2, \ldots$ and $\lambda_1(B) \geq \lambda_2(B) \geq \ldots$ be a basic system of eigenvectors and eigenvalues of B.

(a) *If $\langle Ax, x \rangle \leq \langle Bx, x \rangle$ for all $x \in \mathcal{H}$, then*

$$\lambda_n(A) = \min_{\substack{M \\ \dim M = n-1}} \max_{\substack{\|x\|=1 \\ x \perp M}} \langle Ax, x \rangle \leq \min_{\substack{M \\ \dim M = n-1}} \max_{\substack{\|x\|=1 \\ x \perp M}} \langle Bx, x \rangle = \lambda_n(B).$$

(b) $|\lambda_n(A) - \lambda_n(B)| \leq \|A - B\|.$

To see this, we note that if $\|x\| = 1$, then

$$|\langle Ax, x\rangle - \langle Bx, x\rangle| = |\langle (A - B)x, x\rangle| \leq \|A - B\|.$$

Hence

$$\langle Ax, x\rangle \leq \langle Bx, x\rangle + \|A - B\| \tag{9.4}$$

$$\langle Bx, x\rangle \leq \langle Ax, x\rangle + \|A - B\|. \tag{9.5}$$

It follows from (9.4), (9.5) and Theorem 9.1 that

$$\lambda_n(A) \leq \lambda_n(B) + \|A - B\|$$
$$\lambda_n(B) \leq \lambda_n(A) + \|A - B\|$$

or, equivalently,

$$|\lambda_n(A) - \lambda_n(B)| \leq \|A - B\|.$$

Thus if $\|K_j - K_0\| \to 0$, where each K_j, $j > 0$, is a compact non-negative operator in $\mathcal{L}(\mathcal{H})$, then K_0 is also compact and non-negative. Let $\lambda_1(K_j) \geq \lambda_2(K_j) \geq \ldots$ be the basic system of eigenvalues of K_j, $j = 0, 1, \ldots$. Then from the inequality (b), $\lambda_n(K_j) \to \lambda_n(K_0)$ as $j \to \infty$.

(c) *Since* A *and* B *are compact and non-negative, so is* A $+$ B. *Let* $\lambda_1(A+B) \geq \lambda_2(A + B) \geq \ldots$ *be the basic system of eigenvalues of* A $+$ B. *Then*

$$\lambda_n(A) + \lambda_m(B) \geq \lambda_{n+m-1}(A + B). \tag{9.6}$$

Indeed, from Equations (9.1),

$$\lambda_n(A) = \max_{\substack{\|x\|=1 \\ x \perp \{\varphi_1, \ldots, \varphi_{n-1}\}}} \langle Ax, x\rangle \tag{9.7}$$

$$\lambda_m(B) = \max_{\substack{\|x\|=1 \\ x \perp \{\psi_1, \ldots, \psi_{m-1}\}}} \langle Bx, x\rangle. \tag{9.8}$$

Take $M = \mathrm{sp}\{\varphi_1, \ldots, \varphi_{n-1}, \psi_1, \ldots, \psi_{m-1}\}$. Then $\dim M = n + m - j$ for some $j \geq 2$. Since $M^\perp = \{\varphi_1, \ldots, \varphi_{n-1}\}^\perp \cap \{\psi_1, \ldots, \psi_{m-1}\}^\perp$, it follows from (9.3), (9.4) and Theorem 9.1 that

$$\lambda_n(A) + \lambda_m(B) \geq \max_{\substack{\|x\|=1 \\ x \perp M}} \langle Ax, x\rangle + \max_{\substack{\|x\|=1 \\ x \perp M}} \langle Bx, x\rangle$$

$$\geq \max_{\substack{\|x\|=1 \\ x \perp M}} \langle (A + B)x, x\rangle \geq \lambda_{n+m-j+1}(A + B)$$

$$\geq \lambda_{n+m-1}(A + B).$$

Exercises IV

1. Find all the eigenvectors and eigenvalues of the following operators K.

   (a) $K\xi = (\sum_{j=1}^{\infty} \xi_j (\frac{1}{\sqrt{2}})^{j-1} + \xi_1, \frac{1}{\sqrt{2}}\xi_1, \frac{1}{\sqrt{4}}\xi_1, \ldots); \ K : \ell_2 \rightarrow \ell_2$

   (b) $(K\varphi)(t) = t \int_{-\pi}^{\pi} \varphi(x) \cos x \, dx + \cos t \int_{-\pi}^{\pi} x\varphi(x) dx; \ K : L_2[-\pi, \pi]$
   $\rightarrow L_2[-\pi, \pi]$.

2. Let $K : L_2[-\pi, \pi] \rightarrow L_2[-\pi, \pi]$ be given by $(K\varphi)(t) = \int_{-\pi}^{\pi} k(t-s)\varphi(s)$
   ds. Find all the eigenvectors and eigenvalues of K when K is given by

   (a) $K(t) = |t|$

   (b) $K(t) = \sin t$.

3. What are the eigenvectors and eigenvalues of an orthogonal projection?

4. If $A \in \mathcal{L}(\mathcal{H})$ and $A^2 = 0$, find the eigenvectors and eigenvalues of A.

5. Let P be an orthogonal projection on a Hilbert space $\mathcal{H}$. Find all the eigenvectors and eigenvalues of $I - 2P$.

6. Given an orthogonal projection P on a Hilbert space, prove that $S = I - 2P$
   if and only if $S = S^*$ and $S^2 = I$.

7. Let K be the operator in exercises 1 and 2. Define A on $\mathcal{H} \oplus \mathcal{H}$ by

$$A = \begin{pmatrix} 0 & K \\ K^* & 0 \end{pmatrix}.$$

   Find the eigenvectors and eigenvalues of A.

8. Let K be a compact self adjoint operator on a Hilbert space $\mathcal{H}$. If one
   knows the eigenvalues and eigenvectors of K, what are the eigenvalues and
   eigenvectors of A, defined on $\mathcal{H} \oplus \mathcal{H}$, if

$$A = \begin{pmatrix} 0 & K \\ K & 0 \end{pmatrix}?$$

9. Let K be a compact operator on a Hilbert space. Let $\{\lambda_j\}$ be the eigenvalues
   of K^*K. Find the eigenvalues of

$$A = \begin{pmatrix} 0 & K \\ K^* & 0 \end{pmatrix}.$$

10. Show that every $|\lambda| < 1$ is an eigenvalue of the backward shift operator S_ℓ.

11. Given that $(\alpha I - A)^{-1}$ exists, find x, where

    (a) $(\alpha I - A)x = e_1$ and $A : \ell_2 \rightarrow \ell_2$ is given by $A\xi = (3\xi_1 + \xi_2, \xi_1 + 3\xi_2, 0, 0, \ldots)$.

(b) $(\alpha I - A)x = \cos 2t$, and $A: L_2[-\pi, \pi] \to L_2[-\pi, \pi]$ is given by
$(A\varphi)(t) = \int_{-\pi}^{\pi} (t - s)^2 \varphi(s) \, ds$.

12. Let A, B be two operators on a Hilbert space and let φ be an eigenvector of AB corresponding to an eigenvalue $\lambda \neq 0$. Prove that B_φ is an eigenvector of BA corresponding to λ.

13. Prove that all non zero eigenvalues λ or AB and BA are the same and that

$$B[\mathrm{Ker}(AB - \lambda I)] = \mathrm{Ker}(BA - \lambda I).$$

Is it always correct that $\dim \mathrm{Ker}(AB - \lambda I) = \dim \mathrm{Ker}(BA - \lambda I)$ if $\lambda \neq 0$ is an eigenvalue of AB?

14. What conclusions can one draw in problems 12 and 13 if $\lambda = 0$?

15. Given $\lambda \neq 0$, prove that if one of the two operators $(AB - \lambda I)$ and $(BA - \lambda I)$ is invertible, then so is the other and

$$(AB - \lambda I)^{-1} = \frac{1}{\lambda}[-I + A(BA - \lambda I)^{-1}B].$$

16. Let A be defined on $\mathcal{H} \oplus \mathcal{H}$ by

$$A = \begin{pmatrix} B & 0 & 0 \\ 0 & 0 & C \\ 0 & 0 & 0 \end{pmatrix}$$

Where $B, C \in \mathcal{L}(\mathcal{H})$, B compact and self adjoint. Prove that A^2 is compact and self adjoint.

17. Given $A \in \mathcal{L}(\mathcal{H})$, Suppose A^2 is compact and self adjoint. Show that A has an eigenvalue and that the eigenvectors of A corresponding to λ and μ, where $\lambda^2 \neq \mu^2$, are orthogonal.

18. Let A be a self adjoint operator on a Hilbert space.

(a) Prove that if A^2 is compact, then A is compact.

(b) Generalize this result to the case when A^n is compact for some $n \in \mathbb{N}$.

19. Let A be a positive compact operator on a Hilbert space. Let $\{\varphi_j\}_{j=1}^\infty$ be a basic system of eigenvectors of A corresponding to eigenvalues $\{\lambda_j\}_{j=1}^\infty$ $(\lambda_1 \geq \lambda_2 \geq \ldots)$. Let $L_{j-1} = \mathrm{sp}\{\varphi_1, \ldots, \varphi_{j-1}\}$, $j \geq 1$, where $L_0 = \{0\}$. Verify that

$$\min_{\substack{(Ax,x)=1 \\ x \perp L_{j-1}}} \langle x, x \rangle = \frac{1}{\lambda_j}.$$

20. Let A be a compact positive operator on a Hilbert space. Let $\lambda_1 \geq \lambda_2 \geq \cdots \geq 0$ be the eigenvalues of A. Show that

(a) $\max\limits_{\substack{L \\ \dim L=j-1}} \min\limits_{\substack{(Ax,x)=1 \\ X \perp L}} \langle x, x \rangle = \frac{1}{\lambda_j}$

(b) $\min\limits_{\substack{L \\ \dim L=j}} \max\limits_{\substack{(Ax,x)=1 \\ X \in L}} \langle x, x \rangle = \frac{1}{\lambda_j}.$

21. Let P be an orthogonal projection on a Hilbert space $\mathcal{H}$. Prove that for any $A \in \mathcal{L}(\mathcal{H})$, $A^*PA \leq A^*A$, i.e., $A^*A - A^*PA$ is positive.

22. Define $K \in \mathcal{L}(\mathcal{H})$ by $Kh = \sum_{j=1}^{n} \lambda_j \langle h, \varphi_j \rangle \varphi_j$, where $\{\varphi_j\}_{j=1}^{n}$ is an ortho-normal system and $\lambda_j \geq 0$. Let P be an orthogonal projection on $\mathcal{H}$. Define $K_p h = \sum_{j=1}^{n} \lambda_j \langle h, P\varphi_j \rangle P\varphi_j$. Prove that $\lambda_j(K_p) \leq \lambda_j(K)$.

23. Let P be an orthogonal projection on a Hilbert space $\mathcal{H}$. Prove that if $A \in \mathcal{L}(\mathcal{H})$ is compact, then

$$\lambda_j(PA^*AP) \leq \lambda_j(A^*A).$$

24. Prove or disprove the statement that if P is an orthogonal projection on a Hilbert space $\mathcal{H}$ and $A \in \mathcal{L}(\mathcal{H})$, then $PA^*AP \leq A^*A$.

25. Let A be a compact positive operator on a Hilbert space $\mathcal{H}$. Let the matrix representation of A with respect to some orthonormal basis be $(a_{ij})_{i,j=1}^{\infty}$. Let A_n be the operator corresponding to the orthonormal basis and the matrix $(b_{ij})_{i,j=1}^{\infty}$, where $b_{ij} = a_{ij}, 1 \leq i, j \leq n$, and zero otherwise. Prove that $\lambda_j(A_n) \leq \lambda_j(A), j = 1, 2, \ldots$.

26. Let A be a compact positive operator on a Hilbert space $\mathcal{H}$. Find an operator Z such that Z is compact and $Z^2 = A$. Also, find an operator Z such that Z is compact and $Z^n = A$.

27. Prove that if A is a compact positive operator on a Hilbert space $\mathcal{H}$ and P is an orthogonal projection on $\mathcal{H}$, then

$$\lambda_n(PAP) \leq \lambda_n(A).$$

28. Does the conclusion in exercise 27 remain true if A is not assumed to be positive?

29. Let A be a compact operator or a Hilbert space $\mathcal{H}$ and let $\varphi_1, \ldots, \varphi_n$ be an orthonormal system of vectors in $\mathcal{H}$. Prove that

$$g(A\varphi_1, \ldots, A\varphi_n) \leq \lambda_1(A^*A) \ldots \lambda_n(A^*A),$$

where g denotes the Gram determinant.

30. Prove that for a compact operator A on a Hilbert space $\mathcal{H}$,

$$\max\limits_{\substack{\{\varphi_1, \ldots, \varphi_n\} \\ \text{orthonormal}}} g(A\varphi_1, \ldots, A\varphi_n) = \lambda_1(A^*A) \ldots \lambda_n(A^*A).$$

31. Let A be a compact positive definite operator on a Hilbert space $\mathcal{H}$. Prove that for any orthonormal system of vectors $\{\varphi_i\}_{i=1}^n$ in $\mathcal{H}$,

$$\sum_{i=1}^n \langle A\varphi_i, \varphi_1 \rangle \leq \sum_{i=1}^n \lambda_i(A).$$

32. Prove that under the conditions in Exercise 31,

$$\sum_{j=1}^n \lambda_j(A) = \max \left\{ \sum_{j=1}^n \langle A\varphi_j, \varphi_j \rangle : \{\varphi_j\}_{j=1}^n \text{ orthonormal} \right\}.$$

Chapter V
Spectral Theory of Integral Operators

Using the theory developed in Chapter IV, we now present some fundamental theorems concerning the spectral theory of compact self adjoint integral operators. In general, the spectral series representations of these operators converge in the L_2-norm which is not strong enough for many applications. Therefore we prove the Hilbert-Schmidt theorem and Mercer's theorem since each of these theorems gives conditions for a uniform convergence of the spectral decomposition of the integral operators. As a corollary of Mercer's theorem we obtain the trace formula for positive integral operators with continuous kernel function.

5.1 Hilbert-Schmidt Theorem

We recall that if k is in $L_2([a, b] \times [a, b])$ and $\overline{k(t, s)} = k(s, t)$ a.e., then the integral operator K defined by

$$(Kf)(t) = \int_a^b k(t, s) f(s) ds$$

is a compact self adjoint operator on $L_2([a, b])$. Consequently, there exists a basic system of eigenvectors $\{\varphi_k\}$ and eigenvalues $\{\lambda_k\}$ of K $(K \neq 0)$ such that

$$(Kf)(t) = \sum_k \lambda_k \left(\int_a^b f(s)\overline{\varphi_k(s)}ds \right) \varphi_k(t). \qquad (1.1)$$

Convergence of the series means convergence with respect to the norm on $L_2([a, b])$, i.e.,

$$\lim_{n \to \infty} \int_a^b \left| (Kf)(t) - \sum_{k=1}^n \lambda_k \langle f, \varphi_k \rangle \varphi_k(t) \right|^2 dt = 0.$$

For many purposes this type of convergence is too weak, whereas uniform convergence in (1.1) is most desirable. The problem is to give sufficient, but reasonable, conditions so that for each $f \in L_2([a, b])$, the series in (1.1) converges uniformly on $[a, b]$. The Hilbert-Schmidt theorem which follows now is very useful in this regard.

Theorem 1.1 *Let k be a Lebesgue measurable function on* $[a, b] \times [a, b]$ *such that*

$$\overline{k(t, s)} = k(s, t) \text{ a.e.}$$

and

$$\sup_t \int_a^b |k(t, s)|^2 ds < \infty.$$

Let $\{\varphi_n\}, \{\lambda_n\}$ *be a basic system of eigenvectors and eigenvalues of K, where K is the integral operator with kernel function k. Then for all* $f \in L_2([a, b])$,

$$\int_a^b k(t, s) f(s) ds = \sum_k \lambda_k \left(\int_a^b f(s) \overline{\varphi_k(s)} ds \right) \varphi_k(t) \text{ a.e.}$$

The series converges absolutely and uniformly on $[a, b]$.

Proof: By Schwarz's inequality,

$$\sum_{j=m}^n |\lambda_j \langle f, \varphi_j \rangle \varphi_j(t)|$$

$$\leq \left(\sum_{j=m}^n |\lambda_j \varphi_j(t)|^2 \right)^{1/2} \left(\sum_{j=m}^n |\langle f, \varphi_j \rangle|^2 \right)^{1/2}. \tag{1.2}$$

Now

$$\lambda_j \varphi_j(t) = (K\varphi_j)(t) = \int_a^b k(t, s) \varphi_j(s) \, ds = \langle k_t, \bar{\varphi}_j \rangle,$$

where $k_t(s) = k(t, s)$. Therefore, since k_t is in $L_2([a, b])$, it follows from Bessel's inequality and the hypotheses that

$$\sum_j |\lambda_j \varphi_j(t)|^2 = \sum_j |\langle k_t, \bar{\varphi}_j \rangle|^2 \leq \|k_t\|^2$$

$$= \int_a^b |k(t, s)|^2 \, ds \leq \sup_t \int_a^b |k(t, s)|^2 \, ds$$

$$= C^2 < \infty. \tag{1.3}$$

Let $\varepsilon > 0$ be given. Since $\sum_j |\langle f, \varphi_j \rangle|^2 \leq \|f\|^2$, there exists an integer N such that if $n > m \geq N$, then

$$\sum_{j=m}^n |\langle f, \varphi_j \rangle|^2 \leq \varepsilon^2. \tag{1.4}$$

Thus from (1.2), (1.3), and (1.4)

$$\sum_{j=m}^n |\lambda_j \langle f, \varphi_j \rangle \varphi_j(t)| \leq C\varepsilon, \quad n > m \geq N, \quad t \in [a, b].$$

Hence $\sum_j \lambda_j \langle f, \varphi_j \rangle \varphi_j(t)$ converges absolutely and uniformly on $[a, b]$ by the Cauchy criterion. Since this series also converges to $(Kf)(t)$ with respect to the norm on $L_2([a, b])$, it follows that $(Kf)(t)$ is the limit of the series for almost every t. □

Disregarding the validity of termwise integration of the series in the above theorem, we formally write

$$\int_a^b k(t, s) f(s) \, ds = \sum_k \int_a^b \lambda_k \varphi_k(t) \bar{\varphi}_k(s) f(s) \, ds$$

$$= \int_a^b \sum_k \lambda_k \varphi_k(t) \bar{\varphi}_k(s) f(s) \, ds.$$

Since f is arbitrary in $L_2([a, b])$, it is not unreasonable to expect that under the appropriate conditions,

$$k(t, s) = \sum_k \lambda_k \varphi_k(t) \bar{\varphi}_k(s) \quad \text{a.e.} \tag{1.5}$$

The result we shall prove in Section 3 is Mercer's theorem which states that if k is continuous and each λ_k is non-negative, then the series (1.5) converges uniformly and absolutely on $[a, b] \times [a, b]$ to $k(t, s)$.

Now let us prove another fact. We know from IV.6.1(b) that $L_2([a, b])$ has an orthonormal basis $\varphi_1, \varphi_2, \dots$ consisting of eigenvectors of K. Since $\Phi_{ij}(t, s) = \varphi_i(t) \overline{\varphi_j(s)}$ form an orthonormal basis for $L_2([a, b] \times [a, b])$,

$$k = \sum_{i,j=1}^{\infty} \langle k, \Phi_{ij} \rangle \Phi_{ij} \tag{1.6}$$

and

$$\langle k, \Phi_{ij} \rangle = \int_a^b \int_a^b k(t, s) \bar{\varphi}_i(t) \varphi_j(s) \, ds \, dt$$

$$= \int_a^b \bar{\varphi}_i(t) \left[\int_a^b k(t, s) \varphi_j(s) \, ds \right] dt$$

$$= \langle K\varphi_j, \varphi_i \rangle = \lambda_j \langle \varphi_j, \varphi_i \rangle = \lambda_j \delta_{ji}. \tag{1.7}$$

Thus (1.1) and (1.2) imply

$$\int_a^b \int_a^b |k(t, s)|^2 \, ds \, dt = \|k\|^2 = \sum_{i,j=1}^{\infty} |\langle k, \Phi_{ij} \rangle|^2 = \sum_{j=1}^{\infty} \lambda_j^2.$$

We have therefore obtained the following result.

Theorem 1.2 *Suppose $k \in L_2([a, b] \times [a, b])$ and $\overline{k(t, s)} = k(s, t)$ a.e. If $\{\lambda_k\}$ is the basic system of eigenvalue of K, where K is the integral operator with kernel function k, then*

$$\int_a^b \int_a^b |k(t, s)|^2 \, ds \, dt = \sum_j \lambda_j^2.$$

The formula is the continuous analogue of the fact that if (a_{ij}) is an $n \times n$ self adjoint matrix with eigenvalues $\lambda_1, \ldots, \lambda_n$, counted according to multiplicity, then

$$\sum_{i,j=1}^n |a_{ij}|^2 = \sum_{i=1}^n \lambda_i^2.$$

Here $k(t, s)$ is replaced by $k(i, j) = a_{ij}$ and the intergral is replaced by a sum.

5.2 Preliminaries for Mercer's Theorem

Lemma 2.1 *If k is continuous on $[a, b] \times [a, b]$ and*

$$\int_a^b \int_a^b k(t, s) f(s) \overline{f(t)} \, ds \, dt \geq 0$$

for all $f \in L_2([a, b])$, then

$$k(t, t) \geq 0 \text{ for all } t \in [a, b].$$

Proof: The function $k(t, t)$ is real valued. Indeed, the integral operator with kernel function k is positive and therefore self adjoint. Hence $\overline{k(t, t)} = k(t, t)$. Suppose $k(t_0, t_0) < 0$ for some $t_0 \in [a, b]$. It follows from the continuity of k that $\Re \, k(t, s) < 0$ for all (t, s) in some square $[c, d] \times [c, d]$ containing (t_0, t_0). But then for $g(s) = 1$ if $s \in [c, d]$ and zero otherwise,

$$0 \leq \int_a^b \int_a^b k(t, s) g(s) \overline{g(t)} \, ds \, dt = \Re \int_c^d \int_c^d k(t, s) \, ds \, dt < 0$$

which is absurd. Hence $k(t, t) \geq 0$ for all $t \in [a, b]$.   □

Lemma 2.2 *If k is a continuous complex valued function on $[a, b] \times [a, b]$, then for any $\varphi \in L_2([a, b])$,*

$$h(t) = \int_a^b k(t, s) \varphi(s) \, ds$$

is continuous on $[a, b]$.

Proof: By Schwarz's inequality,

$$|h(t) - h(t_0)| \leq \int_a^b |k(t, s) - k(t_0, s)||\varphi(s)| \, ds$$

$$\leq \|\varphi\| \left(\int_a^b |k(t, s) - k(t_0, s)|^2 \, ds \right)^{\frac{1}{2}}.$$

This last inequality, together with the uniform continuity of k on $[a, b] \times [a, b]$, imply the theorem. □

Dini's Theorem 2.3 *Let $\{f_n\}$ be a sequence of real valued continuous functions on $[a, b]$. Suppose $f_1(t) \leq f_2(t) \leq \ldots$ for all $t \in [a, b]$ and $f(t) = \lim_{n \to \infty} f_n(t)$ is continuous on $[a, b]$. Then $\{f_n\}$ converges uniformly to f on $[a, b]$.*

Proof: Given $\varepsilon > 0$, let

$$F_n = \{t : f(t) - f_n(t) \geq \varepsilon\}, \quad n = 1, 2, \ldots.$$

Then $F_n \supset F_{n+1}$ and it follows from the continuity of $f - f_n$ that F_n is a closed set. Furthermore, the pointwise convergence of $\{f_n\}$ implies that $\cap_{n=1}^\infty F_n = \emptyset$. Hence $[a, b] \subset \cup_{n=1}^\infty F_n'$, where F_n' is the complement of F_n. Since $[a, b]$ is closed and bounded and each F_n' is an open set, there exists an integer N such that $[a, b] \subset \cup_{n=1}^N F_n'$. Therefore

$$F_N = \bigcap_{n=1}^N F_n = \emptyset.$$

This implies that for all $n \geq N$ and all $t \in [a, b]$,

$$|f(t) - f_n(t)| = f(t) - f_n(t) \leq f(t) - f_N(t) < \varepsilon.$$

Hence $\{f_n\}$ converges uniformly to f on $[a, b]$. □

5.3 Mercer's Theorem

We are now ready to prove the series expansion for k which was discussed in Section 1.

Theorem 3.1 *Let k be continuous on $[a, b] \times [a, b]$. Suppose that for all $f \in L_2([a, b])$,*

$$\int_a^b \int_a^b k(t, s) f(s) \bar{f}(t) ds \, dt \geq 0.$$

If $\{\varphi_n\}$, $\{\lambda_n\}$ is a basic system of eigenvectors and eigenvalues of the integral operator with kernel function k, then for all t and s in $[a, b]$,

$$k(t, s) = \sum_j \lambda_j \varphi_j(t)\bar{\varphi}_j(s).$$

The series converges absolutely and uniformly on $[a, b] \times [a, b]$.

Proof: Let K be the integral operator with kernel function k. It follows from the hypotheses that K is compact, positive and $\lambda_j = \langle K\varphi_j, \varphi_j \rangle \geq 0$. Schwarz's inequality applied to the sequences $\{\sqrt{\lambda_j}\varphi_j(t)\}$ and $\{\sqrt{\lambda_j}\varphi_j(s)\}$ yields

$$\sum_{j=m}^{n} |\lambda_j \varphi_j(t)\bar{\varphi}_j(s)| \leq \left(\sum_{j=m}^{n} \lambda_j |\varphi_j(t)|^2\right)^{1/2} \left(\sum_{j=m}^{n} \lambda_j |\varphi_j(s)|^2\right)^{1/2}. \tag{3.1}$$

First we establish that for all t

$$\sum_j \lambda_j |\varphi_j(t)|^2 \leq \max_s k(s, s). \tag{3.2}$$

Let

$$k_n(t, s) = k(t, s) - \sum_{j=1}^{n} \lambda_j \varphi_j(t)\bar{\varphi}_j(s).$$

Since each φ_j is an eigenvector of K, it follows from Lemma 2.2 that φ_j is continuous which implies that k_n is continuous. A straightforward computation verifies that

$$\int_a^b \int_a^b k_n(t, s) f(s) \bar{f}(t) \, ds \, dt = \langle Kf, f \rangle - \sum_{j=1}^{n} \lambda_j |\langle f, \varphi_j \rangle|^2$$

$$= \sum_{j>n} \lambda_j |\langle f, \varphi_j \rangle|^2 \geq 0.$$

Hence we have from Lemma 2.1 that for each t,

$$0 \leq k_n(t, t) = k(t, t) - \sum_{j=1}^{n} \lambda_j |\varphi_j(t)|^2.$$

Since n was arbitrary, inequality (3.2) follows. For fixed t and $\varepsilon > 0$, (3.1) and (3.2) imply the existence of an integer N such that for $n > m \geq N$

$$\sum_{j=m}^{n} \lambda_j |\varphi_j(t)\varphi_j(s)| \leq \varepsilon C, \quad s \in [a, b],$$

where $C^2 = \max_{x \in [a,b]} k(x, x)$. Therefore we have from the uniform Cauchy criterion that $\sum_j \lambda_j \varphi_j(t) \bar{\varphi}_j(s)$ converges absolutely and uniformly in s for each t.

The next step in the proof is to show that the series converges to $k(t, s)$. Once this is done, an application of Dini's theorem will conclude the proof.

Let

$$\tilde{k}(t, s) = \sum_j \lambda_j \varphi_j(t) \bar{\varphi}_j(s).$$

For $f \in L_2([a, b])$ and t fixed, the uniform convergence of the series in s and the continuity of each $\bar{\varphi}_j$ imply that $\tilde{k}(t, s)$ is continuous as a function of s and

$$\int_a^b [k(t, s) - \tilde{k}(t, s)] f(s) \, ds = (Kf)(t) - \sum_j \lambda_j \langle f, \varphi_j \rangle \varphi_j(t). \tag{3.3}$$

If $f \in \operatorname{Ker} K = \operatorname{Im} K^\perp$, then since $\varphi_j = \frac{1}{\lambda_j} K \varphi_j \in \operatorname{Im} K$, we have that $Kf = 0$ and $\langle f, \varphi_j \rangle = 0$. Thus the right side of (3.3) is zero. If $f = \varphi_{i,}$ then the right side of (3.3) is $\lambda_i \varphi_i(t) - \lambda_i \varphi_i(t) = 0$. Thus for each t, $k(t, \ ) - \tilde{k}(t, \ )$ is orthogonal to $L_2([a, b])$ by IV.6.1(a). Hence, $\tilde{k}(t, s) = k(t, s)$ for each t and almost every s. But then $\tilde{k}(t, s) = k(t, s)$ for every t and s since $\tilde{k}(t, \ )$ and $k(t, \ )$ are continuous. We have shown that

$$k(t, s) = \tilde{k}(t, s) = \sum_j \lambda_j \varphi_j(t) \bar{\varphi}_j(s).$$

In particular,

$$k(t, t) = \sum_j \lambda_j |\varphi_j(t)|^2.$$

The partial sums of this series form an increasing sequence of continuous functions which converges pointwise to the continuous function $k(t, t)$. Dini's theorem asserts that this series converges uniformly to $k(t, t)$. Thus given $\varepsilon > 0$, there exists an integer N such that for $n > m \geq N$,

$$\sum_{j=m}^n \lambda_j |\varphi_j(t)|^2 < \varepsilon^2 \quad \text{for all} \quad t \in [a, b].$$

This observation, together with (3.1) and (3.2), imply that for all $n > m \geq N$ and all $(t, s) \in [a, b] \times [a, b]$,

$$\sum_{j=m}^n \lambda_j |\varphi_j(t) \bar{\varphi}_j(s)| \leq \varepsilon C.$$

Hence $\sum_j \lambda_j \varphi_j(t) \bar{\varphi}_j(s)$ converges absolutely and uniformly on $[a, b] \times [a, b]$.

Mercer's theorem is not true if we remove the assumption that K is positive. To see this, let us consider the example in IV.1 where $k(t, s) = h(t - s)$ and h is continuous. If the conclusion in Mercer's theorem were to hold, then h would be the uniform limit of its Fourier series. However, there are well known examples of continuous functions where this is not the case.   □

5.4 Trace Formula for Integral Operators

The trace formula for finite matrices states that if (a_{ij}) is an $n \times n$ matrix with eigenvalues $\lambda_1, \ldots, \lambda_n$, counted according to multiplicity, then $\sum_{j=1}^{n} a_{jj} = \sum_{j=1}^{n} \lambda_j$. The following theorem is the continuous analogue of the trace formula.

Theorem 4.1 *Let k be continuous on $[a, b] \times [a, b]$. Suppose that for all $f \in L_2([a, b])$,*

$$\int_a^b \int_a^b k(t, s) f(s) \bar{f}(t) \, ds \, dt \geq 0.$$

If K is the integral operator with kernel function k and $\{\lambda_j\}$ is the basic system of eigenvalues of K, then

$$\sum_j \lambda_j = \int_a^b k(t, t) \, dt.$$

Proof: Let $\{\varphi_j\}$ be a basic system of eigenvectors of K corresponding to $\{\lambda_j\}$. By Mercer's theorem, the series

$$k(t, t) = \sum_j \lambda_j |\varphi_j(t)|^2$$

converges uniformly on $[a, b]$. Hence

$$\int_a^b k(t, t) \, dt = \sum_j \lambda_j \|\varphi_j\|^2 = \sum_j \lambda_j.$$

□

If, more generally, k is in $L_2([a, b] \times [a, b])$ and $\overline{k(t, s)} = k(s, t)$ a.e., then $\sum_j \lambda_j$ may diverge whereas $\sum_j \lambda_j^2 < \infty$ by Theorem 1.2.

Exercises V

1. Which of the integral operators on $L_2[-\pi, \pi]$ with the following kernel functions are positive and which are not?

   (a) $k(t, s) = \cos(t - s)$
   (b) $k(t, s) = (t - s)^3$
   (c) $k(t, s) = (t^2 - 2s^2)e^{t+s}$.

2. Prove that if a bounded integral operator with kernel function k is positive on $L_2[a, b]$, then it is positive on $L_2[c, d]$, where $[c, d] \subset [a, b]$. Is the converse true?

3. For $c \in (a, b)$, suppose an integral operator K with kernel function k is positive on $L_2[a, c]$ and $L_2[c, b]$. Is it true that k is positive on $L_2[a, b]$?

4. Let k be a compact positive integral operator on $L_2[0, 1]$ with a kernel function defined on $[0, 1] \times [0, 1]$. Let K' be the integral operator with the same kernel function restricted to $[\eta, \xi] \times [\eta, \xi] \subset [0, 1] \times [0, 1]$. Prove that $\lambda_n(K') \leq \lambda_n(K)$, where $\lambda_n(K)$ denotes the nth eigenvalue of K (in the decreasing sequence of eigenvalues).

5. Under what conditions does a kernel $h(t - s)$ defined on $[-1, 1] \times [-1, 1]$ define a positive bounded integral operator on $L_2[0, 1]$? Hint: Example IV.1.

6. Using Mercer's theorem, what can you say about the convergence of the Fourier series of a continuous periodic function with period 2π?

7. Let $k^{(2)}(t, s) = \int_a^b k(t, x)k(x, s)\, dx$, where k is a self adjoint Hilbert-Schmidt kernel, i.e. $\overline{k(t, s)} = k(s, t)$ a.e. and $k \in L_2([a, b] \times [a, b])$. Prove that if $\{\lambda_j\}_{j=1}^\infty$ is the basic system of eigenvalues of the operator $(K\varphi)(t) = \int_a^b k(t, s)\varphi(s)\, ds$, then $\sum_{j=1}^\infty \lambda_j^4 = \int_a^b \int_a^b |k^{(2)}(t, s)|^2\, ds\, dt$.

8. Let g be a continuous complex valued function on $[a, b] \times [a, b]$. Prove that

$$k(t, s) = \int_a^b \overline{g(x, t)}\, g(x, s)\, dx$$

satisfies the hypotheses of Mercer's theorem.

9. Let k be defined as in exercise 8 and let K be the corresponding integral operator. Show that

$$\sum_j \lambda_j(K) = \int_a^b \int_a^b |g(s, t)|^2\, ds\, dt.$$

10. Generalize the results in exercises 8 and 9 to the case where $g(t, s)$ is a Hilbert-Schmidt kernel which is L_2-continuous. This means that for any $\varepsilon > 0$, there exists a $\delta > 0$ such that

$$|t_1 - t_2| < \delta \quad \text{implies} \quad \int_a^b |g(t_1, s) - g(t_2, s)|^2\, ds < \varepsilon \text{ and}$$

$$|s_1 - s_2| < \delta \quad \text{implies} \quad \int_a^b |g(t, s_1) - g(t, s_2)|^2\, dt < \varepsilon.$$

11. Prove that the conclusions of Mercer's theorems remain valid as long as all but a *finite* number of the eigenvalues of the self adjoint operator (with continuous kernel) are positive.

12. Let K be an integral operator on $L_2(a, b)$ with kernel $k(t, s) = \sum_{j=1}^{n} a_j(t) \times b_j(s)$, $a_j, b_j \in L_2([a, b])$, $1 \leq j \leq n$. Prove that K can be written in the form

$$(K\varphi)(t) = \int_a^b A(t) B(s) \varphi(s) \, ds,$$

where A and B are matrix valued functions.

13. Let K be the operator in exercise 12. Using the integral representation of K obtained in the exercise, determine those λ for which $\lambda I - K$ is invertible and find $(\lambda I - K)^{-1}$.

14. Let K be the integral operator on $L_2[0, 1]$ with kernel function $k(t, s) = \min(t, s)$, $0 \leq t, s \leq 1$. Show that K is positive and find its eigenvalues and eigenvectors.

15. Let K be the integral operator on $L_2([0, \infty])$ with kernel function

$$k(x, y) = \sum_{k=0}^{\infty} \frac{1}{(k+1)^2} [\cos (k+1)x \sin ky - \sin (k+1)x \cos ky].$$

Show that K has no eigenvalues.

Chapter VI
Unbounded Operators on Hilbert Space

The theory developed thus far concentrated on bounded linear operators on a Hilbert space which had applications to integral equations. However, differential equations give rise to an important class of unbounded linear operators which are not defined on all of $L_2([a, b])$. In this chapter an introduction to unbounded operators is presented which includes the spectral theorem for the Sturm-Liouville operator. Simple examples of the spectral theory of unbounded self adjoint operators are also given. For a more detailed theory the reader is referred to [G], [GGK1], [K] and [DS2].

6.1 Closed Operators and First Examples

It is indeed fortunate that essentially all the important differential operators form a class of operators for which an extensive theory is developed. These are the closed operators which we now define.

Definition: Let $\mathcal{H}_1$ and $\mathcal{H}_2$ be Hilbert spaces. An operator A with domain $\mathcal{D}(A) \subseteq \mathcal{H}_1$ and range in $\mathcal{H}_2$ is called a *closed operator* if it has the property that whenever $\{x_n\}$ is a sequence in $\mathcal{D}(A)$ satisfying $x_n \to x$ in $\mathcal{H}_1$ and $Ax_n \to y$ in $\mathcal{H}_2$, then $x \in \mathcal{D}(A)$ and $Ax = y$.

Clearly, a bounded linear operator on a Hilbert space is closed.

We now give an example of an unbounded closed operator.

Let $\mathcal{H} = L_2([0, 1])$. Define the differential operator A by $\mathcal{D}(A) = \{f \in \mathcal{H} | f$ is absolutely continuous, $f' \in \mathcal{H}, f(0) = 0\}$, and $Af = f'$.

The operator A is unbounded since $f_n(t) = t^n, n = 1, 2, \ldots$ is in $\mathcal{D}(A)$, $\|f_n\|^2 = \int_0^1 t^{2n} dt = \frac{1}{2n+1} \leq 1$ and

$$\|Af_n\|^2 = \int_0^1 n^2 t^{2n-2} dt = \frac{n^2}{2n - 1} \to \infty.$$

The operator A is closed. To see this, we first note that $\operatorname{Ker} A = \{0\}$ and $\operatorname{Im} A = \mathcal{H}$. Indeed, given $g \in \mathcal{H}$, take $f(t) = \int_0^t g(s) ds$. (Recall that $L_2([0, 1]) \subseteq L_1([0, 1])$). Then $f \in \mathcal{D}(A)$ and $Af = g$. Define $A^{-1}g = f, g \in \mathcal{H}$.

The operator A^{-1} is a bounded linear operator on $\mathcal{H}$ with range $\mathcal{D}(A)$ since the Schwarz inequality gives

$$|(A^{-1}g)(t)| \leq \int_0^1 |g(s)|ds \leq \left(\int_0^1 |g(s)|^2 ds\right)^{1/2} = \|g\|.$$

Therefore

$$\|A^{-1}g\|^2 = \int_0^1 |(A^{-1}g)(t)|^2 dt \leq \int_0^1 \|g\|^2 \, dt = \|g\|^2$$

Hence $\|A^{-1}\| \leq 1$. Suppose

$$f_n \to f, \, f_n \in \mathcal{D}(A) \text{ and } Af_n \to h \in \mathcal{H}.$$

Then $f_n = A^{-1}Af_n \to A^{-1}h$. Hence $f = A^{-1}h \in \mathcal{D}(A)$ and $Af = h$ which shows that A is closed.

Definition: Let $\mathcal{H}_1$ and $\mathcal{H}_2$ be Hilbert spaces and let A be a linear operator with domain $\mathcal{D}(A) \subseteq \mathcal{H}_1$ and range Im $A \subseteq \mathcal{H}_2$. Suppose there exists a bounded linear operator A^{-1} mapping $\mathcal{H}_2$ into $\mathcal{H}_1$ with the properties $AA^{-1}y = y$ for all $y \in \mathcal{H}_2$ and $A^{-1}Ax = x$ for all $x \in \mathcal{D}(A)$. In this case we say that A is *invertible* and A^{-1} is *the inverse* of A. Note that if A is invertible, then Ker $A = \{0\}$ and Im $A = \mathcal{H}_2$. Clearly, A cannot have more than one inverse.

The differential operator in the above example is invertible and its inverse is $(A^{-1}g)(t) = \int_0^t g(s)ds$, $g \in L_2([0, 1])$.

The notation $A(\mathcal{H}_1 \to \mathcal{H}_2)$ signifies that A is a linear operator with domain in Hilbert space $\mathcal{H}_1$ and range in Hilbert space $\mathcal{H}_2$.

Theorem 1.1 *Let $A(\mathcal{H}_1 \to \mathcal{H}_2)$ be invertible. Then A is a closed operator.*

Proof: Suppose $x_n \to x, x_n \in \mathcal{D}(A)$ and $Ax_n \to y$. Since A^{-1} is bounded on $\mathcal{H}_2$, $x_n = A^{-1}Ax_n \to A^{-1}y$. Therefore $x = A^{-1}y \in \mathcal{D}(A)$ and $Ax = AA^{-1}y = y$. $\qquad\square$

We shall show in Theorem XII.4.2 that the converse theorem also holds, i.e., if the operator A is closed with Ker $A = \{0\}$ and Im $A = \mathcal{H}_2$, then A is invertible.

6.2 The Second Derivative as an Operator

One of the main motivations for the development of a theory of integral operators is that certain differential equations with prescribed boundary conditions can be transformed into equivalent integral equations. As an illustration, let us begin with the simple boundary value problem.

$$-y''(x) = f(x) \tag{2.1}$$
$$y(0) = y(1) = 0, \tag{2.2}$$

where f is a function in $L_2([0, 1])$. To find a solution, we integrate both sides of (2.1) twice and obtain

$$y(x) = -\int_0^x \int_0^t f(s)\,ds\,dt + c_1 x + c_2, \tag{2.3}$$

where

$$c_2 = y(0) = 0 \text{ and } c_1 = \int_0^1 \int_0^t f(s)\,ds\,dt. \tag{2.4}$$

Interchanging the order of integration in (2.3) and (2.4) yields

$$y(x) = -\int_0^x \int_s^x f(s)\,dt\,ds + x\int_0^1 \int_s^1 f(s)\,dt\,ds$$
$$= \int_0^x (s - x)f(s)\,ds + \int_0^1 x(1 - s)f(s)\,ds. \tag{2.5}$$

Hence

$$y(x) = \int_0^1 g(x, s)f(s)\,ds, \tag{2.6}$$

where

$$g(x, s) = \begin{cases} s(1 - x), & 0 \le s \le x \\ x(1 - s), & x \le s \le 1. \end{cases} \tag{2.7}$$

Conversely, if y is given by (2.6), a straightforward computation verifies that y satisfies (2.1) and (2.2) a.e. The function g is called the *Green's function* corresponding to the boundary value problem.

Let us consider the above result from the point of view of operator theory. We want to express the differential expression $-y''$ with the boundary conditions (2.2) as a linear operator. The action of the operator is clear. However, we must define its domain. To do this, we note that (2.3) implies that the derivative y' is an indefinite integral or, equivalently, y' is absolutely continuous (cf. [R]). An important property of absolutely continuous functions, which we shall use later, is that the usual "integration by parts" formula holds for the integral of fg, where f is absolutely continuous and g is Lebesgue integrable.

Let the domain $\mathcal{D}(L)$ of L be the subspace of $L_2([0, 1])$ consisting of those complex valued functions y which satisfy (2.2), have first order derivatives which are absolutely continuous on $[0, 1]$ and have second order derivatives which are in $L_2([0, 1])$. Note that $y''(x)$ exists for almost every x since y' is absolutely continuous. Define $Ly = -y''$. It is clear from (2.2) that L is injective.

Take G to be the integral operator with kernel function g defined in (2.7). Since g is continuous on $[0, 1] \times [0, 1]$ and $\overline{g(x, s)} = g(s, x)$, G is a compact self adjoint operator on $L_2([0, 1])$. From the discussion above, $y = Gf$ satisfies (2.1) a.e. for every $f \in L_2([0, 1])$. Thus

$$LGf = f. \tag{2.8}$$

Since L is also injective, we have that L is invertible with $L^{-1} = G$. Thus L is a closed linear operator by Theorem 1.1. If follows that φ is an eigenvector of L with eigenvalue λ if and only if φ is an eigenvector of G with eigenvalue $\frac{1}{\lambda}$. Thus, since G is compact self adjoint and $\mathrm{Ker}\, G = (0)$, $L_2([0, 1])$ has an orthonormal basis consisting of eigenvectors of L. The eigenvalues of L are those real λ for which

$$y'' + \lambda y = 0 \tag{2.9}$$

and

$$y(0) = y(1) = 0 \tag{2.10}$$

has a non-trivial solution. Since the general solution to (2.9) is

$$\begin{aligned}
y &= ax + b, & \lambda &= 0 \\
y &= a \cos \sqrt{\lambda} x + b \sin \sqrt{\lambda} x, & \lambda &> 0 \\
y &= ae^{\sqrt{-\lambda} x} + be^{-\sqrt{-\lambda} x}, & \lambda &< 0
\end{aligned}$$

it follows from the boundary conditions (2.10) that the eigenvalues are $\lambda = n^2\pi^2, n = 1, 2, \ldots$, with $b_n \sin n\pi x, b_n \neq 0$ the corresponding eigenvectors. The eigenvectors $\sqrt{2} \sin n\pi x, n = 1, 2, \ldots$, therefore form an orthonormal basis for $L_2([0, 1])$.

Similarly, if we change the domain of L by replacing the boundary conditions (2.10) by $y(-\pi) = y(\pi), y'(-\pi) = y'(\pi)$, then the eigenvalues of L are those λ for which the boundary value problem

$$y'' + \lambda y = 0$$

$$y(-\pi) = y(\pi), \quad y'(-\pi) = y'(\pi)$$

has a non trivial solution. It follows that $\lambda = n^2, n = 0, 1, \ldots$ are the eigenvalues of L with $a_n \cos nt + b_n \sin nt$ $(|a_n|^2 + |b_n|^2 \neq 0)$ the corresponding eigenvectors. Thus, $\{\frac{1}{\sqrt{2\pi}}, \frac{\cos nt}{\sqrt{\pi}}, \frac{\sin nt}{\sqrt{\pi}}\}_{n=1}^{\infty}$ is an orthonormal system of eigenvectors of L which we know forms an orthonormal basis for $L_2([-\pi, \pi])$.

6.3 The Graph Norm

A closed linear operator in a Hilbert space can be considered as a bounded operator acting between two Hilbert spaces. This is achieved by introducing an appropriate norm on its domain. We need the following preliminary results. Let $\mathcal{H}_1$ and $\mathcal{H}_2$ be Hilbert spaces. The *product space* $\mathcal{H}_1 \times \mathcal{H}_2$ is the set of all ordered pairs (x, y) with $x \in \mathcal{H}_1$ and $y \in \mathcal{H}_2$. The operations of addition $+$ and scalar multiplication $\cdot$ are defined in the usual way by

$$(x_1, y_1) + (x_2, y_2) = (x_1 + x_2, y_1 + y_2)$$
$$\alpha \cdot (x, y) = (\alpha x, \alpha y)$$

Under these operations, $\mathcal{H}_1 \times \mathcal{H}_2$ is a vector space. It is easy to see that $\mathcal{H}_1 \times \mathcal{H}_2$ is a Hilbert space with respect to the inner product

$$\langle (x_1, y_1), (x_2, y_2) \rangle = \langle x_1, x_2 \rangle + \langle y_1, y_2 \rangle \tag{3.1}$$

and corresponding norm

$$\| (x, y) \| = \langle (x, y), (x, y) \rangle^{1/2} = (\|x\|^2 + \|y\|^2)^{1/2}. \tag{3.2}$$

Given operator $A(\mathcal{H}_1 \rightarrow \mathcal{H}_2)$, the *graph* $G(A)$ of A is the subspace of $\mathcal{H}_1 \times \mathcal{H}_2$ consisting of the ordered pairs $(x, Ax), x \in \mathcal{D}(A)$. It follows readily from the definition of the norm on $\mathcal{H}_1 \times \mathcal{H}_2$ that A is a closed operator if and only if its graph $G(A)$ is a closed subspace of $\mathcal{H}_1 \times \mathcal{H}_2$.

On the domain $\mathcal{D}(A)$ we define the inner product $\langle \ , \ \rangle_A$ as follows:

$$\langle u, v \rangle_A = \langle u, v \rangle + \langle Au, Av \rangle, u, v \in \mathcal{D}(A).$$

The corresponding norm $\| \ \|_A$ is

$$\|u\|_A = \langle u, u \rangle^{\frac{1}{2}} = (\|u\|^2 + \|Au\|^2)^{\frac{1}{2}}$$

which is exactly the norm of the pair $(u, Au) \in G(A)$. We call the norm $\| \ \|_A$ the *graph norm* on $\mathcal{D}(A)$. If A is closed, then the inner product space $\mathcal{D}(A)$ with norm $\| \ \|_A$ is complete. Indeed, suppose $\|x_n - x_m\|_A \rightarrow 0$ as $n, m \rightarrow \infty$. Then $\|x_n - x_m\| \rightarrow 0$ and $\|Ax_n - Ax_m\| \rightarrow 0$ as $n, m \rightarrow \infty$. Hence there exists $x \in \mathcal{H}_1$ and $y \in \mathcal{H}_2$ such that $x_n \rightarrow x$ and $Ax_n \rightarrow y$. Since A is closed, $x \in \mathcal{D}(A)$ and $Ax = y$. Therefore

$$\|x_n - x\|_A^2 = \|x_n - x\|^2 + \|Ax_n - Ax\|^2 \rightarrow 0.$$

The operator A mapping $(\mathcal{D}_A : \| \ \|_A)$ into $\mathcal{H}_2$ is bounded since

$$\|Ax\|^2 \leq \|x\|^2 + \|Ax\|^2 = \|x\|_A^2.$$

This graph norm allows one to reduce theorems and problems for closed operators to corresponding results for bounded linear operators. For example, let the closed operator A have the property that it is invertible when considered as a (bounded) map from the Hilbert space $(\mathcal{D}(A) : \| \ \|_A)$ onto $\mathcal{H}_2$. Suppose $B(\mathcal{H}_1 \to \mathcal{H}_2)$ has domain $\mathcal{D}(B) \supseteq \mathcal{D}(A)$ and satisfies

$$\|Bx\| \leq C(\|x\|^2 + \|Ax\|^2)^{1/2} = C\|x\|_A, x \in \mathcal{D}(A).$$

Then for $|\lambda|$ sufficiently small, $A + \lambda B$ is invertible. This follows from Corollary II.8.2 applied to the bounded operators A and B on the Hilbert space $(\mathcal{D}(A) : \| \ \|_A)$.

Let L be the second order differential operator described in Section 2. Then, as we have seen, L is a closed operator. With respect to its graph norm,

$$\|f\|_L = \left(\int_0^1 |f(t)|^2 dt \right)^{1/2} + \left(\int_0^1 |f''(t)|^2 dt \right)^{1/2}, f \in \mathcal{D}(L).$$

The operator L is a bounded linear operator on the Hilbert space $(\mathcal{D}(L) : \| \ \|_L)$ with range $L_2([0, 1])$.

6.4 Adjoint Operators

In this section we extend the concept of the adjoint of a bounded operator to the adjoint of an unbounded operator.

Definition: Let $A(\mathcal{H}_1 \to \mathcal{H}_2)$ have domain $\mathcal{D}(A)$ dense in $\mathcal{H}_1$, i.e., the closure $\overline{\mathcal{D}(A)} = \mathcal{H}_1$. We say that A is *densely defined*. The *adjoint* $A^*(\mathcal{H}_2 \to \mathcal{H}_1)$ is defined as follows:

$$\mathcal{D}(A^*) = \{y \in \mathcal{H}_2| \text{ there exists } z \in \mathcal{H}_1 \text{ such that } \langle Ax, y \rangle = \langle x, z \rangle$$
$$\text{for all } x \in \mathcal{D}(A)\}$$

Obviously $0 \in \mathcal{D}(A^*)$. The vector z with the above property is unique since $\langle x, z \rangle = \langle x, u \rangle$ for all $x \in \mathcal{D}(A)$ implies $(v - z) \perp \overline{\mathcal{D}(A)} = \mathcal{H}_1$. Hence $v - z = 0$. We define $A^*y = z$. Thus

$$\langle Ax, y \rangle = \langle x, z \rangle = \langle x, A^*y \rangle, x \in \mathcal{D}(A), y \in \mathcal{D}(A^*).$$

Note that if A is bounded on $\mathcal{H}_1$, then A^* is the adjoint defined in Section II.11. It is easy to see that A^* is a linear operator.

Examples: Let $\mathcal{H} = L_2([0, 1])$. Define $A(\mathcal{H} \to \mathcal{H})$ by

$$\mathcal{D}(A) = \{f \in \mathcal{H}| f \text{ is absolutely continuous on } [0, 1], f' \in \mathcal{H},$$
$$f(0) = f(1) = 0\}, Af = f'. \tag{4.1}$$

The space $C_0^\infty([0, 1])$ of infinitely differentiable functions which vanish outside a closed subinterval of the open interval $(0, 1)$ is well known to be dense in $L_2([0, 1])$ with respect to the L_2-norm. Since $C_0^\infty([0, 1]) \subseteq \mathcal{D}(A)$, we have that A is densely defined. We shall show that $A^* = S$, where

$$\mathcal{D}(S) = \{g \in \mathcal{H} \,|\, g \text{ is absolutely continuous on } [0, 1], \, g' \in \mathcal{H}\}$$
$$Sg = -g' \tag{4.2}$$

Suppose $g \in \mathcal{D}(A^*)$ and $A^*g = h$. Then for $f \in \mathcal{D}(A)$,

$$\langle Af, g \rangle = \int_0^1 f'(t)\, \bar{g}(t)\, dt = \langle f, A^*g \rangle = \int_0^1 f(t)\, \bar{h}(t)\, dt. \tag{4.3}$$

Since $f \in \mathcal{D}(A)$, $f(0) = f(1) = 0$. Integration by parts yields

$$\int_0^1 f(t)\, \bar{h}(t)\, dt = -\int_0^1 (H(t) + C)\, f'(t)\, dt, \tag{4.4}$$

where C is an arbitray constant and $H(t) = \int_0^t \bar{h}(s)\, ds$. Then by (4.3) and (4.4),

$$0 = \int_0^1 (\bar{g}(t) + H(t) + C)\, f'(t)\, dt, \, f \in \mathcal{D}(A). \tag{4.5}$$

Let

$$f_0(t) = \int_0^t \overline{(\bar{g}(s) + H(s) + C_0)}\, ds, \tag{4.6}$$

where C_0 is chosen so that $f_0(1) = 0$. Then $f_0 \in \mathcal{D}(A)$ and if follows from (4.5) and (4.6), with f replaced by f_0, that

$$0 = \int_0^1 |\bar{g}(t) + H(t) + C_0|^2\, dt$$

Hence $g(t) = -\bar{H}(t) - \bar{C}_0 = -\int_0^t h(s)\, ds - \bar{C}_0$, which shows that g is absolutely continuous on $[0, 1]$ and $g' = -h \in \mathcal{H}$. Hence $g \in \mathcal{D}(S)$ and $\langle Af, g \rangle = \langle f, -g' \rangle = \langle f, Sg \rangle$ by (4.3). Therefore $\mathcal{D}(A^*) \subseteq \mathcal{D}(S)$ and $A^*g = Sg$, $g \in \mathcal{D}(A^*)$.

It remains to prove that $\mathcal{D}(S) \subseteq \mathcal{D}(A^*)$. Given $v \in \mathcal{D}(S)$ and $u \in \mathcal{D}(A)$,

$$\langle Au, v \rangle = \int_0^1 u'(t)\, \bar{v}(t)\, dt = -\int_0^1 u(t)\, \bar{v}'(t)\, dt = \langle u, Sv \rangle$$

Therefore $v \in \mathcal{D}(A^*)$ and $A^*v = Sv = -v'$.

Theorem 4.1 Let $A(\mathcal{H}_1 \to \mathcal{H}_2)$ be densely defined. Then its adjoint A^* is a closed operator.

Proof: Suppose $y_n \to y$, $y_n \in \mathcal{D}(A^*)$ and $A^*y_n \to v$. Then for all $x \in \mathcal{D}(A)$,

$$\langle Ax, y \rangle = \lim_n \langle Ax, y_n \rangle = \lim_n \langle x, A^*y_n \rangle = \langle x, v \rangle$$

Hence $y \in \mathcal{D}(A^*)$ and $A^*y = v$. $\qquad\square$

Since the operator S in the above example is A^*, Theorem 4.1 shows that S is a closed operator.

Theorem 4.2 *Let $A(\mathcal{H}_1 \to \mathcal{H}_2)$ be a densely defined invertible operator. Then A^* is invertible and*

$$(A^*)^{-1} = (A^{-1})^*$$

Proof: Given $v \in \mathcal{D}(A^*)$ and $y \in \mathcal{H}_2$, it follows from the boundedness of A^{-1} that

$$\langle (A^{-1})^* A^* v, y \rangle = \langle A^* v, A^{-1} y \rangle = \langle v, A A^{-1} y \rangle = \langle v, y \rangle$$

Since y is arbitrary in $\mathcal{H}_2$,

$$(A^{-1})^* A^* v = v, \, v \in \mathcal{D}(A^*). \tag{4.7}$$

Now for $x \in \mathcal{D}(A)$ and $w \in \mathcal{H}_1$, it follows from the boundedness of $(A^{-1})^*$ on $\mathcal{H}_1$ that

$$\langle Ax, (A^{-1})^* w \rangle = \langle A^{-1} Ax, w \rangle = \langle x, w \rangle.$$

Thus, by definition of A^*, $(A^{-1})^* w \in \mathcal{D}(A^*)$ and

$$A^* (A^{-1})^* w = w. \tag{4.8}$$

Since $(A^{-1})^* \in L(\mathcal{H}_1, \mathcal{H}_2)$, we have from (4.7) and (4.8) that A^* is invertible and $(A^*)^{-1} = (A^{-1})^* \in L(\mathcal{H}_1, \mathcal{H}_2)$. $\qquad\square$

Definition: A densely defined linear operator $A(\mathcal{H} \to \mathcal{H})$ is called *self adjoint* if $A^* = A$.

The following result is an immediate consequence of Theorem 4.1.

Corollary 4.3 *A self adjoint operator is closed.*

Theorem 4.4 *A densely defined invertible operator $A(\mathcal{H} \to \mathcal{H})$ is self adjoint if and only if its inverse A^{-1} is self adjoint.*

Proof: Suppose A is self adjoint. Then by Theorem 4.2 $(A^{-1})^* = (A^*)^{-1} = A^{-1}$, i.e., A^{-1} is self adjoint. Conversely, if A^{-1} is self adjoint, then $\mathcal{D}(A^*) = \text{Im}(A^*)^{-1} = \text{Im}(A^{-1})^* = \text{Im} A^{-1} = \mathcal{D}(A)$. Given $y \in \mathcal{D}(A^*) = \mathcal{D}(A)$,

$$(A^{-1})^* A^* y = (A^*)^{-1} A^* y = y.$$

and

$$(A^{-1})^* Ay = (A^{-1}) Ay = y.$$

Since $(A^{-1})^*$ is $1 - 1$, $Ay = A^* y$. $\qquad\square$

It was shown that the second order differential operator L discussed in Section 2 is invertible with inverse L^{-1}, a compact self adjoint operator.

Hence L is self adjoint by Theorem 4.4.

6.5 Sturm-Liouville Operators

As an application of Theorem IV.5.1, we consider the following Sturm-Liouville system.

A Sturm-Liouville system is a differential equation of the form

$$\frac{d}{dx}\left(p(x)\frac{dy}{dx}\right) + q(x)y = f(x) \tag{i}$$

together with boundary conditions

$$a_1 y(a) + a_2 y'(a) = 0$$
$$b_1 y(b) + b_2 y'(b) = 0, \tag{ii}$$

where a_i, b_i are real numbers with $a_1^2 + a_2^2 \neq 0$, $b_1^2 + b_2^2 \neq 0$. The system is used to describe, for example, motions of vibrating strings, elastic bars, and membranes. For more details see [CH] Chapter V.

Suppose p, p' and q are continuous real valued functions and $p(x) \neq 0$ for each $x \in [a, b]$. We now give some important properties of the eigenvalues and eigenvectors of the corresponding *Sturm-Liouville operator L*. In order to do so, we use some facts from differential equations and the theory of compact self adjoint operators.

Define the linear differential operator L as follows: The domain $\mathcal{D}(L)$ of L consists of those functions y which satisfy (ii), have first order derivatives which are absolutely continuous on $[a, b]$ and have second order derivatives in $L_2([a, b])$. Let

$$Ly = \frac{d}{dx}\left(p(x)\frac{dy}{dx}\right) + q(x)y.$$

Assume that zero is not an eigenvalue of L, i.e., L is injective. Now for $f = 0$, there exist real valued functions $y_1 \neq 0$ and $y_2 \neq 0$ such that y_1, y_2 satisfy (i), y_1'' and y_2'' are continuous, y_1 satisfies the first condition in (ii), and y_2 satisfies the second condition in (ii) ([DS2]), XIII.2.32). Let

$$W(t) = \det\begin{pmatrix} y_1(t) & y_2(t) \\ y_1'(t) & y_2'(t) \end{pmatrix},$$

which is called the *Wronskian* of (y_1, y_2). A straightforward computation verifies that $(pW)' = 0$. Thus pW is a real valued non-zero constant function. Let $(pW)^{-1} = c$. We shall show that the function g given by

$$g(x, s) = \begin{cases} cy_2(s)y_1(x) & a \leq x \leq s \\ cy_1(s)y_2(x) & s \leq x \leq b \end{cases} \tag{5.1}$$

is the Green's function corresponding to L, i.e., for each $f \in L_2([a, b])$,

$$y(x) = \int_a^b g(x, s)f(s)ds$$

lies in $\mathcal{D}(L)$ and $Ly = f$.

Let G be the integral operator with kernel function g. Since g is a continuous real valued function on $[a, b] \times [a, b]$ and $g(x, s) = g(s, x)$, G is compact and self adjoint. We now prove that

$$LGf = f, \; f \in L_2([a, b]). \tag{5.2}$$

First we show that $y = Gf$ is in $\mathcal{D}(L)$. It is clear from the definition of g that

$$y(x) = y_1(x)Y_2(x) + y_2(x)Y_1(x), \tag{5.3}$$

where

$$Y_1(x) = \int_a^x cy_1(s) f(s) \, ds,$$

$$Y_2(x) = \int_x^b cy_2(s) f(s) \, ds.$$

Differentiation of both sides of (5.3) yields

$$\begin{aligned}
y' &= y_1' Y_2 - y_1 cy_2 f + y_2 cy_1 f + y_2' Y_1 \\
&= y_1' Y_2 + y_2' Y_1 \quad \text{a.e.}
\end{aligned} \tag{5.4}$$

Actually, (5.4) holds for all x. To see this, let $h = y_1' Y_2 + y_2' Y_1$ and let

$$\tilde{y}(x) = y(a) + \int_a^x h(t) \, dt.$$

Now $\tilde{y}' = y'$ a.e. and $\tilde{y}$ and y are absolutely continuous. This follows from the absolute continuity of y_i and Y_i, $i = 1, 2$. Hence

$$(\tilde{y} - y)(x) = \int_a^x (\tilde{y} - y)'(t) \, dt = 0,$$

or $\tilde{y} = y'$. Thus (5.4) holds for all x which implies that y' is absolutely continuous. Moreover,

$$y'' \stackrel{\text{a.e.}}{=} -y_1' cy_2 f + Y_2 y_1'' + y_2' cy_1 f + Y_1 y_2'' \in L_2([a, b]).$$

Finally, since $Y_1(a) = 0$ and y_1 satisfies the first equation in (ii),

$$\begin{aligned}
a_1 y(a) + a_2 y'(a) &= (a_1 y_2(a) + a_2 y_2'(a))Y_1(a) + (a_1 y_1(a) + a_2 y_1'(a))Y_2(a) \\
&= 0.
\end{aligned}$$

Similarly,

$$b_1 y(b) + b_2 y'(b) = 0.$$

We have therefore shown that y is in $\mathcal{D}(L)$. It remains to prove that $Ly = f$. It follows from (5.4), the definition of c and properties of y_1 and y_2 that

$$(py')' = (py_1'Y_2 + py_2'Y_1)' = (py_1')'Y_2 + (py_2')'Y_1 + pc(y_1y_2' - y_2y_1')f$$
$$= -qy_1Y_2 - qy_2Y_1 + f$$
$$= -qy + f \quad \text{a.e.}$$

Thus $Ly = f$. Hence we have established (5.2). Since it is assumed that L is injective, we have from (5.2) that L is invertible with $L^{-1} = G$ which is compact and self adjoint. Each eigenvalue λ of G is *simple*, i.e., $\dim \text{Ker}(\lambda I - G) = 1$. For if u and v are linearly independent in $\text{Ker}(\lambda I - G) = \text{Ker}(\frac{1}{\lambda} - L)$, then by a basic result from differential equations, any solution w to the differential equation

$$\frac{d}{dx}\left[p(x)\frac{dy}{dx}\right] + \left(q(x) - \frac{1}{\lambda}\right)y = 0 \tag{5.5}$$

is a linear combination of u and v. Therefore w satisfies (ii). But there exists a solution to (5.5) with boundary conditions $y(a) = \alpha$, $y'(a) = \beta$, which can be chosen so that (ii) fails to hold. Hence λ is simple.

The following theorem summarizes these results.

Theorem 5.1 *Let L be the Sturm-Liouville operator with $\text{Ker } L = \{0\}$. Then L is a self adjoint invertible operator with compact inverse $L^{-1} = G$, where G is the integral operator on $L_2([a, b])$ with kernel function g defined in (5.1). In addition, $\dim \text{Ker}(\lambda I - L) \leq 1, \lambda \in \mathbb{C}$ and L has infinitely many eigenvalues.*

Corollary 5.2 *The eigenvalues of the Sturm-Liouville operator L form an infinite sequence $\{\mu_k\}$ of real numbers which can be ordered so that*

$$|\mu_1| < |\mu_2| < \cdots \to \infty.$$

If φ_j, $\|\varphi_j\| = 1$, is an eigenvector of L corresponding to μ_j, then $\varphi_1, \varphi_2, \ldots$ is an orthonormal basis for $L_2([a, b])$.

A vector $y \in L_2([a, b])$ is in the domain of L if and only if

$$\sum_{j=1}^{\infty} |\mu_j|^2 |\langle y, \varphi_j \rangle|^2 < \infty.$$

For such vectors,

$$(Ly)(t) = \sum_{j=1}^{\infty} \mu_j \langle y, \varphi_j \rangle \varphi_j(t);$$

convergence is with respect to the norm on $L_2([a, b])$. The series

$$y(t) = \sum_{j=1}^{\infty} \langle y, \varphi_j \rangle \varphi_j(t), y \in \mathcal{D}(L),$$

converges uniformly and absolutely on $[a, b]$.

Proof: If $\mathrm{Ker}\,L = (0)$, then the corollary follows from the theorem above and Theorems V.1.1, IV.8.2.

Suppose zero is an eigenvalue of L. Choose a real number r such that r is not an eigenvalue of L. The existence of such an r may be seen as follows.

Integrating by parts and making use of (ii), a straight forward computation yields

$$\langle Lu, v \rangle = \langle u, Lv \rangle, \quad u, v \in \mathcal{D}(L).$$

Hence eigenvectors corresponding to distinct eigenvalues of L are orthogonal. Thus the set of eigenvalues of L is countable, otherwise $L_2([a, b])$ would contain an uncountable orthonormal set of eigenvectors of L, which contradicts the corollary in Appendix 1. Since the set of real numbers is uncountable, the existence of r is assured.

The corollary now follows from the above results applied to $L - rI$ in place of L.

Since $rI - L$ is self-adjoint for r real and not an eigenvalue of L, it follows from Corollary 4.3 that L is a closed operator.   □

6.6 Self Adjoint Operators with Compact Inverse

Theorem 6.1 *Let $\mathcal{H}$ be an infinite dimensional Hilbert space. Suppose $A(\mathcal{H} \to \mathcal{H})$ is self adjoint and has a compact inverse. Then*

(a) *there exists an orthonormal basis $\{\varphi_1, \varphi_2, \ldots\}$ of $\mathcal{H}$ consisting of eigenvectors of A. If $\mu_1, \mu_2, \ldots$ are the corresponding eigenvalues, then each μ_j is real and $|\mu_j| \to \infty$. The numbers of repetitions of μ_j in the sequence $\mu_1, \mu_2, \ldots$ is finite and equals $\dim \mathrm{Ker}(\mu_j - A)$.*

(b) $\mathcal{D}(A) = \{x \in \mathcal{H} \mid \sum_j |\mu|^2 |\langle x, \varphi_j \rangle|^2 < \infty\}$

(c) $Ax = \sum_j \mu_j \langle x, \varphi_j \rangle \varphi_j, \ x \in \mathcal{D}(A).$

Proof: Let $K = A^{-1}$. Then K is compact and self adjoint by Theorem 4.4. Hence by Theorem IV.5.1 and observations (a)–(d) in IV.6.1, there exists an orthonormal basis $\{\varphi_1, \varphi_2, \ldots\}$ of $\mathcal{H}$ consisting of eigenvectors of K with corresponding non-zero real eigenvalues λ_1, λ_2, which converge to zero. The number of repetitions of λ_j in the sequence $\lambda_1, \lambda_2, \ldots$ is finite and equals $\dim \mathrm{Ker}\,(\lambda_j - K)$. The operators A and K have the same eigenvectors φ since $A^{-1}\varphi = K\varphi = \lambda\varphi$ if and only if $\lambda \neq 0$ and $A\varphi = \frac{1}{\lambda}\varphi$.

Statement (a) follows with $\mu_j = \frac{1}{\lambda_j}$.

Suppose that for some $x \in \mathcal{H}$, $\sum_j |\mu_j|^2 |\langle x, \varphi_g \rangle|^2 < \infty$.

Then the series $\sum_j \mu_j \langle x, \varphi_j \rangle \varphi_j$ converges. Now A^{-1} is bounded and $A^{-1}\varphi_j = \frac{1}{\lambda_j}\varphi_j = \mu_j\varphi_j$. Hence

$$x = \sum_j \langle x, \varphi_j \rangle \varphi_j = \sum_j \mu_j \langle x, \varphi_j \rangle A^{-1}\varphi_j$$

$$= A^{-1}\left(\sum_j \mu_j \langle x, \varphi_j \rangle \varphi_j\right) \in \mathcal{D}(A).$$

On the other hand, if $x \in \mathcal{D}(A)$, then since $K = A^{-1}$ is self adjoint and $K\varphi_j = \lambda_j\varphi_j$, we have

$$Ax = \sum_j \langle Ax, \varphi_j \rangle \varphi_j = \sum_j \mu_j \langle Ax, K\varphi_j \rangle \varphi_j$$

$$= \sum_j \mu_j \langle KAx, \varphi_j \rangle \varphi_j = \sum_j \mu_j \langle x, \varphi_j \rangle \varphi_j$$

which proves (c) as well as (b) since

$$\|Ax\|^2 = \sum_j \mu_j^2 |\langle x, \varphi_j \rangle|^2.$$

$\square$

Exercises VI

1. An operator $A(\mathcal{H} \to \mathcal{H})$ is called *symmetric* if $\langle Au, v \rangle = \langle u, Av \rangle$ for all u, v in $\mathcal{D}(A)$ (A need not be densely defined). Prove the following statements.

   (a) $\|A + ix\|^2 = \|x\|^2 + \|Ax\|^2$

   (b) A is a closed operator if and only if Im $(A + iI)$ is closed.

2. Prove that if $A(\mathcal{H} \to \mathcal{H})$ is densely defined, symmetric and invertible, then A is self adjoint.

3. Let $A(\mathcal{H} \to \mathcal{H})$ be densely defined and let $G(A)$ denote the graph of A with the graph norm. Define V on the Hilbert space $\mathcal{H} \times \mathcal{H}$ by $V(x, y) = (-y, x)$. Prove that $G(A^*) = (VG(A))^\perp$, the orthogonal complement of $VG(A)$ in $\mathcal{H} \times \mathcal{H}$. If, in addition, A is closed, show that $\mathcal{H} \times \mathcal{H} = VG(A) \oplus G(A^*)$.

4. Given $A(\mathcal{H}_1 \to \mathcal{H}_2)$ and $B(\mathcal{H}_1 \to \mathcal{H}_2)$, the operator $A + B$ is defined as follows: its domain

$$\mathcal{D}(A + B) = \mathcal{D}(A) \cap \mathcal{D}(B) \text{ and}$$

$$(A + B)x = Ax + Bx, x \in \mathcal{D}(A + B).$$

Suppose $A + B$ is densely defined. Prove that

(a) $(A + B)^*$ is an extension of $A^* + B^*$, i.e., $\mathcal{D}(A^* + B^*) \subseteq \mathcal{D}(A + B)^*$, $(A^* + B^*)y = (A + B)^*y$, $y \in \mathcal{D}(A^* + B^*)$.

(b) If, in addition, A is bounded on $\mathcal{H}$, then $(A + B)^* = A^* + B^*$.

(c) If B is an extension of A, then A^* is an extension of B^*.

5. Given $S(\mathcal{H} \to \mathcal{H})$ and $T(\mathcal{H} \to \mathcal{H})$, the operater ST is defined as follows:

$$\mathcal{D}(ST) = \{x \in \mathcal{D}(T) | Tx \in \mathcal{D}(S)\}$$
$$STx = S(Tx), x \in \mathcal{D}(ST).$$

Suppose S, T and ST are densely defined. Prove that $(ST)^*$ is an extension of T^*S^*. If, in addition, S is bounded on $\mathcal{H}$, prove that $(ST)^* = T^*S^*$.

6. Set $\mathcal{H} = \ell_2$ and let T be the operator defined as follows: $\mathcal{D}(T) = \mathrm{sp}\{e_k\}_1^\infty$, where $\{e_k\}_1^\infty$, is the standard basis in l_2,

$$T(\alpha_1, \alpha_2, \ldots, \alpha_m, 0, 0, \ldots) = \left(\sum_{j=1}^m j\alpha_j, \alpha_2, \alpha_3, \ldots, \alpha_m, 0, 0, \ldots\right)$$

Prove that $\mathcal{D}(T^*) = \{(0, \beta_1, \beta_2, \ldots) | (\beta_j) \in \ell_2\}$

$$T^*y = y, \quad y \in \mathcal{D}(T^*).$$

7. Let $\mathcal{H} = L_2([0, 1])$. Define operators $T_i(\mathcal{H} \to \mathcal{H})$, $i = 1, 2, 3$ as follows:

$\mathcal{D}(T_1) = \{f \in \mathcal{H} | f$ is absolutely continuous on $[0, 1]$, $f' \in L_2([a, b])\}$
$T_1 f = if'$
$\mathcal{D}(T_2) = \mathcal{D}(T_1) \cap \{f | f(0) = f(1)\}$
$T_2 f = if'$
$\mathcal{D}(T_3) = \mathcal{D}(T_1) \cap \{f | f(0) = f(1) = 0\}.$
$T_3 f = if'$

Prove that

a) $T_1^* = T_3$
b) $T_2^* = T_2$.

8. Suppose $A(\mathcal{H} \to \mathcal{H})$ is densely defined. Prove

a) $\mathrm{Ker}A^* = \mathrm{Im}A^\perp \cap \mathcal{D}(A^*)$

b) If A is closed, then $\mathrm{Ker}A = \mathrm{Im}A^{*\perp} \cap \mathcal{D}(A)$

9. Let $\mathcal{H} = L_2(\mathbb{R})$. Suppose $A(\mathcal{H} \to \mathcal{H})$ is the operator defined by $\mathcal{D}(A) = C_0^\infty(\mathbb{R})$ and $(Ax)(t) = tx(t)$. Is A a closed operator?

10. Find the eigenvalues, eigenvectors and the Green's function for the following Sturm-Liouville operators.

  (a) $Ly = -y''$; $y'(0) = 0$, $y(1) = 0$
  (b) $Ly = -y''$; $y(0) = 0$, $y(1) + y'(1) = 0$
  (c) $Ly = -y''$; $y(0) = 0$, $y'(1) = 0$
  (d) $Ly = -y'' - y$; $y'(0) = 0$, $y(1) = 0$.

11. Given the Sturm-Liouville operator

$$Ly = -y'' - y; \quad y(0) = y\left(\frac{\pi}{2}\right) = 0,$$

express the inverse of L as an integral operator on $L_2([0, \pi/2])$.

12. Consider the boundary value problem

$$\tau y = y^{(n)} + a_0(x)y^{n-1} + \cdots + a_{n-1}(x)y = f(x)$$

$$y(0) = y'(0) = \cdots = y^{n-1}(0) = 0, \qquad (*)$$

where each a_k has a continuous n^{th} order derivative and f is in $L_2[0, 1]$. Define the domain $\mathcal{D}(L)$ of L to be those y such that $y^{(n-1)}$ is absolutely continuous on $[0, 1]$, $y^{(n)}$ is in $L_2[0, 1]$, and y satisfies $(*)$. Let $Ly = \tau y$. Show that the equation $Ly = f$ has a solution if and only if the integral equation

$$\varphi(x) = f(x) + \int_0^x k(x, s)\varphi(s)ds$$

has a solution in $\mathcal{D}(L)$, where

$$k(x, s) = -\sum_{j=0}^{n-1} \frac{a_j(x)}{j!}(x - s)^j.$$

Hint: Let $\varphi = y^{(n)}$ and integrate by parts.

13. In exercise 12, let $n = 2$ and verify that the equation $Ly = f$ has a solution given by

$$y(t) = \int_a^t g(t, s)f(s)ds,$$

where

$$g(t, s) = \frac{y_1(s)y_2(t) - y_1(t)y_2(s)}{y_1(s)y_2'(s) - y_1'(s)y_2(s)}$$

with y_1, y_2 any basis for Ker L.

Chapter VII
Oscillations of an Elastic String

The aim of this chapter is to describe the motion of a vibrating string in terms of the eigenvalues and eigenvectors of an integral operator.

7.1 The Displacement Function

Let us consider an elastic string whose ends are fastened at 0 and ℓ. We assume that the segment 0ℓ is much greater than the natural length of the string. This means that the string is always under large tension.

When a vertical force F is applied to the string, each point x on the string has a displacement $u(x)$ from the segment. We shall determine $u(x)$, first when F is concentrated at a point, then when F is continuously distributed. Finally, we study the motion of the string by regarding acceleration as an inertial force.

A vertical force $F(y)$ is applied at the point y causing a displacement $u(x)$ at each point x of the string. Let $d = u(y)$ (Figure 1). The relative displacement of a point on the string from a fixed end is independent of the point. Thus

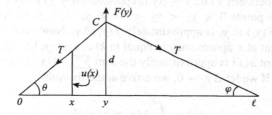

Figure 1

$$\frac{u(x)}{x} = \frac{d}{y}, 0 \leq x \leq y$$

$$\frac{u(x)}{\ell - x} = \frac{d}{\ell - y}, y \leq x \leq \ell. \tag{1.1}$$

Let us suppose that the string is subject to small displacements and that the tension T is constant. Since the sum of F and the vertical components of T is zero (the tension acts in the direction of $C0$ and $C\ell$), we have

$$F = T \sin \theta + T \sin \varphi.$$

From the assumption that the displacements are small, it follows that F is approximately

$$T \tan \theta + T \tan \varphi = T \frac{d\ell}{y(\ell - y)}.$$

Solving for d in (1.1) and substituting this value of d in the above equation yields

$$u(x) = k(x, y)F(y), \tag{1.2}$$

where

$$k(x, y) = \begin{cases} \frac{x(\ell - y)}{T\ell}, & 0 \le x \le y \\ \frac{y(\ell - x)}{T\ell}, & y \le x \le \ell. \end{cases} \tag{1.3}$$

Recall that for $T = \ell = 1$, $k(x, y)$ is the Green's function corresponding to $-\frac{d^2}{dx^2}$ (cf. (2.7) in Section VI.2.). We note that the displacement of the point x arising from the action of a force applied at the point y is equal to the displacement of the point y arising from the same force applied at the point x. In other words, if $F(x) = F(y)$, then $u(x) = u(y)$. This can be seen from (1.2) and (1.3) since $k(x, y) = k(y, x)$ and therefore

$$u(y) = k(y, x)F(x) = k(x, y)F(y) = u(x).$$

Let us consider the more general situation where the string is subjected to a continuously distributed vertical force with density distribution $f(y)$. The force acting on the segment between y and $y + \Delta y$ is approximately $f(y)\Delta y$ (Δy is "small"). Thus if we take points $0 < y_1 < y_2 < \cdots < y_n = \ell$, with $\Delta y = y_{i+1} - y_i$, then the force $F(y_i)$ at y_i is approximately $f(y_i)\Delta y$. Now each $F(y_i)$ gives rise to a displacement at x approximately equal to $k(x, y_i)f(y_i)\Delta y$ (Equation (1.2)). The displacement $u(x)$ is approximately the sum $\sum_{i=1}^{n} k(x, y_i)f(y_i)\Delta y$ of these displacements. If we let $\Delta y \to 0$, we arrive at the formula

$$u(x) = \int_0^\ell k(x, y)f(y)\, dy. \tag{1.4}$$

7.2 Basic Harmonic Oscillations

An important type of oscillation is a *simple harmonic oscillation*. This is one in which each point x on the string, when released from its initial position, moves vertically so that its displacement $u(x, t)$ is a sinusoidal function of time t with frequency ω which is independent of the point, i.e.,

$$u(x, t) = u(x) \sin \omega t, \quad \omega > 0. \tag{2.1}$$

Let us start with the assumption that there is no external force acting on the string. If $\rho^2(x)$ is the density of the string, then by Newton's second law of motion (force is the product of mass and acceleration), the inertial force acting on the section of the string between y and $y + \Delta y$ is

$$-\rho^2(y)\Delta y u_{tt}(y, t). \tag{2.2}$$

It follows from (2.1) and (2.2) that the density distribution $f(y)$ is given by

$$f(y) = \rho^2(y)\omega^2 u(y) \sin \omega t. \tag{2.3}$$

Equations (1.4) and (1.3) yield

$$u(x, t) = u(x) \sin \omega t$$

$$= \left(\omega^2 \int_0^\ell \rho^2(y) k(x, y) u(y) \, dy\right) \sin \omega t.$$

If the density ρ^2 is assumed to be constant, then

$$u(x) = \rho^2 \omega^2 \int_0^\ell k(x, y) u(y) \, dy \tag{2.4}$$

or

$$(I - \rho^2 \omega^2 K)u = 0, \tag{2.5}$$

where K is the integral operator on $L_2[0, \ell]$ with kernel function k defined in (1.3) of Section 1.

In view of (2.5), our immediate goal is to determine the eigenvalues and eigenvectors of K. It follows from the discussion in VI.2 that φ is an eigenvector of K with eigenvalues λ if and only if $\lambda \neq 0$ and

$$\varphi'' + \frac{1}{\lambda T}\varphi = 0$$

$$\varphi(0) = \varphi(\ell) = 0. \tag{2.6}$$

The general form of a solution to (2.6) is given by

$$\varphi(x) = a \cos \frac{x}{\sqrt{T\lambda}} + b \sin \frac{x}{\sqrt{T\lambda}}, \quad \lambda > 0. \tag{2.7}$$

$$\varphi(x) = a e^{\frac{x}{\sqrt{-T\lambda}}} + b e^{\frac{-x}{\sqrt{-T\lambda}}}, \quad \lambda < 0. \tag{2.8}$$

If (2.8) holds and $\varphi(0) = \varphi(\ell) = 0$, then $\varphi = 0$. Thus every eigenvalue of K is positive and therefore K is positive, as noted in the beginning of Section IV.9. In order to determine these positive eigenvalues, we note that (2.6) and (2.7) imply

$$\lambda_n = \frac{1}{T}\left(\frac{\ell}{n\pi}\right)^2, \quad n = 1, 2, \ldots \tag{2.9}$$

are the eigenvalues of K. The eigenvectors corresponding to λ_n are of the form

$$b_n \sin \frac{n\pi x}{\ell}, \quad b_n \neq 0. \tag{2.10}$$

In particular,

$$\varphi_n(x) = \sqrt{\frac{2}{\ell}} \sin \frac{n\pi x}{\ell} \tag{2.11}$$

is an eigenvector of K corresponding to λ_n, and $\|\varphi_n\| = 1$. Moreover, $\varphi_1, \varphi_2, \ldots$ is an orthonormal basis for $L_2([0, \ell])$ as seen in Section VI.2. It follows from (2.5), (2.9) and (2.10) that for $n = 1, 2, \ldots,$

$$u_n(x, t) = u_n(x) \sin \omega_n t = b_n \sin \frac{n\pi x}{\ell} \sin \frac{\sqrt{T}n\pi}{\rho\ell} t$$

is the general form of the displacement of the point x.

Motions for which $u(x, t)$ has the above form are called *basic harmonic motions*. The numbers $\omega_n = \frac{\sqrt{T}n\pi}{\rho\ell}$ are called *characteristic frequencies*.

7.3 Harmonic Oscillations with an External Force

Suppose there is an external simple harmonic force which is continuously distributed and acts vertically on the string during motion. To be specific, let the density distribution of the force be

$$h(y) \sin \omega t. \tag{3.1}$$

Let us assume that the density ρ^2 of the string is constant. Then by (2.3) in Section 2 and (3.1), the density distribution $f(y)$ of the sum of the external force and intertial force is given by

$$f(y) = [h(y) + \rho^2\omega^2 u(y)] \sin \omega t,$$

and from (1.4) in Section 1 we get

$$u(x, t) = u(x) \sin \omega t = \left[g(x) + \rho^2\omega^2 \int_0^\ell k(x, y)u(y) \, dy \right] \sin \omega t,$$

where

$$g(x) = \int_0^\ell k(x, y)h(y) \, dy.$$

Hence

$$(I - \rho^2\omega^2 K)u = g. \tag{3.2}$$

If $(\rho\omega)^{-2}$ is not an eigenvalue of K, i.e., $\omega \neq \sqrt{T}\frac{n\pi}{\rho\ell}$, $n = 1, 2, \ldots$, then by Theorem IV.8.1, Equation (3.2) has the unique solution

$$u(x) = g(x) + \sum_{n=1}^{\infty} \frac{\lambda_n}{(\rho\omega)^{-2} - \lambda_n} \langle g, \varphi_n \rangle \varphi_n(x),$$

where λ_n and φ_n are defined in (2.9) and (2.11) of Section 2. The series converges uniformly and absolutely on $[0, \ell]$. This follows from the uniform and absolute convergence of $\sum_{n=1}^{\infty} \lambda_n \langle g, \varphi_n \rangle \varphi_n(x)$, which is guaranteed by Theorem V.1.1 and the boundedness of the sequence $\{\frac{1}{(\rho\omega)^{-2} - \lambda_n}\}$.

Thus

$$u(x) = g(x) + \sum_{n=1}^{\infty} a_n \frac{\sin n\pi x}{\ell}, \tag{3.3}$$

where

$$a_n = \frac{2}{\ell} \frac{\lambda_n}{(\rho\omega)^{-2} - \lambda_n} \int_0^\ell g(y) \sin \frac{n\pi y}{\ell}\, dy,$$

$$\lambda_n = \frac{1}{T} \left(\frac{\ell}{n\pi} \right)^2, \quad \omega \neq \sqrt{T}\frac{n\pi}{\rho\ell}, \quad n = 1, 2, \ldots.$$

We have seen that harmonic oscillations which arise from an external force, with a density distribution of the form $h(y) \sin \omega t$, is decomposable into basic harmonic oscillations.

Since the operator K with kernel function k, defined in (1.3) of Section 1 was shown to be positive, it follows from Theorem V.3.1 that

$$k(x, y) = \sum_{n=1}^{\infty} \lambda_n \varphi_n(x) \bar{\varphi}_n(y)$$

$$= \frac{2\ell}{T\pi^2} \sum_{n=1}^{\infty} \frac{1}{n^2} \sin \frac{n\pi x}{\ell} \sin \frac{n\pi y}{\ell}. \tag{3.4}$$

The series converges uniformly and absolutely on $[0, \ell] \times [0, \ell]$.

For a thorough and rigorous treatment of the equations describing large vibrations of strings, as well as additional references, we refer the reader to [A].

Chapter VIII
Operational Calculus with Applications

The spectral theory which was studied in the preceding chapters provides a means for the development of a theory of functions of a compact self adjoint operator. We now present this theory with applications to a variety of problems in differential equations.

8.1 Functions of a Compact Self Adjoint Operator

Suppose $A \in \mathcal{L}(\mathcal{H})$ is a compact self adjoint operator with a basic system of eigenvectors $\{\varphi_n\}$ and eigenvalues $\{\lambda_n\}$. Let P_0 be the orthogonal projection from $\mathcal{H}$ onto $\operatorname{Ker} A$. Then for each $x \in \mathcal{H}$,

$$x = P_0 x + \sum_k \langle x, \varphi_k \rangle \varphi_k$$

$$Ax = \sum_k \lambda_k \langle x, \varphi_k \rangle \varphi_k$$

$$A^2 x = \sum_k \lambda_k^2 \langle x, \varphi_k \rangle \varphi_k$$

$$\vdots$$

$$A^n x = \sum_k \lambda_k^n \langle x, \varphi_k \rangle \varphi_k. \tag{1.1}$$

For any polynomial $p(z) = \sum_{k=0}^{n} a_k z^k$, it is natural to define $p(A) = \sum_{k=0}^{n} a_k A^k$. Therefore it follows from (1.1) that

$$p(A)x = p(0) P_0 x + \sum_k p(\lambda_k) \langle x, \varphi_k \rangle \varphi_k. \tag{1.2}$$

Let $\sigma(A)$ be the subset of $\mathbb{C}$ consisting of zero and the eigenvalues of A. Let f be a complex valued function which is bounded on $\sigma(A)$. Guided by (1.2), we define the operator $f(A)$ on $\mathcal{H}$ by

$$f(A)x = f(0) P_0 x + \sum_k f(\lambda_k) \langle x, \varphi_k \rangle \varphi_k. \tag{1.3}$$

The operator $f(A)$ does not depend on the choice of the eigenvectors $\{\varphi_k\}$. Indeed, let $\mu_1, \mu_2, \ldots$ be the distinct non-zero eigenvalues of A and let P_n be the orthogonal projection onto $\mathrm{Ker}(\mu_n I - A)$. Suppose $\varphi_j, \varphi_{j+1}, \ldots, \varphi_k$ is a basis for $\mathrm{Ker}(\mu_n I - A)$. Then

$$P_n x = \sum_{i=j}^{k} \langle x, \varphi_i \rangle \varphi_i$$

and it follows from (1.3) that

$$f(A)x = f(0) P_0 x + \sum_n f(\mu_n) P_n x. \tag{1.4}$$

Next we show that

$$\|f(A)\| = \sup\{|f(\lambda)| : \lambda \text{ an eigenvalue of } A\}. \tag{1.5}$$

Since $P_0 x$ is orthogonal to each φ_k, it follows from (1.3) that

$$\|f(A)x\|^2 = |f(0)|^2 \|P_0 x\|^2 + \sum_k |f(\lambda_k)|^2 |\langle x, \varphi_k \rangle|^2$$

$$\leq \|x\|^2 \sup_k \{f(\lambda_k)|^2 + |f(0)|^2 \|P_0\|^2\} \tag{1.6}$$

and

$$\|f(A)\| \geq |f(A)\varphi_n| = |f(\lambda_n)|,$$
$$\|f(A)x\| \geq |f(0)|\|x\|, \ x \in \mathrm{Ker}\, A. \tag{1.7}$$

It is easy to see that (1.6) and (1.7) imply (1.5).

The operator $f(A) - f(0) P_0 x$ is compact and self adjoint if and only if $\{f(\lambda_k)\}$ is a sequence of real numbers which is either finite or converges to zero.

At first glance the following result does not seem to be important. However, as we shall see, it does have some interesting applications.

Theorem 1.1 *Suppose F and g are complex valued functions which are bounded on $\sigma(A)$. Then*

(i) $(\alpha f + \beta g)(A) = \alpha f(A) + \beta g(A), \alpha, \beta \in \mathbb{C}$.

(ii) $(f.g)(A) = f(A)g(A)$.

(iii) *If $\{f_n\}$ is a sequence of bounded complex valued functions which converges uniformly on $\sigma(A)$ to the bounded function f, then*

$$\|f_n(A) - f(A)\| \to 0.$$

Proof: Let $\{\varphi_n\}$, $\{\lambda_n\}$ be a basic system of eigenvectors and eigenvalues of A. It is clear from the definition of $f(A)$ and $g(A)$ that (i) holds. (ii). By definition,

$$f(A)g(A)x = f(0)P_0g(A)x + \sum_k f(\lambda_k)\langle g(A)x, \varphi_k\rangle\varphi_k. \tag{1.8}$$

Since $\varphi_k \perp \mathrm{Ker}\, A$, $1 \leq k$,

$$\langle g(A)x, \varphi_k\rangle = \left\langle g(0)P_0x + \sum_j g(\lambda_j)\langle x, \varphi_j\rangle\varphi_j, \varphi_k \right\rangle$$

$$= g(\lambda_k)\langle x, \varphi_k\rangle \tag{1.9}$$

and

$$P_0g(A)x = P_0\left[g(0)P_0x + \sum_k g(\lambda_k)\langle x, \varphi_k\rangle\varphi_k \right]$$

$$= g(0)P_0x. \tag{1.10}$$

Property (ii) follows immediately from (1.8), (1.9) and (1.10). (iii). By (i) and Equation (1.5),

$$\|f_n(A) - f(A)\| = \|(f_n - f)(A)\|$$

$$\leq \sup_{\lambda\in\sigma(A)} |f_n(\lambda) - f(\lambda)| \to 0 \text{ as } n \to \infty.$$

$\square$

Corollary 1.2 *Given a compact positive operator $A \in \mathcal{L}(\mathcal{H})$, there exists a unique compact positive operator K such that $K^2 = A$. In fact,*

$$Kx = \sum_n \sqrt{\lambda_n}\langle x, \varphi_n\rangle\varphi_n, \tag{1.11}$$

where $\{\lambda_n\}$, $\{\varphi_n\}$ is a basic system of eigenvalues and eigenvectors of A.

Proof: If K is given by (1.11), then K is compact positive and

$$K^2x = \sum_n \lambda_n\langle x, \varphi_n\rangle\varphi_n = Ax.$$

Suppose B is a compact positive operator such that $B^2 = A$. To prove that $B = K$, it suffices to show that for all $\lambda > 0$,

$$\mathrm{Ker}(\lambda I - B) = \mathrm{Ker}(\lambda I - K). \tag{1.12}$$

But (1.12) is clear since

$$(\lambda I + K)(\lambda I - K) = \lambda^2 I - A = (\lambda I + B)(\lambda I - B)$$

and $-\lambda$ is not an eigenvalue of either B or K.

$\square$

Theorem 1.3 *Suppose $A \in \mathcal{L}(\mathcal{H})$ is compact and self adjoint. If $f(z) = \sum_{k=0}^{\infty} a_k z^k$ is analytic on a closed disc containing $\sigma(A)$, then $f(A) = \sum_{k=0}^{\infty} a_k A^k$. The series converges in norm.*

Proof: Apply Theorem 1.1 to $f_n(z) = \sum_{k=0}^{n} a_k z^k$ and f. $\square$

In particular, if $f(z) = e^z = \sum_{k=0}^{\infty} \frac{z^k}{k!}$, then

$$e^A \equiv f(A) = \sum_{k=0}^{\infty} \frac{A^k}{k!}.$$

The series converges in norm and

$$e^A x = f(A)x = P_0 x + \sum_k e^{\lambda_k} \langle x, \varphi_k \rangle \varphi_k = \sum_k \frac{A^k x}{k!}. \qquad (1.13)$$

Moreover, we have from Equation (1.5) that

$$\|e^A\| = \sup\{|e^\lambda| : \lambda \text{ an eigenvalue of } A\}.$$

A simple application of Theorem 1.1 yields the following results.
Suppose f and $g = \frac{1}{f}$ are bounded on $\sigma(A)$. Then

$$f(A)g(A) = g(A)f(A) = (g.f)(A) = I.$$

Thus $f(A)$ is invertible and $f(A)^{-1} = g(A)$. In particular, if $\mu \notin \sigma(A)$, let $f(\lambda) = \mu - \lambda$. Then f and $g(\lambda) = \frac{1}{\mu - \lambda}$ are bounded on $\sigma(A)$. Hence $\mu I - A$ is invertible and

$$(\mu I - A)^{-1} x = g(A)x = \frac{1}{\mu} P_0 x + \sum_k \frac{1}{\mu - \lambda_k} \langle x, \varphi_k \rangle \varphi_k.$$

By Equation (1.5),

$$\|(\mu I - A)^{-1}\| = \|g(A)\| = \sup \left\{ \frac{1}{|\mu - \lambda|} : \lambda \text{ an eigenvalue of } A \right\}. \qquad (1.14)$$

Examples: 1. Let us consider the operator $B = \beta I + A$. Now $B = f(A)$, where $f(\lambda) = \beta + \lambda$, and φ is an eigenvector of A with eigenvalue λ if and only if φ is an eigenvector of B with eigenvalue $\beta + \lambda$. Therefore, if $\{\eta_1, \eta_2, \ldots\}$ is the set of eigenvalues of B which does not contain β, and Q_j is the orthogonal projection onto Ker $(\eta_j I - B)$, then it follows from Equation (1.4) that $Q_i Q_j = 0, i \neq j$ and

$$Bx = \beta P_0 x + \sum_j \eta_j Q_j x, \qquad (1.15)$$

where P_0 is the orthogonal projection onto Ker $A = \text{Ker}(\beta I - B)$. Furthermore,

$$\|B\| = \sup\{|\beta + \lambda| : \lambda \text{ an eigenvalue of } A\}$$
$$= \sup\{|\mu| : \mu \text{ an eigenvalue of } B\}. \tag{1.16}$$

If $\{\eta_k\}$ is an infinite sequence, then it converges to β since the sequence of eigenvectors of A converges to zero.

While the series in (1.15) converges pointwise to B, it does not converge in norm if $\beta \neq 0$ (and there are infinitely many η_j). Indeed, given any integer N, choose n so large that $n > N$ and $|\eta_n| > \frac{|\beta|}{2}$. Then for $\varphi \in \text{Ker}(\eta_n - B)$, $\|\varphi\| = 1$,

$$\left\| B - \beta P_0 - \sum_{j=1}^{N} \eta_j Q_j \right\| \geq \left\| \left(B - \beta P_0 - \sum_{j=1}^{N} \eta_j Q_j \right) \varphi \right\|$$
$$= \|B\varphi\| = |\eta_n| > \frac{|\beta|}{2}.$$

2. Let K be an integral operator on $\mathcal{H} = L_2([a, b])$ with kernel function $K \in L_2([a, b] \times [a, b])$. Suppose $\{\lambda_n\}$, $\{\varphi_n\}$ is a basic system of eigenvalues and eigenvectors of K. If f is a complex valued function defined at zero and $\sum_n |(\lambda_n)|^2 < \infty$, then for all v in $\mathcal{H}$,

$$(f(K)v)(t) = f(0)(P_0 v)(t) + \int_a^b \widetilde{k}(t, s)\, v(s)\, ds \quad \text{a.e.} \tag{1.17}$$

where

$$\widetilde{k}(t, s) = \sum_n f(\lambda_n)\varphi_n(t)\, \overline{\varphi_n}(s); \tag{1.18}$$

convergence is with respect to the norm on $L_2([a, b] \times [a, b])$ We shall prove the following more general result

Let $\{\psi_n\}$ be an orthonormal basis for $\mathcal{H} = L_2([a, b])$. If $A \in \mathcal{L}(\mathcal{H})$ and $\sum_{i,j} |\langle A\psi_i, \psi_j\rangle|^2 < \infty$, then A is the integral operatoir with kernel function

$$k(t, s) = \sum_{i,j} \langle A\psi_i, \psi_j\rangle \psi_i(t)\, \overline{\psi_j}(s); \tag{1.19}$$

convergence is with respect to the norm on $L_2([a, b] \times [a, b])$.

To prove this, put $C_{ij} = \langle A\psi_i, \psi_j\rangle$ and $\phi_{ij}(t, s) = \psi_i(t)\,\overline{\psi_j}(s)$. Since $\{\phi_{ij}\}$ is orthonormal in $L_2([a, b] \times [a, b])$, $\sum_{i,j} \|C_{ij}\phi_{ij}\|^2 = \sum_{i,j} C_{ij}^2 < \infty$. Hence $\sum_{i,j} C_{ij}\phi_{ij}$ converges to some $k \in L_2([a, b] \times [a, b])$. Let K be the integral operator with kernel function k and let K_n be the integral operator with kernel function $k_n = \sum_{i,j=1}^{n} C_{ij}\phi_{ij}$. Since $\|K_n - K\| \leq \|k_n - k\| \to 0$, a straightforward computation verifies that

$$\langle K\psi_i, \psi_j\rangle = \lim_{n \to \infty} \langle K_n\psi_i, \psi_j\rangle = C_{ij} = \langle A\psi_i, \psi_j\rangle.$$

Thus for all u, v in $\mathcal{H} = \overline{sp}\{\psi_j\}$,

$$\langle Ku, v \rangle = \langle Au, v \rangle,$$

which implies that $A = K$.

Returning to our original integral operator K, we choose $A = f(K) - f(0) P_0$. Our choice of an orthonormal basis for $\mathcal{H}$ is $\{\varphi_n\}$ together with any orthonormal basis for Ker $K = sp\{\varphi_n\}^{\perp}$. Since $Au = 0$, $u \in$ Ker K, and $\langle A\varphi_i, \varphi_j \rangle = f(\lambda_i)\delta_{ij}$, Equations (1.17) and (1.18) follow from (1.19).

3. As a special case, let K be the integral operator on $L_2([-\pi, \pi])$ with kernel function $k(t, s) = h(t - s)$, where h is a continuous complex valued function of period 2π. It was shown in IV.1 that

$$\lambda_n = \int_{-\pi}^{\pi} h(s)e^{-ins} \, ds \quad n = 0, \pm 1, \ldots$$

are the eigenvalues of K with corresponding eigenvectors

$$\varphi_n(t) = \frac{1}{\sqrt{2\pi}} \, e^{int}.$$

Since $\{\varphi_n\}$ is an orthonormal basis for $L_2([-\pi, \pi])$, $\{\varphi_n\}$, $\{\lambda_n\}$ is a basic system of eigenvectors of K and Ker $K = 0$. Hence for any positive integer p, equations (1.17) and (1.18) applied to $f(\lambda) = \lambda^p$ give

$$(K^p v)(t) = \int_a^b \tilde{k}_p(t - s)v(s) \, ds \text{ a.e.,}$$

where

$$\tilde{k}_p(t) = \frac{1}{2\pi} \sum_{n=-\infty}^{\infty} \lambda_n^p \, e^{int} \quad \text{(convergence in } L_2([-\pi, \pi])).$$

8.2 Differential Equations in Hilbert Space

As we shall see in Section 3 and 4, certain integro-differential equations and systems of differential equations give rise to a differential equation of the form

$$y'(t) = Ay(t), \quad -\infty < t < \infty$$
$$y(0) = x \in \mathcal{H}, \tag{2.1}$$

where A is in $\mathcal{L}(\mathcal{H})$, y maps $(-\infty, \infty)$ into $\mathcal{H}$ and

$$y'(t) = \lim_{h \to 0} \frac{y(t + h) - y(t)}{h} \in \mathcal{H}.$$

Though the main results in this section are valid for an arbitrary operator $A \in \mathcal{L}(\mathcal{H})$, we shall assume that A is compact and self adjoint. This enables us to express the solution to (2.1) as a series involving a basic system of eigenvectors of A.

If we consider the one dimensional case that A is a number, then the solution to (2.1) is $y(t) = e^{tA} x$. It turns out that this is also the solution when A is an operator.

Theorem 2.1 *Let $A \in \mathcal{L}(\mathcal{H})$ be compact and self adjoint with a basic system of eigencectors $\{\varphi_k\}$ and eigenvalues $\{\lambda_k\}$. For each $x \in \mathcal{H}$, there exists a unique solution to*

$$y'(t) = Ay(t), \quad -\infty < t < \infty \tag{2.2}$$
$$y(0) = x.$$

The solution is

$$y(t) = e^{tA} x = P_0 x + \sum_k e^{\lambda_k t} (x, \varphi_k) \varphi_k,$$

where P_0 is the orthogonal projection from $\mathcal{H}$ onto Ker A.

Proof: From Theorem 1.1 and Equation 1.13 in Section 1, if follows that if $y(t) = e^{tA} x$, then

$$\left\| \frac{y(t+h) - y(t)}{h} - Ay(t) \right\| = \left\| e^{tA} \left[\left(\frac{e^{hA} - I}{h} - A \right) x \right] \right\|$$

$$\leq \|x\| \|e^{tA}\| \|A\| \sum_{k=1}^{\infty} \frac{|h|^k \|A\|^k}{k!}$$

$$= \|x\| \|e^{tA}\| \|A\| (e^{|h| \|A\|} - 1) \to 0 \text{ as } h \to 0.$$

Thus $y'(t)$ exists and
$$y'(t) = Ay(t).$$

Also,
$$y(0) = e^0 x = x.$$

Suppose $v(t)$ is also a solution to (2.1). Let

$$w(t) = v(t) - y(t)$$
$$w_k(t) = (w(t), \varphi_k).$$

Then

$$\frac{d}{dt} w_k(t) = \lim_{h \to 0} \left(\frac{w(t+h) - w(t)}{h}, \varphi_k \right) = (Aw(t), \varphi_k)$$
$$= (w(t), A\varphi_k) = \lambda_k w_k(t)$$

and

$$w_k(0) = (w(0), \varphi_k) = (0, \varphi_k) = 0.$$

Hence $w_k(t) = 0$. Therefore

$$w'(t) = Aw(t) = \sum_k \lambda_k w_k(t)\varphi_k = 0.$$

But this implies that $y(t) - v(t) = w(t) = 0$. Indeed, for any $z \in \mathcal{H}$,

$$\frac{d}{dt}\langle w(t), z\rangle = \langle w'(t), z\rangle = 0.$$

Hence the complex valued function $\langle w(t), z\rangle$ is a constant function. Since $\langle w(0), z\rangle = \langle 0, z\rangle = 0$, $\langle w(t), z\rangle = 0$ for all t and all z. Thus $w(t) = 0$ for all t. $\square$

8.3 Infinite Systems of Differential Equations

Let us consider the infinite system of differential equations

$$y_1'(t) = a_{11}y_1(t) + a_{12}y_2(t) + \ldots$$
$$y_2'(t) = a_{21}y_1(t) + a_{22}y_2(t) + \ldots$$
$$\vdots \qquad \vdots \qquad \vdots$$

with boundary conditions

$$\{y_i(0)\} = \{x_i\} \in \ell_2.$$

Suppose $\sum_{i,j=1}^{\infty} |a_{ij}|^2 < \infty$ and $\overline{a_{ij}} = a_{ji}$.

In order to solve this system, we define $A \in \mathcal{L}(\ell_2)$ by

$$A(\alpha_1, \alpha_2, \ldots) = (\beta_1, \beta_2, \ldots),$$

where $\beta_i = \sum_{j=1}^{\infty} a_{ij}\alpha_j$. Since A is compact and self adjoint (II.16, Example 2, II.11.1 Example 6), it has a basic system of eigenvectors $\{\varphi_k\}$ and eigenvalues $\{\lambda_k\}$. By Theorem 2.1, if $x = (x_1, x_2, \ldots) \in \ell_2$, then

$$y(t) = P_0 x + \sum_k e^{\lambda_k t}\langle x, \varphi_k\rangle\varphi_k \in \ell_2$$

satisfies

$$y'(t) = Ay(t), \quad -\infty < t < \infty$$
$$y(0) = x.$$

Let $y_j(t)$ be the jth component of $y(t)$. Then $y_j(0) = x_j$ and

$$\left| \frac{y_j(t+h) - y_j(t)}{h} - \sum_k a_{jk}y_k(t) \right| \leq \left\| \frac{y(t+h) - y(t)}{h} - Ay(t) \right\| \to 0$$

as $h \to 0$. Hence $\{y_j(t)\}_{j=1}^{\infty}$ is the desired solution.

8.4 Integro-Differential Equations

Let us consider the integro-differential equation

$$\frac{\partial \varphi}{\partial t}(t, x) = \int_a^b k(x, s)\varphi(t, s)ds, \quad -\infty < t < \infty, \quad a \le x \le b \quad (4.1)$$

with boundary condition

$$\varphi(0, x) = g(x), \quad (4.2)$$

where k is continuous on $[a, b] \times [a, b]$, $\overline{k(x, s)} = k(s, x)$ and g is a given function in $L_2([a, b])$. Equations of this type arise, for instance, in the theory of energy transport.

In order to use Theorem 2.1 to solve the equation, we express (4.1) in the form $y'(t) = Ky(t)$ as follows:

Define $y : (-\infty, \infty) \to L_2([a, b])$ by

$$y(t) = \varphi(t,).$$

That is to say, for each t, $y(t)$ is the function $\varphi(t, x)$ considered as a function of x. Let K be the integral operator on $L_2([a, b])$ with kernel function k. Then (4.1) can be written in the form

$$y'(t) = Ky(t).$$
$$y(0) = g. \quad (4.3)$$

We must be careful here. Equation (4.3) means that

$$\frac{\varphi(t + h,) - \varphi(t,)}{h} = \frac{y(t + h) - y(t)}{h}$$

converges, with respect to the norm on $L_2([a, b])$, to the function $K\varphi(t,)$ as $h \to 0$. Let us temporarily disregard all questions concerning convergence. We know that the solution to (4.3) is

$$\varphi(t, x) = (e^{tK}g)(x) = (P_0 g)(x) + \sum_{k=1}^{\infty} e^{\lambda_k t}\langle g, \varphi_k \rangle \varphi_k(x)$$

$$= g(x) + \sum_{k=1}^{\infty} (e^{\lambda_k t} - 1)\langle g, \varphi_k \rangle \varphi_k(x), \quad (4.4)$$

where $\{\varphi_k\}$, $\{\lambda_k\}$ is a basic system of eigenvectors and eigenvalues of K.

To prove that the series converges and that φ is the desired solution, we recall from the Hilbert-Schmidt theorem that

$$(Kg)(x) = \sum_k \lambda_k \langle g, \varphi_k \rangle \varphi_k(x) \quad (4.5)$$

converges uniformly and absolutely on $[a, b]$. Thus, for any finite interval J, the series in (4.4) converges uniformly on $J \times [a, b]$ since the mean value theorem implies

$$|e^{\lambda_k t} - 1| \leq |\lambda_k| C, \quad 1 \leq k,$$

for some constant C and all $t \in J$. If we differentiate the series in (4.4) termwise with respect to t, we get the series $\sum_{k=1}^{\infty} \lambda_k e^{\lambda_k t} \langle g, \varphi_k \rangle \varphi_k(x)$ with converges uniformly on $J \times [a, b]$. Hence $\frac{\partial \varphi}{\partial t}(t, x)$ exists and

$$\frac{\partial \varphi}{\partial t}(t, x) = \sum_{k=1}^{\infty} \lambda_k e^{\lambda_k t} \langle g, \varphi_k \rangle \varphi_k(x),$$

which is easily seen to equal the right side of (4.1). Obviously, $\varphi(0, x) = g(x)$.

Exercises VIII

1. For the following operators A find e^{iA}, $\cos A$ and $\sin A$.

    (a) $A : \ell_2 \to \ell_2 : A\xi = (\sqrt{2\pi}\xi_2, \sqrt{2\pi}\xi_1, 2\pi\xi_3, 2\pi\xi_4, \ldots)$
    (b) $A : L_2[-\pi, \pi] \to L_2[-\pi.\pi]$;

    $$(A\varphi)(t) = t \int_{-\pi}^{\pi} \varphi(x) \cos x \, dx + \cos t \int_{-\pi}^{\pi} \varphi(x) x \, dx$$

    (c) $A : L_2[-\pi, \pi] \to L_2[-\pi.\pi]$; $(A\varphi)(t) = \int_{-\pi}^{\pi} (t - s)^2 \varphi(s) \, ds$
    (d) $A : L_2[-\pi, \pi] \to L_2[-\pi.\pi]$; $(A\varphi)(t) = \int_{-\pi}^{\pi} \cos \frac{1}{2}(t - s)\varphi(s) \, ds$.

2. For each of the operators A in problem 1, check that $e^{iA} = \cos A + i \sin A$ and $\cos^2 A + \sin^2 A = I$. Which is true, $\sin 2A = 2 \cos A \sin A$ or $\sin 2A = 2 \sin A \cos A$?

3. Solve the following integro-differential equations with the general boundary condition

    $$\varphi(0, x) = f(x) \in L_2[-\pi, \pi].$$

    (a) $\frac{\partial}{\partial t}\varphi(t, x) = \cos x \int_{-\pi}^{\pi} \varphi(t, \xi) \cos \xi \, d\xi$
    (b) $\frac{\partial}{\partial t}\varphi(t, x) = \int_{-\pi}^{\pi} (x - s)^2 \varphi(t, s) \, ds$
    (c) $\frac{\partial}{\partial t}\varphi(t, x) = \int_{-\pi}^{\pi} \varphi(t, s) \cos \frac{1}{2}(x - s) \, ds$.

4. Let A be a positive compact operator on a Hilbert space. Check that the differential equation $\frac{d^2 y}{dt^2} + Ay = 0$ with boundary conditions $y(0) = y_0$, $y'(0) = x_0$ has the solution

    $$y(t) = \cos t\sqrt{A} y_0 + \frac{\sin t\sqrt{A}}{\sqrt{A}} x_0,$$

where $\frac{\sin t\sqrt{A}}{\sqrt{A}}$ is the operator $f(A)$ with $f(\lambda) = \frac{\sin t\sqrt{\lambda}}{\sqrt{\lambda}}$ for $\lambda > 0$ and $f(0) = t$.

5. Solve the following equations with boundary conditions

$$\varphi(0, x) = 0$$
$$\varphi'(0, x) = f(x) \in L_2[-\pi, \pi]$$

(a) $\frac{\partial^2}{\partial t^2}\varphi(t, x) = \cos x \int_{-\pi}^{\pi} \cos \xi \, \varphi(t, \xi) \, d\xi$

(b) $\frac{\partial^2}{\partial t^2}\varphi(t, x) = \int_{-\pi}^{\pi} \cos(x - s) \, \varphi(t, s) \, ds$.

6. Find a formula for a solution to the nonhomogeneous equation

$$\begin{cases} \frac{\partial}{\partial t}\varphi(t, x) = \int_a^b k(x, s)\varphi(t, s) \, ds + h(t, x) \\ \varphi(0, x) = g(x) \in L_2[a, b], \end{cases}$$

where k is continuous on $[a, b] \times [a, b]$ and $k(t, s) = \overline{k(s, t)}$. Hint: Consider a solution to $y' - ay = f$, where a is a number, and generalize as in VI.4.

7. Solve the following nonhomogeneous equations for $\varphi(0, x) = g(x) \in L_2[-\pi, \pi]$.

(a) $\frac{\partial}{\partial t}\varphi(t, x) = \cos x \int_{-\pi}^{\pi} \cos \xi \, \varphi(t, \xi) \, d\xi + \cos x$

(b) $\frac{\partial}{\partial t}\varphi(t, x) = \int_{-\pi}^{\pi} (x - s)^2 \varphi(t, s) \, ds + tx$

(c) $\frac{\partial}{\partial t}\varphi(t, x) = \int_{-\pi}^{\pi} \cos\frac{1}{2}(x - s)\varphi(t, s) \, ds + x$.

Chapter IX
Solving Linear Equations by Iterative Methods

In this chapter we return to the problem of solving the equation $x - Ax = y$. One of the themes of this book is to devise algorithms which generate a sequence of vectors that converges to a solution. This chapter further develops this theme. The main results rely on the spectral theorem.

9.1 The Main Theorem

Let $\mathcal{H}$ be a Hilbert space. Given $A \in \mathcal{L}(\mathcal{H})$ and $y \in \mathcal{H}$, choose any $x_0 \in \mathcal{H}$ and define

$$x_{n+1} = y + Ax_n, \quad n = 0, 1, \ldots. \tag{1.1}$$

If $\{x_n\}$ converges, say to z, then it follows from the continuity of A that $z = y + Az$. Thus z is a solution to the equation

$$x - Ax = y. \tag{1.2}$$

The important question is, when does $\{x_n\}$ converge? Now

$$\begin{aligned}
x_1 &= y + Ax_0 \\
x_2 &= y + Ax_1 = y + Ay + A^2 x_0 \\
&\;\;\vdots \qquad \vdots \\
x_n &= y + Ay + A^2 y + \cdots + A^{n-1} y + A^n x_0.
\end{aligned}$$

Thus, if $\|A\| < 1$, then $A^n x_0 \to 0$ and

$$\lim_{n \to \infty} x_n = \sum_{k=0}^{\infty} A^k y = (I - A)^{-1} y.$$

With no restrictions on the norm of A, a sequence can be constructed which converges to a solution of (1.2). We shall now prove this result under the assumption that A is compact.

Theorem 1.1 *Suppose $K \in \mathcal{L}(\mathcal{H})$ is compact. Assume that for a given $y \in \mathcal{H}$ and a given complex number λ, the equation*

$$\lambda x + Kx = y \tag{1.3}$$

has a solution. For any $x_0 \in \mathcal{H}$, define

$$x_{n+1} = (I - \alpha B^* B)x_n + \alpha B^* y, \quad n = 0, 1, \ldots,$$

where

$$B = \lambda I + K \text{ and } 0 < \alpha < 2\|B\|^{-2}.$$

Then the sequence $\{x_n\}$ converges to a solution of (1.3).
If $\mathcal{H}$ is finite dimensional or if $\lambda \neq 0$ and

$$0 < \alpha < \min\left(2\|B\|^{-2}, 2|\lambda|^{-2}\right),$$

then the convergence is uniform as x_0 ranges over any bounded set.

The proof depends on the following results which are interesting in thier own right.

9.2 Preliminaries for the Proof

Lemma 2.1 *Given $A \in \mathcal{L}(\mathcal{H})$ and $y \in \mathcal{H}$, suppose that the equation $x - Ax = y$ has a solution. Given $x_0 \in \mathcal{H}$, define*

$$x_{n+1} = y + Ax_n, \quad n = 0, 1, \ldots.$$

The sequence $\{x_n\}$ converges (to a solution of $x - Ax = y$) for every $x_0 \in \mathcal{H}$ if and only if $\{A^n x\}$ converges for every $x \in \mathcal{H}$.
The sequence $\{x_n\}$ converges uniformly as x_0 ranges over any bounded set if and only if $\{A^n\}$ converges in norm.

Proof: Let $B_n = \sum_{k=0}^n A^k$. Then

$$x_{n+1} = B_n y + A^{n+1} x_0. \tag{2.1}$$

Suppose $\{x_n\}$ converges for every $x_0 \in \mathcal{H}$. Then for $x_0 = 0$. $\{B_n y\}$ converges and (2.1) implies $\{A^n x_0\}$ converges. Let $Lx = \lim_{n\to\infty} A^n x$, $x \in \mathcal{H}$. If $\{x_n\}$ converges uniformly as x_0 ranges over the l-ball S of $\mathcal{H}$, then it follows from (2.1) that $A^n x \to Lx$ uniformly on S. Thus $\|A_n - L\| \to 0$.

Assume $\{A^n x\}$ converges for every x. The convergence of $\{x_n\}$ will follow from (2.1) once we prove that $\{B^n y\}$ converges. By hypothesis, there exists a $w \in \mathcal{H}$ such that $B_n y = B_n(w - Aw) = w - A^{n+1}w$. Since $\{A^n w\}$ converges, so does $\{B_n y\}$. If $\{A^n\}$ converges in norm, then it follows readily from (2.1) that $\{x_n\}$ converges. uniformly as x_0 ranges over any bounded set. $\qquad \square$

Theorem 2.2 *Let $A = \beta I + K$, where K is a compact self adjoint operator on $\mathcal{H}$ and β is a real number. The sequence $\{A^n x\}$ converges for every $x \in \mathcal{H}$ if and only if $\|A\| \leq 1$ and $\mathrm{Ker}(I + A) = (0)$. In this case, the operator P defined on $\mathcal{H}$ by $Px = \lim_{n\to\infty} A^n x$ is the orthogonal projection onto $\mathrm{Ker}(I - A)$.*
If, in addition, $|\beta| < 1$ or $\mathcal{H}$ is finite dimensional, then $\|A^n - P\| \to 0$.

Proof: Suppose $\{A^n x\}$ converges for each $x \in \mathcal{H}$. If λ is an eigenvalue of A with eigenvector φ, then $\lambda^n \varphi = A^n \varphi$ converges and λ is real since A is self adjoint. Hence $-1 < \lambda \leq 1$. Thus $\mathrm{Ker}(I + A) = (0)$ and, by VIII.1, Equation (1.16),

$$\|A\| = \sup \{|\lambda| : \lambda \text{ an eigenvalue of } A\} \leq 1.$$

Suppose $\|A\| \leq 1$ and $\mathrm{Ker}(I + A) = (0)$. Given $x \in \mathcal{H}$, there exists a unique $u \in \mathrm{Ker}(I - A)$ and a unique $v \in \mathrm{Ker}(I - A)^{\perp}$ such that $x = u + v$. Let P be the orthogonal projection onto $\mathrm{Ker}(I - A)$. Then for $n = 1, 2, \ldots,$

$$A^n x = u + A^n v$$

and

$$\|A^n x - Px\| = \|A^n v\| \leq \|A_M\|^n \|v\| \leq \|A_M\|^n \|x\|, \tag{2.2}$$

where A_M is the restriction of A to $M = \mathrm{Ker}(I - A)^{\perp}$.

The idea of the proof is to show that $\|A_M\| < 1$ if $|\beta| < 1$ and $A^n v \to 0$ for all $v \in M$. When this is accomplished, the theorem is a direct consequence of (2.2). First we note that $AM \subset M$. Indeed, if $v \in M$ and $u \in \mathrm{Ker}(I - A)$, then

$$\langle Av, u \rangle = \langle v, Au \rangle = \langle v, u \rangle = 0.$$

Thus A_M is in $\mathcal{L}(M)$ and $A_M = I_M + K_M$, where I_M and K_M are the restrictions of I and K to M, respectively. Hence it follows from VIII.1, Equation (1.16), that $\|A_M\|$ is $|\beta|$ or $|\mu|$, where μ is eigenvalue of A_M with largest modulus. Now $|\mu| \leq \|A\| \leq 1$ and from the definition of M and the assumption $\mathrm{Ker}(I+A) = (0)$, it follow that $u \neq \pm 1$. Thus $|\mu| < 1$ and therefore $\|A_M\| < 1$ if $|\beta| < 1$. Hence we have from (2.2) that $\|A^n - P\| \to 0$.

If $\mathcal{H}$ is finite dimensional, then A is compact and self adjoint. Thus we may apply the above result to $0I + A$ to conclude that $\|A^n - P\| \to 0$.

Suppose $|\beta| > 1$. Then $-K = \beta I - A$ is invertible since $\|A\| \leq 1$. But then $\mathcal{H}$ is finite dimensional due to the compactness of K. Hence $\|A^n - P\| \to 0$.

Finally, assume $|\beta| = 1$. If $M = (0)$, then $A = I$ and the theorem is trivial. Suppose $M \neq (0)$. There exists an orthonormal set $\{\varphi_k\} \subset M$ and eigenvalues $\{\mu_k\}$ of A_M such that for each $v \in M$ and $n = 1, 2, \ldots,$

$$\|A^n v\|^2 = \left\| \sum_k \mu_k^n \langle v, \varphi_k \rangle \right\|^2 = \sum_k |\mu_k|^{2n} |\langle v, \varphi_k \rangle|^2. \tag{2.3}$$

Let $\varepsilon > 0$ be given. There exists an integer N such that

$$\sum_{k > N} |\langle v, \varphi_k \rangle|^2 < \varepsilon/2. \tag{2.4}$$

We have shown above that the modulus of any eigenvalue of A_M is less than 1. Hence, for all n sufficiently large,

$$\sum_{k=1}^{N} |\mu_k|^{2n} |\langle v, \varphi_k \rangle|^2 < \varepsilon/2. \tag{2.5}$$

Equations (2.2)–(2.5) imply $A^n x \to P x$.   □

9.3 Proof of the Main Theorem

We are now prepared to prove Theorem 1.1. Let $A = I - \alpha B^* B$. The first step in the proof is to verify that $\{A^n x\}$ converges for each x by showing that A satisfies the conditions in Theorem 2.2. It is easy to see that

$$A = I - \alpha B^* B = I - \alpha(\bar{\lambda} I + K^*)(\lambda I + K) = (1 - \alpha |\lambda|^2) I + K_1, \tag{3.1}$$

where K_1 is compact and self adjoint. Given $x \in \mathcal{H}$, $\|x\| = 1$,

$$\langle Ax, x \rangle = \langle x - \alpha B^* B x, x \rangle = 1 - \alpha \|Bx\|^2.$$

If therefore follows from the conditions on α that

$$-1 < \langle Ax, x \rangle \le 1.$$

Hence $\|A\| \le 1$ by Theorem IV.4.1, and

$$\|(I + A)x\| \ge \langle (I + A)x, x \rangle = 1 + \langle Ax, x \rangle > 0,$$

which implies that $\text{Ker}(I + A) = 0$. Moreover, the equation

$$x - Ax = \alpha B^* y \tag{3.2}$$

has a solution. Indeed, by hypothesis, there exists a $w \in \mathcal{H}$ such that $Bw = y$ and

$$w - Aw = \alpha B^* B w = \alpha B^* y.$$

Therefore, we have from Theorem 2.2 and Lemma 2.1 that $\{x_n\}$ converges to a solution v of Equation (3.2). The convergence is uniform as x_0 ranges over any bounded set if $\mathcal{H}$ is finite dimensional or α satisfies the requirements of the theorem. Now

$$\alpha B^* B v = v - Av = \alpha B^* y,$$

hence

$$Bv - y \in \text{Ker } B^* = \text{Im } B^\perp.$$

Since $Bw = y$,

$$Bv - y = B(v - w) \in \text{Im } B \cap \text{Im } B^\perp = (0)$$

or

$$\lambda v + Kv = Bv = y.$$

The iterative method in Theorem 1.1 does not have any specific relation to the spectral theorem, although the proof is strongly based on the theorem.

Even if Equation (1.3) in Theorem 1.1 does not have a solution, nevertheless $\{x_n\}$ converges for an appropriate α when $\lambda \neq 0$.

Lemma 3.1* *If $B = \mu I + K$, where $K \in \mathcal{L}(\mathcal{H})$ is compact and $\mu \neq 0$, then for any $y \in \mathcal{H}$, the equation $B^*Bx = B^*y$ has a solution.*

Proof: It is easy to see that $A = BB^* - |\mu|^2 I$ is compact and self adjoint. Let $\{\lambda_j\}$, $\{\varphi_j\}$ be a basic system of eigenvaluoes and eigenvectors of A. For each $y \in \mathcal{H}$, there exists a $v \in \operatorname{Ker} A$ such that

$$y = v + \sum_j \langle y, \varphi_j\rangle \varphi_j.$$

Nothing that $B^*\varphi_j = 0$ if $j \in N = \{j : |\mu|^2 + \lambda_j = 0\}$, a straightforward computation verifies that $B^*Bx = B^*y$, where

$$x = \frac{1}{|\mu|^2} B^*v + \sum_{j \notin N} \frac{1}{|\mu|^2 + \lambda_j} \langle y, \varphi_j\rangle B^*\varphi_j.$$

□

Theorem 3.2 *Let $\{x_n\}$ be the sequence defined in Theorem 1.1. If $\lambda \neq 0$, then for $0 < \alpha < 2\|B\|^{-2}$, the sequence $\{x_n\}$ converges. For $0 < \alpha < \min(2\|B\|^{-2}, 2|\lambda|^{-2})$, $\{x_n\}$ converges uniformly as x_0 ranges over any bounded set.*

If $v = \lim x_n$, then $(\lambda I + K)v = Py$, where P is the orthogonal projection from $\mathcal{H}$ onto $\operatorname{Im}(\lambda I + K)$. Thus

$$\inf_{x \in \mathcal{H}} \|\lambda x + Kx - y\| = \|\lambda v + Kv - y\|.$$

Proof: The proof of the main theorem relied on the fact that $\alpha B^*Bx = x - Ax = \alpha B^*y$ has a solution. But this we know from Lemma 3.1. An inspection of the proof of the main Theorem 1.1 shows that $\{x_n\}$ converges to some v and $Bv - y \in \operatorname{Im} B^\perp$. Thus

$$(\lambda I + K)v = Bv = Py.$$

The last statement of the theorem follows from the properties of P. □

---

* Lemma 3.1 still holds if, more generally, we only require that Im B be closed. Indeed,

$$\mathcal{H} = \operatorname{Im} B \oplus \operatorname{Im} B^\perp = \operatorname{Im} B \oplus \operatorname{Ker} B^*.$$

Thus $y = Bw + z$ for some $w \in \mathcal{H}$ and $z \in \operatorname{Ker} B^*$. Hence $B^*y = B^*Bw$.

Theorem XV.4.2 shows that every operator of the form $\mu I + K$, where K is compact and $\mu \neq 0$ has a closed range.

9.4 Application to Integral Equations

The iteration procedure in Theorem 1.1 can be readily applied to compact integral operators as follows:

Let K be an integral operator with kernel function $k \in L_2([a, b] \times [a, b])$. Given $g \in L_2([a, b])$, suppose that for some complex number λ, the equation

$$\lambda f(t) + \int_a^b k(t, s) f(s) \, ds = g(t) \quad \text{a.e.} \tag{4.1}$$

has a solution in $L_2([a, b])$.

Now for any $f \in L_2([a, b])$, we have from Theorem II.11.2 that

$$(K^* K f) = \int_a^b \overline{k(s, t)} \left\{ \int_a^b k(s, \xi) f(\xi) \, d\xi \right\} ds$$

$$= \int_a^b \int_a^b \overline{k(s, t)} k(s, \xi) f(\xi) \, d\xi \, ds.$$

Given $f_0 \in L_2([a, b])$, define, for $n = 0, 1, \ldots,$

$$f_{n+1}(t) = (1 - \alpha |\lambda|^2) f_n(t) - \alpha \left[\int_a^b (\bar{\lambda} k(t, s) + \lambda \overline{k(s, t)}) f_n(s) \, ds \right.$$

$$+ \int_a^b \int_a^b \overline{k(s, t)} k(s, \xi) f_n(\xi) \, d\xi \, ds \Big]$$

$$+ \alpha \bar{\lambda} g(t) + \alpha \int_a^b \overline{k(s, t)} g(s) \, ds,$$

where

$$0 < \alpha < 2 \| \lambda I + K \|^{-2}.$$

Since K is compact, it follows from Theorem 1.1 that $\{f_n\}$ converges in $L_2([a, b])$ to a solution of (4.1).

Let $L f_0 = \lim_{n \to \infty} f_n$, i.e., $\| L f_0 - f_n \| \to 0$. Suppose $\lambda \neq 0$. If α satisfies (4.2) and $\alpha < 2|\lambda|^{-2}$, then for each $\varepsilon > 0$, there exists an integer N such that for all f_0 in the r-ball of $\mathcal{L}_2([a, b])$,

$$\| L f_0 - f_n \| < \varepsilon, \quad n \geq N.$$

The results in this chapter are based on the works of Krasnosel'skil [Kra] and Fel'dman [F].

Chapter X
Further Developments of the Spectral Theorem

In this chapter we continue to develop the spectral theory. Simultaneous diagonalization, normal operators, and unitary equivalence are the main topics.

10.1 Simultaneous Diagonalization

Suppose A and B in $\mathcal{L}(\mathcal{H})$ can be diagonalized simultaneously, i.e., there exists an orthonormal system of eigenvectors $\{\varphi_n\}$ of both A and B with corresponding eigenvalues $\{\lambda_n\}$ and $\{\mu_n\}$, respectively, such that for all x,

$$Ax = \sum_n \lambda_n \langle x, \varphi_n \rangle \varphi_n, \quad Bx = \sum_n \mu_n \langle x, \varphi_n \rangle \varphi_n.$$

Clearly, $AB = BA$.

An important result is that the converse holds if A and B are compact and self adjoint.

Theorem 1.1 *Suppose A and B are compact self adjoint operators in $\mathcal{L}(\mathcal{H})$. If $AB = BA$, then A and B are diagonalizable simultaneously.*

Proof: Let $\{\lambda_n\}$ be the basic system of eigenvalues of A. It was shown in the proof of the Spectral Theorem IV.5.1, that if $\varphi_1 \in \mathrm{Ker}\,(\lambda_1 - A)$, $\|\varphi_1\| = 1$ and

$$\varphi_n \in \{\varphi_1, \ldots, \varphi_{n-1}\}^{\perp} \cap \mathrm{Ker}\,(\lambda_n - A), \quad n = 2, 3, \ldots, \tag{1.1}$$

then $\{\varphi_n\}$ is a basic system of eigenvectors of A. It follows from the assumption $AB = BA$ that $\mathrm{Ker}\,(\lambda_n - A)$ is invariant under B. Thus, since B is compact and self adjoint, we could choose φ_n so that φ_n satisfies (1.1) and φ_n is an eigenvector of B, say $B\varphi_n = \mu_n \varphi_n$. Given $x \in \mathcal{H}$, there exists a $u \in \mathrm{Ker}\,A$ such that

$$x = u + \sum_k \langle x, \varphi_k \rangle \varphi_k. \tag{1.2}$$

If $B\,\mathrm{Ker}\,A = (0)$, then

$$Bx = \sum \mu_k \langle x, \varphi_k \rangle \varphi_k. \tag{1.3}$$

Suppose $B \operatorname{Ker} A \neq \{0\}$. Since $\operatorname{Ker} A$ is invariant under B, there exists an orthonormal system $\{\psi_j\} \subset \operatorname{Ker} A$ and a sequence of eigenvalues $\{\eta_j\}$ of B such that for all $u \in \operatorname{Ker} A$,

$$Bu = \sum_j \eta_j \langle u, \psi_j \rangle \psi_j. \tag{1.4}$$

It is clear from (1.2), (1.3) and (1.4) that either $\{\varphi_n\}$ or $\{\psi_1, \varphi_1, \psi_2, \varphi_2, \ldots\}$ diagonalizes both A and B.

The theorem above shows that there exists an orthonormal basis $\{\varphi_j\}$ for $M = \overline{\operatorname{Im} A}$ such that the matrices corresponding to the restrictions A_M, B_M and $\{\varphi_j\}$ are diagonal matrices. $\square$

10.2 Compact Normal Operators

Analogous to the complex numbers, every $A \in \mathcal{L}(\mathcal{H})$ can be expressed in the form

$$A = A_1 + i A_2,$$

where

$$A_1 = \frac{1}{2} (A + A^*), \qquad A_2 = \frac{1}{2i} (A - A^*).$$

The operators A_1 and A_2 are self adjoint and are called the *real* and *imaginary* parts of A, respectively.

It A is compact, then A_1 and A_2 are compact self adjoint. Therefore, A_1 and A_2 are each diagonalizable, but, in general, A is not. However, if $A_1 A_2 = A_2 A_1$, then by Theorem 1.1, A_1 and A_2 are simultaneously diagonalizable and therefore A is diagonalizable.

A simple computation verifies that $A_1 A_2 = A_2 A_1$ if and only if $A A^* = A^* A$. This leads us to the following definition.

Definition: An operator $A \in \mathcal{L}(\mathcal{H})$ is *normal* if $A A^* = A^* A$.

Obviously, every self adjoint operator is normal.

The following result is a generalization of the spectral theorem.

Theorem 2.1 *An operator $A \in \mathcal{L}(\mathcal{H})$ is compact and normal if and only if it has a basic system of eigenvectors and eigenvalues, where the sequence of eigenvalues of A converges to zero if it is infinite.*

Proof: Suppose A is compact and normal. Let A_1 and A_2 be the real and imaginary parts of A. Then by the above discussion, there exists an orthonormal system $\{\varphi_n\}$ and corresponding real eigenvalues $\{\lambda_n\}$ and $\{\mu_n\}$ of A_1 and A_2, respectively such that for each $x \in \mathcal{H}$,

$$Ax = A_1 x + i A_2 x = \sum_n (\lambda_n + i \mu_n) \langle x, \varphi_n \rangle \varphi_n \tag{2.1}$$

and $\lambda_n + i \mu_n \to 0$ if $\{\varphi_n\}$ is an infinite set.

Suppose (2.1) holds and $\eta_k = \lambda_k + i\mu_k \to 0$. Then

$$A_n x = \sum_{k=1}^{n} \eta_k \langle x, \varphi_k \rangle \varphi_k$$

is compact and $\|A_n - A\| \to 0$. Hence A is compact. Moreover,

$$A^* x = \sum_k \bar{\eta}_k \langle x, \varphi_k \rangle \varphi_k$$

since

$$\langle Au, x \rangle = \sum_k \eta_k \langle u, \varphi_k \rangle \langle \varphi_k, x \rangle = \left\langle u, \sum_k \bar{\eta}_k \langle x, \varphi_k \rangle \varphi_k \right\rangle.$$

Therefore,

$$A^* A x = \sum_k |\eta_k|^2 \langle x, \varphi_k \rangle \varphi_k = A A^* x$$

for all $x \in \mathcal{H}$, which means that A is normal. $\qquad\qquad\square$

Application. Let K be an integral operator with kernel function $k \in L_2([a, b] \times [a, b])$. For $f \in L_2([a, b])$,

$$(K^* K f)(t) = \int_a^b \overline{k(s, t)} \left\{ \int_a^b k(s, \xi) f(\xi) \, d\xi \right\} ds$$

$$= \int_a^b \int_a^b \overline{k(s, t)} k(s, \xi) f(\xi) \, d\xi ds$$

and

$$(K K^* f)(t) = \int_a^b \int_a^b k(t, s) \overline{k(\xi, s)} f(\xi) \, d\xi ds.$$

Hence a necessary and sufficient condition that K be normal is that for almost every $(t, \xi) \in [a, b] \times [a, b]$,

$$\int_a^b \overline{k(s, t)} k(s, \xi) \, ds = \int_a^b k(t, s) \overline{k(\xi, s)} \, ds. \qquad (2.2)$$

Since K is compact, we know from Theorem 2.1 that if (2.2) holds, then K has a basic system of eigenvectors $\{\varphi_n\}$ and eigenvalues $\{\lambda_n\}$. Thus

$$\lim_{n \to \infty} \int_a^b \left| \int_a^b k(t, s) f(s) \, ds - \sum_{j=1}^{n} \alpha_j \varphi_j(t) \right|^2 dt = 0,$$

where

$$\alpha_j = \lambda_j \int_a^b f(s) \overline{\varphi_j}(s) \, ds.$$

Moreover, $\lambda_j \to 0$ if K is of infinite rank.

In view of Theorem 2.1, an operational calculus for compact normal operators can be developed in the same way as for compact self adjoint operators.

10.3 Unitary Operators

It was shown in Theorem I.18.1 that any two complex Hilbert spaces $\mathcal{H}_1$, $\mathcal{H}_2$ of the same dimension are equivalent in the sense that there exists a $U \in \mathcal{L}(\mathcal{H}_1, \mathcal{H}_2)$ such that U is surjective and $\|Ux\| = \|x\|$ for all $x \in \mathcal{H}_1$.

An operator U which has the property that $\|Ux\| = \|x\|$ for all x is called an *isometry*.

Isometries have some very special properties as seen in the next theorem.

Theorem 3.1 *Let U be in $\mathcal{L}(\mathcal{H}_1, \mathcal{H}_2)$. The following statements are equivalent.*

   (i) *U is an isometry.*

   (ii) *$U^*U = I_{\mathcal{H}_1}$, the identity operator on $\mathcal{H}_1$.*

   (iii) *$\langle Ux, Uy \rangle = \langle x, y \rangle$ for all x and $y \in \mathcal{H}_1$.*

If U is an isometry and $\mathrm{Im}\, U = \mathcal{H}_2$, then $UU^* = I_{\mathcal{H}_2}$, the identity operator on $\mathcal{H}_2$.

Proof: (i) $\Rightarrow$ (ii). For all $x \in \mathcal{H}_1$,

$$\langle (U^*U - I_{\mathcal{H}_1})x, x \rangle = \|Ux\|^2 - \|x\|^2 = 0.$$

Thus $U^*U - I_{\mathcal{H}_1} = 0$ by Corollary IV.4.2.

   (ii) $\Rightarrow$ (iii).

$$\langle Ux, Uy \rangle = \langle U^*Ux, y \rangle = \langle x, y \rangle.$$

   (iii) $\Rightarrow$ (i).

$$\|Ux\|^2 = \langle Ux, Ux \rangle = \langle x, y \rangle = \|x\|^2.$$

Suppose U is an isometry from $\mathcal{H}_1$ onto $\mathcal{H}_2$. Given x and y in $\mathcal{H}_2$, there exists a $z \in \mathcal{H}_1$ such that $Uz = y$. Hence it follows from (ii) that

$$\langle UU^*x, y \rangle = \langle UU^*x, Uz \rangle = \langle U^*x, U^*Uz \rangle$$
$$= \langle U^*x, z \rangle = \langle x, Uz \rangle = \langle x, y \rangle.$$

Since y is arbitrary, $UU^*x = x$.   □

Definition: A linear isometry which maps $\mathcal{H}$ *onto* $\mathcal{H}$ is called a *unitary operator*.

The above theorem shows that an operator $U \in \mathcal{L}(\mathcal{H})$ is unitary if and only if it is invertible and $U^{-1} = U^*$.

Thus a unitary operator is normal.

We have seen in (20.4) of Section II.20 that of $U \in \mathcal{L}(\mathcal{H})$ is unitary, then the spectrum

$$\sigma(A) \subset \{\lambda \mid |\lambda| = 1\}. \tag{3.1}$$

Examples: 1. Let $a(t)$ be a Lebesgue measurable function on $[a, b]$ such that $|a(t)| = 1$ a.e. The operator U defined on $L_2([a, b])$ by

$$(Uf)(t) = a(t)f(t)$$

is unitary.

2. Given a real number r, let $f_r(t) = f(t + r)$ for all $t \in (-\infty, \infty)$. The operator defined on $L_2[(-\infty, \infty)]$ by $Uf = f_r$ is unitary.

3. The forward shift operator on ℓ_2 is an isometry which is not unitary since it is not surjective.

Unitary operators can be used to identify compact self adjoint operators with each other provided they have certain properties in common. The following theorem is an illustration of this assertion.

Theorem 3.2 *Let A and B be compact self adjoint operators on a separable Hilbert space $\mathcal{H}$. There exists a unitary operator U on $\mathcal{H}$ such that $U^*BU = A$ if and only if*

$$\dim \mathrm{Ker}\,(\lambda I - A) = \dim \mathrm{Ker}\,(\lambda I - B)$$

for all $\lambda \in \mathbb{C}$.

Proof: Suppose $\dim \mathrm{Ker}(\lambda - A) = \dim \mathrm{Ker}(\lambda - B)$ for all $\lambda \in \mathbb{C}$. It follows from IV.6.1(d) that A and B have the same basic system of eigenvalues $\{\lambda_n\}$. Let $\{\varphi_n\}$ and $\{\psi_n\}$ be basic systems of eigenvectors of A and B, respectively, corresponding to $\{\lambda_n\}$. Since $\dim \mathrm{Ker}\,A = \dim \mathrm{Ker}\,B$, there exists, by Theorem I.18.1, a linear isometry U_0 which maps $\mathrm{Ker}\,A$ onto $\mathrm{Ker}\,B$. Define $U \in \mathcal{L}(\mathcal{H})$ as follows. Given $x \in \mathcal{H}$, there exists a unique $u \in \mathrm{Ker}\,A$ such that

$$x = u + \sum_k \langle x, \varphi_k \rangle \varphi_k.$$

Let

$$Ux = U_0 u + \sum_k \langle x, \varphi_k \rangle \psi_k.$$

It is easy to see that U is a linear isometry. Moreover, U is surjective. For given $z \in \mathcal{H}$, there exists a unique $w \in \mathrm{Ker}\,B$ such that

$$z = w + \sum_k \langle z, \psi_k \rangle \psi_k.$$

Clearly

$$U\left(U_0^{-1}w + \sum_k \langle z, \psi_k\rangle\varphi_k\right) = z.$$

Given $x \in \mathcal{H}$,

$$BUx = \sum_k \langle x, \varphi_k\rangle B\psi_k = \sum_k \lambda_k\langle x, \varphi_k\rangle\psi_k$$

$$= U\sum_k \lambda_k\langle x, \varphi_k\rangle\varphi_k = UAx.$$

Since $U^*U = I$, we have $U^*BU = A$.

Conversely, if $U^*BU = A$, where U is unitary, then

$$U^*(\lambda I - B)U = \lambda I - A, \quad \lambda \in \mathbb{C},$$

which implies

$$\dim\operatorname{Ker}(\lambda I - B) = \dim\operatorname{Ker}(\lambda I - A).$$

The operators A and B in the above theorem are said to be *unitarily equivalent*.

□

10.4 Singular Values

Given Hilbert spaces $\mathcal{H}_1$ and $\mathcal{H}_2$, let A be a compact operator in $\mathcal{L}(\mathcal{H}_1, \mathcal{H}_2)$. The operator A^*A is compact and positive since

$$\langle A^*Ax, x\rangle = \|Ax\|^2 \geq 0. \tag{4.1}$$

Therefore it has a basic system of eigenvalues

$$\lambda_1(A^*A) \geq \lambda_2(A^*A) \geq \cdots > 0 \tag{4.2}$$

and eigenvectors $\varphi_1, \varphi_2, \ldots$

In case the sequence in (4.1) is finite, we extend it to an infinite sequence by adding on zeros. We define the *singular values* of A, written $s_j(A)$, by

$$s_j(A) = \sqrt{\lambda_j(A^*A)}, \; j = 1, 2, \ldots$$

note that $s_j(A) \geq s_{j+1}(A)$ and $s_j(A) \to 0$. The importance of singular values is seen in the following characterization of compact operators.

Theorem 4.1 *If $A \in \mathcal{L}(\mathcal{H}_1, \mathcal{H}_2)$ is compact, then there exist orthonormal systems $\{\varphi_j\} \subseteq \mathcal{H}_1$ and $\{\psi_j\} \subseteq \mathcal{H}_2$ such that for all $x \in \mathcal{H}_1$,*

$$Ax = \sum_{j=1}^{\nu(A)} s_j(A)\langle x, \varphi_j\rangle\psi_j, \tag{4.3}$$

where $\nu(A)$ is the number of non-zero singular values of A, counted according to multiplicities. Also,

$$A^*y = \sum_{j=1}^{\nu(A)} s_j(A)\langle y, \psi_j\rangle\varphi_j, \tag{4.4}$$

Proof: If $A = 0$, the theorem is trivial. Suppose $A \neq 0$. Let $\{\lambda_j(A^*A)\}$ and $\{\varphi_j\}$ be the basic system of eigenvalues and eigenvectors of A^*A. For brevity we write $\lambda_j = \lambda_j(A^*A)$ and $s_j = \sqrt{\lambda_j}$, $1 \leq j \leq \nu(A)$. Take $\psi_j = \frac{1}{s_j}A\varphi_j$, $1 \leq j \leq \nu(A)$. The sequence ψ_j is orthonormal since

$$\langle \psi_k, \psi_j\rangle = \frac{1}{s_k s_j}\langle A^*A\varphi_k, \varphi_j\rangle = \frac{s_k^2}{s_k s_j}s_{kj}.$$

Given $x \in \mathcal{H}_1$, there exists, by IV.6.1(a), a $u \in \text{Ker } A^*A$ such that

$$x = u + \sum_{j=1}^{\nu(A)}\langle x, \varphi_j\rangle\varphi_j. \tag{4.5}$$

The equality $\text{Ker } A^*A = \text{Ker } A$ follows from (4.1).

Hence formula (4.5) gives

$$Ax = \sum_{j=1}^{\nu(A)}\langle x, \varphi_j\rangle A\varphi_j = \sum_{j=1}^{\nu(A)} s_j\langle x, \varphi_j\rangle\psi_j.$$

Thus for all $x \in \mathcal{H}_1$ and $y \in \mathcal{H}_2$,

$$\langle Ax, y\rangle = \sum_{j=1}^{\nu(A)} s_j\langle x, \varphi_j\rangle\langle\psi_j, y\rangle = \left\langle x, \sum_{j=1}^{\nu(A)} s_j\langle y, \psi_j\rangle\varphi_j\right\rangle$$

which implies equality (4.4). □

Theorem 4.2 *Every compact operator in $\mathcal{L}(\mathcal{H}_1, \mathcal{H}_2)$ is the limit in norm of a sequence of operators of finite rank.*

Proof: Given a compact operator $A \in \mathcal{L}(\mathcal{H}_1, \mathcal{H}_2)$, it has a representation given by formula (4.3) in Theorem 4.1. For each positive integer n, Define

$$A_n x = \sum_{j=1}^{n} s_j(A)\langle x, \varphi_j\rangle\psi_j.$$

It is easy to see that $\|A_n - A\| \to 0$. □

Corollary 4.3 *Let $\mathcal{H}_1$ and $\mathcal{H}_2$ be Hilbert spaces and let A be a compact operator in $\mathcal{L}(\mathcal{H}_1, \mathcal{H}_2)$. Then the singular values*

$$s_j(A^*) = s_j(A).$$

Proof: The operator A^* is compact by Theorem II.16.2. Hence $s_j(A^*) = \lambda_j^{1/2}(AA^*)$. Let A and A^* have the representations appearing in (4.3) and (4.4). Then for $y \in \mathcal{H}_2$,

$$AA^*y = \sum_{j=1}^{\nu(A)} s_j(A)\langle A^*y, \varphi_j\rangle \psi_j$$

and

$$\langle A^*y, \varphi_j\rangle = \left(\sum_{k=1}^{\nu(a)} s_k(A)\langle y, \psi_k\rangle \langle \varphi_k, \varphi_j\rangle = s_j(A)\langle y, \psi_j\rangle\right),$$

Hence

$$AA^*y = \sum_{j=1}^{\nu(A)} s_j^2(A)\langle y, \psi_j\rangle \psi_j$$

Thus

$$AA^*\psi_k = s_k^2(A)\psi_k = \lambda_k(AA^*)\psi_k, \quad 1 \le k \le \nu(A)$$

and

$$s_k(A^*) = \lambda_k^{1/2}(AA^*) = s_k(A), \quad k = 1, 2, \ldots$$

$\square$

Example: Let $K : L_2([0, 1]) \to L_2([0, 1])$ be the integral operator

$$(Kf)(t) = 2i \int_0^t f(s)\, ds.$$

By taking

$$k(t, s) = \begin{cases} 1, & 0 \le s \le t \le 1 \\ 0, & s > t \end{cases}$$

it follows from example 3 in II.16, that K is a compact operator. Since the kernel function corresponding to K^* is $\overline{k(s, t)}$, we have that

$$(K^*f)(t) = -2i \int_t^1 f(s)\, ds.$$

Hence

$$(K^*Kf)(t) = 4 \int_t^1 \left(\int_0^s f(\eta)d\eta\right) ds.$$

To find the positive eigenvalues λ of K^*K, we set

$$4\int_t^1 \left(\int_0^s f(\eta)d\eta\right) ds = \lambda f(t), \quad \lambda > 0.$$

a straight forward computation shows that

$$\lambda f'' + 4f = 0,$$

$$f(1) = 0, \quad f'(0) = 0. \tag{i}$$

The general solution of the differential equation in (i) is

$$f(t) = ae^{2it/\lambda^{\frac{1}{2}}} + be^{-2it/\lambda^{\frac{1}{2}}}$$

Since

$$f'(t) = a\frac{2i}{\lambda^{1/2}} e^{2it} - b\frac{2i}{\lambda^{1/2}} e^{-2it/\lambda^{1/2}}$$

we have

$$0 = f'(0) = \frac{2i}{\lambda^{1/2}}(a - b).$$

Thus

$$f(t) = a(e^{2it/\lambda^{1/2}} + e^{-2it/\lambda^{1/2}}) = 2a\cos 2t/\lambda^{1/2}$$

and

$$0 = f(1) = 2a\cos 2/\lambda^{1/2}.$$

Hence

$$2/\lambda^{1/2} = (2k + 1)\pi, \quad k = 0, 1, 2, \ldots$$

or

$$s_k(K) = \lambda^{1/2} = \frac{2}{(2k + 1)\pi}, \quad k = 0, 1, 2, \ldots$$

It follows from equality (4.1) and the min-max Theorem IV.9.1 applied to A^*A that

$$s_j(A) = \min_{\substack{M \\ \dim M = j-1}} \max_{\substack{\|x\|=1 \\ X \perp M}} \|Ax\|. \tag{4.6}$$

Another important formula for the singular values is the following:

Theorem 4.4 *Let $\mathcal{H}$ be a Hilbert space and let $A \in \mathcal{L}(\mathcal{H})$ be compact. Then for $n = 1, 2, \ldots$*

$$s_n(A) = \min\{\|A - F\| \mid F \in \mathcal{L}(\mathcal{H}), \text{ rank } F \leq n - 1\}. \tag{4.7}$$

Proof: Suppose rank $F = p \leq n - 1$. Then $\dim(\operatorname{Ker} F)^{\perp} = \dim \operatorname{Im} F^* = \dim \operatorname{Im} F = p$. Hence by formula (4.6), with $M = (\operatorname{Ker} F)^{\perp}$,

$$s_n(A) \leq s_{p+1}(A) \leq \max_{\substack{\|x\|=1 \\ x \in \operatorname{Ker} F}} \|Ax\| = \max_{\substack{\|x\|=1 \\ x \in \operatorname{Ker} F}} \|(A - F)x\| \leq \|A - F\|.$$

Since F was an arbitrary operator in $\mathcal{L}(\mathcal{H})$ of rank $\leq n - 1$,

$$s_n(A) \leq \inf\{\|A - F\| \mid F \in \mathcal{L}(\mathcal{H}), \text{ rank } F \leq n - 1\}.$$

It remains to prove that the infimum above is attained and equals $s_n(A)$. By Theorem 4.1 the operator $A = \sum_{j=1}^{\nu(A)} s_j(A)\langle x, \varphi_j\rangle\psi_j$, where $\{\varphi_j\}$ and $\{\psi_j\}$ are orthonormal systems. For $n < \nu(A) + 1$, define for $n > 1$ the operator $F_n x = \sum_{j=1}^{n-1} s_j(A)\langle x, \varphi_j\rangle\psi_j$, and $F_1 = 0$. Since rank $F_n = n - 1$,

$$\|(A - F_n)x\|^2 = \left\| \sum_n^{\nu(A)} s_j(A)\langle x, \varphi_j\rangle\psi_j \right\|^2 \leq s_n^2(A)\|x\|^2$$

Thus $\|A - F_n\| \leq s_n(A)$. Hence formula (4.7) holds for $n < \nu(A) + 1$. Thus if $m \geq \nu(A) + 1$, then $s_n = 0$ and rank $A \leq n - 1$. Therefore the infimum is attained at $F = A$. $\qquad\square$

The next result shows that the singular values are continuous functions of compact operators in $\mathcal{L}(\mathcal{H})$.

Corollary 4.5 *Let A and B be compact operators in $\mathcal{L}(\mathcal{H})$. Then*

$$|s_n(A) - s_n(B)| \leq \|A - B\|. \tag{4.8}$$

Proof: For any $F \in \mathcal{L}(\mathcal{H})$ of rank at most $n - 1$,

$$s_n(A) \leq \|A - F\| \leq \|A - B\| + \|B - F\|.$$

Hence

$$s_n(A) - \|A - B\| \leq \inf\{\|B - F\| \mid F \in \mathcal{L}(\mathcal{H}),$$
$$\text{rank } F \leq n - 1\} = s_n(B) \tag{4.9}$$

by Theorem 4.4. Interchanging the roles of A and B in (4.9) yields (4.8). $\qquad\square$

10.5 Trace Class and Hilbert-Schmidt Operators

In this section we introduce two important classes of compact operators.

Definition: A bounded linear operator A on a Hilbert space $\mathcal{H}$ is said to be of *trace class* if it is compact and $\sum_{j=1}^{\infty} s_j(A) < \infty$.

Theorem 5.1 *Let k be continuous on $[a, b] \times [a, b]$. Suppose that for all $f \in L_2([a, b] \times [a, b])$,*

$$\int_a^b \int_a^b k(t, s) f(s) \bar{f}(t) \, ds dt \geq 0. \tag{5.1}$$

Then the operator K defined by

$$(Kf)(t) = \int_a^b k(t, s) f(s) \, ds$$

is of trace class and

$$\sum_{j=1}^{\infty} s_j(K) = \int_a^b k(t, t) \, dt \tag{5.2}$$

Proof: The operator K is compact as seen in Example 3 of Section II.16. Moreover K is positive by (5.1). Hence K is self adjoint and therefore the eigenvalues of $K^*K = K^2$ are the squares of the eigenvalues of K. Thus $s_j(K) = \lambda_j(K)$ and by Theorem V.4.1, equality (5.2), holds. $\square$

The compact operator

$$(Kf)(t) = 2i \int_0^t f(t) ds \tag{5.3}$$

defined on $L_2([0, 1])$ was shown in the example in Section 4 to have singular values $s_k(A) = \frac{2}{(2k+1)\pi}, k = 0, 1, \ldots$. Therefore K is not of trace class.

Definition: A bounded linear operator A on a Hilbert space $\mathcal{H}$ is said to be a *Hilbert-Schmidt operator* if A is compact and $\sum_{j=1}^{\infty} s_j^2(A) < \infty$.

Every trace class operator is clearly a Hilbert-Schmidt operator.

Theorem 5.2 *Suppose $k \in L_2([a, b] \times [a, b])$ and $\overline{k(t, s)} = k(s, t)$. Then the operator K defined on $L_2([a, b])$ by*

$$(Kf)(t) = \int_a^b k(t, s) f(s) \, ds$$

is Hilbert-Schmidt and

$$\sum_{j=1}^{\infty} s_j(K)^2 = \int_a^b \int_a^b |k(t,s)|^2 \, ds \, dt. \tag{5.4}$$

Proof: The operator K is compact and self adjoint. Hence $s_j^2(K) = \lambda_j^2(K)$ and formula (5.4) is a consequence of Theorem V.1.2.

$\square$

The compact operator K defined in (5.3) is not of trace class but is a Hilbert-Schmidt operator since

$$\sum_{j=1}^{\infty} s_k^2(K) = \sum_{k=1}^{\infty} \left(\frac{2}{(2k+1)\pi} \right)^2 < \infty.$$

Thorough treatments of trace class and Hilbert-Schmidt operators appear in [GKre], [GGK1] Chapters VII and VIII, and [DS2].

Exercises X

1. Which of the following operators are normal and which are not?
   (a) $K : L_2[0,1] \to L_2[0,1]$; $(K\varphi)(t) = \int_0^1 (t-s)^2 \varphi(s) \, ds$.
   (b) $V : L_2[0,1] \to L_2[0,1]$; $(Vf)(t) = \int_0^t f(s) \, ds$.
   (c) $S_r : \ell_2 \to \ell_2$; $S_r \xi = (0, \xi_1, \xi_2, \ldots)$.
   (d) $A : L_2[0,1] \to L_2[0,1]$; $(Af)(t) = a(t)f(t)$, where a is bounded and Lebesgue measurable on $[0,1]$.

2. Which of the following pairs of operators defined in Exercise 1 commute and which do not?
   (a) S_ℓ and S_r^k, $k > 1$, where S_ℓ is the left shift on ℓ_2.
   (b) A, V.
   (c) K, V.

3. Let A and B be self adjoint compact operators on an infinite dimensional Hilbert space. Let Ker A = Ker B = $\{0\}$ and $AB = BA$. Prove that there exists a bijection f on the natural numbers such that the eigenvalues

$$\lambda_k(\alpha A + \beta B) = \alpha \lambda_k(A) + \beta \lambda_{f(k)}(B).$$

4. Let A be a self adjoint operator on a Hilbert space $\mathcal{H}$. Define on $\mathcal{H} \oplus \mathcal{H} \oplus \mathcal{H}$ an operator

$$B = \begin{pmatrix} 0 & 0 & A \\ A & 0 & 0 \\ 0 & A & 0 \end{pmatrix}$$

Prove that B is normal.

5. Let A be a compact self adjoint operator on a Hilbert space $\mathcal{H}$. Let $\{\lambda_j\}$, $\{\varphi_j\}$ be a basic system of eigenvalues and eigenvectors of A. Find the eigenvalues and the eigenvectors of the operator B defined in Exercise 4.

6. Generalize Exercises 4 and 5 to an operator of the type

$$B = \begin{pmatrix} 0 & 0 \cdots 0 & A \\ A & 0 \cdots 0 & 0 \\ 0 & A \cdots 0 & 0 \\ \vdots & \cdots & \\ 0 & 0 \cdots A & 0 \end{pmatrix}.$$

7. Prove that if $A \in \mathcal{L}(\mathcal{H})$ is self adjoint, then $A \pm iI$ are invertible and $V = (A - iI)(A + iI)^{-1}$ is unitary. Hint: $\|Ax \pm ix\|^2 = \|Ax\|^2 + \|x\|^2$. V is called the *Cayley* transform of A.

8. Let K be a compact normal operator on a Hilbert space. Prove that $I + K$ is unitary if and only if $|1 + \lambda_j(K)| = 1$ for all j.

9. Let A be a compact normal operator on a Hilbert space. Suppose $A = G + iF$, where G and F are self adjoint operators. Prove that

$$e^A = e^G (\cos F + i \sin F).$$

Is this also true in case A is compact but not necessarily normal?

10. Let A and B be compact self adjoint operators on a Hilbert space. Prove that if $AB = BA$, then $e^{\alpha A + \beta B} = e^{\alpha A} \cdot e^{\beta B}$.

11. Does the formula in Exercise 10 hold if $AB \neq BA$?

12. Let A and B be compact self adjoint operators on a Hilbert space. Prove that if $AB = BA$ and two functions $f(A)$ and $f(B)$ are defined, then

$$f(A)g(B) = g(B)f(A).$$

13. Find the real and imaginary parts of the following operators.
    (a) $(K\varphi)(t) = \int_{-\pi}^{\pi} k(t - s)\varphi(s)\, ds$; $K : L_2[-\pi, \pi] \to L_2[-\pi, \pi]$, $k \in L_2[-2\pi, 2\pi]$.
    (b) $(Af)(t) = 2i \int_t^1 f(s)\, ds$; $A : L_2[0, 1] \to L_2[0, 1]$.
    (c) $(Af)(t) = 2 \int_0^t k(t, s) f(s)\, ds$; $A : L_2[0, 1] \to L_2[0, 1]$, $k \in L_2([0, 1] \times [0, 1])$.
    (d) $A = \begin{pmatrix} 0 & A_{12} & A_{13} & \cdots & A_{1n} \\ 0 & 0 & A_{23} & \cdots & A_{2n} \\ \cdot & \cdot & \cdot & & \cdot \\ 0 & 0 & 0 & \cdots & A_{n-1n} \\ 0 & 0 & 0 & \cdots & 0 \end{pmatrix}$, $A_{ij} \in \mathcal{L}(\mathcal{H})$.

14. Let G be a self adjoint invertible operator on a Hilbert space $\mathcal{H}$ and let $B = G^2$. Define on $\mathcal{H}$ a new scalar product $\langle \cdot, \cdot \rangle_B$ by $\langle h, g \rangle_B = \langle Bh, g \rangle$. Prove the $\mathcal{H}$ is a Hilbert space with respect to the new scalar product and that the new norm is equivalent to the old one.

15. Let B and $\langle \cdot, \cdot \rangle_B$ be defined as in exercise 14. For any $A \in \mathcal{L}(\mathcal{H})$, define $A^\times$ to be the adjoint operator of A with respect to the new scalar product $\langle \cdot, \cdot \rangle_B$. Prove that $A^\times = B^{-1}A^*B$ and that A is self adjoint with respect to the new scalar product if and only if $BA = A^*B$.

16. Let B be the operator defined in exercise 14. Let $A \neq 0$ be a compact operator such that $A^*B = BA$. Prove that

    (a) A has only real eigenvalues and at least one eigenvalue of A is not equal to zero.

    (b) If φ_j and φ_k are eigenvectors of A corresponding to different eigenvalues, then φ_j and φ_k are orthogonal with respect to the new scalar product $\langle \cdot, \cdot \rangle_B$.

    (c) A admits a spectral representation

    $$Ah = \sum_j \lambda_j \langle h, \varphi_j \rangle \varphi_j,$$

    where $\{\lambda_j\}$ are the eigenvalues and the φ_j form a system for which $\langle B\varphi_j, \varphi_k \rangle = s_{jk}$.

    (d) Conversely, if A has the representation given in (c), then $BA = A^*B$.

17. Let B be defined on $L_2[a, b]$ by $(B\varphi)(t) = p(t)\varphi(t)$. where p is a positive Lebesgue measurable function such that

    $$0 < \inf_{t \in [a,b]} p(t), \quad \sup_{t \in [a,b]} p(t) < \infty.$$

    Define A on $L_2[a, b]$ by $(A\varphi)(t) = \int_a^b k(t, s)p(s)\varphi(s)ds$, where $k \neq 0$ is a self adjoint Hilbert-Schmidt kernel on $[a, b] \times [a, b]$. Prove that

    (a) $BA = A^*B$.

    (b) A has eigenvalues different from zero.

    (c) The eigenvectors of A corresponding to different eigenvalues are orthogonal in the scalar product $\langle \cdot, \cdot \rangle_B$.

18. Prove that $A \in \mathcal{L}(\mathcal{H})$ is normal if and only if $\|Ax\| = \|A^*x\|$ for all $x \in \mathcal{H}$, in which case $\mathcal{H} = \overline{\operatorname{Im} A} \oplus \operatorname{Ker} A$.

19. Generalize Theorem 1.1 for a finite number of operators.

20. Is it possible to generalize Theorem 1.1 for an infinite set of operators?

21. Prove that if A is a compact normal operator and M is a closed A-invariant subspace, then M is also A^*-invariant and $M^\perp$ is A-invariant. Hint: Let $\{p_n\}$ be a sequence of polynomials which converges uniformly on $\sigma(A)$ to $f(z) = \bar{z}$. Consider $p_n(A)$ and $f(A)$ restricted to M.

22. Let A be a compact normal operator. Prove that a closed subspace is A-invariant if and only if it is $A_\mathcal{R}$ and $A_\mathcal{I}$ invariant, where $A_\mathcal{R}$ and $A_\mathcal{I}$ are the real and imaginary parts of A, respectively.

23. Generalize Theorem 1.1 for two compact normal operators.

24. Generalize Theorem 1.1 for a finite number of compact normal operators.

25. Prove that any bounded finite rank operator on a Hilbert space is a Hilbert-Schmidt operator.

26. Prove that any bounded finite rank operator on a Hilbert space is trace class.

27. Suppose that for some orthonormal basis $\varphi_1, \varphi_2, \ldots$ in Hilbert space $\mathcal{H}$ and for some operator $A \in \mathcal{L}(\mathcal{H})$ the series $\sum_{j=1}^{\infty} \|A\varphi_j\|^2$ converges. Prove that A is Hilbert-Schmidt.

28. Suppose that for some orthonormal basis $\varphi_1, \varphi_2, \ldots$ in Hilbert space $\mathcal{H}$ and for some operator $A \in \mathcal{L}(\mathcal{H})$ the series $\sum_{j=1}^{\infty} |\langle A\varphi_j, \varphi_j \rangle|$ converges. Is the operator A trace-class?

29. If the operator A in problem 28 is known to be positive, is A trace class?

30. How many non zero singular values does a bounded operator of a given rank k have?

31. Give an example of a self adjoint compact operator which is

    (a) not trace class

    (b) not Hilbert-Schmidt

32. Give an example of a compact operator which has only zero as its eigenvalue and is not Hilbert-Schmidt.

Chapter XI
Banach Spaces

A norm can be defined on $\mathbb{C}^n$ in many different ways. Some of them stem from an inner product. An example of a norm on $\mathbb{C}^n$ which does not arise from an inner product is $\|(\alpha_1, \alpha_2, \ldots, \alpha_n)\| = \max_{1 \le i \le m} |\alpha_i|$.

This is due, for example, to the fact that the vectors $(1, 0, \ldots, 0)$ and $(0, 1, 0, \ldots, 0)$ do not satisfy the parallelogram law (cf. I. 3.1) in that norm. As we shall see, all norms on $\mathbb{C}^n$ are equivalent. This need not be the case for infinite dimensional spaces as the situation is more complicated.

Many problems are cast in an infinite dimensional setting with norms which do not have some of the nice geometrical properties of a Hilbert space norm. The choice of the norm plays a vital role in applications and in solutions of certain problems. This leads us to study, in this chapter, Banach spaces which are more general that Hilbert spaces. In subsequent chapters we extend the theory of linear operators to Banach spaces.

11.1 Definitions and Examples

Let us start with the definition of a Banach space.

Definition: Let X be a vector space over $\mathbb{C}$. A *norm* $\| \cdot \|$ on X is a real valued function on X which has the following properties.

    (i) $\|x\| \ge 0$, $\|x\| = 0$ implies $x = 0$.

    (ii) $\|\alpha x\| = |\alpha| \|x\|$, $\alpha \in \mathbb{C}$.

    (iii) $\|x + y\| \le \|x\| + \|y\|$ (triangle inequality).

The space X, together with $\| \cdot \|$, is called a *normed linear space*.

Many of the definitions which appear in the preceding chapters carry over verbatim to normed linear spaces. For convenience we shall not restate most of the definitions.

For example, a sequence $\{x_n\}$ in a normed linear space X is called a Cauchy sequence if $\|x_n - x_m\| \to 0$ as $n, m \to \infty$. X is *complete* if every Cauchy sequence in X converges.

A complete normed linear space is called a *Banach space*.

Examples: 1. Every Hilbert space is a Banach space.

2. Let $C([a, b])$ denote the set of continuous complex valued functions defined on the interval $a \leq x \leq b$. With the usual definitions of addition and scalar multiplication of functions, $C([a, b])$ is a vector space over $\mathbb{C}$ and

$$\|f\| = \max_{t \in [a,b]} |f(t)|$$

defines a norm on this space.

Since $\|f_n - f_m\| \to 0$, as $n, m \to \infty$, if and only if $\{f_n - f_m\}$ converges uniformly to zero on $[a, b]$, it follows from the uniform Cauchy criterion that $\{f_n\}$ converges in $C([a, b])$. Thus $C([a, b])$ is a Banach space.

3. For $1 \leq p < \infty$, let ℓ_p be the set of those sequences $\xi = (\xi_1, \xi_2, \ldots)$ of complex numbers for which $\sum_{n=1}^{\infty} |\xi_n|^p < \infty$. Defining addition and scalar multiplication as for ℓ_2, it can be shown ([TL] Section II.4 and [Sc] Section 2.4) that ℓ_p is a vector space and

$$\|\xi\| = \left(\sum_{n=1}^{\infty} |\xi_n|^p \right)^{1/p}$$

is a norm on ℓ_p. The completeness of ℓ_p can be established by the argument which is used to prove that ℓ_2 is complete (I.2.1).

4. Let ℓ_∞ be the set of all bounded sequences of complex numbers. With respect to the definitions of addition and scalar multiplication as defined on ℓ_2, the set ℓ_∞ is a vector space and

$$\|(\xi_1, \xi_2, \ldots)\| = \sup_n |\xi_n|$$

defines a norm on ℓ_∞. It is easy to show that ℓ_∞ is complete.

5. Let c_0 be the subspace of ℓ_∞ consisting of sequences of complex numbers which converge to z. It is easy to verify that c_0 is a Banach space.

6. Another useful source of examples is the Banach space $\ell_p(Z), 1 \leq p \leq \infty$, of all "doubly infinite" sequences

$$\xi = (\ldots, \xi_{-2}, \xi_{-1}, \xi_0, \xi_1, \xi_2, \ldots)$$

with

$$\|\xi\|_p = \left(\sum_{n=-\infty}^{\infty} |\xi_n|^p \right)^{1/p}, \quad 1 \leq p < \infty,$$

$$\|\xi\|_\infty = \sup_n |\xi_n|, \quad p = \infty.$$

7. $L_p([a, b])$, $1 \leq p < \infty$, is a Banach space with

$$\|f\|_p = \left(\int_a^b |f(t)|^p \, dt \right)^{1/p},$$

and $L_\infty([a, b])$ is a Banach space with

$$\|f\|_\infty = \text{ess sup} |f(t)|.$$

A treatment of these spaces may be found in [R] (see also Appendix 2).

Theorem 1.1 *If X and Y are Banach spaces, then $\mathcal{L}(X, Y)$ is a Banach space.*

Proof: Suppose $\{A_n\}$ is a Cauchy sequence in $\mathcal{L}(X, Y)$. Then for $x \in X$,

$$\|A_n x - A_m x\| \leq \|A_n - A_m\| \|x\| \to 0$$

as $n, m \to \infty$. Since Y is complete, $\{A_n x\}$ converges. Define $Ax = \lim_{n \to \infty} A_n x$. Given $\varepsilon > 0$, there exists an integer N such that $\|A_n - A_m\| \leq \varepsilon$, $n, m \geq N$. Therefore, if $\|x\| = 1$ and $n \geq N$, then

$$\|A_n x - Ax\| = \lim_{m \to \infty} \|A_n x - A_m x\| \leq \lim_{m \to \infty} \|A_n - A_m\| \leq \varepsilon.$$

This implies that $A = A - A_N + A_N \in \mathcal{L}(X, Y)$ and

$$\|A - A_n\| \to 0 \quad n \to \infty.$$

$\square$

Excluding ℓ_2 and $L_2([a, b])$, none of the Banach spaces in Examples 2–8 is a Hilbert space, i.e., there does not exist an inner product $\langle \, , \, \rangle$ on the space such that $\|x\| = \langle x, x \rangle^{1/2}$ (Exercise 8).

There are some very basic properties which Hilbert spaces possess and arbitrary Banach spaces lack. For example, not every closed subspace of a Banach (not Hilbert) space is complemented. Indeed, the closed subspace c_0 of ℓ_∞ is not complemented in ℓ_∞ (cf. [W]). These differences, which we shall point out in subsequent sections, are due to the fact that a Banach space has less structure than a Hilbert space.

11.2 Finite Dimensional Normed Linear Spaces

Given a finite dimensional vector space X over $\mathbb{C}$, there are infinitely many norms which can be defined on X. For example, suppose $x_1, \ldots, x_n$ is a basis for X. Then for $x = \sum_{i=1}^{n} \alpha_i x_i$,

$$\|x\|_1 = \sum_{i=1}^{n} |\alpha_i|, \quad \|x\|_{(r)} = r\|x\|_1, \quad r > 0,$$

$$\|x\|_2 = \left(\sum_{i=1}^{n} |\alpha_i|^2 \right)^{1/2}, \quad \|x\|_\infty = \max_i |\alpha_i| \tag{2.1}$$

all define norms on X. However, we shall now show that *all* norms on X are equivalent in the following sense.

Definition: Two norms $\| \cdot \|$ and $\| \cdot \|_1$, on a vector space X, are called *equivalent* if there exist numbers C and $m > 0$ such that for all $x \in X$,

$$m\|x\|_1 \leq \|x\| \leq C\|x\|_1.$$

It is clear that if $\| \|$ and $\| \|_1$ are equivalent norms on X, then $X_0 = (X, \| \|)$ is complete if and only if $X_1 = (X, \| \|_1)$ is complete. Also, a sequence converges in X_0 if and only if it converges in X_1.

Theorem 2.1 *Any two norms on a finite dimensional vector space are equivalent.*

Proof: Let $x_1, \ldots, x_n$ be a basis for the vector space X. For any norm $\| \cdot \|$ on X and any $\alpha_k \in \mathbb{C}, 1 \leq k \leq n$,

$$\left\| \sum_{k=1}^{n} \alpha_k x_k \right\| \leq \left(\sum_{k=1}^{n} \|x_k\|^2 \right)^{1/2} \left(\sum_{k=1}^{n} |\alpha_k|^2 \right)^{1/2}. \tag{2.2}$$

We shall show that there exists an $m > 0$ such that for $\alpha_k \in \mathbb{C}, 1 \leq k \leq n$,

$$\left\| \sum_{k=1}^{n} \alpha_k x_k \right\| \geq m \left(\sum_{k=1}^{n} |\alpha_k|^2 \right)^{1/2}. \tag{2.3}$$

Define the real valued function f on $S = \{\alpha \in \mathbb{C}^n : \|\alpha\| = 1\}$ by

$$f(\beta_1, \ldots, \beta_n) = \left\| \sum_{k=1}^{n} \beta_k x_k \right\|.$$

Since f is continuous and S is compact, f assumes its minimum m at some point $\xi \in S$. The linear independence of $\{x_1, \ldots, x_n\}$ implies that $m = f(\xi) > 0$. Hence for $0 \neq \alpha = (\alpha_1, \ldots, \alpha_n) \in \mathbb{C}^n$,

$$\left\| \sum_{k=1}^n \alpha_k x_k \right\| = \|\alpha\| f\left(\frac{\alpha}{\|\alpha\|} \right) \geq m \left(\sum_{k=1}^n |\alpha_k|^2 \right)^{1/2}.$$

Since (2.2) and (2.3) are valid for any norm on X, it follows that any two norms on X are equivalent. $\qquad\square$

From the properties of $\mathbb{C}^n$ and Theorem 2.1, we obtain the following result.

Theorem 2.2 *If X is a finite dimensional normed linear space, then*

(a) *X is complete.*

(b) *Every bounded sequence in X has a convergent subsequence.*

Property (a) implies that every finite dimensional subspace of a normed linear space Y is closed in Y.

Property (b) is false if X is infinite dimensional. In order to prove this we use the following lemma.

Lemma 2.3 *If M is a finite dimensional, proper subspace of a normed linear space X, then there exists an $x \in X$ such that*

$$1 = \|x\| = d(x, M),$$

where $d(x, M)$ is the distance from x to M.

Proof: Let z be in X but not in M. There exists a sequence $\{m_k\}$ in M such that $\|z - m_k\| \to d(z, M) > 0$. Since M is finite dimensional and $\{m_k\}$ is bounded, there exists a subsequence $\{m_{k'}\}$ of $\{m_k\}$ and an $m \in M$ such that $m_{k'} \to m$. Hence

$$0 < \|z - m\| = \lim_{k' \to \infty} \|z - m_{k'}\| = d(z, M) = d(z - m, M).$$

Thus for $x = \frac{z-m}{\|z-m\|}$,

$$1 = \|x\| = \frac{d(z - m, M)}{\|z - m\|} = d(x, M). \qquad\square$$

Theorem 2.4 *If every sequence in the 1-sphere of a normed linear space X has a convergent subsequence, then X is finite dimensional.*

Proof: Assume X is infinite dimensional. Choose $x_1 \in X$, $\|x_1\| = 1$. By Lemma 2.3, there exists an $x_2 \in X$ such that

$$1 = \|x_2\| = d(x_2, \text{sp}\{x_1\}).$$

After $\{x_1, \ldots, x_k\}$ has been obtained, choose $x_{k+1} \in X$ such that

$$1 = \|x_{k+1}\| = d(x_{k+1}, M_k),$$

where $M_k = \text{sp}\{x_1, \ldots, x_k\}$. Now $\{x_n\}$ is in the unit sphere of X but the sequence does not have a convergent subsequence since

$$\|x_n - x_k\| \geq d(x_n, M_{n-1}) = 1, \quad n > k.$$

This contradicts the hypothesis of the theorem. Therefore, X is finite dimensional. $\square$

In general, the sum of two closed subspaces of a Banach space need not be closed (cf., Section XII.5) unless one of the subspaces is finite dimensional.

Theorem 2.5 *If M is a closed subspace and N is a finite dimensional subspace of a normed linear space, then $M + N = \{u + v : u \in M, v \in N\}$ is closed.*

Proof: Assume dim $N = 1$, say $N = \text{sp}\{x\}$. The lemma is trivial if $x \in M$. Suppose $x \notin M$ and $z_k \to y$, where

$$z_k = \alpha_k x + u_k, \ u_k \in M, \ \alpha_k \in \mathbb{C}.$$

The sequence $\{\alpha_k\}$ is bounded; otherwise there exists a subsequence $\{\alpha_{k'}\}$ such that $0 < |\alpha_{k'}| \to \infty$. Hence

$$x + \frac{u_{k'}}{\alpha_{k'}} = \frac{z_{k'}}{\alpha_{k'}} \to 0.$$

Thus x is in M since M is closed, which is a contradiction. Consequenctly, $\{\alpha_k\}$ is bounded and therefore it has a subsequence $\{\alpha_{k'}\}$ which converges to some $\alpha \in \mathbb{C}$. Thus

$$u_{k'} = z_{k'} - \alpha_{k'} x \to y - \alpha x.$$

Hence $y - \alpha x$ is in M or $y \in M + N$. The proof of the lemma now follows by induction. $\square$

11.3 Separable Banach Spaces and Schauder Bases

Definition: A normed linear space X is called *separable* if it contains a countable set which is dense in X.

Examples: 1. $C([a, b])$ and $L_p([a, b])$, $1 \leq p < \infty$, are separable Banach spaces since it follows from the Weierstrass approximation theorem that the countable set of polynomials with complex rational coefficients is dense in $C([a, b])$ which, in turn, is dense in $L_p([a, b])$.

2. ℓ_p, $1 \leq p < \infty$, is separable since the countable set of all sequences $\{\alpha_k\}$ of complex rationals, where $\alpha_k = 0$ for all k sufficiently large, is dense in ℓ_p.

3. ℓ_∞ is not separable. For given any countable set $\{x_n\}$ in ℓ_∞, we can always find an $x \in \ell_\infty$ such that $\|x - x_n\| \geq 1$. Indeed, suppose $x_n = (\alpha_1^{(n)}, \alpha_2^{(n)}, \ldots)$. Let $\alpha_n = 0$ if $|\alpha_n^{(n)}| > 1$ and $\alpha_n = 1 + |\alpha_n^{(n)}|$ otherwise. Then $x = (\alpha_1, \alpha_2, \ldots)$ is in ℓ_∞ and

$$\|x - x_n\| \geq |\alpha_n - \alpha_n^{(n)}| \geq 1.$$

We have seen that every separable Hilbert space $\mathcal{H}$ has an orthonormal basis $\{\varphi_n\}$. Thus every $x \in \mathcal{H}$ can be represented by $x = \sum_k \alpha_k \varphi_k$ and the representation is unique, i.e., if $x = \sum_k \beta_k \varphi_k$, then $\alpha_k = \beta_k$.

For a Banach space we have the following definition.

Definition: A sequence $\{x_n\}$ in a Banach space X is called a *Schauder basis* for X if for every $x \in X$, there exists a unique sequence $\{\alpha_n\}$ in $\mathbb{C}$ such that $x = \sum_n \alpha_n x_n$.

The standard basis $\{e_n\}$ is a Schauder basis for ℓ_p, $1 \leq p < \infty$. $C([a, b])$ has a Schauder basis of polynomials and $L_p([-\pi, \pi])$, $1 \leq p < \infty$ has a Schauder basis consisting of $\{1, \cos nx, \sin nx\}_{m=1}^\infty$. These results are difficult to prove.

Many of the separable Banach spaces which one encounters have a Schauder basis. However, there is one without a Schauder basis [E].

11.4 Conjugate Spaces

Throughout the rest of this chapter, X denotes a Banach space.

As we shall see in subsequent sections, it is very useful to study the space of bounded linear functionals on X. In this section we shall describe all bounded linear functionals on a certain X. In contrast with the Hilbert space case, the description depends essentially on the space X. Instead of a generalized Riesz theorem (II.5.2) for Banach spaces, it turns out that we need many theorems, one for each space.

By Theorem 1.1, the space of all bounded linear functionals on X with norm $\|f\| = \sup_{\|x\|=1} |f(x)|$ is a Banach space. This space is called the *conjugate space* of X and is denoted by X'.

We shall now describe the conjugate space ℓ'_p, $1 \leq p < \infty$.

Theorem 4.1 *Given $f \in \ell'_p$, $1 \le p < \infty$, there exists a unique $\eta = (\eta_1, \eta_2, \ldots) \in$ ℓ_q, $\frac{1}{p} + \frac{1}{q} = 1$, ($q = \infty$ when $p = 1$), such that for all $\xi = (\xi_1, \xi_2, \ldots) \in \ell_p$,*

$$f(\xi) = \sum_{k=1}^{\infty} \xi_k \eta_k. \tag{4.1}$$

Moreover, $\|f\| = \|\eta\|$ and $\eta = (f(e_1), f(e_2), \ldots)$, where $\{e_k\}$ is the standard basis for ℓ_p.

Conversely, given $\{\eta_k\} \in \ell_q$, (4.1) defines an $f \in \ell'_p$.

Proof: We shall only prove the theorem for $p = 1$. The proof for $1 < p < \infty$ appears in [TL] p. 143.

Given $f \in \ell'_1$, let $\eta_k = f(e_k)$. For $\xi = (\xi_1, \xi_2, \ldots) \in \ell_1$,

$$f(\xi) = f\left(\sum_{k=1}^{\infty} \xi_k e_k\right) = \sum_{k=1}^{\infty} \xi_k \eta_k. \tag{4.2}$$

and $\eta = (\eta_1, \eta_2, \ldots)$ is in ℓ_∞ since

$$\|\eta\|_\infty = \sup_k |\eta_k| = \sup_k |f(e_k)| \le \|f\|. \tag{4.3}$$

Also,

$$|f(\xi)| \le \sum_{k=1}^{\infty} |\xi_k \eta_k| \le \|\eta\|_\infty \sum_{k=1}^{\infty} |\xi_k| = \|\eta\|_\infty \|\xi\|. \tag{4.4}$$

Thus we have from (4.3) and (4.4) that $\|f\| = \|\eta\|_\infty$.

If there exists $(\beta_1, \beta_2, \ldots) \in \ell_\infty$ such that $f(\xi) = \sum_{k=1}^{\infty} \xi_k \beta_k$ for all $\xi = (\xi_1, \xi_2, \ldots) \in \ell_1$, then $f(e_j) = \beta_j$.

Inequality (4.4) implies that the functional f which is given by (4.1) is in ℓ'_1.

Theorem 4.1 shows that ℓ'_p can be identified with ℓ_q in the following sense.

There exists a linear isometry J which maps ℓ'_p onto ℓ_q. The operator J is defined by $Jf = \{f(e_k)\}$. For $\eta = (\eta_1, \eta_2, \ldots) \in \ell_q$, $J^{-1}\eta = g \in \ell'_p$, where

$$g(\xi) = \sum_{k=1}^{\infty} \xi_k \eta_k.$$

The proof of the following description of $L'_p([a, b])$, $1 \le p < \infty$, appears in [R] p. 246.   □

Theorem 4.2 *For each $F \in L'_p([a, b])$, $1 \le p < \infty$, there corresponds a unique (up to sets of measure zero) $g \in L_q([a, b])$, $\frac{1}{p} + \frac{1}{q} = 1$, such that for all $f \in L_p([a, b])$,*

$$F(f) = \int_a^b f(t)g(t)\, dt. \tag{4.5}$$

Furthermore, $\|F\| = \|g\|$.

Conversely, given $g \in L_q([a, b])$, the functional F defined by (4.5) is in $L'_p([a, b])$.

Additional examples of conjugate spaces may be found in [DS1].

11.5 Hahn-Banach Theorem

One of the fundamental results in the theory of Banach spaces is the Hahn-Banach theorem. This section contains the proof of the theorem together with some very important applications.

Hahn-Banach Theorem 5.1 *If f is a bounded linear functional which is defined on a subspace M of a normed linear space X, then f can be extended to a bounded linear functional F defined on X such that $\|f\| = \|F\|$.*

The theorem is clear if X is a Hilbert space (see Exercise XII-5).

In order to prove the Hahn-Banach theorem, we need the following preliminary results.

A set E is called *partially ordered* if there exists a binary relation, $\leq$, defined for certain pairs $(x, y) \in E \times E$ such that

  (i) $x \leq x$,

  (ii) if $x \leq y$ and $y \leq z$, then $x \leq z$.

Let F be a subset of E. An element $x \in E$ is an *upper bound* for F (with respect to $\leq$) if $y \leq x$ for every $\in F$.

The set F is *totally ordered* if for every y and z in F, either $y \leq z$ or $z \leq y$.

An element $m \in E$ is said to be *maximal* if $m \leq x$ implies $m = x$.

For example, if E is the set of all subsets of a set S, let $A \leq B$ mean $A \subset B$. Then E is partially ordered with respect to $\leq$. If F is a subset of E, then $\cup_{A \in F} A$ is an upper bound for F. If F consists of sets $A_1 \subseteq A_2 \subseteq \ldots$, then F is totally ordered. The set E is the only maximal element in E.

The following lemma, which is equivalent to the action of choice, has many applications in functional analysis and other branches of mathematics. The reader is referred to [DS1] Section I.2 for a further discussion of Zorn's lemma.

Zorn's Lemma. *Let E be a non-empty partially ordered set. If every totally ordered subset of E has an upper bound in E, then E contains a maximal element.*

The following result is a general version of the Hahn-Banach theorem.

Theorem 5.2 *Suppose X is a vector space over the real or complex numbers. Let p be a real-valued function defined on X such that for all x, y in X and numbers α,*

  (a) $p(x + y) \leq p(x) + p(y)$

  (b) $p(\alpha x) = |\alpha| p(x)$.

Suppose f is a linear functional defined on a subspace M of X and

$$|f(z)| \leq p(z) \quad \text{for all } z \in M.$$

Then f can be extended to a linear functional F defined on all of X such that

$$|F(x)| \leq p(x) \quad \text{for all } x \in X.$$

Proof: Let $\mathcal{E}$ be the set of all linear functionals g such that the domain $D(g)$ of g is contained in X, $g = f$ on M, and $|g(x)| \leq p(x)$ $x \in D(g)$. Obviously, f is in $\mathcal{E}$. We partially order $\mathcal{E}$ by letting $g \leq h$ mean that h is an extension of g. The idea of the proof is to use Zorn's lemma to show that $\mathcal{E}$ has a maximal element and that this element is defined on all of X.

Let $\mathcal{J}$ be a totally ordered subset of $\mathcal{E}$. Define the linear functional H by

$$D(H) = \bigcup_{g \in \mathcal{J}} D(g)$$
$$H(x) = g(x) \quad \text{if } x \in D(g).$$

Since $\mathcal{J}$ is totally ordered, it follows that H is in $\mathcal{E}$ and that H is an upper bound of $\mathcal{J}$. Hence $\mathcal{E}$ has a maximal element F by Zorn's lemma. The theorem is proved once we show that $D(F) = X$.

Suppose there exists an $x \in X$ which is not in $D(F)$. Let $X_1 = \text{sp}\{x\} \oplus D(F)$. We shall show that there exists a $G \in \mathcal{E}$ which is an extension of F to X_1, thereby contradicting the maximality of F. It is clear that we should define G on X_1 by

$$G(\alpha x + z) = \alpha G(x) + F(z), \quad z \in D(F), \tag{5.1}$$

with $G(x)$ chosen so that

$$|G(\alpha x + z)| \leq p(\alpha x + z), \quad z \in D(F). \tag{5.2}$$

In order to determine $G(x)$, let us first assume that X is a vector space over the reals. If (5.2) is to hold, then in particular,

$$G(x + z) \leq p(x + z), \quad z \in D(F) \tag{5.3}$$

and

$$G(-x - y) \leq p(-x - y) = p(x + y), \quad y \in D(F). \tag{5.4}$$

Thus, from (5.1), (5.3) and (5.4) we want

$$G(x) \leq p(x + z) - F(z), \quad z \in D(F). \tag{5.5}$$

and

$$G(x) \geq -p(x + y) - F(y), \quad y \in D(F). \tag{5.6}$$

It is easy to check that the right side of (5.5) is greater than or equal to the right side of (5.6) for all z and y in $D(F)$. Thus, if we define

$$G(x) = \inf\{p(x+z) - F(z): \quad z \in D(F)\},$$

it follows that (5.3) and (5.4) hold. But then

$$G(\alpha x + z) \le p(\alpha x + z), \quad z \in D(F). \tag{5.7}$$

To see this, suppose $\alpha > 0$. Then (5.3) implies

$$G(\alpha x + z) = \alpha G(x + z/\alpha) \le \alpha p(x + z/\alpha) = p(\alpha x + z).$$

If $\alpha < 0$, then (5.4) implies

$$G(\alpha x + z) = -\alpha G(-x - z/\alpha) \le -\alpha p(-x - z/\alpha) = p(\alpha x + z).$$

If $\alpha = 0$, then

$$G(z) = F(z) \le p(z).$$

Finally, (5.7) implies (5.2) since

$$-G(\alpha x + z) = G(-\alpha x - z) \le p(-\alpha x - z) = p(\alpha x + z).$$

To summarize, we have constructed a $G \in \mathcal{E}$ such that $F \le G$ but $F \ne G$. This contradicts the maximality of F. Hence $D(F) = X$ and the theorem is proved for X a vector space over the reals.

Suppose that X is a vector space over the complex numbers. We write

$$f(z) = \Re f(z) + i \Im f(z), \quad z \in M,$$

where $\Re f(z)$ and $\Im f(z)$ denote the real and imaginary parts of $f(z)$, respectively. Since $\Im f(z) = -\Re f(iz)$, we have

$$f(z) = \Re f(z) - i \Re f(iz). \tag{5.8}$$

Let X_r be X considered as a vector space over the reals. Now $\Re f$ is a linear functional on M, considered as a subspace of X_r. Moreover,

$$|\Re f(z)| \le |f(z)| \le p(z), \quad z \in M.$$

Hence, by the result we proved for real vector spaces, there exists a linear extension G of $\Re f$ to all of X_r such that $|G(z)| \le p(x), x \in X_r$. Guided by (5.8), we define F on X by

$$F(x) = G(x) - iG(ix).$$

Now G is a linear extension of f to all of X. Writing $F(x) = |F(x)|e^{i\theta}$, we get

$$|F(x)| = F(e^{-i\theta}x) = \Re F(e^{-i\theta}x) = G(e^{-i\theta}x) \le p(e^{-i\theta}x) = p(x).$$

This completes the proof of the theorem. $\qquad\qquad\qquad\qquad\qquad\qquad\quad\square$

Proof of the Hahn-Banach Theorem. Define p on X by $p(x) = \|f\| \|x\|$, and apply the preceding theorem to f and p. $\square$

Now for some important applications of the Hahn-Banach theorem.

Theorem 5.3 *Let M be a subspace of a normed linear space X. Given $x \in X$ such that*

$$\inf_{u \in M} \|x - u\| = d(x, M) > 0,$$

there exists an $f \in X'$ such that

$$\|f\| = 1, \ f(M) = 0 \quad and \quad f(x) = d(x, M).$$

Proof: Since $x \notin M$, every $z \in M_0 = \text{sp}\{x\} + M$ is uniquely represented in the form

$$z = \alpha x + u, \quad \alpha \in \mathbb{C}, \quad u \in M.$$

The function

$$g(\alpha x + u) = \alpha d, \quad d = d(x, M),$$

is linear on M_0 and for $\alpha \neq 0$,

$$\|\alpha x + u\| = |\alpha| \left\| x + \frac{1}{\alpha} u \right\| \geq |\alpha| d = |g(\alpha x + u)|.$$

Thus $\|g\| \leq 1$. On the other hand, there exists a sequence $\{u_k\}$ in M such that $\|x - u_k\| \to d$. Hence

$$d = g(x - u_k) \leq \|g\| \|x - u_k\| \to \|g\| d,$$

which implies that $1 \leq \|g\|$. Therefore, $\|g\| = 1$. The theorem now follows by extending g to $f \in X'$ such that $1 = \|g\| = \|f\|$. $\square$

For $M = (0)$, the above theorem gives the following result.

Corollary 5.4 *Given $x \in X$, there exists an $f \in X'$ such that $\|f\| = 1$ and $f(x) = \|x\|$.*

Corollary 5.5 *For each $x \in X$,*

$$\|x\| = \max\{|f(x)| : f \in X', \ \|f\| = 1\}.$$

Indeed, there exists a $g \in X'$ such that $\|g\| = 1$ and $g(x) = \|x\|$. Thus for all $f \in X', \|f\| = 1$,

$$|f(x)| \leq \|x\| = g(x).$$

Corollary 5.6 *Given a linearly independent set* $\{x_1, \ldots, x_n\} \subset X$, *there exist* $f_1, f_2, \ldots, f_n$ *in* X' *such that*

$$f_j(x_k) = \delta_{jk}, \quad 1 \le j, k \le n.$$

If $x \in \text{sp}\{x_1, \ldots, x_n\}$, *then* $x = \sum_{j=1}^n f_j(x)x_j$.

Proof: Let $M_j = \text{sp}\{x_i : i \ne j\}$, $1 \le j \le n$. Since M_j is a closed subspace of X and $x_j \notin M_j$, Theorem 5.3 ensures the existence of a $g_j \in X'$ such that $g_j(x_j) = d_j \ne 0$ and $g_j M_j = 0$. The functionals $f_j = \frac{1}{d_j} g_j$, $1 \le j \le n$, have the desired properties.

If $x = \sum_{k=1}^n \alpha_k x_k$, then $f_j(x) = \sum_{k=1}^n \alpha_k f_j(x_k) = \alpha_j$.

The system $\{x_1, \ldots, x_n\}$, $\{f_1, \ldots, f_n\}$ is called a *biorthogonal system.* $\square$

Definition: A closed subspace M of X is said to be *complemented* in X if there exists a closed subspace N of X such that

(i) $X = M + N = \{u + v : u \in M, v \in N\}$ and

(ii) $M \cap N = (0)$.

X is called the *direct sum* of M and N and is written $X = M \oplus N$.

It is clear that $X = M \oplus N$ if and only if for each $x \in X$ there exists a *unique* $u \in M$ and a *unique* $v \in N$ such that $x = u + v$.

A closed subspace M of a Hilbert space is complemented by $M^\perp$. However, as we pointed out in Section 1, a closed subspace of a Banach space need not be complemented, unless the subspace is finite dimensional.

Theorem 5.7 *Every finite dimensional subspace of* X *is complemented in* X.

Proof: Let $x_1, \ldots, x_n$ be a basis for a subspace M of X. By Corollary 5.6, there exist $f_1, \ldots, f_n$ in X' such that $f_i(x_j) = \delta_{ij}$. Now $N = \bigcap_{i=1}^n \text{Ker } f_i$ is a closed subspace of X. Furthermore, $M \cap N = (0)$. For if $z \in M \cap N$, then from Corollary 5.6 and the definition of N, we get $z = \sum_{i=1}^n f_i(z)x_i = 0$. Given $x \in X$, let $u = \sum_{i=1}^n f_i(x)x_i \in M$ and $v = x - u$. Then $x = u + v$ and v is in N since

$$f_k(v) = f_k(x) - f_k(u) = f_k(x) - \sum_{i=1}^n f_i(x)f_k(x_i)$$
$$= f_k(x) - f_k(x) = 0, \quad 1 \le k \le n.$$

Hence $X = M \oplus N$. $\square$

The direct complement of a subspace is not unique. For example, $e_1 = (1, 0)$ is complemented in $\mathbb{C}^2$ by $\text{sp}\{v\}$, where $v \notin \text{sp}\{e_1\}$.

Exercises XI

Throughout these exercises, X and Y denote Banach spaces.

1. Check that the following vectors are in ℓ_p, for $1 \le p \le \infty$.
   (a) $\xi = (1, \frac{1}{2}, \frac{1}{4}, \frac{1}{8}, \frac{1}{16}, \dots)$
   (b) $\xi = (1, \frac{1}{4}, \frac{1}{9}, \frac{1}{16}, \frac{1}{25}, \dots)$

2. Let $1 \le p_1 \le p_2 \le \infty$ and $0 \le a \le b < \infty$. Prove or disprove the following:
   (a) $\ell_{p_1} \subseteq \ell_{p_2}$
   (b) $L_{p_1}[a, b] \supseteq L_{p_2}[a, b]$
   (c) $\|\xi\|_{p_1} \le \|\xi\|_{p_2}$ for $\xi \in \ell_{p_1}$
   (d) $\|\varphi\|_{p_1} \ge \|\varphi\|_{p_2}$ for $\varphi \in L_{p_2}[a, b]$.

3. Let $1 \le p_1 \le p_2 \le \infty$. Prove or disprove the following:
   (a) $L_{p_1}(0, \infty) \subseteq L_{p_2}(0, \infty)$
   (b) $L_{p_2}(0, \infty) \subseteq L_{p_1}(0, \infty)$

4. Show that the intersection of the unit ball in ℓ_∞ with the plane

   $$\{(\xi_1, \xi_2, 0, 0, \dots) \mid \xi_1, \xi_2 \in \mathbb{C}\}$$

   is the square

   $$\{(\xi_1, \xi_2) \mid |\xi_1| \le 1, \quad |\xi_2| \le 1\}.$$

5. Find the intersection of the unit ball in ℓ_p, $1 \le p \le \infty$, with the plane

   $$\{(\xi_1, \xi_2, 0, \dots) \mid \xi_1, \xi_2 \in \mathbb{C}\}.$$

6. Prove that the unit ball in a normed linear space is
   (a) convex
   (b) symmetric (if it contains x, then it also contains $-x$)
   (c) closed

7. Suppose that in an n-dimensional normed linear space X there is given a convex, bounded, closed symmetric set S which is not contained in a proper subspace of X. Define a function

   $$\| \cdot \| : X \to R^+ \quad \text{by} \quad \|x\| = \min\left\{ |\lambda| : \frac{x}{\lambda} \in S \right\}.$$

   Prove that $\| \cdot \|$ is a norm and $(X, \| \cdot \|)$ is a Banach space in which S is the unit ball.

8. Prove that none of the following spaces is a Hilbert space, i.e., it is impossible to define an inner product on the space so that $\langle x, x \rangle^{1/2}$ is the original $\|x\|$.

    (a) $\ell_p, p \neq 2$
    (b) $L_p[a, b], p \neq 2$
    (c) $C[a, b]$.

9. Prove that a Banach space X is a Hilbert space if and only if for all $x, y \in X$,

$$\|x + y\|^2 + \|x - y\|^2 = 2\|x\|^2 + 2\|y\|^2.$$

    Hint: $\langle x, y \rangle = \|\frac{1}{2}(x+y)\|^2 - \|\frac{1}{2}(x-y)\|^2 + i\|\frac{1}{2}(x+iy)\|^2 - i\|\frac{1}{2}(x-iy)\|^2$.

10. Prove that a Banach space X is a Hilbert space if and only if the intersection of any plane $\mathrm{sp}\{x, y\}$, $x, y \in X$ with the unit sphere of X is a circle or an ellipse.

11. Given the ellipsoid $\{(x_1, \ldots, x_n) \in \mathbb{C}^n : \sum_{i=1}^{n} \frac{|x_i|^2}{a_i^2} \leq 1\}$, introduce an inner product on $\mathbb{C}^n$ such that this ellipsoid becomes the unit ball.

12. Let F be the set of all $\xi = (\xi_1, \xi_2, \ldots) \in \ell_\infty$ with $\xi_1 = 0$. Check that the vectors y are closest in F to x.

    (a) $x = (2, 0, 0, \ldots)$, y is any element in F with $|y_j| \leq 2$, $j \geq 2$.
    (b) $x = (1, 1, 0, \ldots)$, y is any element in F with $|y_2 - 1| \leq 1$ and $|y_j| \leq 1$, $j = 3, 4, \ldots$.
    (c) $x = (1, \frac{1}{2}, \frac{1}{4}, \ldots)$, y is any element in F with $\left| y_n - \frac{1}{2^{n-1}} \right| \leq 1$, $n \geq 1$.

13. Given two vectors $(1, 1, 0, 0, \ldots)$ and $(\xi_1, \xi_2, \xi_3, \xi_4, 0, 0, \ldots)$ in ℓ_∞, where $0 \leq |\xi_j| \leq 1$, $j = 1, 2, 3, 4$, prove that the distance in ℓ_∞ between those vectors does not exceed 2 and that there exist $\xi_1, \xi_2, \xi_3, \xi_4$ such that the distance is exactly 2.

14. Show that there exist two vectors x and y in ℓ_∞ such that they are linearly independent, $\|x\| = \|y\| = 1$ and $\|x + y\| = 2$.

15. Prove that if the unit sphere of a normed linear space contains a line segment, then there exist vectors x and y such that $\|x + y\| = \|x\| + \|y\|$ and x, y are linearly independent (a line segment is a set of the form $\{au + (1 - a)v: 0 \leq a \leq 1\}$.

16. Prove that if a normed linear space X contains linearly independent vectors x and y such that $\|x + y\| = \|x\| + \|y\|$, then there is a line segment contained in the unit sphere of X.

17. Prove that there are no line segments contained in the unit sphere of a normed linear space if and only if the closest element in a subspace to a given vector is unique.

18. Prove that in ℓ_p, $1 < p < \infty$, there are no line segments contained in the unit sphere.

19. Find the intersection of the unit ball in $C[0, 1]$ with the following subspaces:

    (a) sp$\{t\}$
    (b) sp$\{1, t\}$
    (c) sp$\{1, t, t^2\}$
    (d) sp$\{1 - t, t\}$.

20. Given two spheres $\|x - y_0\| = \|y_0\|$ and $\|x + y_0\| = \|y_0\|$ in a normed linear space, how many points can these spheres have in common?

21. Let $\xi = (\xi_1, \xi_2, \ldots)$ be a vector in c_0 (the subspace of ℓ_∞ consisting of sequences which converge to zero). By renumbering the entries of ξ we obtain ξ^* with $|\xi_1^*| \geq |\xi_2^*| \geq \ldots$. Let $X_1 = \{\xi \in c_0 : \sum_{j=1}^{\infty} \frac{1}{j} |\xi_j^*| < \infty\}$. Define a norm on X_1 by $\|\xi\|_1 = \sum_{j=1}^{\infty} \frac{1}{j} |\xi_j^*|$.

    Prove that X_1 is a Banach space with respect to this norm. Let $X^{(1)} = \{\xi \in c_0 | \sup_n \frac{\sum_{j=1}^{n} |\xi_j^*|}{\sum_{j=1}^{n} \frac{1}{j}} < \infty\}$ and define a norm on $X^{(1)}$ by

$$\|\xi\|^{(1)} = \sup_n \frac{\sum_{j=1}^{n} |\xi_j^*|}{\sum_{j=1}^{n} \frac{1}{j}}.$$

    Prove that $X^{(1)}$ is a Banach space with respect to this norm. Prove that X_1 and $X^{(1)}$ are not Hilbert spaces.

22. Let M be a closed subspace of X. Define the *quotient space* X/M to be the vector space consisting of the cosets $[x] = \{x + z : z \in M\}$ with vector addition and scalar multiplication defined by $[x] + [y] = [x + y]$, $\alpha[x] = [\alpha x]$. Define $\|[x]\|_1 = d(x, M)$, the distance from x to M. Prove that $\| \cdot \|_1$ is a norm on X/M and that $(X/M, \| \cdot \|_1)$ is complete. We shall refer to this space in XII exercise 17.

23. A Banach space is called *uniformly convex* if $\|x_n\| = \|y_n\| = 1$ and $\|\frac{1}{2}(x_n + y_n)\| \to 1$ imply $x_n - y_n \to 0$.

    (a) Prove that every Hilbert space is uniformly convex.

    (b) It can be shown that the following inequalities hold for every f and g in $L_p[a, b]$.

$$\|f + g\|^p + \|f - g\|^p \leq 2^{p-1}(\|f\|^p + \|g\|^p), \quad 2 \leq p < \infty.$$
$$\|f + g\|^q + \|f - g\|^q \leq 2(\|f\|^p + \|g\|^p)^{q-1}, \quad 1 < p \leq 2,$$
$$\frac{1}{p} + \frac{1}{q} = 1.$$

    Prove that $L_p[a, b]$, $1 < p < \infty$, is uniformly convex.

(c) Give an example of a 2-dimensional Banach space which is not uniformly convex.

24. Let S be a bounded closed convex set in the uniformly convex space X. Prove that there exists a unique $s_0 \in S$ smallest norm, i.e., $\|s_0\| < \|s\|$ for all $s \in S$, $s \neq s_0$. Hint: Choose $s_n \in S$ so that $s_n \to \inf_{s \in S} \|s\|$ and show that $\{s_n\}$ is a Cauchy sequence.

25. Let X be uniformly convex. Show that for each $f \neq 0$ in X', there exists a unique $x \in X$ such that $\|x\| = 1$ and $\|f\| = |f(x)|$.

26. Prove that the conjugate space of c_0 is ℓ_1.

27. Prove that a linear functional on X is bounded if and only if its kernel is closed.

28. Prove that if X is separable, then there exists a countable set $f_1, f_2, \ldots$ in X' with the property that if $x \neq 0$, then there exists an f_j such that $f_j(x) \neq 0$.

29. Let $x_1 = (\alpha, \beta, 0, \ldots), x_2 = (0, \alpha, \beta, 0, \ldots), x_3 = (0, 0, \alpha, \beta, 0, \ldots), \ldots,$ where $|\frac{\alpha}{\beta}| > 1$. Show that $\{x_j\}$ is a Schauder basis for ℓ_2.

30. Let $x_1 = (\alpha, 0, \ldots), x_2 = (\alpha, \beta, 0, \ldots), x_3 = (0, \alpha, \beta, 0, \ldots), \ldots,$ where $|\frac{\alpha}{\beta}| < 1$. Show that $\{x_j\}$ is a Schauder basis for ℓ_2.

31. Let $A \in \mathcal{L}(X)$ be invertible. Prove that $\{x_j\}$ is a Schauder basis for X if and only if $\{Ax_j\}$ is a Schauder basis for X.

Chapter XII
Linear Operators on a Banach Space

Many of the definitions, theorems and proofs concerning operators in $\mathcal{L}(\mathcal{H}_1, \mathcal{H}_2)$ carry over verbatim to operators in $\mathcal{L}(X, Y)$, where X and Y are Banach spaces. Chapter II, Sections 1, 3, 8 and Theorems 16.1, 16.3 are illustrations of this assertion. We shall refer to these results within the framework of Banach spaces even though they were stated for Hilbert spaces.

There are also very substantial differences between operators on Banach and Hilbert spaces. For instance, in general, there is no notion of a self adjoint operator on a Banach space which is not a Hilbert space.

12.1 Description of Bounded Operators

The following result characterizes operators in $\mathcal{L}(\ell_1)$ which stem from matrices.

Theorem 1.1 *Let* $(a_{jk})_{j,k=1}^{\infty}$ *be a matrix of complex numbers. Define A on ℓ_1 by*

$$A(\xi_1, \xi_2, \ldots) = (\eta_1, \eta_2, \ldots),$$

where

$$\eta_j = \sum_{k=1}^{\infty} a_{jk} \xi_k, \quad j = 1, 2, \ldots.$$

The operator A is in $\mathcal{L}(\ell_1)$ if and only if

$$m = \sup_k \sum_{j=1}^{\infty} |\alpha_{jk}| < \infty,$$

in which case $\|A\| = m$.

Proof: If A is in $\mathcal{L}(\ell_1)$ and $\{e_k\}$ is the standard basis for ℓ_1, then

$$\|A\| \geq \|Ae_k\| = \sum_{j=1}^{\infty} |a_{jk}|, \quad k = 1, 2, \ldots.$$

Thus $\|A\| \geq m$.

On the other hand, if $m < \infty$, then given $\xi = (\xi_1, \xi_2, \ldots) \in \ell_1$,

$$\|A\xi\| = \sum_{j=1}^{\infty} \left| \sum_{k=1}^{\infty} a_{jk}\xi_k \right| \le \sum_{k=1}^{\infty} \sum_{j=1}^{\infty} |a_{jk}||\xi_k| \le m \sum_{k=1}^{\infty} |\xi_k| = m\|\xi\|.$$

Hence $\|A\| \le m$. $\qquad\qquad\qquad\qquad\qquad\qquad\qquad\qquad\qquad\qquad\qquad\square$

Theorem 1.2 *Given a continuous complex valued function k defined on $[a, b] \times [a, b]$, let K be defined on $X = C([a, b])$ by*

$$(Kf)(t) = \int_a^b k(t, s) f(s) \, ds.$$

The operator K is in $\mathcal{L}(X)$ and

$$\|K\| = \max_t \int_a^b |k(t, s)| \, ds.$$

Proof: It follows from the uniform continuity of k that K is a linear map from X into X. Let

$$m = \max_t \int_a^b |k(t, s)| \, ds.$$

Since

$$|(Kf)(t)| \le \int_a^b |k(t, s)||f(s)| \, ds \le m\|f\|,$$

we have $\|K\| \le m$. Now

$$y(t) = \int_a^b |k(t, s)| \, ds$$

is continuous on $[a, b]$ and therefore y attains its maximum at some point $t_0 \in [a, b]$. Let

$$g(s) = \frac{\overline{k(t_0, s)}}{|k(t_0, s)|} \quad \text{if } k(t_0, s) \ne 0$$

and zero otherwise. Since g is bounded and Lebesgue measurable, it is in $L_1([a, b])$. It follows from the fact that $C([a, b])$ is dense in $L_1([a, b])$ that there exists a sequence $\{g_n\}$ in $C([a, b])$ such that

$$\|g_n\| = \max_{s \in [a,b]} |g_n(s)| \le 1$$

and $\{g_n\}$ converges in $L_1([a, b])$ to g. Hence

$$\|K\| \ge \|Kg_n\| \ge (Kg_n)(t_0) = \int_a^b k(t_0, s)g_n(s) \, ds \to \int_a^b k(t_0, s)g(s) \, ds.$$

Therefore,

$$\|K\| \geq \int_a^b k(t_0, s) g(s) \, ds = \int_a^b |k(t_0, s)| \, ds = m.$$

Let K be an integral operator with kernel function $k \in L_2([a, b] \times [a, b])$. Suppose that $\{\varphi_n\}$, $\{\lambda_n\}$ is a basic system of eigenvectors and real eigenvalues of K. The Hilbert-Schmidt Theorem V.1.1 states that if

$$C = \sup_t \int_a^b |k(t, s)|^2 \, ds < \infty,$$

then

$$(Kf)(t) = \sum_j \lambda_j \langle f, \varphi_j \rangle \varphi_j(t) \quad \text{a.e.,} \tag{1.1}$$

and the series converges uniformly and absolutely on $[a, b]$. We shall give the following simple proof of this result.

Since

$$|(Kf)(t)| \leq \left(\int_a^b |k(t, s)|^2 \, ds \right)^{1/2} \|f\|_2 \leq C \|f\|_2,$$

it follows that K is a bounded linear map from $L_2([a, b])$ into $L_\infty([a, b])$. Now we know that

$$s_n(t) = (P_0 f)(t) + \sum_{j=1}^n \langle f, \varphi_j \rangle \varphi_j(t)$$

converges in $L_2([a, b])$ to f. Therefore,

$$(Ks_n)(t) = \sum_{j=1}^n \lambda_j \langle f, \varphi_j \rangle \varphi_j(t)$$

converges in $L_\infty([a, b])$ to Kf. The uniform convergence of the series in (1.1) follows from the definition of the norm on $L_\infty([a, b])$. The absolute convergence follows from the fact that the argument does not depend on the arrangement of the terms of the series.   □

12.2 Closed Linear Operators

In Chapter VI we introduced closed linear operators and proved some results in the setting of Hilbert spaces. These definitions and the proofs of the corresponding results carry over verbatim to Banach spaces. For example, let X and Y be Banach spaces. A linear operator A with domain $\mathcal{D}(A) \subseteq X$ and range Im $A \subseteq Y$, denoted by $A(X \to Y)$, is called a *closed operator* if it has the property that whenever $\{x_n\}$ is a sequence in $\mathcal{D}(A)$ satisfying $x_n \to x$ in X and $Ax_n \to y \in Y$, then $x \in \mathcal{D}(A)$ and $Ax = y$.

Examples: 1. Let $X = Y = C[0, 1]$ and let A be defined by

$$\mathcal{D}(A) = \{f \in X \,|\, f' \in X\}, \quad Af = f'.$$

The linear operator A is unbounded. For if $f_n(t) = t^n$, then

$$\|f_n\| = \max_{t \in [0.1]} |f_n(t)| = 1 \quad \text{and} \quad \|Af_n\| = \|f_n'\| = n.$$

To see that A is closed, suppose that $f_n \to f$ and $Af_n \to g$. Then it follows from the definition of the norm on $C([0, 1])$ that $\{f_n\}$ and $\{f_n'\}$ converge uniformly on $[0, 1]$ to f and g, respectively. By considering $\int_0^t f'(s)\,ds$, it is easy to verify that f is differentiable on $[0, 1]$ and $Af = f' = g$. Thus A is closed.

2. If A is invertible, then A is closed. The proof of this assertion is exactly the same as the proof of Theorem VI.1.1.

3. Let $X = Y = L_p([0, 1])$, $1 \le p < \infty$ and let A be defined by $\mathcal{D}(A) = \{f \in X \,|\, f$ absolutely continuous on $[0, 1]$, $f' \in X$, $f(0) = 0\}$, $Af = f'$.

A is obviously one-one and $\text{Im } A = X$ since given $g \in X$,

$$(A^{-1}g)(t) = \int_0^t g(s)\,ds.$$

Now by Hölder's inequality,

$$|A^{-1}g(t)| \le \int_0^1 |g(s)|\,ds \le \left(\int_0^1 |g(s)|^p \right)^{1/p} = \|g\|_p.$$

Hence

$$\|A^{-1}g\|_p^p = \int_0^1 |(A^{-1}g)(t)|^p \, dt \le \|g\|_p^p.$$

Thus A is closed by Example 2.

For a through treatment of closed differential operators in a Banach space setting, the reader is referred to [G], [GGK1] and [DS 2].

Define the *graph norm* $\|\ \|_A$ on $\mathcal{D}(A)$ by

$$\|x\|_A = \|x\| + \|Ax\|.$$

If A is closed, then the argument used in Section VI.3 shows that $\mathcal{D}(A)$ with norm $\|\cdot\|_A$ is a Banach space.

The operator A is closed if and only if its graph $G(A) = \{(x, Ax)\,|\,x \in \mathcal{D}(A)\}$ is a closed subspace of $X \times Y$ with norm $\|(x, y)\| = \|x\| + \|y\|$.

12.3 Closed Graph Theorem

The closed graph theorem is another fundamental result in operator theory. The theorem cannot be fully appreciated unless one sees some important applications – which we shall present in the next section.

Closed Graph Theorem 3.1 *A closed linear operator which maps a Banach space into a Banach space is continuous.*

It should be stressed that the closed graph theorem requires that the domain of the operator be *complete*. The differential operator given in Example 1 of the preceding section is closed but unbounded. Its domain is a proper dense subspace of the Banach space $C([0, 1])$.

To prove the theorem we need the Baire category theorem (Theorem 3.2 below) and a lemma. In the proof we use the following notation. Given x_0 in a normed linear space X and $r > 0$,

$$S(x_0, r) = \{x : \|x - x_0\| < r\}$$
$$\overline{S}(x_0, r) = \{x : \|x - x_0\| \leq r\}.$$

For $Z \subset X$ and $\alpha \in \mathbb{C}$

$$\alpha Z = \{\alpha z : z \in Z\}.$$

A vector x_0 is called an *interior point* of a set $Z \subset X$ if $S(x_0, r) \subset Z$ for some $r > 0$.

Baire Category Theorem 3.2 *If a Banach space is the union of a countable number of closed sets, then at least one of the closed sets has an interior point.*

Proof: Let the Banach space $X = \cup_n C_n$, where each C_n is a closed set. Suppose that none of the C_n has an interior point. Choose $x_1 \in C_1$. Since $S(x_1, 1) \not\subset C_1$ and C_1 is closed, there exist x_2 and r_2, $0 < r_2 < \frac{1}{2}$, such that

$$\overline{S}(x_2, r_2) \subset S(x_1, r_1) \quad \text{and} \quad \overline{S}(x_2, r_2) \cap C_1 = \emptyset.$$

Since $S(x_2, r_2) \not\subset C_2$ and C_2 is closed, there exist x_3 and r_3, $0 < r_3 < \frac{1}{3}$, such that

$$\overline{S}(x_3, r_3) \subset S(x_2, r_2) \quad \text{and} \quad \overline{S}(x_3, r_3) \cap C_2 = \emptyset.$$

Continuing in this manner, we obtain sequences $\{x_n\}$ and $\{r_n\}$, $0 < r_n < \frac{1}{n}$, such that

$$\overline{S}(x_{n+1}, r_{n+1}) \subset S(x_n, r_n) \quad \text{and} \quad \overline{S}(x_{n+1}, r_{n+1}) \cap C_n = \emptyset. \tag{3.1}$$

The sequence $\{x_n\}$ is a Cauchy sequence. For if $n > m$, then we have from (3.1) that $x_n \in S(x_m, r_m)$, i.e.,

$$\|x_n - x_m\| < r_m < \frac{1}{m}. \tag{3.2}$$

Hence $\{x_n\}$ converges to some $x \in X$. Fix m and let $n \to \infty$ in (3.2). Then $\|x - x_m\| \le r_m$, i.e., $x \in S(x_m, r_m)$ which is disjoint from C_{m-1}, $m = 2, \ldots$. But this is impossible since $X = \cup_n C_n$. $\qquad\square$

Lemma 3.3 *Suppose C is a convex set in X and $C = (-1)C$. If C has an interior point, then zero is also an interior point of C.*

Proof: Suppose $S(x_0, r) \subset C$. If $\|x\| < 2r$, then

$$x = \left(x_0 + \frac{x}{2}\right) - \left(x_0 - \frac{x}{2}\right) \in S(x_0, r) + (-1)S(x_0, r) \subset C + C.$$

But $C + C = 2C$. Indeed, given u and v in C,

$$u + v = 2\left(\frac{1}{2}u + \frac{1}{2}v\right) \in 2C.$$

Since C is convex. We have shown that $S(0, 2r) \subset 2C$, which implies that $S(0, r) \subset C$. $\qquad\square$

Proof of the Closed Graph Theorem: Let T be a closed linear operator which maps the Banach space X into the Banach space Y. Define $Z = \{x : \|Tx\| < 1\}$. First we prove that the closure $\overline{Z}$ of Z has an interior point.

Since $D(T) = X$ and T is linear, $X = \cup_{n=1}^{\infty} nZ$. It follows from the Baire Category Theorem that there exists a positive integer k such that $k\overline{Z} = \overline{kZ}$ has an interior point. Therefore, $\overline{Z}$ has an interior point. It is easy to verify that $\overline{Z}$ is convex and $\overline{Z} = (-1)\overline{Z}$. By Lemma 3.3 $S(0, r) \subset \overline{Z}$ for some $r > 0$, which implies

$$S(0, \alpha r) \subset \alpha \overline{Z} = \overline{\alpha Z}, \quad \alpha > 0. \tag{3.3}$$

Given $0 < \varepsilon < 1$ and $\|x\| < r$, we have from (3.3) that x is in $\overline{Z}$. Therefore, there exists an $x_1 \subset Z$ such that $\|x - x_1\| < \varepsilon r$. Since $x - x_1 \in S(0, r) \subset \overline{\varepsilon Z}$, there exists an $x_2 \in \varepsilon Z$ such that

$$\|x - x_1 - x_2\| < \varepsilon^2 r.$$

Inductively, there exists a sequence $\{x_n\}$ such that

$$\left\| x - \sum_{k=1}^{n} x_k \right\| < \varepsilon^n r, \quad x_n \in \varepsilon^{n-1} Z. \tag{3.4}$$

Let $s_n = \sum_{k=1}^{n} x_k$. From (3.4) and the definition of Z we get

$$s_n \to x \quad \text{and} \quad \|Tx_n\| < \varepsilon^{n-1}. \tag{3.5}$$

Now $\{Ts_n\}$ is a Cauchy sequence since (3.5) implies that for $n > m$,

$$\|Ts_n - Ts_m\| \le \sum_{k=m+1}^{\infty} \|Tx_k\| \le \sum_{k=m}^{\infty} \varepsilon^k = \frac{\varepsilon^m}{1 - \varepsilon} \to 0$$

as $m \to \infty$. Hence, by the completeness of Y, $Ts_n \to y$ for some $y \in Y$. So, we have $s_n \to x$ and $Ts_n \to y$. Since T is closed, $Tx = y$. Thus

$$\|Tx\| = \|y\| = \lim_{n \to \infty} \|Ts_n\| \le \sum_{k=1}^{\infty} \|Tx_k\| \le \frac{1}{1 - \varepsilon},$$

whenever $\|x\| < r$. In particular, if $\|v\| = 1$, then

$$\left\| T\left(\frac{r}{2}v\right) \right\| < \frac{1}{1 - \varepsilon}.$$

Thus

$$\|Tv\| < \frac{2}{r(1 - \varepsilon)}, \qquad \|v\| = 1,$$

which shows that $\|T\| \le \frac{2}{r(1-\varepsilon)}$. □

12.4 Applications of the Closed Graph Theorem

In this section we present a number of applications of the closed graph theorem. The first concerns invertibility.

Recall that an operator $A \in \mathcal{L}(X, Y)$ is invertible if each of the conditions (a) Ker $A = (0)$ (b) Im $A = Y$ (c) A^{-1} is bounded hold. The next theorem shows that (a) and (b) imply (c).

Theorem 4.1 *Suppose X and Y are Banach spaces. If $A(X \to Y)$ is closed with properties* Ker $A = (0)$ *and* Im $A = Y$, *then A^{-1} is bounded on Y.*

Proof: Since $G(A) = \{(x, Ax) : x \in X\}$ is closed in $X \times Y$, $G(A^{-1}) = \{(Ax, x) : x \in X\}$ is closed in $Y \times X$, i.e., A^{-1} is a closed linear operator mapping Y into X. Therefore, A^{-1} is bounded by the closed graph theorem. □

Corollary 4.2 *Suppose $\| \ \|_1$ and $\| \ \|_2$ are norms on the vector space X such that $(X, \| \ \|_1)$ and $(X, \| \ \|_2)$ are complete. If there exists a number C such that*

$$\|x\|_1 \le C\|x\|_2 \quad \text{for all} \quad x \in X,$$

then $\| \ \|_1$ and $\| \ \|_2$ are equivalent, i.e., there exists $C_1 > 0$ such that $C_1\|x\|_2 \le \|x\|_1 \le C\|x\|_2$.

Proof: Let I be the identity map on X. Now I is a bounded linear map from $(X, \| \ \|_2)$ onto $(X, \| \ \|_1)$ since

$$\|Ix\|_1 = \|x\|_1 \le C\|x\|_2.$$

Hence I^{-1} is bounded by Theorem 4.1, and

$$\frac{1}{C}\|x\|_1 \le \|x\|_2 = \|I^{-1}x\|_2 \le \|I^{-1}\|\|x\|_1.$$

$\square$

We now give two proofs of the following fundamental result. One proof relies on the closed graph theorem.

Theorem 4.3 (Uniform Boundedness Principle). *Given Banach spaces X and Y, suppose F is a subset of $\mathcal{L}(X, Y)$ with the property that for each $x \in X$, $\sup_{A \in F} \|Ax\| < \infty$. Then $\sup_{A \in F} \|A\| < \infty$.*

Proof: Suppose $\sup_{A \in F} \|A\| = \infty$. Then for each positive n there exists an operator $A \in F$, denoted by A_n, such that $\|A_n\| \ge n$. However, the exact proof of Theorem II.14.3 with the Hilbert spaces $\mathcal{H}_1, \mathcal{H}_2$ replaced by the Banach spaces X, Y, respectively, shows that $\sup_n \|A_n\| < \infty$. This contradicts $\|A_n\| \ge n$, $n = 1, 2, \ldots$.

Another proof of Theorem 4.3 is the following.

Let $B(F, Y)$ denote the vector space of functions f which map F into Y and have the property that

$$\|f\| = \sup_{A \in F} \|f(A)\| < \infty.$$

Then $\| \ \|$ is a norm on $B(F, Y)$ and by the argument which was used to prove that $\mathcal{L}(X, Y)$ is complete (Theorem XI.1.1), we have that $B(F, Y)$ is complete. Define $T : X \to B(F, Y)$ by

$$(Tx)A = Ax, \ x \in X.$$

It is not difficult to verify that T is a closed linear operator on X. Hence the boundedness of T is ensured by the closed graph theorem. Thus for all $A \in F$ and $x \in X$,

$$\|Ax\| = \|(Tx)A\| \le \|Tx\| \le \|T\|\|x\|,$$

which shows that

$$\sup_{A \in F} \|A\| \le \|T\|.$$

$\square$

Corollary 4.4 *Let X and Y be Banach spaces. Suppose $\{A_n\} \subset \mathcal{L}(X, Y)$ is a sequence such that $\{A_n x\}$ converges for each $x \in X$. Then the operator*

$$Ax = \lim_{n \to \infty} A_n x, \quad x \in X$$

is linear and bounded and the sequence $\{\|A_n\|\}$ is bounded.

Proof: For each $x \in X$, the sequence $\{A_n x\}$ is bounded since it converges. Thus by the uniform boundedness principle, $\sup_n \|A_n\| = m < \infty$ and

$$\|Ax\| = \lim_{n \to \infty} \|A_n x\| \le m\|x\|,$$

which shows that $\|A\| \le m$. ☐

A simple application of the uniform boundedness principle yields the following result.

Theorem 4.5 *Suppose that S is a subset of a Banach space X such that for each $f \in X'$,*

$$\sup_{x \in S} |f(x)| < \infty.$$

Then S is bounded.

Proof: For each $x \in S$, define the linear functional F_x on the conjugate space X' by $F_x f = f(x)$. Clearly, F_x is linear and by Corollary XI.5.5

$$\|F_x\| = \sup_{\|f\|=1} |F_x(f)| = \sup_{\|f\|=1} |f(x)| = \|x\|. \tag{4.1}$$

Thus F_x is a bounded linear functional on the Banach space X' and, by hypothesis, for each $g \in X'$,

$$\sup_{x \in S} |F_x(g)| = \sup_{x \in S} |g(x)| < \infty.$$

Hence by (4.1) and Theorem 4.3,

$$\sup_{x \in S} \|x\| = \sup_{x \in S} \|F_x\| < \infty.$$

☐

Another application of the uniform boundedness principle gives the following result. If $\beta = \{\beta_1, \beta_2, \ldots\}$ is a sequence of complex numbers such that the series $\sum_{j=1}^{\infty} \beta_j \xi_j$ converges for every $\{\xi_j\} \in \ell_p$, $1 \le p < \infty$, then β is in ℓ_q, $\frac{1}{p} + \frac{1}{q} = 1$. To prove this, define f on ℓ_p by $f(\{\xi_j\}) = \sum_{j=1}^{\infty} \beta_j \xi_j$. Let $f_n(\{\xi_j\}) = \sum_{j=1}^{n} \beta_j \xi_j$, $n = 1, 2, \ldots$. Clearly, f_n is in the conjugate space ℓ'_p and $f_n(x) \to f(x)$ for each $x \in \ell$. Hence f is in ℓ'_p by Corollary 4.4 and, by Theorem XI.4.1, β is in ℓ_q with $\|\beta\|_q = \|f\|$.

Theorem 4.6 *An operator $A \in \mathcal{L}(X, Y)$ is one-one and has a closed range if and only if there exists an $m > 0$ such that*

$$\|Ax\| \ge m\|x\| \quad \text{for all} \quad x \in X.$$

The proof is the same as the proof of Corollary II.14.2.

12.5 Complemented Subspaces and Projections

Just as every closed subspace of a Hilbert space has a projection associated with it, so does every closed, *complemented* subspace of a Banach space.

Definition. Let M be a subspace of X. An operator P is called a *projection* from X onto M if it is a bounded linear map from X onto M and $P^2 = P$.

If x is in M, then $Px = x$. Indeed, there exists a $z \in X$ such that $x = Pz$. Hence $Px = P^2z = Pz = x$.

It is easy to see that if P is a projection then $Q = I - P$ is also a projection and Im $P =$ Ker Q, Ker $P =$ Im Q. Hence Im P is closed.

Theorem 5.1 *A closed subspace M of a Banach space X is complemented in X if and only if there exists a projection from X onto M.*

If M is complemented by the closed subspace N, then there exists a $c > 0$ such that

$$\|u + v\| \geq c\|u\|, \quad \text{for all } u \in M \text{ and } v \in N.$$

Proof: Suppose $X = M \oplus N$. Then given $x \in X$, there exists a unique $u \in M$ and a unique $v \in N$ such that $x = u + v$. The operator P defined by $Px = u$ is a linear map from X onto M and $P^2 = P$. We now prove that P is bounded by showing that it is closed.

Suppose

$$x_k = u_k + v_k \to x, \ u_k \in M, \ v_k \in N$$

and $Px_k = u_k \to y$. Now y is in M since M is closed. Therefore $y = Py$ and $v_k \to x - y$. Since N is also closed, $x - y$ is in N and

$$0 = P(x - y) = Px - Py = Px - y.$$

Thus P is closed. Hence P is bounded and

$$\|P\|\|u + v\| \geq \|P(u + v)\| = \|u\|.$$

Conversely, if P is a projection from X onto M, then $N =$ Ker P is a closed subspace of X. Furthermore, $M \cap N = (0)$ since $x \in M \cap N$ implies $x = Px = 0$. Now $x = Px + (x - Px)$ and $x - Px$ is in N. Thus $X = M \oplus N$. $\square$

Example: Let $\mathcal{H}$ be a Hilbert space. We shall now construct two closed subspaces L_1 and L_2 with the following properties:

$$L_1 \cap L_2 = \{0\}$$

and

$$\overline{L_1 \oplus L_2} = H, \quad L_1 \oplus L_2 \neq H,$$

where the left side means the closure of $L_1 \oplus L_2$.

This construction allows one to construct a linear operator Π_0 such that Π_0 is densely defined, $\Pi_0^2 = \Pi_0$ and Π_0 is not bounded. In other words, Π_0 could be considered as an unbounded projection.

Indeed, let us assume that the subspaces L_1 and L_2 are already constructed. Define the operator Π_0 by the equality

$$\Pi_0(x + y) = x, \; x \in L_1, \; y \in L_2.$$

It is obvious that

$$\Pi_0^2 = \Pi_0.$$

Now we have to prove that Π_0 is unbounded on $L_1 \oplus L_2$. We prove it by contradiction. Assume that Π_0 is bounded. Then

$$\|x\| = \|\Pi_0(x + y)\| \le \|\Pi_0\| \, \|x + y\|,$$

also

$$\|y\| = \|(I - \Pi_0)\,(x + y)\| \le \|I - \Pi_0\| \, \|x + y\|.$$

Let $u_n = x_n + y_n, \, x_n \in L_1, \, y_n \in L_2, n = 1, 2, \ldots$ be a convergent sequence from $L_1 \oplus L_2$. Denote by u its limit (in norm). From the above inequalities it follows that the sequences $\{x_n\}_1^\infty$ and $\{y_n\}_1^\infty$ are also convergent. Denote by $x \in L_1$ and $y \in L_2$ their respective limits. It is clear that $u = x + y$ and hence $L_1 \oplus L_2$ is closed and coincides with H. This contradict the properties of L_1 and L_2. We conclude that Π_0 is unbounded.

Now let us return to the construction of L_1 and L_2.

Let H be a separable Hilbert space, and let $\varphi_1, \varphi_2, \ldots$ be an orthonormal basis in H. Define L_1 to be the closed linear span of the vectors

$$\varphi_1, \varphi_3, \varphi_5, \ldots,$$

and let L_2 be the closed linear span of the vectors

$$\varphi_1 + \frac{1}{2}\,\varphi_2, \; \varphi_3 + \frac{1}{2^2}\,\varphi_4, \; \varphi_5 + \frac{1}{2^3}\,\varphi_6, \ldots.$$

It is clear that the first set of vectors is an orthonormal basis for L_1, while the second set is an orthogonal basis for L_2.

The intersection $L_1 \cap L_2$ consists of zero only. Indeed, let $f \in L_1 \cap L_2$. Then f can be represented as

$$f = \sum_{j=1}^{\infty} \alpha_j \varphi_{2j-1}, \quad f = \sum_{j=1}^{\infty} \gamma_j \left(\varphi_{2j-1} + \frac{1}{2j}\,\varphi_j \right),$$

with the two series converging in the norm of H. From these equalities it follows that

$$\sum_{j=1}^{\infty}(\alpha_j - \gamma_j)\varphi_{2j-1} = \sum_{j=1}^{\infty}\gamma_j\frac{1}{2j}\varphi_{2j}.$$

Notice that in this equality the left hand side is in L_1 while the right hand is orthogonal to L_1. This leads to $\alpha_j = \gamma_j$ and $\gamma_j = 0$ for $j = 1, 2, \ldots$. Thus $f = 0$, and hence $L_1 \cap L_2 = \{0\}$.

Also, $\overline{L_1 \oplus L_2} = \mathcal{H}$. For suppose this is not the case. Then there exists a $u \neq 0$ in $\mathcal{H}$ such that $u \perp L$, and $u \perp L_2$. But then $u \perp \varphi_1, \varphi_2, \ldots$ which implies that $u = 0$. Now assume $L_1 \oplus L_2 = \mathcal{H}$. Take

$$x_n = \varphi_{2n-1}, \ n = 1, 2, \ldots$$

and

$$y_n = \varphi_{2n-1} + \frac{1}{2^n}\varphi_{2n}, \ n = 1, 2, \ldots.$$

It is obvious that $\|x_n\| = 1$, $\|y_n\| = \sqrt{1 + \frac{1}{2^{2n}}}$ and

$$\lim_{n\to\infty}\|x_n - y_n\| = \lim_{n\to\infty}\frac{1}{2^n} = 0.$$

But this is impossible since by Theorem 5.1, there exists a $c > 0$ such that

$$\|x_n - y_n\| \geq C\|x_n\| \geq C, \ n = 1, 2, \ldots$$

Hence $L_1 \oplus L_2 \neq \mathcal{H}$.

12.6 One-Sided Invertiblity Revisited

In this section we extend the results of Section II.15 to Banach spaces X and Y. The definition of one-sided invertibility for an operator on a Banach space is the same as for an operator on a Hilbert space (cf. II.15)

Theorem 6.1 *A necessary and sufficient condition for an operator $A \in \mathcal{L}(X, Y)$ to have a left inverse is that $\mathrm{Ker}\, A = \{0\}$ and $\mathrm{Im}\, A$ is closed and complemented in Y. In this case, the operator AL is a projection onto $\mathrm{Im}\, A$ and $\mathrm{Ker}\, AL = \mathrm{Ker}\, L$, where L is a left inverse of A.*

Proof: The proof of the necessity is exactly the same as the proof of Theorem II.15.2. Suppose $\mathrm{Ker}\, A = \{0\}$ and $Y = \mathrm{Im}\, A \oplus M$, where $\mathrm{Im}\, A$ and M are closed subspaces of Y. Let P be the projection from Y onto $\mathrm{Im}\, A$ with $\mathrm{Ker}\, P = M$ and let A_1 be the operator A considered as a map from X onto the Banach space $\mathrm{Im}\, A$. Then A_1 is invertible by Theorem 4.1 and $A_1^{-1}P$ is a bounded left inverse of A since $A_1^{-1}PAx = A_1^{-1}Ax = x$. $\qquad\square$

Theorem 6.2 *A necessary and sufficient condition for an operator $A \in \mathcal{L}(X, Y)$ to have a right inverse is that* Im $A = Y$ *and* Ker A *is complemented in* X. *In this case,* $I - D^{(-1)}D$ *is a projection onto* Ker A, *where* $D^{(-1)}$ *is a right inverse of* A.

Proof: The proof of the necessity is exactly the same as the proof of Theorem II.15.4. Suppose now that Im $A = Y$ and $X = $ Ker $A \oplus N$, where N is a closed subspace of X. Let P be the projection of X onto N with Ker $P = $ Ker A. Let A_1 be the operator A restricted to N. Then A_1 is injective and $A_1 N = AX = Y$. Hence A_1 is invertible and $AA_1^{-1}y = y$.

The proofs of the following theorems are exactly the same as the proofs of the corresponding Theorems II.15.3 and II.15.5. $\qquad\square$

Theorem 6.3 *Let* $A \in \mathcal{L}(X, Y)$ *have a left inverse* $L \in \mathcal{L}(Y, X)$. *If* B *is an operator in* $\mathcal{L}(X, Y)$ *and* $\|A - B\| < \|L^{-1}\|^{-1}$, *then* B *has a left inverse* L, *given by*

$$L_1 = L(I - (A - B)L)^{-1} = L\left(\sum_{k=0}^{\infty}[(A - B)L]\right)^k. \tag{6.1}$$

Moreover,

$$\text{codim Im } B = \text{codim Im } A. \tag{6.2}$$

Theorem 6.4 *Let* $A \in \mathcal{L}(X, Y)$ *have a right inverse* R. *If* B *is an operator in* $\mathcal{L}(X, Y)$ *and* $\|A - B\| < \|R\|^{-1}$, *then* B *has a right inverse* $R_1 \in \mathcal{L}(Y, X)$ *given by*

$$R_1 = (I - R(A - B))^{-1}R. \tag{6.3}$$

Moreover,

$$\dim \text{Ker } A = \dim \text{Ker } B. \tag{6.4}$$

12.7 The Projection Method Revisited

Definition: Let X and Y be Banach spaces and let $\{P_n\}$, $\{Q_n\}$ be sequences of projections in $\mathcal{L}(X)$ and $\mathcal{L}(Y)$, respectively, with the properties that $P_n x \to x$ for all $x \in X$ and $Q_n y \to y$ for all $y \in Y$. Given an invertible operator $A \in \mathcal{L}(X, Y)$, the *projection method* for A seeks to approximate a solution to the equation $Ax = y$ by a sequence $\{x_n\}$ of solutions to the equations

$$Q_n A P_n x = Q_n y, \; n = 1, 2, \ldots. \tag{7.1}$$

The projection method for A is said to *converge* if there an integer N such that for each $y \in Y$ and $n \geq N$, there exists a *unique* solution x_n to equation (7.1) and, in addition, the sequence $\{x_n\}$ converges to $A^{-1}y$. We denote by $\Pi(P_n, Q_n)$ the set of invertible operators for which the convergent method converges.

The projection method for A was presented in Section II.17 for $X = Y$ a Hilbert space and $Q_n = P_n$.

Theorem 7.1 *Let $A \in \mathcal{L}(X, Y)$ be invertible. Then $A \in \Pi(P_n, Q_n)$ if and only if there exists an integer N such that for $n \geq N$,*

   (i) *the restriction of the operator $Q_n A P_n$ to $\operatorname{Im} P_n$ has a bounded inverse on $\operatorname{Im} Q_n$, denoted by $(Q_n A P_n)^{-1}$, and*

   (ii) $\sup_{n \geq N} \|(Q_n A P_n)^{-1}\| < \infty$.

Proof: By arguing as in the proof of Theorem II.17.1 with $P_n A P_n$ replaced by $Q_n A P_n$, we obtain Theorem 7.1. $\qquad\qquad\qquad\qquad\qquad\qquad\qquad\qquad\qquad\qquad$ $\square$

Corollary 7.2 *Let $\{P_n\}$ be a sequence of projections on a Banach space X with $\|P_n\| = 1$, $n = 1, 2, \ldots$. Assume $P_n x \to x$ for all $x \in X$. Suppose $A \in \mathcal{L}(X)$ with $\|I - A\| < 1$. Then $A \in \Pi_n(P_n, P_n)$.*

The proof of the corollary is exactly the same as the proof of Corollary II.17.2. The proof of the next theorem is analogous to the proof of the corresponding Theorem II.17.6 with $P_n A P_n$ replaced by $Q_n A P_n$.

Theorem 7.3 *Suppose $A \in \Pi(P_n, Q_n)$. There exists a $\gamma > 0$ such that if $B \in \mathcal{L}(X, Y)$ and $\|B\| < \gamma$, then $A + B \in \Pi(P_n, Q_n)$.*

12.8 The Spectrum of an Operator

In this section we extend the notion of the spectrum of an operator (which was introduced in Section II.20 for Hilbert space operators) to a Banach space setting. Throughout this section X is a Banach space.

Definition: Given $A \in \mathcal{L}(X)$, a point $\lambda \in C$ is called a *regular point* of A if $\lambda I - A$ is invertible. The set $\rho(A)$ of regular points is called the *resolvent set* of A. The *spectrum $\sigma(A)$* of A is the complement of $\rho(A)$.

Theorem 8.1 *The resolvent set of any $A \in \mathcal{L}(X)$ is an open set in $\mathbb{C}$ containing $\{\lambda \mid |\lambda| > \|A\|\}$. Hence $\sigma(A)$ is a closed bounded set contained in $\{\lambda \mid |\lambda| \leq \|A\|\}$. Furthermore, for $\lambda \in \partial\sigma(A)$, the boundary of $\sigma(A)$, the operator $\lambda I - A$ is neither left nor right invertible.*

Proof: The proof is exactly the same as the proof of Theorem II.20.1 $\qquad\qquad$ $\square$

Examples: 1. If A is a compact self adjoint operator on an infinite dimensional Hilbert space, then $\sigma(A)$ consists of zero and the eigenvalues of A by Theorem IV.8.1. The formula for $(\lambda I - A)^{-1}$ is given in the same theorem.

   2. Let $X = C([c, d])$. Define $A \in \mathcal{L}(X)$ by

$$(Af)(t) = a(t) f(t),$$

where $a(t) \in X$. For $\lambda \neq a(t), t \in [c, d]$, it is clear that $\lambda \in \rho(A)$ and

$$((\lambda I - A)^{-1}g)(t) = \frac{1}{\lambda - a(t)}g(t).$$

If $\lambda = a(t_0)$ for some $t_0 \in [c, d]$, then $\lambda \in \sigma(A)$. Indeed, suppose $\lambda \in \rho(A)$. Then for some $f \in X$ and all $t \in [c, d]$,

$$1 = ((\lambda I - A)f)(t) = (a(t_0) - a(t))f(t).$$

But this is impossible since $\lim_{t \to t_0}(\lambda - a(t))f(t) = 0$. Hence

$$\sigma(A) = \{a(t), c \leq t \leq d\}.$$

3. For $X = \ell_p, 1 \leq p < \infty$, let $A \in \mathcal{L}(X)$ be the backward shift operator,

$$A(\alpha_1, \alpha_2, \ldots) = (\alpha_2, \alpha_3, \ldots).$$

Clearly, $\|A\| = 1$. Thus if $|\lambda| > 1$, then $\lambda \in \rho(A)$ by Theorem 6.1. If $|\lambda| < 1$, then $x = (1, \lambda, \lambda^2, \ldots)$ is in Ker$(\lambda I - A)$. Therefore, $\lambda \in \sigma(A)$. Since $\sigma(A)$ is a closed set, it follows that $\sigma(A) = \{\lambda : |\lambda| \leq 1\}$.

If $|\lambda| > 1$, then

$$(\lambda I - A)^{-1} = \frac{1}{\lambda}\left(I - \frac{A}{\lambda}\right)^{-1} = \sum_{k=0}^{\infty} \frac{A^k}{\lambda^{k+1}}.$$

Thus given $y = (\beta_1, \beta_2, \ldots) \in X$, $(\lambda I - A)^{-1}y = (\alpha_1, \alpha_2, \ldots)$, where

$$\alpha_j = \sum_{k=0}^{\infty} \frac{\beta_{j+k}}{\lambda^{k+1}}.$$

The spectrum of the backward shift operator on $\ell_p, 1 \leq p < \infty$, was shown above to be independent of p. The following example shows that the spectrum can vary as the space varies.

4. Given a number $q > 0$, let $\ell_1(q)$ be the set of those sequences $x = (\alpha_1, \alpha_2, \ldots)$ of complex numbers such that

$$\|x\|_q = \sum_{j=1}^{\infty} |\alpha_j| q^{-j} < \infty.$$

with the usual definitions of additions and scalar multiplication, $\ell_1(q)$, together with the norm $\| \, \|_q$, is a Banach space.

Let A be the backward shift operator on $\ell_1(q)$. We shall show that $\sigma(A) = \{\lambda : |\lambda| \leq q\}$. If $|\lambda| < q$, then $x = (1, \lambda, \lambda^2, \ldots)$ is in Ker$(\lambda I - A)$. Thus

$\lambda \in \sigma(A)$. Suppose $|\lambda| > q$. Given $y = (\beta_1, \beta_2, \ldots) \in \ell_1(q)$, we wish to determine $x = (\alpha_1, \alpha_2, \ldots)$ such that $(\lambda I - A)x = y$. If this x were to exist, then

$$\alpha_{k+1} = \lambda \alpha_k - \beta_k, \ 1 \le k.$$

In particular,

$$\alpha_2 = \lambda \alpha_1 - \beta_1, \ \alpha_3 = \lambda \alpha_2 - \beta_2 = \lambda^2 \alpha_1 - \lambda \beta_1 - \beta_2.$$

By induction,

$$\alpha_{n+1} = \lambda^n \alpha_1 - \sum_{k=0}^{n-1} \lambda^k \beta_{n-k} \tag{8.1}$$

and

$$\alpha_1 = \lambda^{-n} \alpha_{n+1} + \sum_{k=0}^{n-1} \lambda^{k-n} \beta_{n-k}. \tag{8.2}$$

Since x is to be in $\ell_1(q)$ and $|\lambda| > q$,

$$|\lambda^{-n} \alpha_{n+1}| = |\lambda| \, |\lambda|^{-(n+1)} |\alpha_{n+1}| \le |\lambda| \, |\alpha_{n+1}| q^{-(n+1)} \to 0, \tag{8.3}$$

$$\sum_{k=0}^{n-1} |\lambda^{k-n} \beta_{n-k}| \le \sum_{k=0}^{n-1} |\beta_{n-k}| q^{k-n} \le \sum_{j=0}^{\infty} |\beta_j| q^{-j} < \infty. \tag{8.4}$$

It follows from (8.2)–(8.4) that we should take

$$\alpha_1 = \sum_{j=1}^{\infty} \lambda^{-j} \beta_j \tag{8.5}$$

and α_n, $n > 1$, as defined in (8.1). This argument also shows that for each $y \in \ell_1(q)$ there exists a unique $x \in \ell_1(q)$ such that $(\lambda I - A)x = y$, provided $|\lambda| > q$. Formulas (8.1) and (8.5) give $(\lambda I - A)^{-1}y$. Thus, since $\sigma(A)$ is closed and contains $\{\lambda : |\lambda| < q\}$,

$$\sigma(A) = \{\lambda : |\lambda| \le q\}.$$

We know that if $\|A\| < 1$, then $I - A$ is invertible, i.e., $1 \in \rho(A)$ and $(I - A)^{-1} = \sum_{k=0}^{\infty} A^k$. It is possible for the series to converge even though $\|A\| \ge 1$, in which case $I - A$ is invertible. This may be seen in the next section. Let us start with the following theorem.

Theorem 8.2 *Suppose A is in $\mathcal{L}(X)$ and $\sum_{k=0}^{\infty} A^k$ converges in $\mathcal{L}(X)$. Then $I - A$ is invertible and $(I - A)^{-1} = \sum_{k=0}^{\infty} A^k$.*

Proof: Let $S_n = \sum_{k=0}^{n} A^k$ and $S = \lim_{n\to\infty} S_n$. Then $(I - A)S = \lim_{n\to\infty}$
$(I - A)S_n = \lim_{n\to\infty} S_n(I - A) = S(I - A)$.
   Since $(I - A)S_n = I - A^{n+1} \to I$,

$$(I - A)S = S(I - A) = I.$$

Thus

$$\sum_{k=0}^{\infty} A^k = S = (I - A)^{-1}.$$

$\square$

Corollary 8.3 *Suppose $A \in \mathcal{L}(X)$ and $\|A^p\| < 1$ for some positive integer p.
Then $\sum_{k=0}^{\infty} A^k$ converges, $I - A$ is invertible and $(I - A)^{-1} = \sum_{k=0}^{\infty} A^k$.*

Proof: Let

$$S_k = \sum_{n=0}^{\infty} \|A^{np+k}\|, \quad k = 1, 2, \ldots, p.$$

Each S_k is finite since

$$\sum_{n=0}^{\infty} \|A^{np+k}\| \leq \|A^k\| \sum_{n=0}^{\infty} \|A^p\|^n < \infty.$$

Hence

$$\sum_{j=1}^{\infty} \|A^j\| = S_1 + S_2 + \cdots + S_p < \infty.$$

Thus $\sum_{j=0}^{\infty} A^j$ converges since $\mathcal{L}(X)$ is complete (Theorem XI.1.1) and

$$\left\| \sum_{j=m}^{n} A^j \right\| \leq \sum_{j=m}^{n} \|A^j\| \to 0 \quad \text{as} \quad m, n \to \infty.$$

The corollary now follows from Theorem 8.2.

$\square$

12.9 Volterra Integral Operator

Suppose $k(t, s)$ is continuous on $[0, 1] \times [0, 1]$. We shall show that the equation

$$\lambda f(t) - \int_0^t k(t, s) f(s) \, ds = g(t)$$

has a unique solution in $C([0, 1])$ for every $g \in C([0, 1])$ if $\lambda \neq 0$. In other words, if V is the *Volterra integral operator* on $C([0, 1])$ defined by

$$(Vf)(t) \int_0^t k(t, s) f(s) \, ds,$$

then $\sigma(V) = \{0\}$.

First we note that Vf is in $C([0, 1])$ for each $f \in C([0, 1])$. This follows from the uniform continuity of $k(t, s)$ and the equation

$$(Vf)(t_1) - (Vf)(t) = \int_0^{t_1} [k(t_1, s) - k(t, s)] f(s) \, ds + \int_t^{t_1} k(t, s) f(s) \, ds.$$

For $\mu = \max_{0 \leq t, s \leq 1} |k(t, s)|$,

$$|(Vf)(t)| \leq \int_0^t |k(t, s)| \, |f(s)| \, ds \leq \mu t \|f\|$$

$$|(V^2 f)(t)| \leq \int_0^t |k(t, s)| \, |(Vf)(s)| \, ds \leq \mu^2 \|f\| \int_0^t s \, ds = \frac{\mu^2 t^2}{2!} \|f\|.$$

By induction,

$$|(V^k f)(t)| \leq \frac{\mu^k t^k}{k!} \|f\|.$$

Hence

$$\|V^k\| \leq \frac{\mu^k}{k!} \to 0.$$

Thus it follows from Corollary 8.3 that $I - V$ is invertible and $(I - V)^{-1} = \sum_{k=0}^\infty V^k$.

If we replace $k(t, s)$ by $\frac{1}{\lambda} k(t, s)$, $\lambda \neq 0$, we get the invertibility of $I - \frac{V}{\lambda}$. Thus $\lambda I - V$ is invertible and

$$(\lambda I - V)^{-1} = \frac{1}{\lambda} \left(I - \frac{V}{\lambda} \right)^{-1} = \sum_{k=0}^\infty \frac{V^k}{\lambda^{k+1}}.$$

To find a formula for V^k, let $k_1(t, s) = k(t, s)$, if $a \leq s \leq t$ and zero otherwise. Then

$$(Vf)(t) = \int_0^1 k_1(t, s) f(s) \, ds$$

and V^k can be obtained from II.10, formula (10.3).

Note that $\|\frac{V}{\lambda}\|$ can be made as large as we please, yet $\sum_{k=0}^\infty (\frac{V}{\lambda})^k$ converges.

12.10 Analytic Operator Valued Functions

Definition: An operator valued function $A(\lambda)$ which maps a subset of $\mathbb{C}$ into $\mathcal{L}(X)$ is *analytic* at λ_0 if

$$A(\lambda) = A_0 + (\lambda - \lambda_0)A_1 + (\lambda - \lambda_0)^2 A_2 + \cdots,$$

where each A_k is in $\mathcal{L}(X)$ and the series converges for each λ in some neighborhood of λ_0.

Theorem 10.1 *The function $A(\lambda) = (\lambda I - A)^{-1}$ is analytic at each point in the open set $\rho(A)$.*

Proof: Suppose $\lambda_0 \in \rho(A)$. Now

$$\lambda I - A = (\lambda_0 I - A) - (\lambda_0 - \lambda)I = (\lambda_0 I - A)[I - (\lambda_0 - \lambda)A(\lambda_0)]. \quad (10.1)$$

Since $\rho(A)$ is open, we may choose $\varepsilon > 0$ so that $|\lambda - \lambda_0| < \varepsilon$ implies $\lambda \in \rho(A)$ and $\|(\lambda - \lambda_0)A(\lambda_0)\| < 1$. In this case, it follows from (10.1) that

$$A(\lambda) = [I - (\lambda_0 - \lambda)A(\lambda_0)]^{-1}(\lambda_0 I - A)^{-1} = \sum_{k=0}^{\infty}(\lambda_0 - \lambda)^k A(\lambda_0)^{k+1}. \quad (10.2)$$

From the series representation of $A(\lambda)$, it is not difficult to show that

$$\frac{d}{d\lambda}(\lambda I - A)^{-1} = -(\lambda I - A)^{-2}.$$

The function $A(\lambda) = (\lambda I - A)^{-1}$ is called the *resolvent function* of A, or simply, the *resolvent* of A. $\square$

Theorem 10.2 *The spectrum of any $A \in \mathcal{L}(X)$ is non-empty.*

Proof: Suppose $\sigma(A) = \emptyset$. Then by Theorem 10.1 formula (10.2), with $\lambda_0 = 0$,

$$(\lambda I - A)^{-1} = \sum_{k=0}^{\infty}(-1)^k \lambda^k (A^{-1})^{k+1} \quad \text{for all } \lambda \in \mathbb{C}. \quad (10.3)$$

By Corollary XI.5.4, there eixst $x' \in X'$ and $x \in X$ such that $x'x \neq 0$. From (10.3) it follows that

$$f(\lambda) = x'(\lambda I - A)^{-1}x = \sum_{k=0}^{\infty}(-1)^k \lambda^k x'(A^{-1})^{k+1}x, \ \lambda \in \mathbb{C}.$$

Hence $f(\lambda)$ is a complex valued function which is analytic on $\mathbb{C}$, i.e., f is an entire function. Moreover f is bounded. Indeed, since

$$\|(\lambda I - A)x\| \geq (|\lambda| - \|A\|)\|x\|,$$

it follows that for $|\lambda| > \|A\|$,

$$\|(\lambda I - A)^{-1}x\| \leq \frac{\|x\|}{|\lambda| - \|A\|} \to 0 \text{ as } |\lambda| \to \infty. \tag{10.4}$$

Since $f(\lambda)$ is continuous on $\{\lambda \mid |\lambda| \leq \|A\|\}$, it is bounded on this set. Thus f is entire and bounded on $\mathbb{C}$. By Liouville's theorem, f is a constant function. Therefore $x'x = f(0) = f(\lambda) \to 0$ by (10.3). Hence $x'x = 0$ which contradicts our choice of x and x'. We have shown that $\sigma(A) \neq \emptyset$. $\square$

Exercises XII

Throughout these exercises, X and Y denote Banach spaces.

1. Let $\mathcal{B}$ be a Banach space with a given norm $\|\cdot\|$. Check that the following norms $\|\cdot\|_1$ are equivalent to $\|\cdot\|$.

    (a) $\mathcal{B} = \ell_2$; $\|\xi\|_1^2 = \sum_{j=1}^{\infty} |(1 + \frac{1}{j})\xi_j|^2$.

    (b) $\mathcal{B} = L_2[0, 1]$; $\|f\|_1 = \|(2I + V)f\|$, where $Vf(t) = \int_0^t f(s) \, ds$.

    (c) $\mathcal{B} = \ell_p(\mathbb{Z})$, $1 \leq p \leq \infty$; $U : \ell_p(\mathbb{Z}) \to \ell_p(\mathbb{Z})$ is given by

$$U(\ldots \xi_{-1}, \xi_0, \xi_1, \ldots) = (\ldots \xi_{-2}, \xi_{-1}, \xi_0, \xi_1, \ldots)$$
$$\qquad\qquad\quad \uparrow \qquad\qquad\qquad\qquad \uparrow$$
$$\qquad\quad \text{0-place} \qquad\qquad\qquad \text{0-place}$$
$$\|\xi\|_1 = \|U\xi\|.$$

    (d) Let S be a bounded invertible operator on $\mathcal{B}$ and let $\|x\|_1 = \|Sx\|$.

2. Let $\{x_n\}$ be a Schauder basis for X. For $k = 1, 2, \ldots$, define the coordinate functional f_k on X by $f_k(x) = \alpha_k$, where $x = \sum_n \alpha_n x_n$. Thus $x = \sum_n f_n(x)x_n$, $x \in X$. Prove that each f_k is bounded by establishing the following:

    (a) Given $x = \sum_k \alpha_k x_k$, define $\|x\|_\infty = \sup_n \|\sum_{k=1}^n \alpha_k x_k\|$. Then $\|\cdot\|_\infty$ is a norm and $(X, \|\cdot\|_\infty)$ is complete.

    (b) The norms $\|\cdot\|$ and $\|\cdot\|_\infty$ are equivalent.

    (c) Each f_k is linear and bounded on $(X, \|\cdot\|)$.

3. Let A be a linear operator which maps a Hilbert space $\mathcal{H}$ into $\mathcal{H}$. Prove that if $\langle Au, v \rangle = \langle u, Av \rangle$ for all u, v in $\mathcal{H}$, then A is continuous.

4. Let A be a linear operator which maps X into Y. Prove that A is continuous if and only if $y' \circ A$ is continuous for every $y' \in Y'$.

5. Assuming X is a Hilbert space, prove that the Hahn-Banach Theorem can be generalized as follows: If M is a subspace of X and A is in $L(M, Y)$, then A can be extended to an operator $\bar{A} \in \mathcal{L}(X, Y)$ such that $\|\bar{A}\| = \|A\|$.

6. Let M and N be closed subspaces of X with $M \cap N = (0)$. Prove that $M \oplus N$ is closed if and only if there exists a number $c > 0$ such that $\|u + v\| \geq c\|u\|$ and $\|u + v\| \geq c\|v\|$ for all $u \in M$, $v \in N$.

7. Let Z be a Banach space and let $B \in \mathcal{L}(Y, Z)$ be injective. Suppose A is a linear operator which maps X into Y and has the property that BA is bounded. Prove that A is bounded.

8. Let T be in $\mathcal{L}(X, Z)$ and let A be in $\mathcal{L}(Y, Z)$, where Z is a Banach space. Suppose that for each $x \in X$, there exists a unique $y \in Y$ such that $Tx = Ay$. Define $Bx = y$. Prove that B is in $\mathcal{L}(X, Y)$.

9. Let $\{\beta_k\}$ be a bounded sequence of complex numbers. Define $T : \ell_p \to \ell_p$, $1 \leq p \leq \infty$, by $T(\alpha_1, \alpha_2, \ldots) = (\beta_1\alpha_1, \beta_2\alpha_2, \ldots)$. Determine $\sigma(T)$.

10. Given $A \in \mathcal{L}(X)$, prove that

    (a) $\lim_{|\lambda| \to \infty}(\lambda I - A)^{-1} = 0$ and

    (b) $\lim_{|\lambda| \to \infty} \lambda(\lambda I - A)^{-1} = I$.

11. Suppose $\{A_n\}$ is a sequence of invertible operators in $\mathcal{L}(X)$ which converges to A. Prove that A is invertible if and only if $\{\|A_n^{-1}\|\}$ is bounded, in which case $A_n^{-1} \to A^{-1}$.

12. Given $A \in \mathcal{L}(X)$ and given a polynomial p, prove that $\sigma(p(A)) = \{p(\lambda) : \lambda \in \sigma(A)\} \equiv p\sigma(A)$. Hint: Write $p(\lambda) - \mu = c \Pi_{i=1}^{n}(\lambda - \lambda_i)$ and consider $\Pi_{i=1}^{n}(A - \lambda_i I)$.

13. Let $T : \ell_1 \to \ell_1$ be given by $T(\alpha_1, \alpha_2, \ldots) = (0, \alpha_1, \frac{\alpha_2}{2}, \frac{\alpha_3}{3}, \ldots)$.

    (a) Find $\|T^n\|$, $n = 1, 2, \ldots$

    (b) Determine $\sigma(T)$.

14. Given $A \in \mathcal{L}(X)$, let $A(\lambda) = (\lambda - A)^{-1}$, $\lambda \in \rho(A)$.

    (a) Prove that $A(\lambda) - A(\mu) = (\mu - \lambda)A(\lambda)A(\mu)$, $\lambda, \mu \in \rho(A)$.

    (b) Prove that $\frac{d}{d\lambda}A(\lambda) = -A(\lambda)^2$.

15. Let $A(\lambda)$ and $B(\lambda)$ be analytic operator valued functions defined on a set $S \subset \mathbb{C}$. Find the derivatives of $A(\lambda)^{-1}$, $A(\lambda)B(\lambda)$ and $A(\lambda)^{-1}B(\lambda)$ at the points where the derivatives exist.

16. Suppose $A \in \mathcal{L}(X)$ is injective. Prove that the range of A is closed if and only if there exists a $\gamma > 0$ such that $\|Ax\| \geq \gamma\|x\|$ for all $x \in X$.

17. For $A \in \mathcal{L}(X, Y)$ define $\hat{A} : X/\text{Ker}\, A \to Y$ by $\hat{A}[x] = Ax$ (cf. IX, exercise 22, for the definition of X/M).

    (a) Prove that $\hat{A}$ is linear, injective and $\|\hat{A}\| = \|A\|$.

    (b) Prove that the range of A is closed if and only if there exists a $\gamma > 0$ such that $\|Ax\| \geq \gamma d(x, \text{Ker}\, A)$ for all $x \in X$.

18. (a) Let T be in $L(X, Y)$. Suppose there exists a closed subspace N of Y such that the direct sum $\text{Im}(T) \oplus N$ is closed. Prove that $\text{Im}\, T$ is closed. Hint: Define T_0 on $X \times N$ by $T_0(x, z) = Tx + z$; then use exercise 17(b).

    (b) Prove that if $A \in \mathcal{L}(X, Y)$ and codim $\text{Im}\, A < \infty$, then $\text{Im}\, A$ is closed.

Chapter XIII
Compact Operators on Banach Spaces

In this chapter, we present some basic properties of compact operators in $\mathcal{L}(X, Y)$, where X and Y are Banach spaces. The results enable us to determine conditions under which certain integral equations have solutions in $L_p([a, b])$, $1 < p < \infty$ or $C([a, b])$.

An operator $K \in \mathcal{L}(X, Y)$ is *compact* if for every sequence $\{x_n\}$ in X, $\|x_n\| = 1$, the sequence $\{Kx_n\}$ has a convergent subsequence.

Theorems II.16.1, II.16.3 and the corresponding proofs remain the same when the Hilbert spaces are replaced by Banach spaces.

13.1 Examples of Compact Operators

1. Every $K \in \mathcal{L}(X, Y)$ which is of finite rank is compact. For if $\{x_n\}$ is a sequence in the 1-sphere of X, then $\{Kx_n\}$ is a bounded sequence in the finite dimensional space. Im K. Hence $\{Kx_n\}$ has a convergent subsequence by Theorem XI.2.2.

2. Let $\{\beta_k\}$ be a sequence of complex numbers which converges to zero. Define $K \in \mathcal{L}(\ell_p)$, $1 \leq p < \infty$ by $K(\{\alpha_k\}) = \{\beta_k \alpha_k\}$. Let $K_n(\alpha_1, \alpha_2, \ldots) = (\beta_1, \beta_2, \ldots, \beta_n, 0, 0, \ldots)$. Since K_n is a bounded linear operator on ℓ_p of finite rank, it is compact. Moreover,

$$\|K - K_n\| \leq \sup_{k > n} |\beta_k| \to 0 \quad \text{as } n \to \infty.$$

Hence K is compact.

3. Suppose k is continuous on $[a, b] \times [a, b]$. Let $K : C([a, b]) \to C([a, b])$ be the linear integral operator defined by

$$(Kf)(t) = \int_a^b k(t, s) f(s) \, ds.$$

We shall now prove that K is compact.

It follows from the uniform continuity of K that Im $K \subset C([a, b])$. K is bounded since

$$\|Kf\| = \max_{a \leq t \leq b} |(Kf)(t)| \leq \|f\| (b - a) \max_{a \leq s, t \leq b} |k(t, s)|. \tag{1.1}$$

Since k is continuous, there exists a sequence $\{P_n\}$ of polynomials in t and s which converges uniformly to k on $[a, b] \times [a, b]$([R], p. 174). Define K_n on $C([a, b])$ by

$$K_n f = \int_a^b P_n(t, s) f(s) \, ds.$$

It is easy to see that K_n is a bounded linear operator of finite rank on $C([a, b])$. Hence K_n is compact. Replacing k by $k - P_n$ in (1.1) gives

$$\|K - K_n\| \leq (b - a) \max_{a \leq s, t \leq b} |k(t, s) - P_n(t, s)| \to 0$$

as $n \to \infty$. Therefore, K is compact.

4. For $1 < p, q < \infty$, let $k(t, s)$ be in $L_r([0, 1] \times [0, 1])$, where $r = \max(p', q)$, $\frac{1}{p} + \frac{1}{p'} = 1$. We shall now show that the integral operator K defined by

$$(Kf)(t) = \int_0^1 k(t, s) f(s) \, ds$$

is a compact linear map from $L_p([0, 1])$ into $L_q([0, 1])$.

Let $\frac{1}{r'} = 1 - \frac{1}{r}$. Then $r' \leq p$ and for $f \in L_p([0, 1])$, Hölder's inequality applied to $|f|^{r'} \in L_{p/r'}([0, 1])$ and the constant function 1 implies $\|f\|_{r'} \leq \|f\|_p$ and

$$\|Kf\|_q^q = \int_0^1 \left| \int_0^1 k(t, s) f(s) \, ds \right|^q dt$$

$$\leq \|f\|_{r'}^q \int_0^1 \left(\int_0^1 |k(t, s)|^r ds \right)^{q/r} dt. \tag{1.2}$$

Let

$$g(t) = \int_0^1 |k(t, s)|^r ds. \tag{1.3}$$

Since k is in $L_r([0, 1] \times [0, 1])$, g is in $L_1([0, 1])$. Therefore, it follows from Hölder's inequality applied to $|g|^{q/r} \in L_{r/q}$ and the constant function 1 that

$$\int_0^1 |g(t)|^{q/r} dt \leq \left(\int_0^1 |g(t)| \right)^{q/r}. \tag{1.4}$$

From (1.2), (1.3), (1.4), and the observation $\|f\|_{r'} \leq \|f\|_p$, we obtain

$$\|Kf\|_q \leq \|f\|_p \left(\int_0^1 \int_0^1 |k(t, s)|^r ds \, dt \right)^{1/r}. \tag{1.5}$$

Thus K is a bounded linear map from $L_p([0, 1])$ into $L_q([0, 1])$.

The set of polynomials in s and t is dense in $L_r([0, 1] \times [0, 1])$. Thus there exists a sequence $\{P_n(t, s)\}$ of polynomials which converges in $L_r([0, 1] \times [0, 1])$ to $k(t, s)$. Define $k_n : L_p([0, 1]) \to L_q([0, 1])$ by

$$(K_n f)(t) = \int_0^1 P_n(t, s) f(s) \, ds.$$

If we replace k in (1.5), first by P_n and then by $k - P_n$, we see that K_n is a bounded linear operator of finite rank and

$$\|K - K_n\|_q^r \leq \int_0^1 \int_0^1 |k(t, s) - P_n(t, s)|^r \, ds \, dt \to 0.$$

Hence K is compact.

It follows from the example that the Volterra operator V defined in XII.9 is compact linear map from $L_p([0, 1])$ into $L_q([0, 1])$, $1 < p, q < \infty$.

If the restriction $1 < p, q < \infty$ is removed, then the result in Example 4 is false. In [G], p. 90, a bounded Lebesgue measurable function defined on $[0, 1] \times [0, 1]$ is constructed such that the corresponding integral operator is not compact on $L_1([0, 1])$.

5. The Lebesgue integral operator

$$(Kf)(t) = \int_a^b \frac{B(t, s)}{|t - s|^\alpha} f(s) \, ds$$

is compact on $C([a, b])$, where $B(t, s)$ is continuous on $[a, b] \times [a, b]$ and $\alpha < 1$. Also,

$$\|K\| = \max_{t \in [a,b]} \int_a^b \frac{|B(t, s)|}{|t - s|^\alpha} \, ds. \tag{1.6}$$

First we show that Kf is in $C([a, b])$ for each $f \in C[a, b]$. Let $M = \max_{s,t \in [a,b]} |B(t, s)|$ and $\|f\|_\infty = \max_{t \in [a,b]} |f(t)|$.
Then

$$|(Kf)(t)| \leq M \|f\|_\infty \int_a^b |t - s|^{-\alpha} \, ds \leq \frac{2M}{1 - \alpha} (b - a)^{1-\alpha} \|f\|_\infty.$$

Next we show that K is compact on $C([a, b])$. Let $k(t, s) = \frac{B(t,s)}{|t-s|^\alpha}$ and define for $n = 1, 2, \ldots$

$$k_n(t, s) = \begin{cases} 0, |t - s| \leq \frac{1}{2n} \\ (2n|t - s| - 1)k(t, s), \frac{1}{2n} \leq |t - s| \leq \frac{1}{n} \\ k(t, s), |t - s| \geq \frac{1}{n} \end{cases} . \tag{1.7}$$

The function k_n is continuous on $[a, b] \times [a, b]$ and therefore the operator

$$(K_n f)(t) = \int_a^b k_n(t, s) f(s) ds$$

is compact on $C([a, b])$ by Example 3. Let $S_n = \{s \mid |t - s| \leq \frac{1}{n}\}$.
  Since

$$|k(t, s) - k_n(t, s)| = \begin{cases} 0, & s \notin S_n \\ \leq |k(t, s)|, & s \in S_n, \end{cases}$$

we have

$$|((K - K_n)f)(t)| \leq \int_a^b |k(t, s) - k_n(t, s)| |f(s)| ds$$

$$\leq \|f\|_\infty \int_{S_n} |k(t, s)| ds \leq M \|f\|_\infty \int_{t-\frac{1}{n}}^{t+\frac{1}{n}} |t - s|^{-\alpha} ds$$

$$\leq M \|f\|_\infty \frac{2}{1 - \alpha} \left(\frac{2}{n}\right)^{1-\alpha}. \tag{1.8}$$

It follows that $Kf \in C([a, b])$ and $\|K - K_n\| \leq M \frac{2}{1-\alpha} (\frac{2}{n})^\alpha \to 0$.
  Thus K is compact on $C([a, b])$. Finally, we prove equality (b). The function
$F(t) = \int_a^b \frac{|B(t,s)|}{|t-s|^\alpha} ds$ is in $C([a, b])$. This follows from the above result with
$|B(t, s)|$ in place of $B(t, s)$ and $F(t) = \int_a^b \frac{|B(t,s)|}{|t-s|^\alpha} f(s) ds$, where f is identically
1 on $[a, b]$. Thus

$$\max_t \int_a^b |k(t, s)| ds = \int_a^b |k(t_0, s)| ds \tag{1.9}$$

for some $t_0 \in [a, b]$. Since $|k_n(t, s)| \leq |k(t, s)|$ for each n and $s \neq t$, we have
from (1.9) and Theorem XII.1.2 that

$$\|K\| = \lim_{n \to \infty} \|K_n\| = \lim_{n \to \infty} \max_t \int_a^b |k_n(t, s)| ds \leq \max_t \int_a^b |k(t, s)| ds$$

$$= \int_a^b |k(t_0, s)| ds = \lim_{n \to \infty} \int_a^b |k_n(t_0, s)| ds \leq \lim_{n \to \infty} \|K_n\| = \|K\|.$$

Thus (1.6) is established. $\square$

13.2 Decomposition of Operators of Finite Rank

The following theorem is very useful when dealing with operators of finite rank.

Theorem 2.1 *If K is an operator in $\mathcal{L}(X)$ of finite rank, then there exist subspaces
N and Z of X such that N is finite dimensional, Z is closed,*

$$X = N \oplus Z, \quad KN \subset N \text{ and } KZ = (0).$$

Proof: Let $\{Kv_1, \ldots, Kv_n\}$ be a basis for Im K. By Corollary XI.5.6, there exist $f_1, \ldots, f_n$ in X' such that $f_i Kv_j = \delta_{ij}$, $1 \le i, j \le n$, and for each $x \in X$,

$$Kx = \sum_{i=1}^{n} f_i(Kx)Kv_i. \tag{2.1}$$

Let $N = \mathrm{sp}\{v_1, \ldots, v_n, Kv_1, \ldots, Kv_n\}$. Clearly, $KN \subset \mathrm{Im}\, K \subset N$ and from (2.1),

$$x - \sum_{i=1}^{n} f_i(Kx)v_i \in \mathrm{Ker}\, K.$$

Hence,

$$X = N + \mathrm{Ker}\, K. \tag{2.2}$$

Since N is finite dimensional, it follows from Theorem XI.5.7, that there exists a closed subspace Z such that

$$\mathrm{Ker}\, K = (N \cap \mathrm{Ker}\, K) \oplus Z. \tag{2.3}$$

Equations (2.2) and (2.3) imply $X = N \oplus Z$ and $KZ = (0)$. □

Every $K \in \mathcal{L}(X)$ of finite rank can be written in the form

$$Kx = \sum_{i=1}^{n} g_i(x)w_i, \quad g_i \in X', \ w_i \in \mathrm{Im}\, K.$$

Indeed, from (2.1) in the proof of Theorem 2.1, we see that we can choose $g_i(x) = f_i(Kx)$ and $w_i = Kv_i$.

13.3 Approximation by Operators of Finite Rank

It was shown in Theorem X.4.2 that every compact linear operator which maps a Hilbert space into a Hilbert space is the limit, in norm, of a sequence of operators of finite rank. While this result need not hold for Banach spaces, we do have the following results.

Theorem 3.1 *Suppose* $\{T_n\}$ *is a sequence of operators in* $\mathcal{L}(Y)$ *such that* $T_n y \to Ty$ *for all* $y \in Y$. *If* K *is a compact operator in* $\mathcal{L}(X, Y)$, *then*

$$\|T_n K - TK\| \to 0.$$

The proof is exactly the same as the proof of Lemma II.17.8

Corollary 3.2 *Suppose there exists a sequence* $\{P_n\}$ *of operators of finite rank in* $\mathcal{L}(Y)$ *with the property* $P_n y \to y$ *for every* $y \in Y$. *Then every compact operator* $K \in \mathcal{L}(X, Y)$ *is the limit, in norm, of operators of finite rank. In fact,*

$$\|P_n K - K\| \to 0.$$

The operators P_n defined on ℓ_p, $1 \le p < \infty$, by $P_n(\alpha_1, \alpha_2, \ldots) = (\alpha_1, \ldots, \alpha_n,$ $0, 0, \ldots)$, $n = 1, 2, \ldots$ satisfy the hypotheses of the theorem. Hence if $K \in$ $\mathcal{L}(X, \ell_p)$ is compact, then $\|P_n K - K\| \to 0$.

Looking at this another way, let $\{e_j\}$ be the standard basis for ℓ_p. Given $x = (\alpha_1, \alpha_2, \ldots) \in \ell_p$, $x = \sum_{j=1}^{\infty} \alpha_j e_j$. Define $f_j(x) = \alpha_j$ and $P_n(x) = \sum_{j=1}^{n} f_j(x) e_j$. Since $\{P_n\}$ satisfies the hypotheses of Corollary 3.2, $\|P_n K - K\| \to 0$.

More generally, if $\{y_i\}$ is a Schauder basis for Y, it can be shown (Exercise XII-2) that the linear functional f_j defined on Y by

$$f_j(y) = \alpha_j, \text{ where } y = \sum_j \alpha_j y_j$$

is in Y'. Hence, defining $P_n y = \sum_{j=1}^{n} f_j(y) y_j$, it is clear that $\{P_n\}$ satisfies the hypotheses of Corollary 3.2. Thus we have following result.

Corollary 3.3 *If the Banach space Y has a Schauder basis, then every compact operator in $\mathcal{L}(X, Y)$ is the limit, in norm, of a sequence of operators of finite rank.*

Corollary 3.4 *Let $(a_{jk})_{j,k=1}^{\infty}$ be a matrix of complex numbers. The operator A defined on ℓ_1 by*

$$A(\alpha_1, \alpha_2, \ldots) = (\beta_1, \beta_2, \ldots), \beta_j = \sum_{k=1}^{\infty} a_{jk} \alpha_k,$$

is a compact operator in $\mathcal{L}(\ell_1)$ if and only if for each $\varepsilon > 0$, there exists an integer N such that

$$\sup_k \sum_{j>N} |a_{jk}| < \varepsilon. \tag{3.1}$$

Indeed, suppose (3.1) holds. For each positive integer n define $P_n \in \mathcal{L}(\ell_1)$ by $P_n(\alpha_1, \alpha_2, \ldots) = (\alpha_1, \ldots, \alpha_n, 0, 0, \ldots)$. It follows from Theorem XII.1.1 and (3.1) that

$$\|A - P_n A\| = \sup_k \sum_{j>n} |a_{jk}| \to 0 \text{ as } n \to \infty.$$

Conversely, assume A is compact. Then by Corollary 3.2, there exists, for each $\varepsilon > 0$, an integer N such that

$$\sup_k \sum_{j>N} |a_{jk}| = \|P_N A - A\| < \varepsilon. \tag{3.2}$$

In Section XII.7 the convergence of the projection method for an invertible operator in $\mathcal{L}(X, Y)$ was discussed. Recall that $A \in \Pi(P_n, Q_n)$ if A is invertible and the projection method for A converges. Using an argument analogous to the proof of Theorem II.17.7, where $P_n A P_n$ is replaced by $Q_n A P_n$, we obtain the following stability theorem.

Theorem 3.5 *Suppose $A \in \Pi(P_n, Q_n)$ and suppose K is a compact operator in $\mathcal{L}(X, Y)$ with the property that $A + K$ is invertible. Then $A + K \in \Pi(P_n, Q_n)$.*

13.4 First Results in Fredholm Theory

We recall from linear algebra that if A is a linear operator defined on a finite dimensional vector space X, then

$$\dim \operatorname{Ker} A = \operatorname{codim} \operatorname{Im} A \qquad (4.1)$$

This useful result is proved as follows.

Let $\{x_1, \ldots, x_K\}$ be a basis for Ker A. Extend it to a basis $\{x_1, \ldots, x_n\}$ for X. Then $\{Ax_{K+1}, \ldots, Ax\}$ is a basis for Im A. Now $X = M \oplus \operatorname{Im} A$ for some k- dimensional subspace M of X.

As a generalization we have the following theroem.

Theorem 4.1 *Given $K \in \mathcal{L}(X)$, suppose there exists a finite rank operator $K_0 \in \mathcal{L}(X)$, such that $\| K - K_0 \| < 1$. Then $I - K$ has a closed range and*

$$\dim \operatorname{Ker}(I - K) = \operatorname{codim} \operatorname{Im}(I - K) < \infty.$$

In particular, the equation

$$(I - K)x = y$$

has a unique solution for every $y \in X$ if and only if the equation

$$(I - K)x = 0$$

has only the trivial solution $x = 0$.

Proof: First, let us assume that K is of finite rank. By Theorem 2.1, there exist closed subspaces N and Z of X such that N is finite dimensional, $X = N \oplus Z$, $KN \subset N$ and $KZ = \{0\}$. Let $(I - K)_N$ be the restriction of $I - K$ to N. Then

$$\operatorname{Ker}(I - K) = \operatorname{Ker}(I - K)_N \qquad (4.2)$$

$$\operatorname{Im}(I - K) = \operatorname{Im}(I - K)_N \oplus Z. \qquad (4.3)$$

To see this, let $x \in X$ be given. There exist $u \in N$ and $z \in Z$ such that $x = u + z$. Hence $(I - K)x = (I - K)u + z$ and $(I - K)u$ is in N. Thus (4.3) follows. If $(I - K)x = 0$, then $z = u - ku \in N \cap Z = (0)$. Thys $x = u \in N$ and $(I - K)x = (I - K)_N x$.

Since $(I - K)_N$ is in $\mathcal{L}(N)$ and N is finite dimensional, equations (4.1), (4.2) and (4.3) imply

$$\infty > \dim \operatorname{Ker}(I - K) = \dim \operatorname{Ker}(I - K)_N = \operatorname{codim} \operatorname{Im}(I - K)_N$$
$$= \operatorname{codim} \operatorname{Im}(I - K).$$

Also, it follows from Theorem XI.2.5 and (4.3) that $\operatorname{Im}(I - K)$ is closed.

More generally, assume $\|K - K_0\| < 1$, where K_0 is of finite rank. Now $B = I - (K - K_0)$ is invertible and

$$I - K = B - K_0 = (I - K_0 B^{-1})B.$$

Hence

$$\operatorname{Im}(I - K) = \operatorname{Im}(I - K_0 B^{-1}), \tag{4.4}$$

$$B \operatorname{Ker}(I - K) = \operatorname{Ker}(I - K_0 B^{-1}). \tag{4.5}$$

Since $K_0 B^{-1}$ is of finite rank, it follows from what we have shown, together with (4.4) and (4.5), that $\operatorname{Im}(I - K)$ is closed and

$$\infty > \dim \operatorname{Ker}(I - K) = \dim \operatorname{Ker}(I - K_0 B^{-1})$$
$$= \operatorname{codim} \operatorname{Im}(I - K_0 B^{-1}) = \operatorname{codim} \operatorname{Im}(I - K).$$

$\square$

Corollary 4.2 *If $K \in \mathcal{L}(X)$ is the limit, in norm, of a sequence of operators of finite rank, then $I - K$ has a closed range and*

$$\dim \operatorname{Ker}(I - K) = \operatorname{codim} \operatorname{Im}(I - K) < \infty.$$

Corollary 4.3 *If X is a Banach space with a Schauder basis and $K \in \mathcal{L}(X)$ is compact, then $I - K$ has a closed range and*

$$\dim \operatorname{Ker}(I - K) = \operatorname{codim} \operatorname{Im}(I - K) < \infty.$$

Proof: Corollaries 3.3 and 4.2.

The last statement of Theorem 4.1 is referred to as the *Fredholm alternative*.

In Chapter XV Theorem 4.2 we shall show that Corollary 4.2 is valid when it is only assumed that K is compact $\qquad\qquad\qquad\qquad\qquad\qquad\qquad\qquad\square$

13.5 Conjugate Operators on a Banach Space

Corresponding to the adjoint of an operator defined on a Hilbert space, we introduce the concept of the conjugate of an operator defined on a Banach space.

Definition: Given $A \in \mathcal{L}(X, Y)$, the *conjugate* $A' : Y' \to X'$ of A is the operator defined by $A'f = f \circ A$, $f \in Y'$.

It is clear that A' is linear. Furthermore, $\|A'\| = \|A\|$. The proof of this assertion is as follows.

$$\|A'f\| = \|f \circ A\| \le \|f\| \|A\|.$$

Hence $\|A'\| \le \|A\|$. On the other hand, by Corollary XI.5.5

$$\|Ax\| = \max_{\|f\|=1} |f(Ax)| = \max_{\|f\|=1} |(A'f)x| \le \|A'\| \|x\|.$$

Thus $\|A\| \le \|A'\|$.

It is easy to verify that if A and B are in $\mathcal{L}(X, Y)$, then

$$(A + B)' = A' + B' \text{ and } (\alpha A)' = \alpha A'.$$

If C is in $\mathcal{L}(Y, Z)$, then $(CA)' = A'C' \in \mathcal{L}(Z', X')$.

Examples: 1. Suppose $K \in \mathcal{L}(X, Y)$ is of finite rank. Let $y_1, \ldots, y_n$ be a basis for $\mathrm{Im}\, K$. For each j, define the linear functional f_j on X by $f_j(x) = \alpha_j$, where $Kx = \sum_{j=1}^n \alpha_j y_j$. Then for $g \in Y'$,

$$(K'g)x = gKx = \sum_{j=1}^n f_j(x)g(y_j) = \left(\sum_{j=1}^n g(y_j)f_j \right) x.$$

Thus

$$K'g = \sum_{j=1}^n g(y_j)f_j.$$

Choosing $g_k \in Y'$ such that $g_k y_j = \delta_{kj}$, $1 \le j, k \le n$, we obtain $f_k = K'g_k \in X'$.

2. Let K be the integral operator in Section 1, Example 4. If we identify $L_p'([a, b])$ with $L_{p'}([a, b])$, $\frac{1}{p} + \frac{1}{p'} = 1$, as in Theorem XI.4.2, then for $F \in L_q'$ and $g \in L_p([a, b])$,

$$(K'F)g = F(Kg) = \int_a^b F(t) \left\{ \int_a^b k(t, s)g(s)ds \right\} dt$$

$$= \int_a^b g(s) \left\{ \int_a^b k(t, s)F(t)dt \right\} ds.$$

Hence

$$(K'F)(s) = \int_a^b k(t, s)F(t)dt. \tag{5.1}$$

To be more precise, we have shown that if $\widetilde{F}$ in $L_q'([a, b])$ corresponds to $F \in L_{q'}([a, b])$, Then $K'\widetilde{F} \in L_p'([a, b])$ corresponds to the function defined by (5.1).

Analogous to the relationships between the ranges and kernels of an operator
and its adjoint, we have the following results for the conjugate operator.

Definition: For M a subset of X and N a subset of X', define

$$M^\perp = \{f \in X' : f(M) = 0\}$$

and

$$^\perp N = \{x \in X : g(x) = 0 \text{ for all } g \in N\}.$$

Theorem 5.1 *For $A \in \mathcal{L}(X, Y)$,*

    (i) $(\operatorname{Im} A)^\perp = \operatorname{Ker} A'$

    (ii) $\overline{\operatorname{Im} A} =\,^\perp (\operatorname{Ker} A')$

    (iii) $\operatorname{Ker} A =\,^\perp (\operatorname{Im} A')$

    (iv) $(\operatorname{Ker} A)^\perp \supset \overline{\operatorname{Im} A'}$.

Proof: We shall only prove (ii). The proofs of the remaining relationships are
similar. These we leave to the reader.

Suppose $f \in \operatorname{Ker} A'$. For any $x \in X$,

$$f(Ax) = (A'f)x = 0.$$

It therefore follows from the conitnuity of f that $f(\overline{\operatorname{Im} A}) = 0$. Thus
$\overline{\operatorname{Im} A} \subset\,^\perp(\operatorname{Ker} A')$. Assume $x \in\,^\perp(\operatorname{Ker} A')$ but $x \notin \overline{\operatorname{Im} A}$. Then by Theorem XI.5.3
there exists a $g \in X'$ such that

$$g(x) \neq 0 \quad \text{and} \quad g(\operatorname{Im} A) = 0. \tag{5.2}$$

Hence for all $z \in X$,

$$(A'g)z = gAz = 0.$$

Thus g is in $\operatorname{Ker} A'$ and since x is in $^\perp(\operatorname{Ker} A')$, $g(x) = 0$. But this contradicts
(5.2). Hence $^\perp(\operatorname{Ker} A'), \subset \overline{\operatorname{Im} A}$. $\qquad\square$

Theorem 5.2 *Given $A \in \mathcal{L}(X, Y)$, suppose $\operatorname{Im} A$ is closed and codim $\operatorname{Im} A < \infty$.*
Then

$$\dim \operatorname{Ker} A' = \operatorname{codim} \operatorname{Im} A.$$

Proof: By hypothesis, there exists a finite dimensional subspace N such that
$y = \operatorname{Im} A \oplus N$. Let $y_1, \ldots, y_n$ be a basis for N. Since $M_j = \operatorname{sp}\{y_i\}_{i \neq j} \oplus \operatorname{Im} A$
is closed and $y_j \notin M_j$, Theorem XI.5.3 ensures the existence of an $f_j \in Y'$ such
that

$$f_j y_j = 1 \quad \text{and} \quad f_j M_j = 0, \quad 1 \leq j \leq n.$$

Hence

$$f_i y_j = \delta_{ij} \quad \text{and} \quad f_i \in (\text{Im } A)^\perp = \text{Ker } A', \quad 1 \leq i \leq n. \tag{5.3}$$

We need only show that $\{f_1, \ldots, f_n\}$ is a basis for Ker A'. It follows readily from (5.3) that the f_i are linearly independent. Given $f \in \text{Ker } A'$ and $y \in Y$, there exists a $u \in \text{Im } A$ and a $v \in N$ such that $y = u + v$. Since f is in $(\text{Im } A)^\perp$ and $\{y_1, \ldots, y_n\}$ is a basis for N, (5.3) implies that

$$f(y) = f(v) = f\left(\sum_{i=1}^n f_i(v) y_i\right) = \sum_{i=1}^n f_i(v) f(y_i) = \left(\sum_{i=1}^n f(y_i) f_i\right)(y).$$

Thus $f = \sum_{i=1}^n f(y_i) f_i$, which concludes the proof that $f_1, \ldots, f_n$ is a basis for Ker A'. $\quad\square$

Exercise XII-18 shows that codim Im $A < \infty$ implies Im A is closed.

Since $(I - K)' = I - K'$, the next result is an immediate consequence of Theorems 4.1 and 5.2.

Theorem 5.3 *Given $K \in \mathcal{L}(X)$, suppose there exsist a finite rank operator $K_0 \in \mathcal{L}(X)$ such that $\|K - K_0\| < 1$. Then*

$$\infty > \dim \text{Ker } (I - K) = \text{codim Im}(I - K) = \dim \text{Ker}(I - K').$$

We shall now prove the main result in this section, namely, that the conjugate of a compact operator is compact. The proof relies on the following theorem.

Theorem 5.4 *Let W be a subset of a Banach space X. The following statements are equivalent.*

(a) *Every sequence in W has a convergent subsequence.*

(b) *For each $\varepsilon > 0$ there exists a finite set $\{w_1, w_2, \ldots, w_n\}$ of vectors in W with the property that given $w \in W$, there exists a w_k such that $\|w - w_k\| < \varepsilon$.*

Proof: Defining $S(z, r) = \{x \in X | \|x - z\| < r\}$, statement (b) means that $W \subset \cup_{k=1}^n S(w_k, \varepsilon)$. We call $S(z, r)$ a *ball* of radius r.

Assume (a). If (b) is not valid, then for some $\varepsilon > 0$, $W \nsubseteq S(u, \varepsilon)$ where u_1 is any vector in W. Thus there is a $u_2 \in W$, $u_2 \notin S(u_1, \varepsilon)$, i.e., $\|u_2 - u_1\| \geq \varepsilon$. Since $W \nsubseteq S(u_1, \varepsilon) \cup S(u_2, \varepsilon)$, there is a $u_3 \in W$, $u_3 \notin S(u_1, \varepsilon) \cup S(u_2, \varepsilon)$, i.e., $\|u_3 - u_i\| \geq \varepsilon$, $i = 1, 2$. Continuing, we obtain a sequence $\{u_k\} \subseteq W$ such that $\|u_m - u_n\| \geq \varepsilon, m \neq n$. Thus $\{u_n\}$ does not have a convergent subsequence. This contradicts (a).

Assume (b). Let $\{v_k\}$ be a sequence in W. To prove that $\{v_k\}$ has a convergent subsequence, it suffices to assume that $v_m \neq v_n, m \neq n$. Since $V_1 = \{v_1, v_2, \ldots\}$ is an infinite subset of W and W is contained in the union of a finite number of balls of radius $\frac{1}{2}$, at least one of these balls contains an infinite subset $V_2 \subseteq V_1$. Since V_2 is an infinite subset of $V_1 \subseteq W$, it is contained in the union of a finite number of balls of radius $\frac{1}{3}$. Hence there is an infinite subset $V_3 \subseteq V_2$ which is contained in a ball of radius $\frac{1}{3}$. Continuing in this manner, we obtain subsets $V_1 \supseteq V_2 \supseteq V_3 \supseteq \ldots$, and each V_k is an infinite subset of a ball of radius $\frac{1}{k}$. Thus there exists $v_{n_k} \in V_k, n_k > n_{k-1}$.

Since $n_j > n_k$ implies v_{n_j} and v_{n_k} are in V_k and V_k is contained in a ball of radius $\frac{1}{k}$, it follows that $\{v_{n_k}\}$ is a Cauchy sequence which therefore converges in X. $\qquad\square$

Theorem 5.5 *If $K \in \mathcal{L}(X, Y)$ is compact, then $K' \in \mathcal{L}(Y', X')$ is compact.*

Proof: Define $S = \{x \in X \,|\, \|x\| \le 1\}$. Since K is compact, every sequence in KS has a convergent subsequence.

Thus, by Theorem 5.4, for each $\varepsilon > 0$, there exists a finite set $\{Kx_1, Kx_2, \ldots, Kx_n\} \supseteq KS$ with the property that given $x \in S$, there exists a Kx_i such that

$$\|Kx - Kx_i\| < \varepsilon/3. \tag{5.4}$$

Let A be the map from Y' to $\mathbb{C}^n$ given by

$$Ay' = (y'Kx_1, \ldots, y'Kx_n).$$

Since Y' is complete and A is compact, it follows from Theorem 5.4 that there exist y_1', y_2', y_m' in $S_{Y'} = \{y' \in Y' \,|\, \|y'\| \le 1\}$ such that for each $y' \in S_{Y'}$, there is a y_j' for which

$$\|A'y' - A'y_j'\| < \varepsilon/3.$$

Thus,

$$|y'Kx_i - y_j'Kx_i| \le \|A'y - A'y_j'\| < \varepsilon/3, \quad 1 \le i \le n. \tag{5.5}$$

For $x \in X$, $\|x\| = 1$, we have from (5.4) and (5.5),

$$
\begin{aligned}
|K'y'x - K'y_j'x| &\le |y'Kx - y'Kx_i| + |y'Kx_i - y_j'Kx_i| + |y_j'Kx_i - y_j'Kx| \\
&\le \|Kx - Kx_i\| + \varepsilon/3 + \|Kx_i - Kx\| < \varepsilon.
\end{aligned}
$$

Hence $\|K'y' - K'y_j'\| < \varepsilon$. The compactness of K' is now a consequence of Theorem 5.4.

13.6 Spectrum of a Compact Operator

As we pointed out in Section XII.8 the spectrum of an operator $A \in \mathcal{L}(X)$, where X is finite dimensional, consists of a finite number of points, each of which is an eigenvalue of A. Before proving the following extension of this result, we note that *eigenvectors $v_1, \ldots, v_n$ corresponding to distinct eigenvalues $\lambda_1, \ldots, \lambda_n$ of a linear operator T are linearly independent.* For, if not, there exists a v_k

such that $v_1, \ldots, v_{k-1}$ are linearly independent and $v_k \in \mathrm{sp}\{v_1, \ldots, v_{k-1}\}$, say $v_k = \sum_{j=1}^{k-1} \alpha_j v_j$. Since

$$0 = \lambda_k v_k - T v_k = \sum_{j=1}^{k-1} \alpha_j (\lambda_k - \lambda_j) v_j,$$

We have $\alpha_j = 0, 1 \le j \le k - 1$. But then $v_k = 0$ which is impossible. □

Theorem 6.1 *Let X be infinite dimensional, and let $T \in \mathcal{L}(X)$ be the limit in norm of a sequence of operators of finite rank. Then the spectrum of T is a countable set $\{\lambda_1, \lambda_2, \ldots\}$ which includes $\lambda = 0$. If $\lambda_i \neq 0$, then it is an eigenvalue of T. If $\{\lambda_i\}$ is an infinite set, then $\lambda_i \to 0$.*

Proof: Since X is infinite dimensional, $0 \in \sigma(T)$, otherwise $I = T T^{-1}$ is compact. Suppose $0 \neq \lambda \in \sigma(T)$. Then by Corollary 4.2,

$$\mathrm{codim}(\lambda I - T) = \dim \mathrm{Ker}(\lambda I - T) = \dim \mathrm{Ker}\left(I - \frac{T}{\lambda}\right) \neq (0),$$

i.e., λ is an eigenvalue of T. To prove the rest of the theorem, it suffices to show that for each $\varepsilon > 0$, the set $S = \{\lambda \in \sigma(T) : |\lambda| \ge \varepsilon\}$ is finite. Suppose that for some $\varepsilon > 0$, S contains an infinite set $\lambda_1, \lambda_2, \ldots$. Let v_j be an eigenvector of T with eigenvalue λ_j and let $M_n = \mathrm{sp}\{v_1, \ldots, v_n\}$. Since the v_i's are linearly independent, M_{n-1} is a proper subspace of M_n. Hence by Lemma XI.2.3, there exists a $w_n \in M_n$ such that

$$1 = \|w_n\| = d(w_n, M_{n-1}). \tag{6.1}$$

Now

$$\|T w_n - T w_m\| = \|\lambda_n w_n - (\lambda_n w_n - T w_n + T w_m)\|. \tag{6.2}$$

It is easy to see that $\lambda_n w_n - T w_n$ and $T w_m$ are in M_{n-1}, provided $n > m$. Hence it follows from (6.1) and (6.2) that for $n > m$,

$$\|T w_n - T w_m\| \ge |\lambda_n| d(w_n, M_{n-1}) = |\lambda_n| \ge \varepsilon.$$

But this is impossible since $\{T w_n\}$ has a convergent subsequence. □

In Chapter XV, Theorem 4.4, the theorem above will be extended to any compact operator $T \in \mathcal{L}(X)$ without the assumption that T is the limit in norm of operators of finite rank.

The above proof of Theorem 6.1 is short, but it does not explain the analytical motivation which is behind the theorem. It turns out that the proof of this theorem can be derived from the well known fact from complex analysis that if a non-zero analytic function vanishes on a compact set Z, then Z is finite.

We shall prove Theorem 6.1 under the assumption that for each $\varepsilon > 0$ there exists a $K \in \mathcal{L}(X)$ of finite rank such that $\|T - K\| < \varepsilon$. As before, it suffices to show that $S = \{\lambda \in \sigma(T) : |\lambda| \geq \varepsilon\}$ is finite. This is done by constructing a non-zero analytic function which vanishes on S. For $A = T - K$,

$$(\lambda I - T) = \lambda I - A - K \tag{6.3}$$

and $\lambda I - A$ is invertible whenever $\lambda \in S$ since $\left\|\frac{A}{\lambda}\right\| < 1$. Thus $S \subset \rho(A)$.

Let φ be an eigenvector of T corresponding to $\lambda \in S$ and let $A(\lambda) = (\lambda I - A)^{-1}$. From (6.1) we get

$$0 = \lambda\varphi - T\varphi = A(\lambda)[(\lambda I - T)\varphi] = \varphi - A(\lambda)K\varphi. \tag{6.4}$$

Since Im K is finite dimensional, we know from the remark following Theorem 2.1 that there exist $g_1, \ldots, g_n \in X'$ and a basis $y_1, \ldots, y_n$ for Im K such that for all $x \in X$, $Kx = \sum_{j=1}^{n} g_j(x)y_j$. Thus it follows from (6.4) that

$$\varphi = \sum_{j=1}^{n} g_j(\varphi)A(\lambda)y_j. \tag{6.5}$$

Hence

$$g_k(\varphi) = \sum_{j=1}^{n} g_j(\varphi)g_k(A(\lambda)y_j), \quad 1 \leq k \leq n.$$

Writing

$$\xi_k = g_k(\varphi) \quad and \quad a_{kj}(\lambda) = g_k(A(\lambda)y_j), \tag{6.6}$$

we get the system of equations

$$\xi_k - \sum_{j=1}^{n} \xi_j a_{kj}(\lambda) = 0, \quad 1 \leq k \leq n. \tag{6.7}$$

Not all ξ_j are zero; otherwise $\varphi = 0$ by (6.5), which is impossible since φ is an eigenvector. Hence, the system of Equations (6.7) has a non trivial solution, namely $\xi_j = g_j(\varphi)$. Therefore,

$$h(\lambda) = \det(\delta_{kj} - a_{kj}(\lambda)) = 0, \quad \lambda \in S.$$

Conversely, if $h(\lambda) = 0$, there exists $\xi_1, \ldots, \xi_n$, not all zero, such that (6.7) holds. Guided by (6.5) and (6.6), we define $\varphi = \sum_{j=1}^{n} \xi_j A(\lambda)y_j$. Since $y_1, \ldots, y_n$ are linearly independent and $A(\lambda)$ is invertible, it follows that $\varphi \neq 0$. With the use of (6.7), a straightforward computation verifies that $(\lambda I - T)\varphi = 0$.

Thus, $h(S) = 0$ and h is not identically zero since $|\xi| > \|A\| + \|T\|$ implies $\xi \in \rho(A) \cap \rho(T)$, whence $h(\xi) \neq 0$. Moreover, h is analytic on $\rho(A)$. To see this, we recall from Theorem XII.10.1 that $A(\lambda)$ is an analytic operator valued function on $\rho(A)$. It follows readily from the linearity and continuity of g_k that $a_{kj}(\lambda) = g_k(A(\lambda)y_j)$ is a complex valued function which is analytic on $\rho(A)$. Hence h is analytic on $\rho(A)$. A basic result in complex analysis states that if a non-zero analytic function vanishes on a compact set Z, then Z is finite. Since S is compact and $h(S) = 0$, S is finite. This completes the proof of the theorem.

13.7 Applications

In Section 1 it was shown that the operators appearing below are limits in norm of a sequence of operators of finite rank. Hence the conclusions below follow from Corollary 4.2 and Theorem 6.1.

1. Suppose k is continuous on $[a, b] \times [a, b]$. The equation

$$\lambda f(t) - \int_a^b k(t, s) f(s) \, ds = g(t), \quad \lambda \neq 0 \qquad (7.1)$$

has a unique solution in $C([a, b])$ for each $g \in C([a, b])$ if and only if the homogeneous equation

$$\lambda f(t) = \int_a^b k(t, s) f(s) \, ds = 0, \quad \lambda \neq 0, \qquad (7.2)$$

has only the trivial solution in $C([a, b])$.

Except for a countable set of λ, which has zero as the only possible limit point, Equation (7.1) has a unique solution for every $g \in C([a, b])$. For $\lambda \neq 0$, the equation (7.2) has at most a finite number of linear independent solutions.

2. For $1 < p < \infty$, let $k(t, s)$ be in $L_r([0, 1] \times [0, 1])$, where $r = \max(p, p')$, $\frac{1}{p} + \frac{1}{p'} = 1$. The conclusions in the example above remain valid when we replace $C([a, b])$ by $L_p([0, 1])$.

3. Let $(a_{jk})_{j,k=1}^\infty$ be a matrix of complex numbers such that

$$\lim_{n \to \infty} \sup_k \sum_{j=n}^\infty |a_{jk}| = 0.$$

The infinite system of equations

$$\lambda x_j - \sum_{k=1}^\infty a_{jk} x_k = y_j, \quad j = 1, 2, \ldots, \quad \lambda \neq 0 \qquad (7.3)$$

has a unique solution $(x_1, x_2, \ldots) \in \ell_1$ for every $(y_1, y_2, \ldots) \in \ell_1$ if and only if the homogeneous system of equations

$$\lambda x_j - \sum_{k=1}^\infty a_{jk} x_k = 0, \quad j = 1, 2, \ldots, \quad \lambda \neq 0 \qquad (7.4)$$

has only the trivial solution in ℓ_1.

Except for a countable set of λ, which has zero as the only possible limit point, Equation (7.3) has a unique solution $(x_1, x_2, \ldots) \in \ell_1$ for every $(y_1, y_2, \ldots) \in \ell_1$. For $\lambda \neq 0$, Equation (7.4) has at most a finite number of linear independent solutions.

4. If $\sum_{j,k=1}^\infty |a_{jk}|^2 < \infty$, the conclusions in the above example remain valid if we replace ℓ_1 by ℓ_2. Here we also need Example 2 following Theorem II.16.3.

Exercises XIII

Throughout these exercises, X and Y denote Banach spaces.

1. Let $\{\beta_n\}$ be a bounded sequence of complex numbers. Define $T : \ell_p \to \ell_p$, $1 \leq p < \infty$, by $T(\alpha_1, \alpha_2, \ldots) = (\beta_1\alpha_1, \beta_2\alpha_2, \ldots)$. Find T'.

2. Let h be a bounded complex valued Lebsegue measurable function on $[a, b]$. Define $A : L_p[a, b] \to L_p[a, b]$, $1 \leq p < \infty$, by $(Af)(t) = h(t)f(t)$. Find A'.

3. Let $T : L_p[0, \infty) \to L_p[0, \infty)$, $1 \leq p < \infty$, be given by $(Tf)(x) = f(x + 1)$. Find T'.

4. Suppose $P \in \mathcal{L}(X)$ is a projection from X onto M. Prove that P' is a projection with kernel $M^\perp$. What is the range of P'?

5. Prove that if $T \in \mathcal{L}(X, Y)$ is invertible, then T' is also invertible. What is the relationship between $(T')^{-1}$ and $(T^{-1})'$?

6. Suppose $A \in \mathcal{L}(X, Y)$ and Im $A = Y$. Prove that A is an isometry if and only if A' is an isometry and Im $A' = X'$.

7. Prove that a projection is compact if and only if it is of finite rank.

8. Let $K \in L(X)$ be compact. Suppose that $I - K$ is either left or right invertible. Prove that $I - K$ is invertible and $I - (I - K)^{-1}$ is compact.

9. Suppose $\{f_n\} \subset X'$, $\{y_n\} \subset Y$ and $\sum_{n=1}^\infty \|f_n\|\|y_n\| < \infty$. Prove that the operator $K : X \to Y$ given by $Kx = \sum_{n=1}^\infty f_n(x)y_n$ is compact. In this case K is called a *nuclear* operator.

10. Let $K \in L(X, Y)$ be compact. Suppose $\{x_n\}$ is a sequence in X with the property that there exists an $x \in X$ such that $f_n x \to f(x)$ for all $f \in X'$.

    (a) Prove that $Kx_n \to Kx$.

    (b) Give an example which shows that (a) need not hold if K is not compact.

11. Suppose $K \in \mathcal{L}(X)$ is compact and $I - K$ is injective. Prove that if $S \subset X$ and $(I - K)S$ is a bounded set, then S is bounded.

12. Let $K \in \mathcal{L}(X)$ be compact. Given $\lambda \neq 0$, prove that $K_\lambda = \lambda I - K$ has a closed range by establishing the following:

    (a) Ker K_λ is finite dimensional (Hint: $K = \lambda I$ on Ker K_λ). Therefore Ker K_λ is complemented by a closed subspace M.

    (b) Given $y \in \overline{\text{Im } K_\lambda}$ there exists a sequence $\{v_n\}$ in M such that $K_\lambda v_n \to y$.

    (c) The sequence $\{v_n\}$ is bounded.

    (d) The sequence $\{v_n\}$ has a subsequence which converges to some $x \in X$ and $K_\lambda x = y$.

13. Let $\varphi_1, \varphi_2, \ldots$ be an orthonormal basis for a Hilbert space $\mathcal{H}$. Define $T \in \mathcal{L}(\mathcal{H})$ by $Tx = \langle x, \varphi_3 \rangle \varphi_2 + \langle x, \varphi_5 \rangle \varphi_4 + \langle x, \varphi_7 \rangle \varphi_6 + \ldots$. Show that T is not compact but $T^2 = 0$.

14. Suppose $A \in \mathcal{L}(X)$ and A^p is compact for some positive integer p. Prove that the Fredholm Theorems XI-4.1, 5.3 and 6.1 hold for A. Hint: Choose n roots of unity $1, \xi_1, \ldots, \xi_{n-1}$ for some $n \geq p$ so that $\xi_i \in \rho(A)$, $1 \leq i \leq n - 1$ (why do such ξ_i exist?) Then $I - A^n = (I - A)(\xi_1 - A) \ldots (\xi_{n-1} - A)$.

Chapter XIV
Poincaré Operators: Determinant and trace

A linear operator P defined in the standard basis of $\ell_p(1 \leq p < \infty)$ by a matrix of the form

$$P = (p_{jk})_{j,k=1}^{\infty},$$

with $p_{jk} \in \mathbb{C}$ and

$$\sum_{j,k=1}^{\infty} |p_{jk}| < \infty$$

we call a Poincaré operator. H. Poincaré [P] introduced and studied this class in connection with problems in celestial mechanics. One of the important advantages of this class is that for these operators P it is possible to introduce a trace and for operators $I - P$ a determinant that are natural generalizations of the trace and determinant of finite matrices. To keep the presentation of the material in this chapter simpler we restrict ourselves to the Hilbert space ℓ_2. However it can be generalized for many Banach spaces, in particular, for $\ell_p(1 \leq p < \infty)$.

14.1 Determinant and Trace

Denote by $\mathcal{P}$ the set of all Poincaré operators on ℓ_2. Recall that these operators P are defined in the standard basis of ℓ_2 by the matrix

$$P = (p_{jk})_{j,k=1}^{\infty},$$

with the property

$$\sum_{j,k=1}^{\infty} |p_{jk}| < \infty.$$

It is easy to see that those operators can be represented in the form

$$P = \sum_{j,k=1}^{\infty} p_{jk} U_{jk},$$

where

$$U_{jk}x = \langle x, e_k \rangle e_j \quad (x \in \ell_2)$$

and $e_1, e_2, \ldots$ is the standard basis in ℓ_2. Taking into account that

$$\|P\| \leq \sum_{j,k=1}^{\infty} |p_{jk}| \|U_{jk}\|$$

and

$$\|U_{jk}\| = 1,$$

we obtain that P is a bounded operator and

$$\|P\| \leq \|P\|_{\mathcal{P}},$$

where

$$\|P\|_{\mathcal{P}} = \sum_{j,k=1}^{\infty} |p_{jk}|.$$

For any operator $P = (p_{jk})_{j,k=1}^{\infty} \in \mathcal{P}$ we denote by P_n an operator defined on $\mathbb{C}^n$ by the equality:

$$P_n = (p_{jk})_{j,k=1}^{n}.$$

We call this operator the n-th *section* of P. Denote by $P^{(n)} \in \mathcal{P}$ the operator

$$P^{(n)} = (p_{jk}^{(n)})_{j,k=1}^{\infty},$$

where $p_{jk}^{(n)} = p_{jk}$ for $j, k = 1, 2, \ldots, n$ and $p_{jk}^{(n)} = 0$ for the rest of the indices j and k. Note that the following relations hold

$$\|P - P^{(n)}\| \leq \|P - P^{(n)}\|_{\mathcal{P}} = \|P\|_{\mathcal{P}} - \|P^{(n)}\|_{\mathcal{P}}.$$

The last difference converges to zero when $n \to \infty$. From here it follows that P is a compact operator on ℓ_2.

For any operator $P \in \mathcal{P}$ the following inequality holds:

$$|\det(I - P_{n+k}) - \det(I - P_n)| \leq \Pi(P_{n+k}) - \Pi(P_n), \qquad (1.1)$$

where

$$\Pi(P_n) = \prod_{j=1}^{n} \left(1 + \sum_{k=1}^{n} |p_{jk}|\right)$$

Indeed, let

$$V = (v_{jk})_{j,k=1}^{n}$$

be a matrix with entries $v_{jk} \in \mathbb{C}$. According to the definition of the determinant

$$\det V = \sum \gamma_{j_1, j_2, \ldots, j_n} v_{j_1}{}^{1}, v_{j_2}{}^{2}, \ldots, v_{j_n}{}^{n},$$

where the coefficient $\gamma_{j_1, j_2, \ldots, j_n}$ is equal to one of three values $+1, -1$ and 0.

Let the entries v_{jk} have the form

$$v_{jk} = \delta_{jk} - u_{jk}.$$

It is clear that

$$|\det V| \leq \Pi(V),$$

where

$$\Pi(V) = \prod_{j=1}^{n} (1 + |u_{j1}| + |u_{j2}| + \cdots + |u_{jn}|).$$

Let some set $\sum$ of entries u_{jk} in the matrix V be replaced by zeros and the rest is not changed. Denote the new matrix by W. It is easy to see that

$$|\det V - \det W| \leq \Pi(V) - \Pi(W). \tag{1.2}$$

From this inequality follows inequality (1.1). Indeed, introduce $\sum_n$ as the set of u_{jk} for which at least one of the indices j or k is $> n$. Replace u_{jk} by 0, for $j, k \in \sum_n$, then the inequality (1.1) follows from (1.2).

We are now in a position to define the determinant $\det_{\mathcal{P}}(I - P)$ for any $P \in \mathcal{P}$. Let $P_n (n = 1, 2, \ldots)$ be the n-th finite section of P:

$$P_n = (p_{jk})_{j,k=1}^{n}$$

From the relation (1.1) follows that the sequence of complex numbers

$$\det(I_n - P_n) \quad (n = 1, 2, \ldots)$$

is convergent. The limit we denote by $\det_{\mathcal{P}}(I - P)$ and call it the *Poincaré determinant* of the operator $I - P$:

$$\det_{\mathcal{P}}(I - P) = \lim_{n \to \infty} \det(I_n - P_n).$$

Also note that

$$|\det_{\mathcal{P}}(I - P)| \leq \prod_{j=1}^{\infty} (1 + |p_{1j}| + |p_{2j}| + \cdots).$$

Using the inequality $1 + x \leq e^x$ we obtain also

$$|\det_{\mathcal{P}}(I - P)| \leq \exp \left(\sum_{j,k=1}^{\infty} |p_{jk}| \right) = \exp \|P\|_{\mathcal{P}}, \tag{1.3}$$

where $P = (p_{jk})_{j,k=0}^{\infty}$.

The *trace* is introduced for operators $P \in \mathcal{P}$ by the formula

$$\text{tr}_{\mathcal{P}} P = \sum_{j=1}^{\infty} p_{jj},$$

where $P = (p_{jk})_{j,k=1}^{\infty}$. The trace is a linear functional on $\mathcal{P}$ and

$$\text{tr}_{\mathcal{P}} P = \lim_{n \to \infty} \text{tr} P_n.$$

The trace is continuous in $\mathcal{P}$ in the sense

$$|\text{tr}_{\mathcal{P}} P| \le \|P\|_{\mathcal{P}},$$

and hence it is a Lipschitz function on all of $\mathcal{P}$:

$$|\text{tr}_{\mathcal{P}} P' - \text{tr}_{\mathcal{P}} P''| \le \|P' - P''\|_{\mathcal{P}}.$$

Example: 1. Let the operator $V \in \mathcal{P}$ be given by the infinite matrix

$$V = \begin{pmatrix} 0 & u_1 & u_2 & \cdots & u_{n-1} & u_n \cdots \\ -v_1 & 0 & 0 & \cdots & 0 & 0 \cdots \\ 0 & -v_2 & 0 & \cdots & 0 & 0 \cdots \\ 0 & 0 & -v_3 & \cdots & 0 & 0 \cdots \\ \cdot & \cdot & \cdot & \cdots & \cdot & \cdots \end{pmatrix}$$

We shall assume that the series

$$\sum_{j=1}^{\infty} |u_j| \quad \text{and} \quad \sum_{j=1}^{\infty} |v_j|$$

are convergent. It is obvious that $V \in \mathcal{P}$. Now we shall calculate $\det_{\mathcal{P}} (I - V)$. The starting point is the calculation of the determinant of the n-th section

$$\det(I - V_n).$$

If we expand this determinant along the last column we obtain

$$\det (I - V_n) = (-1)^{n+1} u_n v_1 v_2, \ldots, v_n + \det (I - V_{n-1}).$$

Hence

$$\det (I - V_n) = (-1)^{n+1} u_n v_1 v_2 \ldots v_n +$$
$$(-1)^n u_{n-1} v_1 v_2 \ldots v_{n-1} + \cdots + u_1 v_1 + 1.$$

From here it follows that

$$\det_{\mathcal{P}} (I - V) = \sum_{n=1}^{\infty} (-1)^{n+1} u_n v_1 v_2 \ldots v_n + 1$$

where the series converges absolutely. Note that

$$\det_{\mathcal{P}} (I - \mu V) = \sum_{n=1}^{\infty} \mu^{n+1} (-1)^{n+1} u_n v_1, v_2 \ldots v_n + 1$$

and

$$\text{tr}_{\mathcal{P}} V = 0.$$

Note also that the operator V is not of finite rank, generally speaking.

14.2 Finite Rank Operators, Determinants and Traces

Let $\mathcal{F}(\ell_2)$ denote the space of bounded operators of finite rank on ℓ_2. Due to Theorem XIII.2.1, for any operator $F \in \mathcal{F}(\ell_2)$ there exist closed subspaces M_F and N_F of ℓ_2 with the following properties:

a) $M_F \oplus N_F = \ell_2$

b) $F M_F \subseteq M_F$, $N_F \subseteq \text{Ker } F$

c) $\dim M_F < \infty$.

With respect to the decomposition a), the operator F can be represented by

$$F = \begin{pmatrix} F_1 & 0 \\ 0 & 0 \end{pmatrix},$$

where $F_1 = F|_{M_F}$. Hence

$$I - F = \begin{pmatrix} I_1 - F_1 & 0 \\ 0 & I_2 \end{pmatrix}$$

where I, I_1, I_2 stand for the identity in ℓ_2, M_F, N_F respectively. Since M_F is a finite dimensional space, the functionals tr F_1 and det $(I_1 - F_1)$ are well defined. We use this to define the tr F and det $(I - F)$ by the equalities

$$\text{tr } F = \text{tr } F_1 \quad \text{and} \quad \det(I - F) = \det(I_1 - F_1).$$

Note that these definitions do not depend on the particular choice of the subspace M_F. To see this we recall from linear algebra that

$$\text{tr } F_1 = \sum_{j=1}^{m} \lambda_j(F_1) \quad \text{and} \quad \det(I_1 - F_1) = \prod_{j=1}^{m} (1 - \lambda_j(F_1))$$

where $m = \dim M_F$ and $\lambda_1(F_1), \lambda_2(F_1), \ldots, \lambda_m(F_1)$ are the eigenvalues of F_1 counted according to their algebraic multiplicities. In the two identities the zero eigenvalues do not give any contribution. It is also easy to see that the operators F and F_1 have the same nonzero eigenvalues counting multiplicities. Thus

$$\text{tr } F = \sum_j \lambda_j(f) \quad \text{and} \quad \det(I - F) = \prod_j (1 - \lambda_j(F)). \tag{2.1}$$

Now we shall prove the following four properties of the trace and determinant. For any two operators F and $K \in \mathcal{F}(\ell_2)$ the following equalities hold

$$\text{tr}(\alpha F + \beta K) = \alpha \text{ tr } F + \beta \text{ tr } K, \alpha, \beta \in \mathbb{C}$$

$$\text{tr } FK = \text{tr } KF$$

$$\det((I - F)(I - K)) = \det(I - F) \det(I - K)$$

$$\det(I - FK) = \det(I - KF)$$

We shall show that all these properties follow from the corresponding properties in linear algebra. Suppose

$$Fx = \sum_{j=1}^{p} \langle x, \varphi_j \rangle \psi_j \quad \text{and} \quad Kx = \sum_{j=1}^{q} \langle x, f_j \rangle g_j, x \in \ell_2.$$

We set

$$M = \text{span}\{\psi_j, \varphi_j, i = 1, 2, \ldots, p; \ f_k, g_k, k = 1, 2, \ldots, q\}$$

and

$$N = M^{\perp}.$$

Then

$$F(M) \subset M, \quad K(M) \subset M$$

$$N \subset \text{Ker } F, \quad N \subset \text{Ker } K.$$

Now let

$$F_M = F|_M \quad \text{and} \quad K_M = K|_M.$$

The subspace M has a finite dimension and hence all four properties for F_M and K_M follow from linear algebra. This statement is proved in the same way as the first statement in this section. The next theorem gives a useful representation of the trace and determinant.

Theorem 2.1 *Let $F \in \mathcal{F}(\ell_2)$ be given in the form*

$$Fx = \sum_{k=1}^{m} \langle x, \varphi_k \rangle f_k, \quad x \in H$$

Then

$$\operatorname{tr} F = \sum_{k=1}^{m} \langle f_k, \varphi_k \rangle \tag{2.2}$$

and

$$\det(I - F) = \det(\delta_{jk} - \langle f_k, \varphi_j \rangle)_{j,k=1}^{m} \tag{2.3}$$

Proof: Since $\operatorname{tr} F$ is a linear functional, it is enough to prove the first equality for the case $m = 1$. In this case the operator is

$$Fx = \langle x, \varphi_1 \rangle f_1.$$

Let

$$\langle x, \varphi_1 \rangle f_1 = \lambda x \quad (\lambda \neq 0).$$

Then

$$x = \frac{\gamma}{\lambda} f_1, \quad \text{where} \quad \gamma = \langle x, \varphi_1 \rangle$$

and

$$\gamma = \frac{\gamma}{\lambda} \langle f_1, \varphi_1 \rangle.$$

Taking $\gamma \neq 0$ we obtain that the only non-zero eigenvalue of F is

$$\lambda = \langle f_1, \varphi_1 \rangle.$$

Now let us prove the second equality. Consider two operators $X \in \mathcal{L}(\ell_2, \mathbb{C}^m)$ and $Y \in \mathcal{L}(\mathbb{C}^m, \ell_2)$ defined by the equalities:

$$X\varphi = (\langle \varphi, \varphi_1 \rangle, \langle \varphi, \varphi_2 \rangle, \ldots, \langle \varphi, \varphi_m \rangle),$$

$$Y(\alpha_1, \alpha_2, \ldots, \alpha_m) = \sum_{j=1}^{m} \alpha_j f_j.$$

It is clear that $F = YX$. Hence

$$\det(I - F) = \det(I - YX).$$

Note that

$$XY \begin{pmatrix} \alpha_1 \\ \alpha_2 \\ \vdots \\ \alpha_m \end{pmatrix} = \begin{pmatrix} \sum_{k=1}^{m} \alpha_k \langle f_k, \varphi_1 \rangle \\ \vdots \\ \sum_{k=1}^{m} \alpha_k \langle f_k, \varphi_m \rangle \end{pmatrix}.$$

Hence $XY = (\langle f_k, \varphi_j \rangle)_{j,k=1}^{n}.$

Now it remains to prove that

$$\det(I_m - XY) = \det(I - YX).$$

Introduce the following finite rank operator $\begin{pmatrix} 0 & X \\ Y & 0 \end{pmatrix}$ and take

$$A = I + \begin{pmatrix} 0 & X \\ Y & 0 \end{pmatrix},$$

which acts on $\mathbb{C}^m \oplus \ell_2$. Using the following two Schur decompositions of A we obtain

$$A = \begin{pmatrix} I_2 & X \\ Y & I_1 \end{pmatrix} = \begin{pmatrix} I_2 & 0 \\ Y & I_1 \end{pmatrix} \begin{pmatrix} I_2 & 0 \\ 0 & I_1 - YX \end{pmatrix} \begin{pmatrix} I_2 & X \\ 0 & I_1 \end{pmatrix}$$

and

$$A = \begin{pmatrix} I_2 & X \\ Y & I_1 \end{pmatrix} = \begin{pmatrix} I_2 & X \\ 0 & I_1 \end{pmatrix} \begin{pmatrix} I_2 - XY & 0 \\ 0 & I_1 \end{pmatrix} \begin{pmatrix} I_2 & 0 \\ Y & I_1 \end{pmatrix}.$$

Computing the determinants from both sides of these equalities we get

$$\det(I_m - XY) = \det(I - YX).$$

This ends the proof. $\qquad\qquad\qquad\qquad\qquad\qquad\qquad\qquad\qquad\qquad\qquad$ □

The next theorem gives a representation of the trace and the determinant for the case when the finite rank operator is given by an infinite matrix.

Theorem 2.2 *Let the operator $F \in \mathcal{F}(\ell_2)$ be defined in the standard basis $e_1, e_2, \ldots$ of ℓ_2 by the following matrix*

$$F = (a_{jk})_{j,k=1}^{\infty}.$$

Then

$$\operatorname{tr} F = \sum_{k=1}^{\infty} a_{kk}, \qquad\qquad\qquad (2.4)$$

where the series converges absolutely and

$$\det(I - F) = \lim_{n \to \infty} \det(\delta_{jk} - a_{jk})_{j,k=1}^{n} \qquad\qquad (2.5)$$

Proof: The operator F can be represented in the form

$$Fx = \sum_{r=1}^{m} \langle x, \varphi_r \rangle f_r.$$

Hence

$$a_{jk} = \langle Fe_k, e_j \rangle = \sum_{r=1}^{m} \langle e_k, \varphi_r \rangle \langle f_r, e_j \rangle, \quad k, j = 1, 2, \ldots. \quad (2.6)$$

Taking into consideration that

$$f_r = \sum_{j=1}^{\infty} \langle f_r, e_j \rangle e_j, \varphi_s = \sum_{k=1}^{\infty} \langle \varphi_s, e_k \rangle e_k$$

we obtain

$$\langle f_r, \varphi_s \rangle = \sum_{j=0}^{\infty} \langle f_r, e_j \rangle \overline{\langle \varphi_s, e_j \rangle} \quad (2.7)$$

This series converges absolutely since

$$\sum_{j=1}^{\infty} |\langle f_r, e_j \rangle| \, |\langle \varphi_s, e_j \rangle| \leq \|f_r\| \|\varphi_s\|.$$

From formulas (2.6) and (2.7) follows that

$$\sum_{k=1}^{\infty} a_{kk} = \sum_{k=1}^{\infty} \sum_{r=1}^{m} \langle f_r, e_k \rangle \overline{\langle \varphi_r, e_k \rangle}$$

$$= \sum_{r=1}^{m} \sum_{k=1}^{\infty} \langle f_r, e_k \rangle \overline{\langle \varphi_r, e_k \rangle} = \sum_{r=1}^{m} \langle f_r, \varphi_r \rangle.$$

In Theorem 2.1 we proved that

$$\operatorname{tr} F = \sum_{r=1}^{m} \langle f_r, \varphi_r \rangle$$

and hence

$$\operatorname{tr} F = \sum_{k=1}^{\infty} a_{kk}.$$

The first part of the theorem is proved. Now we proceed with the second part.
Denote

$$b_{rs}^{(\nu)} = \sum_{k=1}^{\nu} f_{rk} \overline{\varphi_{sk}}$$

where

$$f_{rk} = \langle f_r, e_k \rangle \quad \text{and} \quad \varphi_{sk} = \langle \varphi_s, e_k \rangle.$$

When ν tends to infinity $b_{rs}^{(\nu)}$ tends to $\langle f_r, \varphi_s \rangle$:

$$\lim_{\nu \to \infty} b_{rs}^{(\nu)} = \langle f_r, \varphi_s \rangle, r, s = 1, \ldots, m.$$

The determinant of finite matrices of fixed size is a continuous function of its entries. From this follows that

$$\lim_{v \to \infty} \det(\delta_{rs} - b_{rs}^v)_{r,s=1}^m = \det(\delta_{rs} - \langle f_r, \varphi_s \rangle)_{r,s=1}^m \qquad (2.8)$$

In Theorem 2.1 it was proved that

$$\det(I - F) = \det(\delta_{rs} - \langle f_r, \varphi_s \rangle)_{r,s=1}^m$$

and this means

$$\det(I - F) = \lim_{v \to \infty} \det(I - B^{(v)}).$$

where

$$B^{(v)} = (b_{rs}^{(v)})_{r,s=1}^m.$$

It is easy to see that the matrix $B^{(v)}$ is a product of two matrices.

$$B^{(v)} = X_{f,v} Y_{\varphi,v}^*,$$

where

$$X_{f,v} = (f_{rk})_{r=1, k=1}^{m, v}$$

and

$$Y_{\varphi,v} = (\varphi_{sj})_{s=1, j=1}^{m, v}.$$

In the proof of Theorem 2.1 it was proved that for two finite matrices $X_{f,v}$ and $Y_{\varphi,v}$

$$\det(I - X_{f,v} Y_{\varphi,v}^*) = \det(I - Y_{\varphi,v}^* X_{f,v}).$$

The matrix $Y_{\varphi,v}^* X_{f,v}$ has the size $v \times v$. Denote

$$Y_{\varphi,v}^* X_{f,v} = (v_{jk})_{j,k=1}^v,$$

where

$$v_{jk} = \sum_{r=1}^m \overline{\varphi}_{jr} f_{rk} = \sum_{r=1}^m \langle e_j, \varphi_r \rangle \langle f_r, e_k \rangle, \quad j, k = 1, 2, \ldots, v.$$

It is clear from (2.6) that

$$v_{jk} = a_{kj} \quad j, k = 1, 2, \ldots, v.$$

It is also now clear that

$$\det(I - B^{(v)}) = \det(\delta_{jk} - a_{jk})_{j,k=0}^v.$$

Now we go to the limit from both sides when $v \to \infty$ and taking into consideration the earlier results, we obtain

$$\det(I - F) = \lim_{v \to \infty} \det(I - B^{(v)})$$

which finishes the proof.   □

Note that so far we have introduced for Poincaré operators $F \in \mathcal{F}(\ell_2)$ two definitions of determinants for $I - F$ and two definitions of trace for F.

Namely, in the first section we defined

$$\det{}_{\mathcal{P}}(I - F) = \lim_{n \to \infty} \det(I - F_n)$$

and

$$\mathrm{tr}{}_{\mathcal{P}} F = \lim_{n \to \infty} \mathrm{tr} \, F_n,$$

where F_n is the n-th section of F. In this section the determinant and the trace were defined via the formulas

$$\det(I - F) = \prod_{r=1}^{m}(1 - \lambda_j(F))$$

and

$$\mathrm{tr} \, F = \sum_{r=1}^{m} \lambda_j(F).$$

As a corollary of Theorem 2.2 we obtain that these two definitions coincide:

$$\det{}_{\mathcal{P}}(I - F) = \det(I - F), \mathrm{tr}{}_{\mathcal{P}} \, F = \mathrm{tr} \, F.$$

In particular

$$\det{}_{\mathcal{P}}(I - F) = \prod_{r=1}^{m}(I - \lambda_j(F)),$$

$$\mathrm{tr}{}_{\mathcal{P}}(I - F) = \sum_{r=1}^{m} \lambda_j(F).$$

In this connection we will drop, in subsequent sections, the subscript $\mathcal{P}$ in the Poincaré determinant.

In conclusion we mention that it can happen that an operator has finite rank and is not a Poincaré operator; see exercise 7.

14.3 Theorems about the Poincaré Determinant

In this section we shall discuss a number of important properties of the determinant. We start with the analysis of continuity of the function $\det (I - X)$, $X \in \mathcal{P}$. Note that $\mathcal{P}$ is a linear space. It is not difficult to see that it is a Banach space in the $\mathcal{P}$-norm $\| \cdot \|_{\mathcal{P}}$.

Theorem 3.1 *The function $\det (I - X)$, $X \in \mathcal{P}$ is Lipschitz continuous. Namely, for any operators $X, Y \in \mathcal{P}$ the following inequality holds:*

$$|\det(I - X) - \det(I - Y)| \leq [\exp(\|X\|_{\mathcal{P}} + \|Y\|_{\mathcal{P}} + 1)]\|X - Y\|_{\mathcal{P}}. \quad (3.1)$$

Proof: Let us first prove the theorem for operators X, Y of finite rank. Introduce the function

$$g(\lambda) = \det\left(I - \frac{1}{2}(X + Y) + \lambda(X - Y)\right).$$

It is obvious that $g(\lambda)$ is an entire function and

$$g\left(\frac{1}{2}\right) = \det(I - Y), \quad g\left(-\frac{1}{2}\right) = \det(I - X).$$

Therefore

$$|\det(I - X) - \det(I - Y)| = \left|g\left(\frac{1}{2}\right) - g\left(-\frac{1}{2}\right)\right| \le \sup_{-\frac{1}{2} \le t \le \frac{1}{2}} |g'(t)|.$$

From complex analysis we use the following Cauchy formula

$$g'(t) = \frac{1}{2\pi} \int_{|\zeta|=\rho} \frac{g(\zeta + t)}{\zeta^2} \, d\zeta,$$

where ρ is a positive number that will be chosen later and $-\frac{1}{2} < t < \frac{1}{2}$.
    Hence

$$|g'(t)| \le \frac{1}{\rho} \sup_{|\zeta|=\rho} |g(t + \zeta)|$$

and

$$|\det(I - X) - \det(I - Y)| \le \frac{1}{\rho} \sup_{|\lambda| \le \rho + \frac{1}{2}} |g(\lambda)|.$$

Put $\rho = \|X - Y\|_{\mathcal{P}}^{-1}$ and $|\lambda| < \rho + \frac{1}{2}$. Taking into account that

$$\left\|\frac{1}{2}(X + Y) + \lambda(X - Y)\right\|_{\mathcal{P}} \le \frac{1}{2}(\|(X + Y)\|_{\mathcal{P}} + \|(X - Y)\|_{\mathcal{P}}) + 1,$$

we obtain

$$\left\|\frac{1}{2}(X + Y) + \lambda(X - Y)\right\|_{\mathcal{P}} \le \|X\|_{\mathcal{P}} + \|Y\|_{\mathcal{P}} + 1.$$

Using now the formula (1.3) is Section 1, we have

$$|g(\lambda)| = \left|\det\left(I - \frac{1}{2}(X + Y) + \lambda(X - Y)\right)\right| \le \exp(\|X\|_{\mathcal{P}} + \|Y\|_{\mathcal{P}} + 1).$$

Now for any finite sections X_n and Y_n of X and Y, respectively. we have

$$|\det(I - X_n) - \det(I - Y_n)| \le \exp(\|X_n\|_{\mathcal{P}} + \|Y_n\|_{\mathcal{P}} + 1)\|X_n - Y_n\|_{\mathcal{P}}.$$

Passing to the limit we obtain (3.1). $\square$

Corollary 3.2 *The function* $\det(I - X)$ *is a continuous function in* $\mathcal{P}$ *in the following sense: For any* $X \in \mathcal{P}$ *and* $\varepsilon > 0$, *there exists a* $\delta > 0$ *such that for any* $Y \in \mathcal{P}$ *which satisfies the inequality*

$$\|X - Y\|_{\mathcal{P}} < \delta,$$

the following inequality holds

$$|\det(I - X) - \det(I - Y)| < \varepsilon.$$

Let us remark that the functions $\operatorname{tr} X$ and $\det(I - X)$ are not continuous on $\mathcal{P}$ in the operator norm. Indeed, define operators

$$F_n = (\alpha_j^{(n)} \delta_{jk})_{j,k=1}^{\infty}, \quad (n = 1, 2, \ldots)$$

where

$$\alpha_j^{(n)} = \frac{-1}{\sqrt{n}}, \ j = 1, 2, \ldots, n$$

and

$$\alpha_j^{(n)} = 0, \ j = n + 1, n + 2, \ldots$$

It is obvious that $F_n \in \mathcal{P}$ and

$$\lim_{n \to \infty} \|F_n\| = 0.$$

Consider $\det(I - F_n)$ and $\operatorname{tr} F_n$. It is clear that

$$\det(I - F_n) = \left(1 + \frac{1}{\sqrt{n}}\right)^n \to \infty$$

and

$$\operatorname{tr} F_n = -\sqrt{n} \to -\infty$$

for $n \to \infty$.

In subsequent sections we will need the following properties of the determinant.

Theorem 3.3 *The set* $\mathcal{P}$ *has the following properties: If* $G^{(1)}$ *and* $G^{(2)} \in \mathcal{P}$ *then* $G = G^{(1)} G^{(2)} \in \mathcal{P}$ *and*

$$\|G\|_{\mathcal{P}} \le \|G^{(1)}\|_{\mathcal{P}} \|G^{(2)}\|_{\mathcal{P}}.$$

Also the following equality holds:

$$\det[(I - G^{(1)})(I - G^{(2)})] = \det(I - G^{(1)}) \det(I - G^{(2)}).$$

Proof: The first property follows from the relations

$$g_{jk} = \sum_{r=1}^{\infty} g_{jr}^{(1)} g_{rk}^{(2)}$$

and

$$|g_{jk}| \le \sum_{r=1}^{\infty} |g_{jr}^{(1)}| \sum_{r=1}^{\infty} |g_{rk}^{(2)}|,$$

Hence

$$\|G\|_{\mathcal{P}} = \sum_{j,k=1}^{\infty} |g_{jk}| \le \|G^{(1)}\|_{\mathcal{P}} \|G^{(2)}\|_{\mathcal{P}}.$$

To deduce the second relation we first remark that

$$G^{(0)} = (I - G^{(1)})(I - G^{(2)}) - I = -G^{(1)} - G^{(2)} + G^{(1)}G^{(2)} \in \mathcal{P}.$$

Consider the equality

$$\det(I - G_n^{(1)}) \det(I - G_n^{(2)}) = \det(I - G_n^{(1)} - G_n^{(2)} + G_1^{(1)}G_n^{(2)}).$$

Here $G_n^{(i)}$ is the n-th section of $G^{(i)}, i = 1, 2$. Now let us take the limit when $n \to \infty$ from both sides of this equality. The limit of the left hand side is equal to

$$\det(I - G^{(1)}) \det(I - G^{(2)})$$

and the limit of the operator $G^{(3)}(n) = G_n^{(1)} + G_n^{(2)} - G_n^{(1)}G_n^{(2)}$ converges to $G^{(0)}$ in the following sense

$$\lim_{n \to \infty} \|G^{(3)}(n) - G^{(0)}\|_{\mathcal{P}} = 0.$$

From the continuity of the function $\det_{\mathcal{P}}(I - X)$ it follows that

$$\lim_{n \to \infty} \det(I - G^{(3)}(n)) = \det(I - G^{(0)})$$

This ends the proof. $\square$

14.4 Determinants and Inversion of Operators

In this section we discuss the role of the determinant in the inversion of operators of the form $I - G, G \in \mathcal{P}$.

Theorem 4.1 *Let $A \in \mathcal{P}$. Then the operator $I - A$ is invertible if and only if* $\det(I - A) \neq 0$. *If the last condition holds, then* $(I - A)^{-1} - I \in \mathcal{P}$.

Proof: Due to Theorem 3.1 the function $\det(I - X)$, $X \in \mathcal{P}$ is uniformly continuous in some neighborhood of $X = 0$. This implies that there exists a $\delta > 0$ such that for all $X \in \mathcal{P}$ with $\|X\|_{\mathcal{P}} < \delta$, the inequality $\det(I - X) \neq 0$ holds. The operator A can be represented in the form $A = A^{(n)} + R_n$, where $\|R_n\|_{\mathcal{P}} < \delta$ and $A^{(n)}$ is the n-th section of A which has finite rank. We shall assume that $\delta < 1$. Then $\|R_n\| \leq \|R_n\|_{\mathcal{P}} < 1$ and the operator $I - R_n$ is invertible. Taking into consideration that

$$(I - R_n)^{-1} = I + \sum_{j=1}^{\infty} R_n^j$$

The series

$$\hat{R} = \sum_{j=1}^{\infty} R_n^j$$

is convergent in the sense that

$$\left\| \hat{R} - \sum_{j=1}^{m} R_n^j \right\|_{\mathcal{P}} \to 0, \quad m \to \infty.$$

This follows from the fact that

$$\lim_{k,m \to \infty} \sum_{j=k+1}^{k+m} \|R_n\|_{\mathcal{P}}^j = 0.$$

Since $\mathcal{P}$ is complete, the operator $\hat{R} \in \mathcal{P}$. Represent $I - A$ in the form

$$I - A = I - A^{(n)} - R_n = (I - R_n)(I - (I - R_n)^{-1} A^{(n)}).$$

The operator

$$(I - R_n)^{-1} A^{(n)} = (I + \hat{R}) A^{(n)} = A^{(n)} + \hat{R} A^{(n)} \in \mathcal{P}$$

and from Theorem 3.3 follows

$$\det(I - A) = \det(I - R_n) \det(I - (I + \hat{R}) A^{(n)}). \tag{4.1}$$

Since $A^{(n)}$ is of finite rank, the operator $(I + \hat{R}) A^{(n)}$ is of finite rank and the operator $I - (I + \hat{R}) A^{(n)}$ is invertible if and only if $\det(I - (I + \hat{R}) A^{(n)}) \neq 0$. Now it is easy to conclude that the operator $I - A$ is invertible if and only if

$\det(I - A) \neq 0$. Note also that if the last condition, holds then in the decomposition $\ell_2 = \operatorname{Im} P_n \oplus \operatorname{Im}(I - P_n)$, where P_n is the orthogonal projection onto the subspace spanned by the first basis vectors $e_1, \ldots, e_n$, the following equality holds

$$I - (I + \hat{R})A^{(n)} = \left(I - \left(I + \begin{pmatrix} \hat{R}_{11} & \hat{R}_{12} \\ \hat{R}_{21} & \hat{R}_{22} \end{pmatrix} \right) \right) \begin{pmatrix} A_n & 0 \\ 0 & 0 \end{pmatrix}$$

$$= I - \begin{pmatrix} (I + \hat{R}_{11})A_n & 0 \\ \hat{R}_{21}A_n & 0 \end{pmatrix}.$$

Hence

$$I - (I + \hat{R})A^{(n)} = \begin{pmatrix} Z_{11} & 0 \\ -Z_{21} & I \end{pmatrix},$$

where

$$Z_{11} = I - (I + \hat{R}_{11})A_n, \qquad Z_{21} = \hat{R}_{21}A_n.$$

From here it follows that

$$(I - (I + \hat{R})A^{(n)})^{-1} = \begin{pmatrix} Z_{11}^{-1} & 0 \\ Z_{21}Z_{11}^{-1} & I \end{pmatrix}.$$

This implies

$$(I - A)^{-1} - I = \left(\begin{pmatrix} Z_{11}^{-1} - I & 0 \\ Z_{21}Z_{11}^{-1} & 0 \end{pmatrix} + I \right) (I - R_n) - I$$

$$= -R_n + \begin{pmatrix} Z_{11}^{-1} - I & 0 \\ Z_{21}Z_{11}^{-1} & 0 \end{pmatrix} - \begin{pmatrix} Z_{11}^{-1} - I & 0 \\ Z_{21}Z_{11}^{-1} & 0 \end{pmatrix} R_n.$$

Taking into consideration that

$$\begin{pmatrix} Z_{11}^{-1} - I & 0 \\ Z_{21}Z_{11}^{-1} & 0 \end{pmatrix} \in \mathcal{P} \quad \text{and} \quad R_n \in \mathcal{P}$$

we obtain $(I - A)^{-1} - I \in \mathcal{P}$. The theorem is proved. $\square$

Corollary 4.2 *Let $A \in \mathcal{P}$ and $\Delta(v) = \det(I - vA)$. The function $\Delta(v)$ is entire in $\mathbb{C}$ and $\Delta(v_0) = 0$ if and only if $v_0 \neq 0$ and $\lambda_0 = 1/v_0$ is an eigenvalue of A. Moreover if κ is the multiplicity of the zero v_0 of the analytic function $\Delta(v)$, then κ is the algebraic multiplicity of the eigenvalue $\lambda_0 = 1/v_0$ of the operator A.*

Proof: All claims except the last follow from the previous theorem. For the last statement we need knowledge about the multiplicity of an eigenvalue. This material is outside the scope of this book. The proof of this statement can be found in [GGKr], Theorem II.6.2. $\qquad\square$

Now we shall use the determinant $\det(I - G)$, $G \in \mathcal{P}$ to obtain formulas for the inverse operator $(I - G)^{-1}$. Let us assume that $G \in \mathcal{P}$ and $\det(I - G) \neq 0$. Then by Theorem 4.1 the operator $I - G$ is invertible and $(I - G)^{-1} - I \in \mathcal{P}$. Represent $G^{(n)}$ on the entire space $\ell_2 = \operatorname{Im} P_n \oplus \operatorname{Im} (I - P_n)$ by the matrix

$$\begin{pmatrix} G_n & 0 \\ 0 & 0 \end{pmatrix},$$

where $P_n x = \{x_1, x_2, \ldots, x_n, 0, 0, \ldots\}$ and G_n is the n-th section of G. It is clear that $I - G^{(n)}$ is invertible for n sufficiently large. We shall now prove that

$$\lim_{n \to \infty} \|(I - G)^{-1} - (I - G^{(n)})^{-1}\|_{\mathcal{P}} = 0.$$

We start with the equalities

$$\begin{aligned} I - G^{(n)} &= (I - G)(I - (I - G)^{-1}(G^{(n)} - G)) \\ &= (I - G)(I - [(I - G)^{-1} - I] + I)(G^{(n)} - G) \end{aligned}$$

Introduce the operator

$$X_n = (I - G)^{-1}(G^{(n)} - G) = ([(I - G)^{-1} - I] + I)(G^{(n)} - G)$$

which is in $\mathcal{P}$ and

$$\|X_n\|_{\mathcal{P}} \leq (\|(I - G)^{-1} - I\|_{\mathcal{P}} + 1)\|G^{(n)} - G\|_{\mathcal{P}}.$$

Hence for n large enough, $\|X_n\|_{\mathcal{P}} < 1$ and

$$(I - G^{(n)})^{-1} = (I - X_n)^{-1}(I - G)^{-1} = \sum_{j=0}^{\infty} X_n^j (I - G)^{-1}.$$

From here it follows that

$$\|(I - G)^{-1} - (I - G^{(n)})^{-1}\|_{\mathcal{P}} \leq (\|(I - G)^{-1} - I\|_{\mathcal{P}} + 1)\frac{\|X_n\|_{\mathcal{P}}}{1 - \|X_n\|_{\mathcal{P}}} \underset{n \to \infty}{\longrightarrow} 0.$$

As it is well known that for $A_n = I - G^{(n)}$,

$$A_n^{-1} = \frac{1}{\det A_n} R_n,$$

where

$$R_n = (r_{jk}^{(n)})_{j,k=1}^{\infty}$$

is defined as follows. The matrix $(r_{jk}^{(n)})_{j,k=1}^n$ is the transposed matrix of algebraic complements of the entries of $I_n - G_n$;

$$r_{kk}^{(n)} = 1 \quad \text{for} \quad k = n+1, \ldots$$

and $r_{jk}^{(n)} = 0$ for the remaining indices. It is obvious that

$$(\det A_n) A_n^{-1} = R_n.$$

Taking into consideration that

$$\lim_{n \to \infty} \det A_n = \det A \quad \text{and} \quad \lim_{n \to \infty} \|A_n^{-1} - A^{-1}\| = 0,$$

where $A = I - G$, we obtain that R_n converges to an operator $R \in \mathcal{L}(\ell_2)$ which we shall call *the operator of algebraic complements* of A:

$$\lim_{n \to \infty} \|R - R_n\| = 0.$$

In particular

$$\lim_{n \to \infty} r_{jk}^{(n)} = r_{jk},$$

where $R = (r_{jk})_{j,k=1}^\infty$.

It is obvious that

$$A = \frac{1}{\det A} R.$$

This equality generalizes, for operators of the form $A = I - G$, $G \in \mathcal{P}$, Cramer's rule of inversion of finite matrices. Now we shall give another representation of the inverse $(I - \mu G)^{-1}$ in terms of a series.

Theorem 4.3 *Let $G \in \mathcal{P}$. Then the following equality holds:*

$$(I - \mu G)^{-1} = I + \frac{\mu}{\det(I - \mu G)} D(G, \mu),$$

where $D(G; \mu)$ is an entire operator valued function

$$D(G; \mu) = \sum_{m=0}^{\infty} d_m(G)\mu^m,$$

of which the coefficients $d_m = d_m(G)$ are operators from $\mathcal{P}$ which are uniquely determined by the recurence relations

$$d_0 = G, \quad d_m = G(d_{m-1} + \Delta_m) \quad m = 1, 2, \ldots.$$

with

$$\det(I - \mu G) = \sum_{j=0}^{\infty} \Delta_j \mu^j.$$

Proof: There exists a $\delta > 0$ such that for $|\mu| < \delta$ the operator $I - \mu G$ is invertible:

$$(I - \mu G)^{-1} = I + \sum_{j=1}^{\infty} \mu^j G^j,$$

where the series converges in the norm of $\mathcal{P}$. Consider the operator $I - \mu G^{(n)}$, where the sequence $\{G^{(n)}\}$ converges to G:

$$\lim_{n \to \infty} \|G^{(n)} - G\|_{\mathcal{P}} = 0.$$

There exists a number $\delta_1 > 0$ $(\delta_1 < \delta)$ such that for $|\mu| \leq \delta_1$,

$$(I - \mu G_0^{(n)})^{-1} = I + \sum_{j=1}^{\infty} \mu^j (G_0^{(n)})^j.$$

Using the computations preceding this theorem we come to the conclusion that

$$\lim_{n \to \infty} \|(I - \mu G^{(n)})^{-1} - (I - \mu G)^{-1}\|_{\mathcal{P}} = 0,$$

uniformly in the disc $|\mu| \leq \delta_1$. We start now with the formula

$$(I - \mu G^{(n)})^{-1} = \frac{1}{\det(I - \mu G^{(n)})} R_n(\mu)$$

where $R_n(\mu)$ is the matrix of algebraic complements for $I - \mu G_0^{(n)}$ (see the definition preceding Theorem 4.3). It is clear that $R_n(\mu)$ is a polynomial in μ with operator valued coefficients. After using the formula

$$(I - \mu G^{(n)})^{-1} = I + \mu G^{(n)}(I - \mu G^{(n)})^{-1}$$

we have

$$(I - \mu G^{(n)})^{-1} = I + \frac{1}{\det(I - \mu G^{(n)})} \mu G^{(n)} R_n(\mu)$$

Denote $G^{(n)} R_n(\mu)$ by $D^{(n)}(\mu)$. Hence

$$D^{(n)}(\mu) = (I - \mu G^{(n)})^{-1} G^{(n)} \det(I - \mu G^{(n)}).$$

Therefore

$$\lim_{n \to \infty} \|D^{(n)}(\mu) - (I - \mu G)^{-1} G \det(I - \mu G)\|_{\mathcal{P}} = 0$$

uniformly on the disk $|\mu| \leq \delta_1$.

Since $D^{(n)}(\mu)$ is a polynomial in μ we may conclude that $D^{(n)}(\mu)$ converges uniformly on $|\mu| \leq r$, where r can be taken arbitrarily large. Thus on $\mathbb{C}$ there exists an operator valued entire function $D(\mu)$ such that

$$\lim_{n \to \infty} \|D(\mu) - D^{(n)}(\mu)\|_{\mathcal{P}} = 0,$$

uniformly in any disk $|\mu| < r$. It is obvious that

$$(I - \mu G)^{-1} = I + \frac{\mu}{\det(I - \mu G)} D(\mu).$$

Since

$$(I - \mu G)D(\mu) = G \det(I - \mu G)$$

and

$$D(\mu) = \sum_{j=0}^{\infty} \mu^j d_j; \ \det(I - \mu G) = \sum_{n=0}^{\infty} \mu^n \Delta_n,$$

we obtain

$$\sum_{n=0}^{\infty} \mu^n d_n - \sum_{n=0}^{\infty} \mu^{n+1} G d_n = G \sum_{n=0}^{\infty} \mu^n \Delta_n.$$

Thus, comparing coefficients, we get

$$d_0 = G; \quad \text{and} \quad d_n = G d_{n-1} + G \Delta_n$$

or

$$d_0 = G \quad \text{and} \quad d_n = G(\Delta_n + d_{n-1}I).$$

Now we come to the conclusion that $D(G; \mu) = D(\mu)$. This ends the proof. $\square$

We note that the last recurrence relation can be further simplified. More than that, it can be solved in closed form as a difference equation. For these results we refer to [GGK1], Ch VII Section 7 and [GGKr] ChVIII Section 1.

14.5 Trace and Determinant Formulas for Poincaré Operators

The main theorem in this section gives formulas for the trace and determinant of a Poincaré operator in terms of its eigenvalues. Here we rely, in part, on complex function theory. In particular, we shall use the following theorem, of which the proof can be found in [Ah] Chapter 5 sec. 2.3 and [GGK1] Theorem VII.5.1. It is a special case of Hadamard's factorization theorem.

Theorem 5.1 *Let $a_1, a_2, \ldots$ be the zeros of an entire function f ordered according to increasing absloute values with multiplicities taken into account. Assume that*

$$\sum_{j} \frac{1}{|a_j|} < \infty.$$

Suppose $f(0) = 1$ and for each $\varepsilon > 0$ there exists a constant $C(\varepsilon)$ such that

$$|f(\lambda)| \leq C(\varepsilon)e^{|\lambda|\varepsilon}, \quad \lambda \in \mathbb{C}.$$

Then f admits the representation

$$f(\lambda) = \prod_j \left(1 - \frac{\lambda}{a_j}\right)$$

which converges uniformly on compact subsets of $\mathbb{C}$.

We know from Corollary 4.2 that for $A \in \mathcal{P}$, the function $\det(I - \lambda A)$ is entire and $\det(I - \lambda_0 A) = 0$ if and only if $\lambda_0 \neq 0$ and $1/\lambda_0$ is an eigenvalue of A. The order of the zero λ_0 of $\det(I - \lambda A)$ is called the *algebraic multiplicity* of $1/\lambda_0$ as an eigenvalue of A.

The following result, which we do not prove, appears in [GKre],Theorem I.4.2.

Theorem 5.2 *Suppose that $\{A_n\}$ is a sequence of compact operators in $\mathcal{L}(\mathcal{H})$ ($\mathcal{H}$ a Hilbert space) which converges in the uniform norm to an operator $A \in \mathcal{L}(\mathcal{H})$. Then for an appropriate enumeration of the eigenvalues $\lambda_j(A_n)$ of the operators A_n, we have*

$$\lim_{n \to \infty} \lambda_j(A_n) = \lambda_j(A), \quad j = 1, 2, \ldots.$$

For the proof of the main theorem about trace and determinant we will need the following properties of the eigenvalues of Poincaré operators.

Lemma 5.3 * *Let $A = (a_{jk})_{j,k=1}^n$ be an $n \times n$ matrix and let $\lambda_1, \lambda_2, \ldots, \lambda_n$ be the eigenvalues of A, counting multiplicities. Then*

$$\sum_{j=1}^n |\lambda_j| \leq \sum_{j,k=1}^n |a_{jk}|. \tag{5.1}$$

Proof: Let $U = (u_{jk})_{j,k=1}^n$ be a unitary matrix such that UAU^* is upper triangular with $\lambda_1, \lambda_2, \ldots, \lambda_n$ on the diagonal. Take B to be the diagonal matrix $\operatorname{diag}(t_1, t_2, \ldots, t_n)$ where $\lambda_j t_j = |\lambda_j|$, $|t_j| = 1$, $1 \leq j \leq n$. Then for $S = BUAU^* = (s_{jk})_{j,k=1}^\infty$ it is clear that

$$\sum_{j=1}^n |\lambda_j| = \sum_{j=1}^n s_{jj}. \tag{5.2}$$

Now for $U^* = (v_{jk})_{j,k=1}^n$,

$$\sum_{j=1}^n s_{jj} = \sum_{j=1}^n \sum_{q=1}^n \left(\sum_{p=1}^n t_j u_{jp} a_{pq}\right) v_{qj}.$$

---

*) The authors are indebted to N. Krupnik for this lemma.

Summing first with respect to j we get

$$\sum_{j=1}^{n} s_{jj} \le \sum_{p,q=1}^{n} |a_{pq}| \left(\sum_{j=1}^{n} |u_{jp}| |v_{qj}| \right)$$

$$\le \sum_{p,q=1}^{n} |a_{pq}| \left(\sum_{j=1}^{n} |u_{jp}|^2 \right)^{1/2} \left(\sum_{j=1}^{n} |v_{qj}|^2 \right)^{1/2}. \qquad (5.3)$$

Since U is unitary, $\sum_{j=1}^{n} |u_{jp}|^2 = \sum_{j=1}^{n} |v_{qj}|^2 = 1$.
Hence inequality (5.1) follows from (5.2) and (5.3). $\qquad\qquad\qquad\qquad \square$

Theorem 5.4 *Let $A \in \mathcal{P}$ and let $\lambda_1(A), \lambda_2(A), \ldots$ be the non-zero eigenvalues of A counted according to their algebraic multiplicities. Then*

$$\sum_{j} |\lambda_j(A)| \le \|A\|_{\mathcal{P}}.$$

Proof: Let A_n be the nth-finite section of $A = (a_{jk})_{j,k=1}^{\infty}$ and let $\lambda_1(A_n)$, $\lambda_2(A_n), \ldots, \lambda_n(A_n)$ be the non-zero eigenvalues of A_n counted according to their multiplicities. Then

$$\lim_{n \to \infty} |\lambda_j(A_n)| = |\lambda_j(A)|, \quad j = 1, 2, \ldots$$

by Theorem 5.2. From Lemma 5.3 we have for $n > k$,

$$\sum_{j=1}^{k} |\lambda_j(A_n)| \le \sum_{j,k=1}^{n} |a_{jk}| \le \|A\|_{\mathcal{P}}.$$

Hence

$$\sum_{j=1}^{k} |\lambda_j(A)| \le \|A\|_{\mathcal{P}}.$$

Since k was arbitrary,

$$\sum_{j} |\lambda_j(A)| \le \|A\|_{\mathcal{P}}.$$

$$\square$$

For the application of Theorem 5.1 to the function $\det(I - \lambda A)$ we need the following theorem.

Theorem 5.5 *Let $A \in \mathcal{P}$. Given $\varepsilon > 0$, there exists a constant $C(\varepsilon)$ such that*

$$|\det(I - \lambda A)| \le C(\varepsilon) e^{\varepsilon |\lambda|}, \quad \lambda \in \mathbb{C},$$

Proof: Since $A \in \mathcal{P}$, we have from Section 1 that

$$|\det(I - \lambda A)| \leq \prod_{j=1}^{\infty} \left(1 + |\lambda| \sum_{k=1}^{\infty} |a_{jk}| \right), \tag{5.4}$$

where $\sum_{j=1}^{\infty} \sum_{k=1}^{\infty} |a_{jk}| < \infty$. Given $\varepsilon > 0$, there exists a positive integer N such that $\sum_{j=N+1}^{\infty} \sum_{k=1}^{\infty} |a_{jk}| < \varepsilon/2$. Let $C_j = \sum_{k=1}^{\infty} |u_{jk}|$, $j = 1, 2, \ldots$. Since $1 + x \leq e^x$, $x \geq 0$, it follows from (5.4) that

$$|\det(I - \lambda A)| \leq \prod_{j=1}^{N} (1 + |\lambda| C_j) \prod_{j=N+1}^{\infty} (1 + |\lambda| C_j) \tag{5.5}$$

$$\leq \prod_{j=1}^{N} (1 + |\lambda| C_j) \exp \left(|\lambda| \sum_{j=N+1}^{\infty} C_j \right) \leq \prod_{j=1}^{N} (1 + |\lambda| C_j) e^{\frac{\varepsilon}{2} |\lambda|}. \tag{5.6}$$

Now the function $(1 + C_j x) e^{-\varepsilon/2 x}$ is bounded on $[0, \infty)$. Hence it follows that there exists a constant $C(\varepsilon)$ such that $\prod_{j=1}^{N} (1 + |\lambda| C_j) e^{-|\lambda| \varepsilon/2} \leq C(\varepsilon)$. Thus from (5.5) we get

$$|\det(I - \lambda A)| \leq C(\varepsilon) e^{|\lambda| \varepsilon/2} e^{|\lambda| \varepsilon/2} = C(\varepsilon) e^{\varepsilon |\lambda|}.$$

$\square$

Now we are ready to prove the main theorem of this section.

Theorem 5.6 *Let $A = (a_{jk})_{j,k=1}^{\infty} \in \mathcal{P}$ and let $\lambda_1(A), \lambda_2(A), \ldots$ be the sequence of the non-zero eigenvalues of A counted according to their algebraic multiplicities. Then*

$$\det(I - \lambda A) = \prod_{j} (1 - \lambda \lambda_j(A)). \tag{i}$$

The convergence of the product is uniform on compact subset of $\mathbb{C}$. Furthermore,

$$\operatorname{tr} A := \sum_{k=1}^{\infty} a_{kk} = \sum_{j} \lambda_j(A). \tag{ii}$$

Proof: (i) From the remark preceding Theorem 5.2, we know that the zeros of the entire function $\det(I - \lambda A)$ are $\frac{1}{\lambda_j(A)}$ of order, by definition, the algebraic multiplicity of $\lambda_j(A)$ for A. Statement (i) now follows from Theorem 5.1, applied to $f(\lambda) = \det(I - \lambda A)$. The conditions of this theorem are met by Theorems 5.4 and 5.5.

(ii) Since the function $\det(I - \lambda A)$ is entire, $\det(I - \lambda A) = \sum_{n=0}^{\infty} c_n \lambda^n$, $\lambda \in \mathbb{C}$. Let A_n be the nth-finite section of A. Now

$$\det(I - \lambda A_n) = \prod_{j=1}^{n} (1 - \lambda \lambda_j(A_n)) = \sum_{k=0}^{n} c_{kn} \lambda^k,$$

where

$$c_{1n} = -\sum_{k=1}^{n} \lambda_k(A_n) = -\sum_{k=1}^{n} a_{kk}.$$

Since $\det(I - \lambda A_n)$ converges uniformly on compact subsets of $\mathbb{C}$ to $\det(I - \lambda A)$, it follows that the sequence $\{c_{1n}\}$ converges to c_1, i.e.,

$$c_1 = -\sum_{k=1}^{\infty} a_{kk}. \tag{5.7}$$

On the other hand, we have from (i) that the sequence of polynomials $\prod_{j=1}^{n}(1 - \lambda\lambda_j(A))$ denoted by $\sum_{j=0}^{n} d_{jn}\lambda^j$ converges uniformly on compact sets to $\det(I - \lambda A)$. Hence the sequence $\{d_{1n}\}$ converges to c_1. Since

$$d_{1n} = -\sum_{j=1}^{n} \lambda_j(A),$$

we obtain

$$-\sum_{j} \lambda_j(A) = c_1.$$

By (5.6)

$$c_1 = -\sum_{k=1}^{\infty} a_{kk}.$$

$\square$

Much of the material in this chapter has its origin in the book [GGKr], Chapters I–III, as well as in the books [GGK1] and [GKre]. Further developments of the theory of traces, determinants and inversion can be found in these books also along with a list of other sources on this subject.

Exercises XIV

1. Let X and Y be any operators in $\mathcal{L}(\mathbb{C}^n)$, and suppose $C_1(n)$ and $C_2(n)$ are positive constants such that

   (a) $|\text{tr } X - \text{tr } Y| \le C_1(n)\|X - Y\|$

   (b) $|\det(I - X) - \det(I - Y)| \le C_2(n)\|X - Y\|$

   Prove that
   $$\lim_{n\to\infty} C_1(n) = \infty \quad \text{and} \quad \lim_{n\to\infty} C_2(n) = \infty,$$

   for any operator norm.

2. Prove that there exists a number $r > 0$ such that for $F, G \in \mathcal{F}(\ell_2)$ with

$$\|F\|_{\mathcal{P}} < r, \quad \|G\|_{\mathcal{P}} < r.$$

The following inequality holds:

$$\|F^n - G^n\|_{\mathcal{P}} \leq n\, r^{n-1} \|F - G\|_{\mathcal{P}}.$$

3. Prove that in a small enough neighborhood $|\lambda| < \delta$ the equality

$$\det(I - \lambda F) = \exp \sum_{n=1}^{\infty} \frac{1}{n} \operatorname{tr}(F^n)\lambda^n.$$

holds, where $F \in \mathcal{P}$.

4. Prove that for any $F \in \mathcal{F}(\ell_2)$ and $\lambda \in \mathbb{C}$,

$$\det(I - \lambda F) = 1 + \sum_{n=1}^{\operatorname{rank} F} (-1)^n \frac{C_n(F)}{n!}\lambda^n,$$

where

$$C_n(F) = \det \begin{pmatrix} \operatorname{tr} F & n-1 & 0 & \cdots & 0 \\ \operatorname{tr} F^2 & \operatorname{tr} F & n-2 & \cdots & 0 \\ \cdot & \cdot & \cdot & \cdot & \cdot \\ \operatorname{tr} F^n & \operatorname{tr} F^{n-1} & \operatorname{tr} F^{n-2} & \cdots & \operatorname{tr} F \end{pmatrix}$$

5. Prove that $\operatorname{tr}(X)$, $X \in \mathcal{P}$, is the only linear functional, continuous in $\|\cdot\|_{\mathcal{P}}$, with the properties

$$\operatorname{tr}(XY) = \operatorname{tr}(XY) \quad X, Y \in \mathcal{P}$$

and

$$\operatorname{tr} \begin{pmatrix} 1 & 0 & \cdot & \cdot & \cdot \\ 0 & 0 & \cdot & \cdot & \cdot \\ \cdot & \cdot & \cdot & \cdot & \cdot \end{pmatrix} = 1.$$

6. Prove that $\det(I - X)$ is the only multiplicative functional, continuous in $\|\cdot\|_{\mathcal{P}}$, with the property that

$$\det[(I - X)(I - Y)] = \det(I - X)\det(I - Y)$$

for $X, Y \in \mathcal{P}$.

7. Construct a finite rank operator $F \in \mathcal{F}(H)$ such that $F \notin \mathcal{P}$.

8. Let $\{\varphi_j\}_1^\infty$ be an orthonormal basis in Hilbert space H. Denote by $\mathcal{P}_\varphi$ the set of operators in $\mathcal{L}(H)$ such that their matrices in the basis $\{\varphi_j\}_1^\infty$ belong to $\mathcal{P}$. Let $\{\psi_j\}_1^\infty$ be another orthonormal basis in H. Find an operator $X \in \mathcal{P}_\varphi$ such that $X \notin \mathcal{P}_\psi$.

9. Let $K \in \mathcal{L}(L_2(a, b))$ be defined by the equality

$$(K\varphi)(t) = \int_a^b k(t, s)\varphi(s)\, ds,$$

with the kernel

$$k(t, s) = \sum_{j=1}^n f_j(t)g_k(s),$$

where $f_j, g_j \in L_2(a, b)$, $j = 1, 2, \ldots, n$.

Prove that

$$\sum_j \lambda_j(K) = \int_a^b k(s, s)\, ds$$

10. Denote by HS_φ the set of all operators A in $\mathcal{L}(H)$ such that their matrices $(a_{jk})_{j,k=1}^\infty$ in the orthonormal basis $\{\varphi_j\}_1^\infty$ satisfy the condition

$$\sum_{j,k=1}^\infty |a_{jk}|^2 < \infty.$$

Let $\{\psi_j\}_1^\infty$ be another orthonormal basis in H. Find an operator $X \in HS_\varphi$ such that $X \notin HS_\psi$.

11. (a) Compute the determinants of the following matrix

$$\begin{pmatrix} x_1 & a_2b_1 & a_3b_1 & \cdots \\ a_1b_2 & x_2 & a_3b_2 & \cdots \\ \cdot & \cdot & \cdot & \cdot \end{pmatrix}$$

under the conditions that

$$\sum_{j=1}^\infty |a_j|^2 < \infty, \quad \sum_{j=1}^\infty |b_j|^2 < \infty$$

and

$$\sum_{j=0}^\infty |x_j - 1| < \infty.$$

(b) The same problem under the same conditions for the special case when

$$x_j = 1 - a_jb_j.$$

In this case the last condition follows from the previous ones.

12. Let $\{e_j\}_{j=-\infty}^{\infty}$ be the standard orthonormal basis of $\ell_2(\mathbb{Z})$ and denote by $\mathcal{P}(\mathbb{Z})$ the set of all operators $A \in \mathcal{L}(\ell_2(\mathbb{Z}))$ such that their matrices $(a_{jk})_{j,k=-\infty}^{\infty}$ in the basis $\{e_j\}_{j=-\infty}^{\infty}$ satisfy the condition

$$\|A\|_{\mathcal{P}(\mathbb{Z})} = \sum_{j,k=-\infty}^{\infty} |a_{jk}| < \infty.$$

Denote by $A^{(-n,m)}$ the finite section

$$(a_{jk})_{j,k=-n}^{m}.$$

(a) Let $A \in \mathcal{P}(\mathbb{Z})$, then

$$\lim_{n,m \to \infty} \det(I - A^{(-n,m)})$$

exists. Denote this limit by

$$\det(I - A).$$

(b) Prove that $\det(I - X)$, $X \in \mathcal{P}(\mathbb{Z})$ is a Lipschitz function in the norm $\|\ \|_{\mathcal{P}(\mathbb{Z})}$.

(c) Prove that for $A \in \mathcal{P}(\mathbb{Z})$ the operator $I - A$ is invertible if and only if $\det(I - A) \neq 0$. If this condition holds, then $(I - A)^{-1} - I \in \mathcal{P}(\mathbb{Z})$.

13. Let $F \in \mathcal{F}(\ell_2(\mathbb{Z}))$. We assume that $(f_{jk})_{j,k=-\infty}^{\infty}$ is the matrix of F in the standard basis $\{e_j\}_{j=-\infty}^{\infty}$ of $\ell_2(\mathbb{Z})$. Denote by $F^{(-n,m)}$ the operator defined by the section

$$(f_{jk})_{j,k=-n}^{m}.$$

Define tr F and $\det(I - F)$ in the same way as these quantities are defined for $F \in \mathcal{F}(\ell_2)$; see Section 2.

(a) Prove that

$$\operatorname{tr} F = \lim_{n,m \to \infty} \operatorname{tr} F^{(-n,m)} = \sum_{k=-\infty}^{\infty} f_{kk}$$

and the series converges absolutely.

(b) Prove that

$$\det(I - F) = \lim_{n,m \to \infty} \det(I - F^{(-n,m)}).$$

14. Compute the determinant $\det(I - V)$, where the operator V is defined in the standard basis of $\ell_2(\mathbb{Z})$ by one of the following matrices:

(a)

$$V = \begin{pmatrix} \ddots & \vdots & \vdots & \vdots & \vdots & \vdots & & \\ \ddots & b_{-4} & 0 & 0 & 0 & 0 & 0 & \cdots \\ \cdots & 0 & b_{-3} & 0 & 0 & 0 & 0 & \cdots \\ \cdots & 0 & 0 & b_{-2} & 0 & 0 & 0 & \cdots \\ \cdots & a_{-3} & a_{-2} & a_{-1} & b_{-1} & 0 & 0 & \cdots \\ \cdots & 0 & 0 & 0 & a_0 & a_1 & a_2 & \cdots \\ \cdots & 0 & 0 & 0 & b_1 & 0 & 0 & \cdots \\ \cdots & 0 & 0 & 0 & 0 & b_2 & 0 & \cdots \\ & \vdots & \vdots & \vdots & \vdots & & \ddots & \ddots \end{pmatrix}$$

$$\sum_{j=-\infty}^{\infty} |a_j| < \infty; \qquad \sum_{j=-\infty}^{\infty} |b_j| < \infty$$

(b) $V = (a_{jk})_{j,k=-\infty}^{\infty}$, where $a_{jk} = a_j b_k$ for $j \neq k$ and

$$a_{jj} = x_j \quad j = 0, \pm 1, \ldots$$

under the condition

$$\sum_{j=-\infty}^{\infty} |a_j|^2 < \infty \quad \sum_{j=-\infty}^{\infty} |b_j|^2 < \infty,$$

and

$$\sum_{j=-\infty}^{\infty} |x_j| < \infty.$$

(c) $V = (b_{jk})_{j,k=-\infty}^{\infty}$, where

$$b_{jk} = a_j b_k \quad j, k = 0, \pm 1, \ldots$$

under the same conditions as in b)

15. Construct a compact operator $A \in \mathcal{L}(\ell_2)$ defined in the standard basis of ℓ_2 by a matrix $(a_{jk})_{j,k=1}^{\infty}$ such that neither of the following limits

$$\lim_{n \to \infty} \sum_{k=1}^{n} a_{kk}, \qquad \lim_{n \to \infty} \det(\delta_{jk} - a_{jk})_{j,k-1}^{n}$$

exists.

16. Let $\mathcal{P}(\mathbb{Z})$ be the set of operators defined in problem 12. Suppose $A = (a_{jk})_{j,k=-\infty}^{\infty} \in \mathcal{P}(\mathbb{Z})$ with its sequence of eigenvalues $\lambda_1(A), \lambda_2(A), \dots$ counted according to their algebraic multiplicities. Prove that

$$\det(I - \lambda A) = \prod_j (1 - \lambda\lambda_j(A)).$$

17. Prove that in problem 16,

$$\mathrm{tr}\, A = \sum_j \lambda_j(A).$$

Chapter XV
Fredholm Operators

Fredholm operators are operators that have a finite dimensional kernel and an image of finite codimension. This class includes all operators acting between finite dimensional spaces and operators of the form two-sided invertible plus compact. Fredholm operators appear in a natural way in the theory of Toeplitz operators. The main properties of Fredholm operators, the perturbation theorems and the stability of the index, are presented in this chapter. The proofs are based on the fact that these operators can be represented as finite rank perturbations of one-sided invertible operators.

15.1 Definition and Examples

Let X and Y be Banach spaces. An operator $A \in \mathcal{L}(X, Y)$ is called a *Fredholm operator* if the numbers $n(A) = \dim \operatorname{Ker} A$ and $d(A) = \operatorname{codim} \operatorname{Im} A$ are finite. In this case the number

$$\operatorname{ind} A = n(A) - d(A)$$

is called the *index* of A.

Examples: 1. As a special case of Theorem XIII.4.1, we have that an operator $I - F$ with $F \in \mathcal{L}(X)$ of finite rank is Fredholm with index zero.

2. Given any positive integers r and k, there exists an operator A on ℓ_2 with $n(A) = k$ and $d(A) = r$. Indeed, define A on ℓ_2 by

$$A(\alpha_1, \alpha_2, \ldots) = (\underbrace{0, 0, \ldots, 0}_{r}, \alpha_{k+1}, \alpha_{k+2}, \ldots).$$

Then $\operatorname{Ker} A = \operatorname{sp}\{e_1, \ldots, e_k\}$ and $\ell_2 = \operatorname{Im} A \oplus \operatorname{sp}\{e_1, \ldots, e_r\}$, where $e_1, e_2, \ldots$ is the standard orthonormal basis in ℓ_2. Thus

$$n(A) = k, \quad d(A) = r, \quad \operatorname{ind} A = k - r.$$

15.2 First Properties

Theorem 2.1 *The range of a Fredholm operator $A \in \mathcal{L}(X, Y)$ is closed.*

Proof: Since $n(A) < \infty$ and $d(A) < \infty$, there exist, by Theorem XI.5.7, closed subspaces M and N such that

$$X = M \oplus \text{Ker } A, \quad Y = \text{Im } A \oplus N.$$

Let $M \times N$ be the Banach space with norm $\|(u, v)\| = \|u\| + \|v\|$. Define the operator $\hat{A}$ on $M \times N$ by $\hat{A}(u, v) = Au + v$. Since $\hat{A}$ is an injective bounded linear map from $M \times N$ onto Y, we have that $\hat{A}$ is invertible by Theorem XII.4.1. Thus for any $u \in M$,

$$\|\hat{A}^{-1}\|\|Au\| \geq \|\hat{A}^{-1}Au\| = \|(u, 0)\| = \|u\|. \tag{2.1}$$

Hence Im $A = \text{Im } A|M$ is closed by Theorem XII.4.6. $\qquad\qquad\square$

Theorem 2.2 *Suppose $A \in \mathcal{L}(X, Y)$ and $B \in \mathcal{L}(Y, Z)$ are Fredholm operators. Then BA is a Fredholm operator with*

$$\text{ind } (BA) = \text{ind } B + \text{ind } A.$$

Proof: Since Ker A is a finite dimensional subspace of Ker BA, there exists a closed subspace M of Ker BA such that

$$\text{Ker } BA = \text{Ker } A \oplus M. \tag{2.2}$$

It is easy to verify that the restriction of A to M is a one to one map from M onto Im $A \cap \text{Ker } B$. Hence

$$\dim M = \dim (\text{Im } A \cap \text{Ker } B) \leq n(B) < \infty.$$

Thus by (2.2),

$$n(BA) = n(A) + n_1, \quad n_1 = \dim N_1, \quad N_1 = \text{Im } A \cap \text{Ker } B. \tag{2.3}$$

Since N_1 is a subspace of the finite dimensional space Ker B, there exists a subspace N_2 of Ker B such that

$$\text{Ker } B = N_1 \oplus N_2. \tag{2.4}$$

Then

$$n(B) = n_1 + n_2, \quad n_2 = \dim N_2. \tag{2.5}$$

Since $d(A)$ and $\dim N_2$ are finite and Im $A \cap N_2 = (0)$, there exists a finite dimensional subspace N_3 such that

$$Y = \text{Im } A \oplus N_2 \oplus N_3 \tag{2.6}$$

and

$$d(A) = n_2 + n_3 < \infty, \quad n_3 = \dim N_3. \tag{2.7}$$

From
$$\text{Ker } B = N_1 \oplus N_2 \subset \text{Im } A \oplus N_2$$
and equalities (2.4) and (2.6), it follows that B is one to one on N_3 and
$$\text{Im } B = \text{Im } BA \oplus BN_3. \tag{2.8}$$
Since $Z = \text{Im } B \oplus W$ for some finite dimensional subspace W, we have from (2.8) that
$$Z = \text{Im } BA \oplus BN_3 \oplus W,$$
which implies that
$$d(BA) = \dim BN_3 + \dim W = \dim N_3 + d(B) = n_3 + d(B) < \alpha. \tag{2.9}$$
From (2.3), (2.5), (2.7) and (2.9), we have that the operator BA is a Fredholm operator with
$$\begin{aligned}
\text{ind } (BA) &= n(A) + n_1 - d(B) - n_3 = n(A) + n(B) - n_2 - d(B) - n_3 \\
&= n(A) + \text{ind } (B) - d(A) = \text{ind } (A) + \text{ind } B.
\end{aligned}$$

$\square$

Theorem 2.3 *An operator $A \in \mathcal{L}(X, Y)$ is a Fredholm operator if and only if A has the representation*
$$A = D + F, \tag{2.10}$$
where F has finite rank and D is a one-sided invertible Fredholm operator.

Proof: Let A have the representation (2.10). Assume that $D^{(-1)}$ is a right inverse of D. Then
$$A = D(I + D^{(-1)}F).$$
Notice that by Theorem XIII.4.1 the operator $I + D^{(-1)}F$ is Fredholm and has index zero. Hence A is Fredholm and ind $A = $ ind D by Theorem 2.2. A similar argument shows that A is Fredholm if D has a left inverse.

Suppose that A is Fredholm. There exists a linear independent set $\{y_1, y_2, \ldots, y_n\}$ and a closed subspace M such that
$$Y = \text{Im } A \oplus \text{sp}\{y_1, y_2, \ldots, y_n\}, \tag{2.11}$$
and
$$X = M \oplus \text{Ker } A.$$
Let $\{x_1, x_2, \ldots x_m\}$ be a basis for Ker A. By Corollary XI.5.6, linear functionals $f_1, f_2, \ldots, f_m$, which are bounded on X, may be chosen so that $f_j(x_k) = \delta_{jk}$, $j, k = 1, 2, \ldots, m$, where δ_{jk} is the Kronecker delta and for every $x \in \text{Ker } A$,
$$x = \sum_{j=1}^{m} f_j(x)x_j.$$

Define a finite rank operator $F \in \mathcal{L}(X, Y)$ by

$$F(x) = \sum_{j=1}^{\min(m,n)} f_j(x) y_j. \tag{2.12}$$

Setting $D = A - F$, we have

$$\text{Im } D = \text{Im } A \oplus \text{Im } F. \tag{2.13}$$

Indeed, it is obvious from (2.11) and (2.12) that $\text{Im } A \cap \text{Im } F = (0)$. Clearly $\text{Im } D \subset \text{Im } A \oplus \text{Im } F$. Given $Ax + \sum_{j=1}^{p} \alpha_j y_j \in \text{Im } A \oplus \text{Im } F$, where $p = \min(m.n)$, it follows from $F(x_k) = y_k$, $k = 1, 2, \ldots, p$ and $x_k \in \text{Ker } A$ that

$$Du = Ax + \sum_{j=1}^{p} \alpha_j y_j \in \text{Im } A \oplus \text{Im } F,$$

where $u = x - \sum_{j=1}^{p} f_j(x) x_j - \sum_{j=1}^{p} \alpha_j x_j$. Thus equality (2.13) holds. If $n \leq m$, then from (2.13), (2.11) and $F(x_k) = y_k$, $1 \leq k \leq n$, we have

$$Y = \text{Im } D, \quad n \leq m.$$

It is easy to see that $\text{Ker } D = \text{sp}\{x_{n+1}, \ldots, x_m\}$ for $n < m$ and

$$\text{Ker } D = \{0\}, \quad n \geq m. \tag{2.14}$$

Thus $\text{Ker } D$ is complemented in X. These results combined with Theorem XII.6.2 show that D is *right invertible* when $n \leq m$, i.e., $d(A) \leq n(A)$. For $n \geq m$, i.e., $d(A) \geq n(A)$, we have from (2.13), (2.14) and Theorem XII.6.1 that $n(D) = 0$ and $d(D) \leq d(A) < \infty$, i.e., D is a left invertible Fredholm operator. Thus (2.10) follows.   □

Summarizing the above results, we have the following corollary.

Corollary 2.4 *Suppose* $A \in \mathcal{L}(X, Y)$ *is a Fredholm operator. Then* A *has the representation*

$$A = D + F$$

where F *has finite rank and* D *is a one-sided invertible Fredholm operator. Moreover,*

  (i) *if* $\text{ind } A \geq 0$, *then* D *is right invertible;*

  (ii) *if* $\text{ind } A \leq 0$, *then* D *is left invertible;*

  (iii) *if* $\text{ind } A = 0$, *then* D *is invertible.*

Lemma 2.5 *Given $A \in \mathcal{L}(X, Y)$ and $B \in \mathcal{L}(Y, Z)$, where X, Y and Z are Banach spaces, suppose that $AB = F$ and $BA = G$ are Fredholm operators. Then A and B are Fredholm operators and*

$$\text{ind}\,(AB) = \text{ind}\,A + \text{ind}\,B.$$

Proof: Since $\text{Im}\,F \subset \text{Im}\,A$ and $\text{Ker}\,A \subset \text{Ker}\,G$, it follows that $d(A) \leq d(F) < \infty$ and $n(A) \leq n(G) < \infty$. Thus A is a Fredholm operator. An analogous argument shows that B is a Fredholm operator. Now apply Theorem 2.2. □

Theorem 2.6 *Suppose $A \in \mathcal{L}(X, Y)$ is a Fredholm operator. Then*

   (a) *the conjugate operator A' is a Fredholm operator,*

   (b) $\text{Im}\,A' = (\text{Ker}\,A)^{\perp}, \quad \text{Im}\,A = {}^{\perp}(\text{Ker}\,A'),$

   (c) $n(A') = d(A), \quad d(A') = n(A),$

   (d) $\text{ind}\,A' = -\text{ind}\,A.$

Proof: (a) Since A is a Fredholm operator, $A = D + F$, where D is a one-sided invertible Fredholm operator and F has finite rank. We shall show that the conjugate operator D' is a Fredholm operator. Let $D^{(-1)}$ denote a one-sided inverse of D. If $D^{(-1)}$ is a left inverse of D, then $D^{(-1)}D = I$ and $I - DD^{(-1)}$ is a projection onto a subspace complementing $\text{Im}\,D$ by Theorem XII.6.1. Since $d(D) < \infty$, it follows that $I - DD^{(-1)} = F_1$ is a finite rank operator. Taking conjugates we see that $D'(D^{(-1)})' = I$ and $(D^{(-1)})'D' = I - F_1'$. Since F_1 has finite rank, so does F_1' by Example 1 in Section XIII.5. Hence $D'(D^{(-1)})'$ and $(D^{(-1)})'D'$ are Fredholm operators by Theorem XIII.4.1. Lemma 2.5 shows that D' is a Fredholm operator. If $D^{(-1)}$ is a right inverse of D, then $DD^{(-1)} = I$ and by Theorem XII.6.2, $I - D^{(-1)}D = F_2$ is a finite rank operator with range the finite dimensional space $\text{Ker}\,D$. Arguing as above, we see that D' is a Fredholm operator. Since $(D^{(-1)})'$ is a one-sided inverse of D', it follows from Theorem 2.3 that $A' = D' + F'$ is a Fredholm operator.

   (b) Since A and A' are Fredholm operators, there exist finite dimensional subspaces Y_0, W and a closed subspace X_0 such that

$$X = X_0 \oplus \text{Ker}\,A, \quad Y = \text{Im}\,A \oplus Y_0, \quad X' = \text{Im}\,A' \oplus W.$$

Theorem XIII.5.1 shows that $\text{Im}\,A = {}^{\perp}(\text{Ker}\,A')$ and $\text{Im}\,A' \subseteq (\text{Ker}\,A)^{\perp}$. To prove (b) it remains to show that $(\text{Ker}\,A)^{\perp} \subset \text{Im}\,A'$. Suppose $x' \in (\text{Im}\,A)^{\perp}$. Define the functional f on $\text{Im}\,A = AX_0$ by

$$f(Ax_0) = x'(x_0), \quad x_0 \in X_0.$$

Since A is one to one on X_0 and AX_0 is closed, it follows from Theorem XII.4.6 that there exists $m > 0$ such that $\|Ax_0\| \geq m\|x_0\|$ for $x_0 \in X_0$. Thus

$$|f(Ax_0)| \leq \|x'\|\|x_0\| \leq \frac{1}{m}\|x'\|\|Ax_0\|.$$

Hence f is a bounded linear functional on Im A. By the Hahn-Banach theorem, f has a bounded linear extension g to all of Y. Now for $x_0 \in X_0$ and $z \in \text{Ker } A$,

$$A'g(x_0 + z) = g(Ax_o) = f(Ax_0) = x'(x_0) = x'(x + z).$$

Therefore $x' = A'g \in \text{Im } A'$ and (b) is proved.

(c) Equality $n(A') = d(A)$ appears in Theorem XIII.5.2. We have

$$X' = \text{Im } A' \oplus W = (\text{Ker } A)^{\perp} \oplus W. \qquad (2.15)$$

Define $\varphi : W \to (\text{Ker } A)'$ by

$$\varphi(w) = w_R,$$

where w_R is the restriction of the functional w to Ker A. The functional φ is linear and one to one, for if $\varphi(w) = w_R = 0$, then $w \in \text{Ker } A^{\perp} \cap W = (0)$. Moreover, φ maps W onto $(\text{Ker } A)'$. Indeed, given $h \in (\text{Ker } A)'$, there exists a bounded linear extension g' of h to X, by the Hahn-Banach theorem. From equality (2.15), the functional $g \in X'$ has the representation

$$g = u + v, \quad u \in (\text{Ker } A)^{\perp}, \quad v \in W.$$

Hence $g - u = v \in W$ and for $x \in \text{Ker } A$,

$$\varphi(v) = v_R, \quad v_R(x) = g(x) - u(x) = g(x) = h(x).$$

Thus $\varphi(v) = h$. Hence, since Ker A is finite dimensional,

$$d(A') = \dim W = \dim \varphi(W) = \dim (\text{Ker } A)' = \dim \text{Ker } A = n(A).$$

Thus (c) is proved. The index formula (d) is immediate.   □

The following corollary follows directly from Theorem 2.6

Corollary 2.7 *Let $A \in \mathcal{L}(X, Y)$ be a Fredholm operator. Then $Ax = y$ has a solution if and only if $A'\varphi = 0$ implies $\varphi(y) = 0$. Also, the equation $A'\psi = f$ has a solution if and only if $Au = 0$ implies $f(u) = 0$.*

15.3 Perturbations Small in Norm

Theorem 3.1 *Given a Fredholm operator $A \in \mathcal{L}(X, Y)$, there exists a $\gamma > 0$ such that if $B \in \mathcal{L}(X, Y)$ and $\|B\| < \gamma$, then $A + B$ is a Fredholm operator with*

$$\text{ind } (A + B) = \text{ind } A.$$

Furthermore,

$$n(A + B) \leq n(A), \quad d(A + B) \leq d(A).$$

Proof: Since A is Fredholm, we have from Theorem 2.3 that A has the representation $A = D + F$, where D is a one-sided invertible Fredholm operator and F has finite rank. Let $D^{(-1)}$ denote a one-sided inverse of D. We know from Theorems XII.6.3 and XII.6.4, that for $\| B \| < \| D^{(-1)} \|^{-1}$, $D + B$ is invertible from the same side as D and

$$n(D + B) = n(D) < 0, \quad d(D + B) = d(D) < \infty. \tag{3.1}$$

Suppose $D^{(-1)}$ is a right inverse of D. Then

$$A + B = D + B + F = (D + B)(I + (D + B)^{-1} F). \tag{3.2}$$

Hence by Theorem 2.2 and equalities (3.1) and (3.2),

$$\text{ind}\,(A + B) = \text{ind}\,(D + B) = \text{ind}\,D = \text{ind}\,(D(I + D^{-1}F)) = \text{ind}\,(A). \tag{3.3}$$

A similar argument shows that (3.3) holds if $D^{(-1)}$ is a left inverse of D.

Next we show that $n(A + B) \leq n(A)$. Suppose $\text{ind}\,(A) \leq 0$. We see from (2.12) and (2.13) in Section 2 that the operator F may be chosen so that

$$\dim \text{Im}\, F = n(A). \tag{3.4}$$

We have from Corollary 2.4 that the operator D is left invertible and, as noticed above, $D + B$ is also left invertible. Thus $\text{Ker}\,(D + B) = \{0\}$ and

$$A + B = D + B + K = (I + F(D + B)^{(-1)})(D + B). \tag{3.5}$$

Since $\text{Ker}\,(I + F(D + B)^{(-1)}) \subset \text{Im}\,(F(D + B)^{(-1)})$, it follows from (3.4) and (3.5) that

$$\begin{aligned}
n(A + B) &= \dim \text{Ker}\,(A + B) \\
&\leq \dim \text{Ker}\,(I + F(D + B)^{(-1)}) + \dim \text{Ker}\,(D + B) \\
&= \dim \text{Ker}\,(I + F(D + B)^{(-1)}) \leq \dim \text{Im}\,(F(D + B^{(-1)})) \\
&\leq \dim \text{Im}\, F = n(A). \tag{3.6}
\end{aligned}$$

Since $\text{ind}\,(A + B) = \text{ind}\,A$, we have that $d(A + B) \leq d(A)$. Now suppose $\text{ind}\,A \geq 0$. Then $\text{ind}\,A' = -\text{ind}\,A \leq 0$ by Theorem 2.6. Hence the above result applied to A', together with Theorem 2.6, yield

$$d(A + B) = n(A' + B') \leq n(A') = d(A). \tag{3.7}$$

Since $\text{ind}\,(A + B) = \text{ind}\,A$, inequality (3.7) implies $n(A + B) \leq n(A)$. $\square$

The above theorem yields the following corollary.

Corollary 3.2 *The spectrum of the forward shift S on ℓ_p, $1 \leq p \leq \infty$, is equal to the closed unit disc. Furthermore, for $|\lambda| < 1$ the operator $\lambda I - S$ is Fredholm with*

$$n(\lambda I - S) = 0, \quad d(\lambda I - S) = 1, \quad \text{ind}\,(\lambda I - S) = -1, \qquad (3.8)$$

and $\lambda I - S$ is not Fredholm for $|\lambda| = 1$.

Proof: Since $\|S\| = 1$, we have $\sigma(S) \subset \{\lambda \mid |\lambda| \leq 1\}$. To prove equality, it suffices to show that $\lambda I - S$ is not two-sided invertible for $|\lambda| < 1$. To do this, let A be the backward shift on ℓ_p. Then $AS = I$ and $\|A\| = 1$. Now take $|\lambda| < 1 = \|A\|^{-1}$. Thus, by Theorem XII.6.3, the operator $\lambda I - S$ is left invertible and

$$\text{codim Im}\,(\lambda I - S) = \text{codim Im}\, S = 1.$$

It follows that $\lambda I - S$ is not two-sided invertible. Also we have proved (3.8).

Fix $|\lambda| = 1$. Suppose $\lambda I - S$ is Fredholm. Then, by the previous theorem, for $|\lambda - \beta|$ sufficiently small the operator $\beta I - S$ is Fredholm and

$$\text{ind}\,(\lambda I - S) = \text{ind}\,(\beta I - S).$$

But this is impossible because ind $(\beta I - S) = -1$ for $|\beta| < 1$ and ind $(\beta I - S) = 0$ for $|\beta| > 1$. Thus $\lambda I - S$ is not Fredholm for $|\lambda| = 1$. □

Examples: 1. As a further illustration of Theorem 3.1 consider the backward shift A on ℓ_p, $1 \leq p \leq \infty$, i.e., $A(\alpha_1, \alpha_2, \ldots) = (\alpha_2, \alpha_3, \ldots)$. Then, as was shown in Example 3 of Section XII.8, the spectrum $\sigma(A) = \{\lambda \mid |\lambda| \leq 1\}$. If $|\lambda| = 1$, then $A - \lambda I$ is not a Fredholm operator. To see this, suppose $A - \lambda I$ were Fredholm. Then if $|\beta| > 1$ and β "close" to λ we have $\beta \in \rho(A)$ and

$$0 = \text{ind}\,(A - \beta I) = \text{ind}\,(A - \lambda I). \qquad (3.9)$$

If $|\eta| < 1$ and η "close" to λ, then it is easy to see that $n(A - \eta I) = 1$ and $\beta(A - \eta I) = 0$. Hence

$$1 = \text{ind}\,(A - \eta I) = \text{ind}\,(A - \lambda I).$$

which contradicts equality (3.9).

2. Let $X = C([0, 1])$. Define $A \in \mathcal{L}(X)$ by $(Af)(t) = t f(t)$. It was shown in XII.8, Example 2 that the spectrum of A is given by

$$\sigma(A) = [0, 1].$$

If $\lambda \in [0, 1]$, then $A - \lambda I$ is not a Fredholm operator. Indeed, suppose $A - \lambda I$ is a Fredholm operator. Then for $\beta \notin [0, 1]$ and β "close" to λ, β is in $\rho(A)$ and $0 = \text{ind}\,(A - \beta I) = \text{ind}\,(A - \lambda I)$. It is clear that $A - \lambda I$ is one to one. Hence $d(A - \lambda I) = \text{ind}\,(A - \lambda I) = 0$. But this shows that $\lambda \in \rho(A)$, which is a contradiction.

15.4 Compact Perturbations

The main result in this section is the following theorem.

Theorem 4.1 *Let $A \in \mathcal{L}(X, Y)$ be a Fredholm operator and let $K \in \mathcal{L}(X, Y)$ be compact. Then $A + K$ is a Fredholm operator with*

$$\operatorname{ind}(A + K) = \operatorname{ind} A$$

The basic step in the proof of the theorem relies on the following result.

Theorem 4.2 *If $K \in \mathcal{L}(X, Y)$ is compact, then $I - K$ is Fredholm with index zero.*

Proof: The subspace $\operatorname{Ker}(I - K)$ is finite dimensional. Indeed, if $\{x_k\}$ is a sequence in the 1-ball of $\operatorname{Ker}(I - K)$, then $\{x_k\} = \{Kx_k\}$ has a convergent subsequence and therefore $\operatorname{Ker}(I - K)$ is finite dimensional by Theorem XI.2.4. Thus $X = \operatorname{Ker}(I - K) \oplus M$, where M is a closed subspace of X. We now prove that $\operatorname{Im}(I - K) = (I - K)M$ is closed. It suffices, by Theorem XII.4.6, to show the existence of $m > 0$, such that

$$\|(I - K)u\| \geq m\|u\| \text{ for all } u \in M. \tag{4.1}$$

If (4.1) fails for every $m > 0$, there exists a sequence $\{u_n\} \subset M$ such that $\|u_n\| = 1$ and $(I - K)u_n \to 0$. Since K is compact, there exists a subsequence $\{u_{n'}\}$ of $\{u_n\}$ such that $\{Ku_{n'}\}$ converges. Hence $\lim_{n \to 0} u_{n'} = \lim_{n \to \infty} Ku_{n'} = u$ for some $u \in M$. Thus $\|u\| = 1$ and

$$(I - K)u = \lim_{n \to \infty} (I - K)u_{n'} = 0.$$

But this is impossible since $I - K$ is one to one on M and $u \neq 0$.

Next we show that $d(I - K) < \infty$. Assume this is not the case. Then for any positive integer n, there exist linearly independent vectors $y_1, y_2, \ldots, y_n$ such that $\operatorname{Im}(I - K) \cap \operatorname{sp}\{y_1, y_2, \ldots, y_n\} = \{0\}$. As we saw in the proof of Theorem XI.5.3, there exist functionals $f_1, f_2, \ldots, f_n$ in $\operatorname{Im}(I - K)^{\perp}$ such that $f_i(y_j) = \delta_{ij}$. Now the set $\{f_1, f_2, \ldots, f_n\}$ is linearly independent since $0 = \sum_{j=1}^{n} \alpha_j f_j$ implies $0 = \left(\sum_{j=1}^{n} \alpha_j f_j\right)(y_k) = \alpha_k, \; k = 1, 2, \ldots, n$. Hence $n \leq \dim \operatorname{Im}(I - K)^{\perp} = \dim \operatorname{Ker}(I - K')$ by Theorem XIII.5.1. Since n was arbitrary, $\dim \operatorname{Ker}(I - K') = \infty$. But this is impossible, for K' is compact and, as we observed above, $\operatorname{Ker}(I - K')$ is finite dimensional. Hence $d(I - K) < \infty$.

We have shown that $I - K$ is Fredholm. Finally we prove that $\operatorname{ind}(I - K) = 0$. For each $\lambda \in \mathbb{C}$, we have seen that $I - \lambda K$ is a Fredholm operator. Define the mapping φ on the interval $[0, 1]$ by $\varphi(\lambda) = \operatorname{ind}(I - \lambda K)$. It follows from Theorem 3.1 that φ is continuous. Since φ is integer valued, φ must be a constant function. Hence

$$0 = \varphi(0) = \varphi(1) = \operatorname{ind}(I - K).$$

This completes the proof.   □

Theorem 4.2 is the full generalization of Corollary XIV.4.2.

Proof of Theorem 4.1: Since A is Fredholm, $A = D + F$, where D is a one-sided invertible Fredholm operator and F has finite rank. Assume that D has a right inverse $D^{(-1)}$. Then

$$A + K = D + F + K = D(I + D^{(-1)}(F + K)). \qquad (4.2)$$

Hence $A + K$ is Fredholm and

$$\text{ind } A + K = \text{ind } D = \text{ind } (D(I + D^{(-1)}F)) = \text{ind } (D + F) = \text{ind } A$$

by Theorems 2.2 and 4.2 and equality (4.2). A similar argument shows that ind $(A + K) = \text{ind } A$ if D has a left inverse.

As a consequence of Theorems 2.3 and 4.1 we have the following result.

Corollary 4.3 *Let A be in $\mathcal{L}(X, Y)$. The following statements are equivalent:*

(a) *A is a Fredholm operator*

(b) *A has the representation $A = D + F$, where F has finite rank and D is a one-sided invertible Fredholm operator*

(c) *A has the representation $A = D + K$, where K is compact and D is a one-sided invertible Fredholm operator.*

Using Theorem 4.2, the proof of the following theorem is the same as the proof of Theorem XIII.6.1.

Theorem 4.4 *Let T be a compact operator in $\mathcal{L}(X)$, where X is an infinite dimensional Banach space. Then for the spectrum $\sigma(T)$, one of the following equalities hold:*

a) $\sigma(T) = \{0\}$

b) $\sigma(T) = \{\lambda_1, \lambda_2, \ldots, \lambda_N; 0\}$

c) $\sigma(T) = \{\lambda_1, \lambda_2, \ldots, ; 0\}$.

All λ_j are different from zero and are eigenvalues of T. In case c) we have $\lim_{j \to \infty} \lambda_j = 0$.

15.5 Unbounded Fredholm Operators

Theorems 3.1 and 4.1 can be extended to closed operators in the following manner. First we extend the definition of a Fredholm operator.

A linear operator $T(X \to Y)$ is called a *Fredholm operator* if the numbers $n(T) = \dim \text{Ker } T$ and $d(T) = \text{codim Im } T$ are both finite. In this case the number ind $T = n(T) - d(T)$ is called the *index* of T. The Sturm-Liouville operator discussed in Section VI.5 is a closed Fredholm operator.

If T is a closed operator, we introduce the *graph norm* $\| \cdot \|_T$ on the domain $\mathcal{D}(T)$ of T by

$$\|x\|_T = \|x\| + \|Tx\|.$$

By the argument given in Section VI.3 we see that $\mathcal{D}(T)$ with norm $\| \cdot \|_T$ is a Banach space which we denote by $\mathcal{D}_T$. Obviously T is bounded on $\mathcal{D}_T$.

Theorem 5.1 *Let* $T(X \to Y)$ *be a closed Fredholm operator. There exists a* $\gamma > 0$ *such that for any operator* $B(X \to Y)$ *with* $\mathcal{D}(B) \supset \mathcal{D}(T)$ *and*

$$\|Bx\| \leq \gamma(\|x\| + \|Tx\|), \quad x \in \mathcal{D}(T), \tag{5.1}$$

the operator $T + B$ *is closed and Fredholm with*

(a) $\mathrm{ind}\,(T + B) = \mathrm{ind}\,(T)$

(b) $n(T + B) \leq n(T)$

(c) $d(T + B) \leq d(T)$

Proof: Notice that $\mathcal{D}(T + B) = \mathcal{D}(T)$. Since T is a bounded Fredholm operator on $\mathcal{D}_T$, Theorem 3.1 shows that there exists $\gamma > 0$ such that (5.1) implies that statements (a), (b) and (c) hold. Without loss of generality we may assume that $\gamma < 1$.

It remains to show that $T + B$ is closed. Given $x \in \mathcal{D}(T) = \mathcal{D}(T + B)$, we have

$$\|(T + B)x\| \leq \gamma\|x\| + (1 + \gamma)\|Tx\| \tag{5.2}$$

and

$$\|(T + B)x\| \geq \|Tx\| - \|Bx\| \geq (1 - \gamma)\|Tx\| - \gamma\|x\|.$$

Hence

$$\|Tx\| \leq \frac{1}{1 - \gamma}(\gamma\|x\| + \|(T + B)x\|). \tag{5.3}$$

Suppose $x_n \to x$, $x_n \in \mathcal{D}(T + B)$, and $(T + B)x_n \to y$. Then from (5.3) we see that $\{Tx_n\}$ is a Cauchy sequence which therefore converges to some $v \in Y$. Since T is closed, $x \in \mathcal{D}(T)$ and $Tx = \lim_{n\to\infty} Tx_n = v$. Hence by (5.2),

$$\|(T + B)(x_n - x)\| \leq \gamma\|x_n - x\| + (1 + \gamma)\|T(x_n - x)\| \to 0.$$

Therefore, $(T + B)x = y$ which proves that $T + B$ is a closed operator.   □

Theorem 5.2 *Let* $T(X \to Y)$ *be a closed Fredholm operator, and let* $K(X \to Y)$ *be a compact operator on* $\mathcal{D}_T$. *Then* $T + K$ *is a closed Fredholm operator with*

$$\mathrm{ind}\,(T + K) = \mathrm{ind}\,T.$$

Proof: Since T is a bounded Fredholm operator on $\mathcal{D}_T$ and K is compact on $\mathcal{D}_T$, the operator $T + K$ is Fredholm with ind $(T + K) =$ ind T by Theorem 4.2. It remains to prove that $T + K$ is closed. Since $n(T + K) < \infty$, there exists a closed subspace M of $\mathcal{D}_T$ such that $\mathcal{D}_T = M \oplus \mathrm{Ker}\ (T + K)$. Now the restriction of $T + K$ to M is bounded on $\mathcal{D}_T$, is one-one and has range Im $(T + K)$ which is closed because $T + K$ is a Fredholm operator. Thus, by Theorem XII.4.6, there exists an $m > 0$ such that

$$\|(T + K)v\| \geq m\|v\|_T, \quad v \in M. \tag{5.4}$$

To see that $T + K$ is closed, suppose $x_n \to x$ and $(T + K)x_n \to y$. Now $x_n = v_n + z_n$, $v_n \in M$, $z_n \in \mathrm{Ker}\ (T + K)$ and $(T + K)v_n = (T + K)x_n \to y$. It follows from (5.4) that the sequence $\{\|v_n\|_T\}$ is bounded. Since K is compact on $\mathcal{D}_T$, there is a subsequence $\{v_{n'}\}$ of $\{v_n\}$ such that $K v_{n'} \to w$ for some $w \in Y$. Hence $T x_{n'} \to y - w$. Since T is closed, $x \in \mathcal{D}(T)$ and $Tx = y - w$. Moreover, $w = Kx$. Indeed,

$$\|x_{n'} - x\|_T = \|x_{n'} - x\| + \|Tx_{n'} - Tx\| \to 0.$$

Thus $Kx_{n'} \to Kx = w$ and $Tx = y - Kx$ or $(T + K)x = y$. Hence $T + K$ is a closed operator. $\qquad\square$

Example: Let $\mathcal{H} = L_2([0, 1])$. Define $A(\mathcal{H} \to \mathcal{H})$ by

$$\mathcal{D}(A) = \{f \in \mathcal{H} \mid f \text{ is absolutely continuous on}[0, 1],\ f(0) = 0,\ f' \in \mathcal{H}\},$$

$$Af = f'$$

Let b be a bounded Lebesgue measurable function on $[0, 1]$. Define $B \in \mathcal{L}(\mathcal{H})$ by

$$(Bg)(t) = b(t)g(t).$$

Then B is A-compact and $A + B$ is a Fredholm operator with ind $(A + B) =$ ind $A = 0$. Indeed, we have from Section VII.1 that A is an invertible operator with inverse

$$(A^{-1}g)(t) = \int_0^t g(t)\, ds.$$

Thus A^{-1} is a compact operator on $\mathcal{H}$. To see that B is A-compact, suppose $\{g_n\}$ is an A-bounded sequence, i.e., $\|g_n\| + \|Ag_n\| \leq M$, $n = 1, 2, \ldots$. Then $\{Ag_n\}$ is a bounded sequence in $\mathcal{H}$, and since A^{-1} is compact, $\{g_n\} = \{A^{-1}Ag_n\}$ has a convergent subsequence. Hence $\{Bg_n\}$ has a convergent subsequence which shows that B is A-compact. From Theorem 5.2 we get that $A + B$ is a closed Fredholm operator with

$$\text{ind}\ (A + B) = \text{ind}\ A = 0.$$

For a thorough treatment of perturbation theory, the reader is referred to the paper [GKre1], and the books [G], [GGK1] and [K]. Applications of Theorems 5.1 and 5.2 to differential operators appear in [G].

Exercises XV

1. Let U be the operator on ℓ_p $(1 \leq p < \infty)$ defined by

$$U(x_0, x_1, x_2, \ldots) = (\underbrace{0, 0, \ldots, 0}_{r}, x_k, x_{k+1}, \ldots).$$

As we have seen in Section 1, the operator U is Fredholm and

$$n(U) = k, \quad d(U) = r, \quad \text{ind } U = k - r.$$

(a) Prove that for the operator

$$U^+(x_0, x_1, x_2, \ldots) = (\underbrace{0, 0, \ldots, 0}_{k}, x_r, x_{r+1}, \ldots)$$

the operators $U^+U - I$ and $UU^+ - I$ define projections on ℓ_p and compute their action.

(b) Prove that for $|\lambda| > 1$ the operator $\lambda I - U$ is invertible and that it is a Fredholm operator with ind $(\lambda I - U) = k - r$ for $|\lambda| < 1$.

(c) Let $k \neq r$. Show that for $|\lambda| = 1$ the operator $\lambda I - U$ is not Fredholm. What happens in this case when $k = r$?

(d) Compute $n(\lambda I - U)$ and $d(\lambda I - U)$ for $|\lambda| < 1$.

(e) Determine the spectrum of $\lambda I - U$ for each λ.

(f) Represent U as the sum of a one-sided invertible operator and a finite rank one.

2. Let U and U^+ be as in the previous exercise with $k \neq r$, and consider the operator

$$A = aU^+ + bI + cU.$$

(a) Show that A is Fredholm if and only if

$$\alpha(\lambda) = a\lambda^{-1} + b + c\lambda \neq 0 \quad (|\lambda| = 1).$$

(b) If the latter condition on α holds, show that ind $A = (k - r)m$, where m is the winding number relative to zero of the oriented curve $t \mapsto \alpha(e^{it})$, with t running from $-\pi$ to π.

Determine the Fredholm properties of the operator A when $k = r$.

3. Find the spectrum of the operator A defined in the previous exercise (with $k \neq r$).

4. Let U and U^+ be as in the first exercise with $k \neq r$, and consider the operator

$$B = \sum_{j=1}^{N} a_{-j}(U^+)^j + a_0 I + \sum_{j=1}^{M} a_j U^j,$$

where $a_{-N}, \ldots, a_M$ are complex numbers.

(a) Show that B is Fredholm if and only if

$$\beta(\lambda) = \sum_{j=-N}^{M} a_j \lambda^j \neq 0 \quad (|\lambda| = 1).$$

(b) If the latter condition on β holds, show that ind $B = (k - r)m$, where m is the winding number relative to zero of the oriented curve $t \to \beta(e^{it})$, with t running from $-\pi$ to π.

5. Find the spectrum of the operator B defined in the previous exercise.

6. If $I - A^n$ is a Fredholm operator, prove that $I - A$ is Fredholm.

7. Let the Banach space X be a direct sum of the closed subspaces X_1 and X_2, i.e.,

$$X = X_1 \oplus X_2.$$

In that case a bounded linear operator A on X is represented by a 2×2 operator matrix

$$A = \begin{bmatrix} A_{11} & A_{12} \\ A_{21} & A_{22} \end{bmatrix}, \qquad (*)$$

where A_{ij} is a bounded linear operator from X_j into X_i $(i, j = 1, 2)$. In fact, $(*)$ means that for $x_1 \in X_1$ and $x_2 \in X_2$ we have $A(x_1 + x_2) = y_1 + y_2$, where

$$y_i = A_{i1}x_1 + A_{i2}x_2 \in X_i \quad (i = 1, 2).$$

(a) If A is invertible and the operators A_{12} and A_{21} are compact, prove that A_{11} and A_{22} are Fredholm and

$$\text{ind } A_{11} = -\text{ind } A_{22}.$$

(b) Is the previous statement true if only one of the operators A_{12} and A_{21} is compact?

(c) If the answer to the previous question is negative, what conclusions can nevertheless be made?

(d) If the operator A_{11} on X_1 is Fredholm, show that there exist operators A_{12}, A_{21} and A_{22} such that A_{12} and A_{21} have finite rank and the operator A given by $(*)$ is invertible.

(e) Let A_{11} on X_1 and A_{22} on X_2 be Fredholm operators, and assume that ind $A_{11} = -\text{ind } A_{22}$. Show that there exist finite rank operators A_{12} and A_{21} such that the operator A given by $(*)$ is invertible.

Chapter XVI
Toeplitz and Singular Integral Operators

In this chapter we develop the theory of Laurent and Toeplitz operators for the case when the underlying space is ℓ_p with $1 \leq p < \infty$. To keep the presentation as simple as possible we have chosen a special class of symbols, namely those that are analytic in an annulus around the unit circle. For this class of symbols the results are independent of p. We prove the theorems about left and right invertibility and derive the Fredholm properties. Also, the convergence of the finite section method is analyzed. In the proofs factorization is used systematically. The chapter also contains extensions of the theory to pair operators and to a simple class of singular integral operators on the unit circle.

16.1 Laurent Operators on $\ell_p(\mathbb{Z})$

In this section we study the invertibility of *Laurent operators* on the Banach space $\ell_p(\mathbb{Z})$, where $1 \leq p < \infty$. Such an operator L assigns to an element $x = (x_j)_{j \in \mathbb{Z}}$ in $\ell_p(\mathbb{Z})$ an element $y = (y_j)_{j \in \mathbb{Z}}$ via the rule

$$Lx = y, \quad y_n = \sum_{k=-\infty}^{\infty} a_{n-k} x_k \quad (n \in \mathbb{Z}). \tag{1.1}$$

In what follows we assume that there exist constants $c \geq 0$ and $0 \leq \rho < 1$ such that

$$|a_n| \leq c \rho^{|n|} \quad (n \in \mathbb{Z}). \tag{1.2}$$

From (1.2) it follows that the operator L in (1.1) is well-defined and bounded. To see this, let V be the bilateral (forward) shift on $\ell_p(\mathbb{Z})$, that is, V is the operator given by

$$V((x_j)_{j \in \mathbb{Z}}) = (x_{j-1})_{j \in \mathbb{Z}}. \tag{1.3}$$

Notice that V is invertible and $\|V^n\| = 1$ for each $n \in \mathbb{Z}$. We claim that the action of the operator L in (1.1) is also given by

$$Lx = \sum_{\nu=-\infty}^{\infty} a_\nu V^\nu x, \quad x \in \ell_p(\mathbb{Z}). \tag{1.4}$$

Indeed, the series in the right hand side of (1.4) converges in $\ell_p(\mathbb{Z})$ to a vector which we shall denote by $\tilde{L}x$. Since the map $(x_j)_{j \in \mathbb{Z}} \to x_n$ is a continuous linear functional on $\ell_p(\mathbb{Z})$, we have

$$(\tilde{L}x)_n = \sum_{\nu=-\infty}^{\infty} a_\nu (V^\nu x)_n = \sum_{\nu=-\infty}^{\infty} a_\nu x_{n-\nu} = \sum_{k=-\infty}^{\infty} a_{n-k} x_k = (Lx)_n.$$

This holds for each n, and thus $\tilde{L}x = Lx$, and (1.4) is proved. Now,

$$\|Lx\| \le \sum_{\nu=-\infty}^{\infty} |a_\nu| \|V^\nu x\| \le \left(\sum_{\nu=-\infty}^{\infty} |a_\nu| \right) \|x\|. \tag{1.5}$$

Thus L is bounded and $\|L\| \le \sum_{\nu=-\infty}^{\infty} |a_\nu|$.

With the operator L in (1.1) we associate the complex function

$$\alpha(\lambda) = \sum_{n=-\infty}^{\infty} a_n \lambda^n, \quad \lambda \in \mathbb{T}. \tag{1.6}$$

Notice that condition (1.2) is equivalent to the requirement that α is analytic in an annulus containing the unit circle. Furthermore, we can recover the coefficients a_n from (1.6) via the formula

$$a_n = \frac{1}{2\pi} \int_{-\pi}^{\pi} \alpha(e^{it}) e^{-int} \, dt, \quad n \in \mathbb{Z}. \tag{1.7}$$

We shall call α the *symbol* of the Laurent operator L.

In the sequel we let $\mathcal{A}$ denote the set of all complex-valued functions α on the unit circle that are analytic on some (depending on α) annulus containing the unit circle. Notice that the set $\mathcal{A}$ is closed under the usual addition and multiplication of functions. Also, if $\alpha \in \mathcal{A}$ and α does not vanish on the unit circle, then α^{-1}, where $\alpha^{-1}(\lambda) = 1/\alpha(\lambda)$ for each $\lambda \in \mathbb{T}$, belongs to $\mathcal{A}$.

Theorem 1.1 *Let L_α be the Laurent operator on $\ell_p(\mathbb{Z})$ with symbol $\alpha \in \mathcal{A}$. Then L_α is invertible if and only if α does not vanish on the unit circle, and in that case*

$$(L_\alpha)^{-1} = L_{\alpha^{-1}}. \tag{1.8}$$

Proof: We split the proof into two parts. In the first part we assume that α does not vanish on $\mathbb{T}$ and we prove (1.9). The second part concerns the necessity of the condition on α.

Part 1. It will be convenient to show first that for each α and β in $\mathcal{A}$ have

$$L_\alpha L_\beta = L_{\alpha\beta}. \tag{1.9}$$

To prove (1.9) it suffices to show that $L_\alpha L_\beta e_j = L_{\alpha\beta} e_j$ for each vector $e_j = (\delta_{jn})_{n \in \mathbb{Z}}$, where δ_{jn} is the Kronecker delta. Notice that $L_\beta e_j = (b_{n-j})_{n \in \mathbb{Z}}$, where

$$b_n = \frac{1}{2\pi} \int_{-\pi}^{\pi} \beta(e^{it}) e^{-int} \, dt \quad (n \in \mathbb{Z}).$$

Let a_n be given by (1.7). Then according to (1.1) we have

$$(L_\alpha L_\beta e_j)_n = \sum_{k=-\infty}^{\infty} a_{n-k} b_{k-j}. \tag{1.10}$$

The right hand side of (1.10) is equal to the coefficient γ_{n-j} of λ^{n-j} in the Laurent series expansion of $\gamma(\lambda) = \alpha(\lambda)\beta(\lambda)$. Thus the right hand side of (1.10) is also equal to $(L_{\alpha\beta} e_j)_n$. This holds for each n, and thus (1.9) is proved.

Now assume that α does not vanish on $\mathbb{T}$, and put $\beta = \alpha^{-1}$. Then (1.9) yields

$$L_{\alpha^{-1}} L_\alpha = I, \qquad L_\alpha L_{\alpha^{-1}} = I,$$

where I is the identity operator on $\ell_p(\mathbb{Z})$. Thus (1.8) holds.

Part 2. Assume L_α is invertible. We have to show that $\alpha(\lambda) \neq 0$ for each $\lambda \in \mathbb{T}$. This will be done by contradiction using arguments similar to those used in the second part of the proof of Theorem 4.1 in Chapter III. So assume $\alpha(\lambda_0) = 0$ for some $\lambda_0 \in \mathbb{T}$, and put $\varepsilon = \|L_\alpha^{-1}\|^{-1}$. By Chapter XII, Theorem 6.3 an operator $\tilde{L}$ on $\ell_p(\mathbb{Z})$ will be invertible whenever

$$\|\tilde{L} - L_\alpha\| < \varepsilon. \tag{1.11}$$

Condition (2) implies that $\sum_{n=-\infty}^{\infty} |a_n|$ is convergent. We choose a positive integer N so that $\sum_{|n|>N} |a_n| < \varepsilon/2$. Put

$$\alpha_N(\lambda) = \sum_{n=-N}^{N} a_n \lambda^n,$$

and let L_{α_N} be the Laurent operator with symbol α_N. Then

$$\|L_\alpha - L_{\alpha_N}\| = \sum_{|n|>N} |a_n| < \varepsilon/2, \tag{1.12}$$

$$|\alpha(\lambda) - \alpha_N(\lambda)| \leq \sum_{|n|>N} |a_n| < \varepsilon/2 \quad (\lambda \in \mathbb{T}). \tag{1.13}$$

Put $\tilde{L} = L_{\alpha_N} - \alpha_n(\lambda_0) I$. From (1.13) and $\alpha(\lambda_0) = 0$ we see that $|\alpha_N(\alpha_0)| < \varepsilon/2$. By combining this with (1.12) we get that $\|L_\alpha - \tilde{L}\| < \varepsilon$, and hence $\tilde{L}$ is invertible. Now, put

$$\omega(\lambda) = (\lambda - \lambda_0)^{-1}(\alpha_N(\lambda) - \alpha_N(\lambda_0)).$$

Notice that ω is a trigonometric polynomial, and hence ω is analytic in an annulus containing $\mathbb{T}$. Let L_ω be the corresponding Laurent operator. Since $\tilde{L}$ is the Laurent operator with symbol $\alpha_N(\lambda) - \alpha_N(\lambda_0)$, formula (1.9) and the fact that $(\lambda - \lambda_0)\omega(\lambda) = \alpha_N(\lambda) - \alpha_N(\lambda_0)$ imply that

$$\tilde{L} = (V - \lambda_0 I)L_\omega = L_\omega(V - \lambda_0 I). \tag{1.14}$$

Since $\tilde{L}$ is invertible, we conclude that $V - \lambda_0 I$ is both left and right invertible, and thus λ_0 is not in the spectrum of V. But this contradicts the fact that $|\lambda_0| = 1$ because $\sigma(V) = \mathbb{T}$. Thus $\alpha(\lambda) \neq 0$ for all $\lambda \in \mathbb{T}$. $\qquad\square$

The arguments used in the second part of the proof of the previous theorem also yield the following result.

Theorem 1.2 *Let L be the Laurent operator on $\ell_p(\mathbb{Z})$ with symbol $\alpha \in \mathcal{A}$. Then L is invertible if and only if L is Fredholm.*

Proof: We only have to establish the "if part" of the theorem. So assume L is Fredholm. In view of Theorem 1.1 it suffices to show that $\alpha(\lambda) \neq 0$ for each $\lambda \in \mathbb{T}$. Assume not. So $\alpha(\lambda_0) = 0$ for some $\lambda_0 \in \mathbb{T}$. By the stability theorems for Fredholm operators there exists $\varepsilon > 0$ such that an operator $\tilde{L}$ on $\ell_p(\mathbb{Z})$ will be Fredholm whenever

$$\|L - \tilde{L}\| < \varepsilon.$$

Now, using this ε, construct $\tilde{L}$ and ω as in the second part of the proof of Theorem 1.1. Then $\tilde{L}$ is Fredholm and (1.14) holds. This implies that the operator $V - \lambda_0 I$ is Fredholm, which is impossible because $|\lambda_0| = 1$. Thus $\alpha(\lambda) \neq 0$ for $\lambda \in \mathbb{T}$, and hence L is invertible. $\qquad\square$

16.2 Toeplitz Operators on ℓ_p

In this section we study the invertibility of *Toeplitz operators* on the Banach space ℓ_p, where $1 \leq p < \infty$. Such an operator T assigns to an element $x = (x_j)_{j=0}^\infty$ in ℓ_p an element $y = (y_j)_{j=0}^\infty$ via the rule

$$Tx = y, \quad y_n = \sum_{k=0}^\infty a_{n-k}x_k, \quad n = 0, 1, 2, \ldots. \tag{2.1}$$

Throughout we assume that the coefficients a_n are given by

$$a_n = \frac{1}{2\pi}\int_{-\pi}^{\pi} \alpha(e^{it})e^{-int}\,dt, \quad n \in \mathbb{Z}, \tag{2.2}$$

where $\alpha \in \mathcal{A}$, that is, α is analytic in an annulus containing the unit circle $\mathbb{T}$, or, equivalently, for some $c \geq 0$ and $0 \leq \rho < 1$ we have

$$|a_n| \leq c\rho^{|n|} \quad (n \in \mathbb{Z}). \tag{2.3}$$

The function α is called the *symbol* of T.

From (2.3) it follows that T is a bounded linear operator on ℓ_p. To prove this, notice that ℓ_p may be identified with the subspace of $\ell_p(\mathbb{Z})$ consisting of all $(x_j)_{j \in \mathbb{Z}}$ such that $x_j = 0$ for $j < 0$. Let P be the projection of $\ell_p(\mathbb{Z})$ onto ℓ_p defined by

$$P(\ldots, x_{-1}, x_0, x_1, \ldots) = (\ldots, 0, 0, x_0, x_1, \ldots). \tag{2.4}$$

Thus $\|P\| = 1$, and $T = PL|\text{Im } P$, where L is the Laurent operator on $\ell_p(\mathbb{Z})$ with symbol α. Thus

$$\|T\| \le \|L\| \le \sum_{n=-\infty}^{\infty} |a_n| < \infty. \tag{2.5}$$

Let $e_0, e_1, e_2, \ldots$ be the standard basis of ℓ_p. Then (2.1) means that

$$\begin{bmatrix} a_0 & a_{-1} & a_{-2} & \cdots \\ a_1 & a_0 & a_{-1} & \cdots \\ a_2 & a_1 & a_0 & \\ \vdots & \vdots & & \ddots \end{bmatrix} \begin{bmatrix} x_0 \\ x_1 \\ x_2 \\ \vdots \end{bmatrix} = \begin{bmatrix} y_0 \\ y_1 \\ y_2 \\ \vdots \end{bmatrix},$$

with the usual matrix column multiplication. Therefore, instead of (2.1) we sometimes simply write

$$T = \begin{bmatrix} a_0 & a_{-1} & a_{-2} & \cdots \\ a_1 & a_0 & a_{-1} & \cdots \\ a_2 & a_1 & a_0 & \\ \vdots & \vdots & & \ddots \end{bmatrix}.$$

To study the invertibility of Toeplitz operators we shall use the notion of a Wiener-Hopf factorization for a function from $\mathcal{A}$. So let $\alpha \in \mathcal{A}$. A factorization

$$\alpha(\lambda) = \alpha_-(\lambda)\lambda^\kappa \alpha_+(\lambda), \quad \lambda \in \mathbb{T}, \tag{2.6}$$

is called a *Wiener-Hopf factorization* of α if κ is an integer, the functions α_- and α_+ both belong to $\mathcal{A}$ and do not vanish on $\mathbb{T}$, and

(i) $\alpha_+(\cdot)^{\pm 1}$ extends to a function which is analytic on the unit disk,

(ii) $\alpha_-(\cdot)^{\pm 1}$ extends to a function which is analytic outside the unit disk including infinity.

Since $\alpha \in \mathcal{A}$, the conditions (i) and (ii) imply that the functions $\alpha_+(\cdot)^{\pm 1}$ and $\alpha_-(\cdot)^{\pm 1}$ are also in $\mathcal{A}$. Furthermore, from (i) and (ii) we see that $\alpha_+(\cdot)^{\pm 1}$ and $\alpha_-(\cdot)^{\pm 1}$ admit series expansions of the following form:

$$\alpha_+(\lambda) = \sum_{j=0}^{\infty} \alpha_j^+ \lambda^j, \qquad \alpha_+(\lambda)^{-1} = \sum_{j=0}^{\infty} \gamma_j^+ \lambda^j \qquad (|\lambda| \le 1),$$

$$\alpha_-(\lambda) = \sum_{j=0}^{\infty} \alpha_{-j}^- \lambda^{-j}, \qquad \alpha_-(\lambda)^{-1} = \sum_{j=0}^{\infty} \gamma_{-j}^- \lambda^{-j} \qquad (|\lambda| \ge 1).$$

Thus the Toeplitz operators with symbols $\alpha_+(\cdot)$ and $\alpha_+(\cdot)^{-1}$ are represented by lower triangular matrices, and those with symbols $\alpha_-(\cdot)$ and $\alpha_-(\cdot)^{-1}$ by upper triangular matrices.

For a Wiener-Hopf factorization to exist it is necessary that α does not vanish on $\mathbb{T}$. The next theorem shows that this condition is also sufficient.

Theorem 2.1 *A function α in $\mathcal{A}$ admits a Wiener-Hopf factorization if and only if α does not vanish on $\mathbb{T}$, and in that case the integer κ in (2.6) is equal to the winding number of α relative to zero.*

Proof: We have already seen that $\alpha(\lambda) \neq 0$ for $\lambda \in \mathbb{T}$ is necessary. To prove that this condition is also sufficient, assume that it is satisfied, and let κ be the winding number of α relative to zero. Put $\omega(\lambda) = \lambda^{-\kappa}\alpha(\lambda)$. Then ω is analytic in an annulus containing the unit circle, ω does not vanish on $\mathbb{T}$, and the winding number of ω relative to zero is equal to zero. It follows from complex function theory that $\omega(\lambda) = \exp f(\lambda)$ for some function f that is analytic on an annulus containing $\mathbb{T}$. In fact, we can take $f = \log \omega$ for a suitable branch of the logarithm. Thus for some $\delta > 0$ the function f admits the following Laurent series expansion:

$$f(\lambda) = \sum_{n=-\infty}^{\infty} f_n \lambda^n, \quad 1 - \delta < |\lambda| < 1 + \delta.$$

Now put

$$f_+(\lambda) = \sum_{n=0}^{\infty} f_n \lambda^n, \quad |\lambda| < 1 + \delta,$$

$$f_-(\lambda) = \sum_{n=-\infty}^{-1} f_n \lambda^n, \quad |\lambda| > 1 - \delta,$$

and set

$$\alpha_+(\lambda) = \exp f_+(\lambda), \quad \alpha_-(\lambda) = \exp f_-(\lambda).$$

Then $\omega(\lambda) = \alpha_-(\lambda)\alpha_+(\lambda)$, and hence

$$\alpha(\lambda) = \alpha_-(\lambda)\lambda^\kappa \alpha_+(\lambda), \quad \lambda \in \mathbb{T}. \tag{2.7}$$

From the analytic properties of f_+ and f_- it immediately follows that (2.7) is a Wiener-Hopf factorization of α.

We conclude by showing that in a Wiener-Hopf factorization of α the integer κ is uniquely determined by α, and thus by the result of the previous paragraph κ will always be equal to the winding number of α relative to zero. So consider the Wiener-Hopf factorization of α in (2.6), and let $\alpha(\lambda) = \tilde{\alpha}_-(\lambda)\lambda^\mu \tilde{\alpha}_+(\lambda)$ be a second Wiener-Hopf factorization. Assume $\mu > \kappa$. Then

$$\tilde{\alpha}_-(\lambda)^{-1}\alpha_-(\lambda)\lambda^{\kappa-\mu} = \tilde{\alpha}_+(\lambda)\alpha_+(\lambda)^{-1}. \tag{2.8}$$

The right hand side of (2.8) extends to a function which is analytic on $|\lambda| < 1 + \delta_+$ for some $\delta_+ > 0$, and the left hand side of (2.8) extends to a function which is analytic on $|\lambda| > 1 - \delta_-$ for some $\delta_- > 0$. Moreover, the left hand side of (2.8) is analytic at infinity and its value at infinity is zero. By Liouville's theorem (from complex function theory), both sides of (2.8) are identically zero, which is impossible. Thus μ cannot be strictly larger than κ. In a similar way one shows that κ cannot be strictly larger than μ. Thus $\kappa = \mu$, and κ is uniquely determined by α. $\square$

We are now ready to state the main result of this section.

Theorem 2.2 *Let T be the Toeplitz operator on $\ell_p, 1 \leq p < \infty$, with symbol $\alpha \in \mathcal{A}$. Then T is left or right invertible if and only if $\alpha(e^{it}) \neq 0$ for $-\pi \leq t \leq \pi$. Assume the latter condition is satisfied and let*

$$\alpha(\lambda) = \alpha_-(\lambda)\lambda^\kappa\alpha_+(\lambda), \quad \lambda \in \mathbb{T}, \tag{2.9}$$

be a Wiener-Hopf factorization of α. Then

(i) *T is left invertible if and only if $\kappa \geq 0$, and in that case codim Im $T = \kappa$,*

(ii) *T is right invertible if and only if $\kappa \leq 0$, and in that case dim Ker $T = -\kappa$,*

and in both cases a left or right inverse is given by

$$T^{(-1)} = T_{\alpha_+^{-1}} S^{(-\kappa)} T_{\alpha_-^{-1}}, \tag{2.10}$$

where $T_{\alpha_+^{-1}}$ and $T_{\alpha_-^{-1}}$ are the Toeplitz operators with symbols $1/\alpha_+(\lambda)$ and $1/\alpha_-(\lambda)$, respectively, and

$$S^{(n)} = \begin{cases} S^n, & n = 0, 1, 2, \ldots, \\ (S^{(-1)})^{-n}, & n = -1, -2, \ldots, \end{cases} \tag{2.11}$$

with S the forward shift and $S^{(-1)}$ the backward shift on ℓ_p.

Formula (1.9) in the previous section for the product of two Laurent operators does not hold for Toeplitz operators. This fact complicates the study of Toeplitz operators. However, we do have the following intermediate result.

Theorem 2.3 *Let T_α and T_β be the Toeplitz operators on $\ell_p, 1 \leq p < \infty$, with symbols α and β from $\mathcal{A}$. If α extends to a function which is analytic outside the unit disk including infinity or β extends to a function which is analytic on the unit disk, then*

$$T_{\alpha\beta} = T_\alpha T_\beta. \tag{2.12}$$

Proof: Let L_α and L_β be the Laurent operators on $\ell_p(\mathbb{Z})$ with symbols α and β, and let P be the projection of $\ell_p(\mathbb{Z})$ defined by (2.4). By identifying Im P with ℓ_p we see that

$$T_\alpha = PL_\alpha|\text{Im } P, \quad T_\beta = PL_\beta|\text{Im } P.$$

Put $Q = I - P$. Since $L_{\alpha\beta} = L_\alpha L_\beta$, we have

$$PL_{\alpha\beta}P = PL_\alpha L_\beta P = PL_\alpha(P + Q)L_\beta P$$
$$= (PL_\alpha P)(PL_\beta P) + (PL_\alpha Q)(QL_\beta P).$$

Now observe that α extends to a function which is analytic outside the unit disk including infinity if and only if

$$a_n = \frac{1}{2\pi} \int_{-\pi}^\pi \alpha(e^{it})e^{-int}\, dt = 0, \quad n = 1, 2, \ldots.$$

It follows that α has this analytic property if and only if $PL_\alpha Q = 0$. Similarly, β extends to a function which is analytic on the unit disk if and only if $QL_\beta P = 0$. Thus our hypotheses imply that the operator $(PL_\alpha Q)(QL_\beta P)$ is equal to zero. Thus $PL_{\alpha\beta}P = (PL_\alpha P)(PL_\beta P)$, which yields (2.12). $\qquad\square$

The order of the factors in (2.12) is important, because in general $T_\alpha T_\beta \neq T_\beta T_\alpha$. However, as we shall see later (Theorem 2.5 below), for α and β in $\mathcal{A}$ the difference $T_\alpha T_\beta = T_\beta T_\alpha$ is always compact. We are now ready to prove Theorem 2.2.

Proof of Theorem 2.2: We split the proof into two parts. In the first part we prove the necessity of the condition $\alpha(\lambda) \neq 0$ for $\lambda \in \mathbb{T}$. The second part concerns the reverse implication and the proof of (2.10).

Part 1. Assume T is left or right invertible, and let $\alpha(\lambda_0) = 0$ for some $\lambda_0 \in \mathbb{T}$. We want to show that these assumptions are contradictory. To do this we use the same line of reasoning as in the second part of the proof of Theorem 1.1. Since T is left or right invertible, we know from Section XII.6 that there exists $\varepsilon > 0$ such that $\tilde{T}$ on ℓ_p is left or right invertible whenever $\|T - \tilde{T}\| < \varepsilon$. Choose a positive integer N such that

$$\sum_{|n|>N} |a_n| < \varepsilon/2, \tag{2.13}$$

and set

$$T_{\alpha_N} = \sum_{n=-N}^N a_n S^{(n)}, \quad \alpha_N(\lambda) = \sum_{n=-N}^N a_n \lambda^n. \tag{2.14}$$

Here $S^{(n)}$ is defined as in (2.11). Then

$$\|T - T_{\alpha_N}\| = \left\| \sum_{|n|>N} a_n S^{(n)} \right\| < \varepsilon/2,$$

and

$$\left| \alpha_N(\lambda_0) \right| = \left| \alpha(\lambda_0) - \alpha_N(\lambda_0) \right| = \left| \sum_{|n|>N} a_n \lambda_0^n \right| < \varepsilon/2.$$

Put $\tilde{T} = T_{\alpha_N} - \alpha_N(\lambda_0)I$. Then $\| T - \tilde{T} \| < \varepsilon$, and hence $\tilde{T}$ is left or right invertible, depending on T being left or right invertible. Put

$$\omega_1(\lambda) = \frac{1}{\lambda - \lambda_0}(\alpha_N(\lambda) - \alpha_N(\lambda_0)), \quad \omega_2(\lambda) = \lambda \omega_1(\lambda).$$

Then

$$\alpha_N(\lambda) - \alpha_N(\lambda_0) = \omega_1(\lambda)(\lambda - \lambda_0) = (1 - \lambda_0 \lambda^{-1})\omega_2(\lambda). \tag{2.15}$$

Both ω_1 and ω_2 are trigonometric polynomials, and hence ω_1 and ω_2 belong to $\mathcal{A}$. Let T_{ω_1} and T_{ω_2} be the corresponding Toeplitz operators. Notice that $\tilde{T}$ is the Toeplitz operator with symbol $\alpha_N(\cdot) - \alpha_N(\lambda_0)$. Thus (2.15) and Theorem 2.3 yield

$$\tilde{T} = T_{\omega_1}(S - \lambda_0 I), \quad \tilde{T} = (I - \lambda_0 S^{(-1)})T_{\omega_2}. \tag{2.16}$$

We have already proved that $\tilde{T}$ is left or right invertible. Thus (2.16) shows that $S - \lambda_0 I$ is left invertible or $I - \lambda_0 S^{(-1)}$ is right invertible. Both are impossible because $\lambda_0 \in \mathbb{T}$. Indeed, the spectra of S and $S^{(-1)}$ are equal to the closed unit disc, and according to the stability results of Section XII.6 the operators $\lambda I - S$ and $\lambda I - S^{(-1)}$ are not one-sided invertible for any $\lambda \in \partial\sigma(T) = \mathbb{T}$. We conclude that for T to be left or right invertible it is necessary that $\alpha(\lambda) \neq 0$ for each $\lambda \in \mathbb{T}$.

Part 2. We assume that α does not vanish on $\mathbb{T}$. So, by Theorem 2.1, the function α admits a Wiener-Hopf factorization. Let this factorization be given by (2.9). From the properties of the factors α_- and α_+ in the right hand side of (2.9) we may conclude (using Theorem 2.3) that

$$T = T_{\alpha_-} S^{(\kappa)} T_{\alpha_+}. \tag{2.17}$$

From Theorem 2.3 and the fact that $\alpha_-^{\pm 1}$ extends to a function which is analytic outside the unit disk including infinity it also follows that

$$T_{\alpha_-} T_{\alpha_-^{-1}} = I, \quad T_{\alpha_-^{-1}} T_{\alpha_-} = I.$$

Thus T_{α_-} is invertible and $(T_{\alpha_-})^{-1} = T_{\alpha_-^{-1}}$. Similarly, T_{α_+} is invertible and $(T_{\alpha_+})^{-1} = T_{\alpha_+^{-1}}$. The fact that in the right hand side of (2.17) the factors T_{α_-} and T_{α_+} are invertible, allows us to obtain the invertibility properties of T from those of $S^{(\kappa)}$.

From (2.11) we see that

$$S^{(-\kappa)} S^{(\kappa)} = I \; (\kappa \geq 0), \quad S^{(\kappa)} S^{(-\kappa)} = I \; (\kappa \leq 0). \tag{2.18}$$

Therefore T is left invertible if and only if $\kappa \geq 0$, and in that case

$$\dim \operatorname{Ker} T = \dim \operatorname{Ker} S^{\kappa} = \kappa.$$

This proves item (i) of the theorem; item (ii) is proved in a similar way. Finally, from (2.17) and (2.18) it follows that the operator $T^{(-1)}$ defined by (2.10) is a left or right inverse of T.

The following theorem is the second main result of this section.

Theorem 2.4 *Let T be the Toeplitz operator on ℓ_p, $1 \leq p < \infty$, with symbol $\alpha \in \mathcal{A}$. Then T is Fredholm on ℓ_p if and only if its symbol α does not vanish on the unit circle, and in that case* ind T *is equal to the negative of the winding number of α relative to zero.*

For the proof of Theorem 2.4 we need the following addition to Theorem 2.3.

Theorem 2.5 *Let T_α and T_β be the Toeplitz operators on ℓ_p with symbols α and β from $\mathcal{A}$. Then the commutator $T_\alpha T_\beta - T_\beta T_\alpha$ is compact.*

Proof: Let L_α and L_β be the Laurent operators on $\ell_p(\mathbb{Z})$ with symbols α and β, respectively, and let P be the projection of $\ell_p(\mathbb{Z})$ defined by (2.4). Put $Q = I - P$. We have already seen in the proof of Theorem 2.3 that

$$PL_\alpha L_\beta P = (PL_\alpha P)(PL_\beta P) + (PL_\alpha Q)(QL_\beta P).$$

By interchanging the order of α and β we get

$$PL_\beta L_\alpha P = (PL_\beta P)(PL_\alpha P) + (PL_\beta Q)(QL_\alpha P).$$

According to formula (1.9) in the previous section, $L_\alpha L_\beta = L_\beta L_\alpha$, and hence

$$\begin{aligned}
(PL_\alpha P)&(PL_\beta P) - (PL_\beta P)(PL_\alpha P) \\
&= (PL_\beta Q)(QL_\alpha P) - (PL_\alpha Q)(QL_\beta P).
\end{aligned} \tag{2.19}$$

Recall that Im P may be identified with ℓ_p, and then

$$T_\alpha = PL_\alpha | \operatorname{Im} P, \qquad T_\beta = PL_\beta | \operatorname{Im} P.$$

Thus in order to prove that $T_\alpha T_\beta - T_\beta T_\alpha$ is compact it suffices to show that the right hand side of (2.19) is compact. In fact, to complete the proof it suffices to show that for $\alpha \in \mathcal{A}$ the operators $PL_\alpha Q$ and $QL_\alpha P$ are compact.

Put $\alpha_N(\lambda) = \sum_{n=-N}^{N} a_n \lambda^n$, where a_n is given by (2.2), and let L_{α_N} be the Laurent operator on $\ell_p(\mathbb{Z})$ with symbol α_N. Then

$$\|L_\alpha - L_{\alpha_N}\| \leq \sum_{|n|>N} |a_n| \to 0 \quad (N \to \infty).$$

It follows that for $N \to \infty$ we have

$$\|PL_\alpha Q - PL_{\alpha_N} Q\| \to 0, \quad \|QL_\alpha P - QL_{\alpha_N} P\| \to 0.$$

Thus, in order to prove that $PL_\alpha Q$ and $QL_\alpha P$ are compact, it is sufficient to show that $PL_{\alpha_N} Q$ and $QL_{\alpha_N} P$ are finite rank operators. But the latter is a consequence of the fact that α_N is a trigonometric polynomial. Indeed, for each $x = (x_j)_{j\in\mathbb{Z}}$ in $\ell_p(\mathbb{Z})$ we have

$$(PL_{\alpha_N} Qx)_n = \sum_{k=1}^{N-n} a_{n+k}x_{-k}, \quad n = 0, \dots, N-1$$

and $(PL_{\alpha_N} Qx)_n = 0$ otherwise. Hence the rank of $PL_{\alpha_N} Q$ is at most N. Similarly, rank $QL_{\alpha_N} P \le N$. $\qquad\square$

Proof of Theorem 2.4: Assume α does not vanish on the unit circle. Then we can use Theorem 2.2 (i) and (ii) to show that T is Fredholm, and that

$$\text{ind } T = \dim \text{Ker } T - \text{codim } T = -\kappa,$$

where κ is the integer appearing in (2.9). According to the second part of Theorem 2.1 this integer κ is equal to the winding number of α with respect to zero. So it remains to show that T is Fredholm implies that α does not vanish on $\mathbb{T}$. This will be proved by contradiction using the same line of arguments as in the first part of the proof of Theorem 2.2

Assume T is Fredholm and $\alpha(\lambda_0) = 0$ for some $\lambda_0 \in \mathbb{T}$. According to Theorem 3.6 there exists $\varepsilon > 0$ such that $\tilde{T}$ on ℓ_p is Fredholm whenever $\|T - \tilde{T}\| < \varepsilon$. Choose a positive integer N such that (2.13) holds, and define T_{α_N} and α_N as in (2.14). Let $\tilde{T}$ be the Toeplitz operator with symbol $\alpha_N - \alpha_N(\lambda_0)$. Then $\|T - \tilde{T}\| < \varepsilon$, and hence $\tilde{T}$ is Fredholm. Put

$$\omega(\lambda) = (\lambda - \lambda_0)^{-1}(\alpha_N(\lambda) - \alpha_N(\lambda_0)), \qquad (2.20)$$

and let T_ω be the Toeplitz operator with symbol ω. From Theorem 2.3 and (2.20) we see that $\tilde{T} = T_\omega(S - \lambda_0 I)$, where S is the forward shift on ℓ_p. Thus $T_\omega(S - \lambda_0 I)$ is Fredholm. Theorem 2.5 tells us that $(S - \lambda_0 I)T_\omega$ is a compact perturbation of $T_\omega(S - \lambda_0 I)$. According to Theorem 4.1 in the previous chapter this yields that $(S - \lambda_0 I)T_\omega$ is also Fredholm. Thus both $T_\omega(S - \lambda_0 I)$ and $(S - L_0 I)T_\omega$ are Fredholm. But then we can use Lemma XV.2.5 to show that $S - \lambda_0 I$ is Fredholm. However, by Corollary XV.3.2, the latter is impossible because $\lambda_0 \in \mathbb{T}$. So we reached a contradiction, and hence $\alpha(\lambda) \ne 0$ for each $\lambda \in \mathbb{T}$ whenever T is Fredholm. $\qquad\square$

The following theorems show that Theorem 2.4 remains true for Toeplitz operators with continuous symbols provided we take $p = 2$. The result is a further addition to Theorem III.4.1.

Theorem 2.6 *Let T be the Toeplitz operator on ℓ_2 with continuous symbol α. Then T is Fredholm on ℓ_2 if and only if its symbol α does not vanish on the unit circle, and in that case* ind T *is equal to the negative of the winding number of α relative to zero.*

Proof: In view of Theorem III.4.1 we only have to show that T is Fredholm implies that its symbol α does not vanish on $\mathbb{T}$. We shall prove this by following the same line of arguments as in Part 2 of the proof of Theorem III.4.1.

Let T be Fredholm, and assume $\alpha(\lambda_0) = 0$ for some $\lambda_0 \in \mathbb{T}$. By Theorem 3.1 from the previous chapter there exists $\varepsilon > 0$ such that the operator $\hat{T}$ on ℓ_2 is Fredholm whenever $\|\hat{T} - T\| < \varepsilon$. Now use the second Weierstrass approximation theorem (see Section I.13) to pick a trigonometric polynomial $\tilde{\alpha}$ such that

$$|\alpha(e^{it}) - \tilde{\alpha}(e^{it})| < \frac{1}{2}\varepsilon, \quad -\pi \le t \le \pi, \tag{2.21}$$

and let $\hat{T}$ be the Toeplitz operator with symbol $\hat{\alpha} = \tilde{\alpha} - \tilde{\alpha}(\lambda_0)$. From (2.21) we see that $|\tilde{\alpha}(\lambda_0)| < \frac{1}{2}\varepsilon$, and hence using Theorem III.2.2 we see that

$$\|\hat{T} - T\| = \max_{-\pi \le t \le \pi} |\tilde{\alpha}(e^{it}) - \tilde{\alpha}(\lambda_0) - \alpha(e^{it})| < \varepsilon.$$

Thus $\hat{T}$ is Fredholm. Notice that $\hat{\alpha} \in \mathcal{A}$. Hence, Theorem 2.4 applied to $\hat{T}$ and $\hat{\alpha}$, shows that $\hat{\alpha}$ does not vanish on $\mathbb{T}$, which contradicts the fact $\hat{\alpha}(\lambda_0) = \tilde{\alpha}(\lambda_0) - \tilde{\alpha}(\lambda_0) = 0$. Therefore T is Fredholm implies $\alpha(\lambda) \ne 0$ for all $\lambda \in \mathbb{T}$. $\square$

16.3 An Illustrative Example

In this section we use the theory developed in the previous section to solve in ℓ_p the following infinite system of equations:

$$\sum_{k=0}^{\infty} e^{i\rho|j-k|}x_k - \lambda x_k = q^j \quad (j = 0, 1, 2, \ldots). \tag{3.1}$$

Here λ is a spectral parameter, q and ρ are complex numbers with $|q| < 1$ and $\Im\rho > 0$.

The above equations appear in a natural way in the study of propagation of electromagnetic waves in a medium of periodic structure; see [Kre], §13 and the references given there. In this particular case $1 - \lambda$ is purely imaginary and $q = e^{i\rho}$.

In the sequel we set $c = e^{i\rho}$. We have $c \ne 0$ and $|c| < 1$ because $\Im\rho > 0$. Let T be the Toeplitz operator on ℓ_p given by

$$T = \begin{bmatrix} 1 & c & c^2 & \cdots \\ c & 1 & c & \cdots \\ c^2 & c & 1 & \\ \vdots & \vdots & & \ddots \end{bmatrix}. \tag{3.2}$$

Since $|c| < 1$, the symbol α of T is given by

$$\alpha(\zeta) = \sum_{j=0}^{\infty} c^j \zeta^j + \sum_{j=1}^{\infty} c^j \zeta^{-j} = \frac{1}{1 - c\zeta} + \frac{c\zeta^{-1}}{1 - c\zeta^{-1}}, \quad \zeta \in \mathbb{T}.$$

Notice that α is analytic in an annulus around the unit circle $\mathbb{T}$, and hence the operator T is well-defined and bounded on ℓ_p $(1 \le p < \infty)$.

Using the operator T the equation (3.1) can be rewritten as

$$Tx - \lambda x = y, \tag{3.3}$$

where $y = (1, q, q^2, \ldots)$ and $x = (x_0, x_1, x_2, \ldots)$. To solve (3.1) or (3.3) we distinguish between $\lambda = 0$ and $\lambda \ne 0$.

Theorem 3.1 *Let $\lambda = 0$. Then equation (3.1) has a unique solution in ℓ_p $(1 \le p < \infty)$, namely,*

$$x_0 = \frac{e^{-i\rho} - q}{e^{-i\rho} - e^{i\rho}}, \qquad x_j = x_0(q - e^{i\rho})q^{j-1}, \quad j = 1, 2, \ldots. \tag{3.4}$$

Proof: First let us prove that the operator T is invertible. To do this, notice that its symbol α can be rewritten as

$$\alpha(\zeta) = \frac{c}{\zeta - c} - \frac{c^{-1}}{\zeta - c^{-1}} = \frac{(c - c^{-1})\zeta}{(\zeta - c)(\zeta - c^{-1})}, \quad \zeta \in \mathbb{T}.$$

It follows that α admits the factorization

$$\alpha(\zeta) = \alpha_-(\zeta)\alpha_+(\zeta), \quad \zeta \in \mathbb{T}, \tag{3.5}$$

with the factors being given by

$$\alpha_-(\zeta) = (c - c^{-1})\frac{\zeta}{\zeta - c} = (c - c^{-1})(1 - c\zeta^{-1})^{-1}, \tag{3.6}$$

$$\alpha_+(\zeta) = \frac{1}{\zeta - c^{-1}} = -c(1 - c\zeta)^{-1}. \tag{3.7}$$

Recall that c is inside the unit disc. Thus $\alpha_+(\cdot)^{\pm 1}$ extends to a function which is analytic on the unit disc, and $\alpha_-(\cdot)^{\pm 1}$ extends to a function which is analytic outside the unit disc including infinity. Thus (3.5) is a Wiener-Hopf factorization with $\kappa = 0$. But then we can apply Theorem 2.2 to show that T is invertible. Moreover,

$$T^{-1} = T_{\alpha_+^{-1}} T_{\alpha_-^{-1}}. \tag{3.8}$$

Now, let us use (3.8) to compute the solution of (3.1). From (3.6) and (3.7) we see that

$$T_{\alpha_-^{-1}} = \frac{1}{c - c^{-1}}(I - cS'), \qquad T_{\alpha_+^{-1}} = -\frac{1}{c}(I - cS),$$

where S' and S are the backward shift and the forward shift on ℓ_p, respectively. It follows that

$$T^{-1} = \frac{1}{1-c^2}(I - cS - cS' + c^2 SS'),$$

and hence the matrix of T^{-1} with respect to the standard basis of ℓ_p is tridiagonal:

$$T^{-1} = \frac{1}{1-c^2} \begin{bmatrix} 1 & -c & 0 & 0 & \cdots \\ -c & 1+c^2 & -c & 0 & \cdots \\ 0 & -c & 1+c^2 & -c & \\ 0 & 0 & -c & 1+c^2 & \ddots \\ \vdots & \vdots & & & \ddots & \ddots \end{bmatrix}.$$

Put $y = (1, q, q^2, \ldots)$, and let $x = (x_0, x_1, x_2, \ldots)$ be $T^{-1}y$. Then

$$x_0 = \frac{1}{1-c^2}(1 - cq) = \frac{c^{-1} - q}{c^{-1} - c},$$

and for $k \geq 1$ we have

$$x_j = \frac{1}{1-c^2}(-cq^{j-1} + (1+c^2)q^j - cq^{j+1})$$

$$= \frac{1}{1-c^2}(-c + (1+c^2)q - cq^2)q^{j-1}$$

$$= \frac{(q-c)(1-cq)}{1-c^2}q^{j-1} = \frac{c^{-1} - q}{c^{-1} - c}(q - c)q^{j-1}.$$

Since $c = e^{i\rho}$, this yields (3.4).   □

Now fix $\lambda \neq 0$. Then the symbol ω of the Toeplitz operator $T - \lambda I$ is given by

$$\omega(\zeta) = \alpha(\zeta) - \lambda = \frac{(c - c^{-1})\zeta}{(\zeta - c)(\zeta - c^{-1})} - \lambda$$

$$= -\frac{\lambda\zeta^2 - \lambda(c + c^{-1})\zeta - (c - c^{-1})\zeta + 1}{(\zeta - c)(\zeta - c^{-1})}$$

$$= -\frac{\lambda\zeta^2 - 2(\lambda\cos\rho + i\sin\rho)\zeta + \lambda}{(\zeta - c)(\zeta - c^{-1})}, \qquad \zeta \in \mathbb{T}.$$

Here we used the convention that for $c = e^{i\rho}$ we have

$$c + c^{-1} = 2\cos\rho, \qquad c - c^{-1} = 2i\sin\rho. \tag{3.9}$$

In the above expression for $\omega(\zeta)$ the numerator is a quadratic polynomial and the product of its two roots z_1 and z_2 is equal to one. So $z_2 = z^{-1}$, and we can assume that $|z_1| \le 1$. Thus for $\lambda \ne 0$ the symbol ω of $T - \lambda I$ is of the form

$$\omega(\zeta) = -\lambda \frac{(\zeta - z_1)(\zeta - z_1^{-1})}{(\zeta - c)(\zeta - c^{-1})}, \qquad \zeta \in \mathbb{T}, \tag{3.10}$$

where $0 < |z_1| \le 1$.

We claim that for $\lambda \ne 0$ the operator $T - \lambda I$ is invertible if and only if $|z_1| < 1$. Indeed, if $T - \lambda I$ is invertible, then $\omega(e^{it}) \ne 0$ for each t by Theorem 2.2, and hence we see from (3.10) that $|z_1| = 1$ is impossible. To prove the converse implication, assume that $|z_1| < 1$. The function ω in (3.10) admits the factorization

$$\omega(\zeta) = \omega_-(\zeta)\omega_+(\zeta), \qquad \zeta \in \mathbb{T}, \tag{3.11}$$

with the factors being given by

$$\omega_-(\zeta) = -\lambda \frac{\zeta - z_1}{\zeta - c} = -\lambda \frac{1 - z_1 \zeta^{-1}}{1 - c\zeta^{-1}}, \tag{3.12}$$

$$\omega_+(\zeta) = \frac{\zeta - z_1^{-1}}{\zeta - c^{-1}} = z_1^{-1} c \frac{1 - z_1 \zeta}{1 - c\zeta}. \tag{3.13}$$

Recall that z_1 and c are in the open unit disc. Thus $\omega_+(\cdot)^{\pm 1}$ extends to function which is analytic on the unit disc, and $\omega_-(\cdot)^{\pm 1}$ extends to a function which is analytic outside the unit disc including infinity. Thus (3.11) is a Wiener-Hopf factorization with $\kappa = 0$, and we can apply Theorem 2.2 to show that $T - \lambda I$ is invertible. Moreover,

$$(T - \lambda I)^{-1} = T_{\omega_+^{-1}} T_{\omega_-^{-1}}, \tag{3.14}$$

where $T_{\omega_+^{-1}}$ and $T_{\omega_-^{-1}}$ are the Toeplitz operators with symbols $1/\omega_+(\lambda)$ and $1/\omega_-(\lambda)$, respectively.

Theorem 3.2 *Let $\lambda \ne 0$, and assume $|z_1| < 1$. Then the equation (3.1) has a unique solution in ℓ_p $(1 \le p < \infty)$, and for $q = e^{i\rho}$ this solution is given by*

$$x_j = \frac{2i \sin \rho}{\lambda(1 - e^{i\rho} z_1)} z_1^{j+1} \qquad (j = 0, 1, 2, \ldots). \tag{3.15}$$

Proof: It remains to prove (3.15). Write $y = (1, q, q^2, \ldots)$, where $q = e^{i\rho}$. Then, according to (3.14), the unique solution $x = (x_0, x_1, x_2, \ldots)$ in ℓ_p $(1 \le p < \infty)$ of (3.1) is given by

$$x = T_{\omega_+^{-1}} T_{\omega_-^{-1}} y. \tag{3.16}$$

Recall that $T_{\omega_-^{-1}}$ is an upper triangular Toeplitz operator. Hence, using (3.12), we have

$$T_{\omega_-^{-1}} y = \omega_- \left(\frac{1}{q}\right)^{-1} y = -\lambda^{-1} \frac{1 - cq}{1 - z_1 q} y. \tag{3.17}$$

Next, since $T_{\omega_+^{-1}}$ is lower triangular,

$$
T_{\omega_-^{-1}} y = T_{\omega_+^{-1}} \begin{bmatrix} 1 \\ q \\ q^2 \\ \vdots \end{bmatrix} = \begin{bmatrix} a_0 \\ a_1 \\ a_2 \\ \vdots \end{bmatrix}, \tag{3.18}
$$

where $a_0, a_1, a_2, \ldots$ are the Taylor coefficients at zero of the function $\omega_+(\zeta)^{-1} \times (1 - q\zeta)^{-1}$.

Now, take $q = e^{i\rho} \,(= c)$. Then we see from (3.13) that

$$
\omega_+(\zeta)(1 - q\zeta)^{-1} = z_1 c^{-1}(1 - z_1\zeta)^{-1},
$$

and hence $a_j = c^{-1} z_1^{j+1}$. By combining this with (3.16)–(3.18) we obtain that

$$
x_j = -\lambda^{-1} \frac{1 - c^2}{1 - z_1 c} c^{-1} z_1^{j+1} = \frac{1}{\lambda} \frac{c - c^{-1}}{1 - z_1 c} z_1^{j+1}, \quad j = 0, 1, 2, \ldots.
$$

Together with the second identity in (3.9) this yields (3.15).   □

Theorem 3.3 *Let $c = e^{i\rho}$ with $\Im\rho > 0$. Then for each $1 \le p < \infty$, the spectrum of the operator T in (3.2) is given by*

$$
\sigma(T) = \left\{ \frac{-i \sin\rho}{\tau - \cos\rho} \mid -1 \le \tau \le 1 \right\}. \tag{3.19}
$$

More precisely, $\sigma(T)$ is an arc on a circle, the endpoints of the arc are

$$
\lambda_{-1} = \frac{i \sin\rho}{1 + \cos\rho}, \qquad \lambda_{+1} = \frac{-i \sin\rho}{1 - \cos\rho},
$$

and it passes through the point

$$
\lambda_0 = \frac{i \sin\rho}{\cos\rho}.
$$

Proof: From the proof of Theorem 3.1 we know that T is invertible. Thus $\lambda \in \sigma(T)$ if and only if $\lambda \ne 0$ and the roots $z_1 = z_1(\lambda)$ and $z_2 = z_1(\lambda)^{-1}$ of the quadratic polynomial

$$
\lambda \zeta^2 - (\lambda(c + c^{-1}) + (c - c^{-1}))\zeta + \lambda
$$

are both on the unit circle. For $\lambda \ne 0$ put

$$
\tau(\lambda) = \frac{z_1(\lambda) + z_1(\lambda)^{-1}}{2} = \frac{1}{2}\left(c + c^{-1} + \frac{c - c^{-1}}{\lambda} \right).
$$

Notice that a nonzero complex number z lies on the unit circle if and only if $(z + z^{-1})/2 \in [-1, 1]$. We conclude that $\lambda \in \sigma(T)$ if and only if $\lambda \neq 0$ and $\tau(\lambda) \in [-1, 1]$, and in that case

$$\lambda = \frac{c - c^{-1}}{2\tau(\lambda) - c - c^{-1}}.$$

Here we used that $|c| < 1$, and hence $(c + c^{-1})/2 \notin [-1, 1]$.

Next, notice that $\tau(\lambda)$ runs through $[-1, 1]$ when λ runs through $\sigma(T)$. Indeed, if $\tau \in [-1, 1]$, then $2\tau \neq c + c^{-1}$, and

$$\lambda := \frac{c - c^{-1}}{2\tau - c - c^{-1}} \neq 0.$$

For this λ we have $\tau(\lambda) = \tau \in [-1, 1]$, and hence $\lambda \in \sigma(T)$. Finally, using (3.9) we see that (3.19) is proved.

From what we have proved so far we see that $\sigma(T)$ is the image of the interval $[-1, 1]$ under the map

$$\tau \mapsto \lambda_\tau = \frac{c - c^{-1}}{2\tau - c - c^{-1}}.$$

Such a linear fractional map transforms an interval into an arc of a circle (or an interval). By taking $\tau = -1$ and $\tau = 1$ we see that $[-1, 1]$ is transformed into the arc which begins at the point λ_{-1} and ends at λ_1, and which passes through λ_0. These three points uniquely determine the circle on which this arc lies. $\quad\square$

16.4 Applications to Pair Operators

Throughout this section (except in some of the remarks) we assume that α and β are functions from the class $\mathcal{A}$, that is, α and β are analytic in an annulus containing the unit circle $\mathbb{T}$. With these two functions we associate an operator $M_{\alpha,\beta}$ on $\ell_p(\mathbb{Z})$, where $1 \leq p \leq \infty$ is fixed. The definition is as follows. Given $x = (x_j)_{j \in \mathbb{Z}}$ the element $y = M_{\alpha,\beta}x$ is the vector $y = (y_j)_{j \in \mathbb{Z}}$ with

$$y_n = \sum_{k=0}^{\infty} a_{n-k}x_k + \sum_{k=-\infty}^{-1} b_{n-k}x_k, \tag{4.1}$$

where for each n

$$a_n = \frac{1}{2\pi} \int_{-\pi}^{\pi} \alpha(e^{it})e^{-int}\, dt, \qquad b_n = \frac{1}{2\pi} \int_{-\pi}^{\pi} \beta(e^{it})e^{-int}\, dt. \tag{4.2}$$

We shall refer to $M_{\alpha,\beta}$ as the *pair operator* defined by α and β.

Notice that for $\alpha = \beta$ the operator $M_{\alpha,\beta}$ is just the Laurent operator with symbol α. More generally, if L_α and L_β denote the Laurent operators on $\ell_p(\mathbb{Z})$ with symbols α and β, respectively, then

$$M_{\alpha,\beta} = L_\alpha P + L_\beta Q, \tag{4.3}$$

where P is the orthogonal projection of $\ell_p(\mathbb{Z})$ onto the subspace consisting of all vectors $(x_j)_{j \in \mathbb{Z}}$ such that $x_j = 0$ for $j < 0$, and $Q = I - P$. From (4.3) it immediately follows that $M_{\alpha,\beta}$ is a bounded operator on $\ell_p(\mathbb{Z})$.   □

Theorem 4.1 *Let α, β be in $\mathcal{A}$. The operator $M_{\alpha,\beta}$ on $\ell_p(\mathbb{Z})$ is Fredholm if and only if the functions α and β do not vanish on $\mathbb{T}$, and in that case* ind $M_{\alpha,\beta}$ *is the negative of the winding number of $\gamma = \alpha/\beta$ relative to zero.*

Before we prove the theorem it will be convenient first to analyze the operator $M_{\alpha,\beta}$ a bit better. Let P and Q be the complementary projections appearing in (4.3). Then $\ell_p(\mathbb{Z}) = \operatorname{Im} Q \oplus \operatorname{Im} P$, and relative to this decomposition we can write $M_{\alpha,\beta}$ as a 2×2 operator matrix,

$$M_{\alpha,\beta} = \begin{bmatrix} QL_\beta|\operatorname{Im} Q & QL_\alpha|\operatorname{Im} P \\ PL_\beta|\operatorname{Im} Q & PL_\alpha|\operatorname{Im} P \end{bmatrix} : \operatorname{Im} Q \oplus \operatorname{Im} P \to \operatorname{Im} Q \oplus \operatorname{Im} P. \quad (4.4)$$

Here each element of the 2×2 matrix is viewed as an operator between the corresponding spaces. For instance, $QL_\alpha|\operatorname{Im} P$ acts from $\operatorname{Im} P$ into $\operatorname{Im} Q$. Recall that $\operatorname{Im} P$ consists of all vector $(x_j)_{j \in \mathbb{Z}}$ in $\ell_p(\mathbb{Z})$ such that $x_j = 0$ for $j < 0$, and hence $\operatorname{Im} P$ can be identified with ℓ_p. Using this identification we have

$$PL_\alpha|\operatorname{Im} P = T_\alpha, \quad (4.5)$$

where T_α is the Toeplitz operator on ℓ_p with symbol α. In Section 2 (see the proof of Theorem 2.5) we have shown that the operator $QL_\alpha P$ is compact, and hence $QL_\alpha|\operatorname{Im} P$ is a compact operator from $\operatorname{Im} P$ into $\operatorname{Im} Q$. Similarly, the operator $PL_\beta|\operatorname{Im} Q$ from $\operatorname{Im} Q$ into $\operatorname{Im} P$ is compact. It remains to analyze $QL_\beta|\operatorname{Im} Q$. Let J be the operator from ℓ_p into $\operatorname{Im} Q$ that transforms the vector $(x_j)_{j=0}^\infty$ in ℓ_p into the vector $(y_j)_{j \in \mathbb{Z}}$, where

$$y_j = \begin{cases} x_{-j-1}, & j = -1, -2, -3, \ldots, \\ 0, & \text{otherwise.} \end{cases} \quad (4.6)$$

Then $J : \ell_p \to \operatorname{Im} Q$ is one-one and onto, and J is norm preserving. A straightforward calculation shows that

$$J^{-1}(QL_\beta|\operatorname{Im} Q)J = T_{\beta^\#}, \quad (4.7)$$

where $T_{\beta^\#}$ is the Toeplitz operator on ℓ_p with symbol $\beta^\#$ given by

$$\beta^\#(\lambda) = \beta\left(\frac{1}{\lambda}\right), \qquad \lambda \in \mathbb{T}. \quad (4.8)$$

Indeed, let b_ν be the coefficient of λ^ν in the Laurent series expansion of β. Then for $x = (x_j)_{j=0}^\infty$ in ℓ_p and $Jx = y = (y_j)_{j\in\mathbb{Z}}$, where y_j is given by (4.6), we have for $n \geq 0$

$$(J^{-1}(QL_\beta|\text{Im } Q)Jx)_n = (J^{-1}QL_\beta y)_n = (QL_\beta y)_{-n-1}$$

$$= \sum_{k=-\infty}^\infty b_{-n-1-k}y_k = \sum_{k=-\infty}^{-1} b_{-n-1-k}x_{-k-1}$$

$$= \sum_{\ell=0}^\infty b_{-n-\ell}x_\ell = \sum_{\ell=0}^\infty b_{n-\ell}^\#x_\ell,$$

where $b_\nu^\#$ is the coefficient of λ^ν in the Laurent series expansion of $\beta^\#$. Notice that $\beta^\#$ is also analytic in an annulus containing the unit circle, that is, $\beta^\# \in \mathcal{A}$. We are now ready to prove Theorem 4.1

Proof of Theorem 4.1: Consider the representation (4.4). Since the operators $QL_\alpha|\text{Im } P$ and $PL_\beta|\text{Im } Q$ are compact, $M_{\alpha,\beta}$ and

$$\begin{bmatrix} QL_\beta|\text{Im } Q & 0 \\ 0 & PL_a|\text{Im } P \end{bmatrix} \tag{4.9}$$

differ by a compact operator. Thus, by Theorem 3.1 in the previous chapter, the operator $M_{\alpha,\beta}$ is Fredholm if and only if the operator (4.9) is Fredholm. But the latter happens if and only if $PL_\alpha|\text{Im } P$ and $QL_\beta\text{Im } Q$ are Fredholm. Now use (4.5) and (4.7). It follows that $M_{\alpha,\beta}$ is Fredholm if and only if T_α and $T_{\beta^\#}$ are Fredholm. Now apply Theorem 2.4 to T_α and $T_{\beta^\#}$. Notice that $\beta^\#$ does not vanish on $\mathbb{T}$ if and only if the same holds true for β. We conclude that $M_{\alpha,\beta}$ is Fredholm if and only if α and β do not vanish on $\mathbb{T}$.

Now, assume α and β do not vanish on $\mathbb{T}$, and put $\gamma = \alpha/\beta$. By applying Theorem 1.1 to L_β we see that L_β is invertible, and thus, using formula (9) in Section 1, we have

$$M_{\alpha,\beta} = L_\beta(L_\gamma P + Q) = L_\beta \begin{bmatrix} I_{\text{Im } Q} & QL_\gamma|\text{Im } P \\ 0 & PL_\gamma|\text{Im } P \end{bmatrix}.$$

Again the operator $QL_\gamma|\text{Im } P$ is compact, and, by Theorem 2.4, the operator $PL_\gamma|\text{Im } P = T_\gamma$ is Fredholm and its index is equal to $-\kappa$, where κ is the winding number of γ relative to zero. By Theorem 3.1 in the previous chapter,

$$\text{ind } M_{\alpha,\beta} = \text{ind } \begin{bmatrix} I_{\text{Im } Q} & 0 \\ 0 & PL_\gamma|\text{Im } P \end{bmatrix} = \text{ind } T_\gamma = -\kappa,$$

which completes the proof. $\qquad\qquad\square$

For the case when $p = 2$ Theorem 4.1 above is also true if α and β are just continuous on the unit circle. To see this recall (see Section III.1) that for α and β

continuous on $\mathbb{T}$ the Laurent operators L_α and L_β are well-defined on $\ell_2(\mathbb{Z})$. In fact, in the terminology of Section III.1 the operators L_α and L_β are the Laurent operators defined by the functions $a(t) = \alpha(e^{it})$ and $b(t) = \beta(e^{it})$, respectively. Also, the operators $QL_\alpha P$ and $PL_\beta Q$ are compact. For instance, to prove that $QL_\alpha P$ is compact, we choose a sequence $\alpha_1, \alpha_2, \ldots$ of trigonometric polynomials which converges uniformly to α on $B\mathbb{T}$. Such a sequence exists because of the second Weierstrass approximation theorem (see Section I.13). Now let L_{α_n} be the Laurent operator defined by $a_n(t) = \alpha_n(e^{it})$. Then $QL_{\alpha_n} P$ is an operator of finite rank (see the last paragraph of the proof of Theorem 2.5), and

$$\|QL_\alpha P - QL_{\alpha_n} P\| \le \|L_\alpha - L_{\alpha_n}\| = \|L_{\alpha - \alpha_n}\|$$
$$= \max_{-\pi \le t \le \pi} |\alpha(e^{it}) - \alpha_n(e^{it})| \to 0 \quad (n \to \infty).$$

Here we used formula (1.3) of Section III.1. Thus $QL_\alpha P$ is the limit in the operator norm of a sequence of finite rank operators, and therefore $QL_\alpha P$ is compact. In a similar way one shows that $PL_\beta Q$ is compact. We can now repeat in precisely the same way all the arguments in the proof of Theorem 4.1 with the exception of three. The first two concern the two references to Theorem 2.4 which have to be replaced by references to Theorem 2.6. The third concerns the reference to Theorem 1.1 which has to be replaced by a reference to Theorem III.1.2. With these changes the proof of Theorem 4.1 also works for the case when α and β are continuous on $\mathbb{T}$ provided $p = 2$.

Theorem 4.2 *Let α, β be in $\mathcal{A}$. The operator $M_{\alpha,\beta}$ is left or right invertible if and only if α and β do not vanish on the unit circle. Assume the latter condition holds, and let κ be the winding number of $\gamma = \alpha/\beta$ relative to zero. Then*

(i) *$M_{\alpha,\beta}$ is left invertible if and only if $\kappa \ge 0$, and in that case codim Im $M_{\alpha,\beta} = \kappa$,*

(ii) *$M_{\alpha,\beta}$ is right invertible if and only if $\kappa \le 0$ and in that case dim Ker $M_{\alpha,\beta} = -\kappa$.*

Furthermore, if γ admits a Wiener-Hopf factorization

$$\gamma(\lambda) = \gamma_-(\lambda)\lambda^\kappa \gamma_+(\lambda), \qquad \lambda \in \mathbb{T}, \tag{4.10}$$

then a left or right inverse of $M_{\alpha,\beta}$ is given by

$$M_{\alpha,\beta}^{(-1)} = (L_{\gamma_+^{-1}} P + L_{\gamma_-} Q)(V^{-\kappa} P + Q) L_{\gamma_-^{-1}} L_{\beta^{-1}}. \tag{4.11}$$

Here L_ω denotes the Laurent operator on $\ell_p(\mathbb{Z})$ with symbol ω, and V is the bilateral forward shift on $\ell_p(\mathbb{Z})$.

Proof: We split the proof in three parts. The first part contains some general facts on the function space $\mathcal{A}$.

Part 1. Consider $\alpha \in \mathcal{A}$. Let a_n be the coefficient of λ^n in the Laurent series expansion of α. We know that there exist constants $c \geq 0$ and $0 \leq \rho < 1$ such that

$$|a_n| \leq c\rho^{|n|} \qquad (n \in \mathbb{Z}).$$

Take $\rho < r < 1$, and let $\alpha_r(\lambda) = \sum_{n=-\infty}^{\infty} a_n r^n \lambda^n$. Then α_r also belongs to $\mathcal{A}$, because

$$|a_n r^n| \leq c\rho^{|n|} r^n \leq c\tilde{\rho}^{|n|} \qquad (n \in \mathbb{Z}),$$

where $\tilde{\rho} = \rho/r < 1$. Next, let L_α and L_{α_r} be the Laurent operators on $\ell_p(\mathbb{Z})$ with symbols α and α_r, respectively. Then

$$\|L_\alpha - L_{\alpha_r}\| \to 0 \qquad (\rho < r \uparrow 1) \tag{4.12}$$

Indeed,

$$\|L_\alpha - L_{\alpha_r}\| \leq \sum_{n=-\infty}^{\infty} |a_n - a_n r^n| = c \sum_{n=-\infty}^{\infty} \rho^{|n|} |1 - r^n|$$

$$\leq c \sum_{n=0}^{\infty} (\rho^n - \rho^n r^n) + c \sum_{n=1}^{\infty} \left(\left(\frac{\rho}{r}\right)^n - \rho^n \right)$$

$$= c\left(\frac{1}{1-\rho} - \frac{1}{1-\rho r}\right) + c\left(\frac{\rho}{r-\rho} - \frac{\rho}{1-\rho}\right) \to 0 \; (\rho < r \uparrow 1).$$

Part 2. In this part we assume that $M_{\alpha,\beta}$ is left or right invertible. We want to show that α and β do not vanish on $\mathbb{T}$. Since $M_{\alpha,\beta}$ is left or right invertible, there exists $\varepsilon > 0$ such that $\|M - M_{\alpha,\beta}\| < \varepsilon$ implies that M is left or right invertible and

$$\dim \text{Ker } M = \dim \text{Ker } M_{\alpha,\beta}, \qquad \text{codim Im } M = \text{codim Im } M_{\alpha,\beta}. \tag{4.13}$$

The left or right invertibility of $M_{\alpha,\beta}$ also implies that the functions α and β are not identically equal to zero. This together with the fact that α and β are analytic in an annulus containing $\mathbb{T}$ shows that the zeros of α and β are isolated. So we can find $0 < r' < 1$ such that for $0 < r' < 1$ the functions α_r and β_r are in $\mathcal{A}$ and do not vanish on $\mathbb{T}$. By definition,

$$M_{\alpha,\beta} = L_\alpha P + L_\beta Q, \qquad M_{\alpha_r,\beta_r} = L_{\alpha_r} P + L_{\beta_r} Q,$$

and hence (4.12) yields

$$\|M_{\alpha,\beta} - M_{\alpha_r,\beta_r}\| \to 0 \qquad (r' < r \uparrow 1).$$

But then we can use (4.13) to show that for $r' < r < 1$ and $1 - r$ sufficiently small we have

$$\dim \text{Ker } M_{\alpha_r,\beta_r} = \dim \text{Ker } M_{\alpha,\beta}, \qquad \text{codim Im } M_{\alpha_r,\beta_r} = \text{codim Im } M_{\alpha,\beta}. \tag{4.14}$$

However, since α_r and β_r do not vanish on $\mathbb{T}$ for $r' < r < 1$, the operator M_{α_r,β_r} is Fredholm (by Theorem 4.1), and hence the identities in (4.14) show that $M_{\alpha,\beta}$ is Fredholm too. But then, again using Theorem 4.1, we can conclude that α and β do not vanish on $\mathbb{T}$.

Part 3. In this part we assume that α and β do not vanish on $\mathbb{T}$, and we prove the statements (i) and (ii) and the inversion formula (4.11). Our assumption on α and β imply that

$$M_{\alpha,\beta} = L_\beta(L_\gamma P + Q), \tag{4.15}$$

where $\gamma = \alpha/\beta$. The function γ does not vanish on $\mathbb{T}$, and hence it admits a Wiener-Hopf factorization as in (4.10), where κ is the winding number of γ relative to zero. Let $L_{\gamma-}$ and $L_{\gamma+}$ be the Laurent operators with symbols γ_- and γ_+. The fact that γ_- and γ_+ do not vanish on $\mathbb{T}$ yields

$$L_{\gamma_-}^{-1} = L_{\gamma_-^{-1}}, \qquad L_{\gamma_+}^{-1} = L_{\gamma_+^{-1}}, \tag{4.16}$$

and from the analyticity properties of γ_- and γ_+ we see that

$$L_{\gamma_-}^{\pm 1} Q = Q L_{\gamma_-}^{\pm 1} Q, \qquad L_{\gamma_+}^{\pm 1} P = P L_{\gamma_+}^{\pm 1} P. \tag{4.17}$$

Using the factorization (4.10) we see that (4.15) can be rewritten as

$$M_{\alpha,\beta} = L_\beta(L_{\gamma_-} V^\kappa L_{\gamma_+} P + Q).$$

where V is the bilateral forward shift on $\ell_p(\mathbb{Z})$. Hence, by (4.16) and (4.17),

$$M_{\alpha,\beta} = L_\beta L_{\gamma_-}(V^\kappa P + Q)(L_{\gamma_+} P + Q)(P + L_{\gamma_-}^{-1} Q). \tag{4.18}$$

Moreover, all factors in the right hand side of (4.18) are invertible with the possible exception of $V^\kappa P + Q$. Now, notice that $V^\kappa P + Q$ is left invertible if $\kappa \geq 0$ and right invertible if $\kappa \leq 0$, and in both cases a left or right inverse of $V^\kappa P + Q$ is given by

$$(V^\kappa P + Q)^{(-1)} = V^{-\kappa} P + Q. \tag{4.19}$$

also

$$\text{codim Im } (V^\kappa P + Q) = \text{codim Im } T_{\lambda^\kappa} = \kappa \quad (k \geq 0), \tag{4.20}$$

$$\dim \text{Ker } (V^\kappa P + Q) = \dim \text{Ker } T_{\lambda^{-\kappa}} = -\kappa \quad (k \leq 0). \tag{4.21}$$

From (4.18)–(4.21) and the invertibility of all factors in the right hand side of (4.18) different from $V^\kappa P + Q$ we see that statements (i) and (ii) hold true, and a left or right inverse of $M_{\alpha,\beta}$ is given by

$$M_{\alpha,\beta}^{(-1)} = (P + L_{\gamma_-}^{-1} Q)^{-1}(L_{\gamma_+} P + Q)^{-1}(V^\kappa P + Q)L_{\gamma_-}^{-1} L_\beta^{-1}.$$

The latter formula together with (4.16) and (4.17) yield (4.11), which completes the proof.   □

We conclude this section with some further information about the pair operator $M_{\alpha,\beta}$ for the case when this operator acts on $\ell_2(\mathbb{Z})$ and the functions α and β are not required to belong to the class $\mathcal{A}$ but are merely continuous. In that case the first part of Theorem 4.2 still holds true, that is, the following theorem holds.

Theorem 4.3 *Let $M_{\alpha,\beta}$ be the pair operator on $\ell_2(\mathbb{Z})$, with α and β being continuous on the unit circle. Then $M_{\alpha,\beta}$ is left or right invertible if and only if α and β do not vanish on $\mathbb{T}$. Assume the latter condition holds, and let κ be the winding number of $\gamma = \alpha/\beta$ relative to zero. Then*

    (i) *$M_{\alpha,\beta}$ is left invertible if and only if $\kappa \geq 0$, and in that case codim Im $M_{\alpha,\beta} = \kappa$,*

    (ii) *$M_{\alpha,\beta}$ is right invertible if and only if $\kappa \leq 0$, and in that case dim Ker $M_{\alpha,\beta} = -\kappa$.*

In particular, $M_{\alpha,\beta}$ is two-sided invertible if and only if $\kappa = 0$.

Proof: We split the proof into two parts. In the first part we prove the necessity of the condition on α and β. The second part concerns the reverse implication and statements (i) and (ii).

Part 1. Let $M_{\alpha,\beta}$ have a left or right inverse, and assume $\alpha(\lambda_0) = 0$ for some $\lambda_0 \in \mathbb{T}$. By the results of Section XII.6 there exists $\varepsilon > 0$ such that the operator $\hat{M}$ on $\ell_2(\mathbb{Z})$ is left or right invertible whenever $\|\hat{M} - M\| < \varepsilon$. Now, pick a trigonometric polynomial $\tilde{\alpha}$ such that $|\tilde{\alpha}(e^{it}) - \alpha(e^{it})| < \varepsilon/4$ for each $-\pi \leq t \leq \pi$. Put $\hat{\alpha} = \tilde{\alpha} - \tilde{\alpha}(\lambda_0)$. Then $\hat{\alpha}(\lambda_0) = 0$ and

$$|\hat{\alpha}(e^{it}) - \alpha(e^{it})| < \frac{1}{2}\varepsilon, \quad -\pi \leq t \leq \pi. \tag{4.22}$$

Next, choose a trigonometric polynomial $\hat{\beta}$ such that

$$|\hat{\beta}(e^{it}) - \beta(e^{it})| < \frac{1}{2}\varepsilon, \quad -\pi \leq t \leq \pi. \tag{4.23}$$

Then, by formula (1.3) in Section III.1, we have

$$\|M_{\hat{\alpha},\hat{\beta}} - M_{\alpha,\beta}\| = \|(L_{\hat{\alpha}} - L_\alpha)P + (L_{\hat{\beta}} - L_\beta)Q\|$$
$$\leq \|L_{\hat{\alpha}} - L_\alpha\| + \|L_{\hat{\beta}} - L_\beta\| < \varepsilon.$$

Thus $M_{\hat{\alpha},\hat{\beta}}$ is left or right invertible. Both $\hat{\alpha}$ and $\hat{\beta}$ belong to class $\mathcal{A}$. Thus Theorem 4.2 implies that $\hat{\alpha}$ does not vanish on $\mathbb{T}$ which contradicts the fact that $\hat{\alpha}(\lambda_0) = 0$. Therefore, $M_{\alpha,\beta}$ is left or right invertible implies that α does not vanish on $\mathbb{T}$. The analogous result for β is proved in the same way.

Part 2. In this part we assume that α and β do not vanish on $\mathbb{T}$. Let κ be the winding number of $\gamma = \alpha/\beta$ relative to zero. Notice that

$$M_{\alpha,\beta} = L_\beta(PL_\gamma P + Q)(I + QL_\gamma P). \tag{4.24}$$

The operator L_β is invertible by Theorem III.1.2. Since $(QL_\gamma P)^2$ is the zero operator, we have that $I + QL_\gamma P$ is invertible too. Next, recall (see formula (5)) that we can identify the operator $PL_\gamma|\text{Im } P$ with the Toeplitz operator T_γ. But then we can use Theorem III.4.1 to show that $PL_\gamma P + Q$ is left or right invertible. The invertibility of the factors L_β and $I + QL_\gamma P$ in (4.24) now yields that $M_{\alpha,\beta}$ is also left or right invertible. Furthermore, by applying (i) and (ii) in Theorem III.4.1 to $T = T_\gamma$, formula (4.24) also proves the statements (i) and (ii) of the present theorem. $\qquad\square$

We conclude this section with a remark about operators that can be considered as conjugates of pair operators. Fix $1 \le p \le \infty$, and let α and β belong to the class $\mathcal{A}$. Given $x = (x_k)_{k \in \mathbb{Z}}$ in $\ell_p(\mathbb{Z})$ we define

$$(K_{\alpha,\beta}x)_j = \begin{cases} \sum_{k=-\infty}^{\infty} a_{j-k}x_k, & j = 0, 1, 2, \ldots, \\ \sum_{k=-\infty}^{\infty} b_{j-k}x_k, & j = -1, -2, \ldots. \end{cases}$$

Here a_n and b_n are given by (4.2). It is straightforward to check that $K_{\alpha,\beta}x \in \ell_p(\mathbb{Z})$, and that $K_{\alpha,\beta}$ is a bounded linear operator on $\ell_p(\mathbb{Z})$. In fact,

$$K_{\alpha,\beta} = PL_\alpha + QL_\beta, \tag{4.25}$$

where L_α and L_β are the Laurent operators with symbols α and β, the operator P is the orthogonal projection of $\ell_p(\mathbb{Z})$ onto the subspace consisting of all vectors $(x_j)_{j \in \mathbb{Z}}$ with $x_j = 0$ for $j < 0$, and $Q = I - P$. We shall refer to $K_{\alpha,\beta}$ as the *associate pair operator* defined by α and β.

From the representations (4.25) and (4.3) it follows that the conjugate of $K_{\alpha,\beta}$ is the operator

$$K'_{\alpha,\beta} = L_{\alpha^\#} P + L_{\beta^\#} Q = M_{\alpha^\#,\beta^\#},$$

where $\alpha^\#(\lambda) = \alpha(\lambda^{-1})$ and $\beta^\#(\lambda) = \beta(\lambda^{-1})$. This connection allows us to conclude that with appropriate modifications, Theorems 4.1–4.3 remain true if the pair operator $M_{\alpha,\beta}$ is replaced by the associate pair operator $K_{\alpha,\beta}$. For instance, $K_{\alpha,\beta}$ is Fredholm if and only if the functions α and β do not vanish. Moreover, in that case $\text{ind } K_{\alpha,\beta} = -\text{ind } M_{\alpha^\#,\beta^\#}$, and hence $\text{ind } K_{\alpha,\beta}$ is again equal to the negative of the winding number of α/β relative to zero. In a similar way, one derives the invertibility properties (one- or two-sided) of $K_{\alpha,\beta}$.

16.5 The Finite Section Method Revisited

In this section we study the convergence of the finite section method of the Laurent operators, Toeplitz operators and pair operators considered in the previous sections. We begin with the Toeplitz operators on ℓ_p. Throughout, $1 \le p < \infty$. For each $n \ge 0$ let P_n be the projection of ℓ_p defined by

$$P_n(x_0, x_1, x_2, \ldots) = (x_0, \ldots, x_n, 0, 0, \ldots). \tag{5.1}$$

Notice that $\|P_n\| = 1$, and $P_n x \to x$ $(n \to \infty)$ for each $x \in \ell_p$. Here we use that $p < \infty$. Indeed, for $x = (x_0, x_1, x_2, \ldots)$ we have

$$\|x - P_n x\| = \left(\sum_{j=n+1}^{\infty} |x_j|^p \right)^{1/p} \to 0 \quad (n \to 0).$$

Now let T be an invertible operator on ℓ_p. We say that the *finite section method converges* for T if for n sufficiently large, $n \geq n_0$ say, the operator $P_n T P_n$ on Im P_n is invertible and for each $y \in \ell_p$ the vector

$$x(n) = (x_0(n), \ldots, x_n(n), 0, 0, \ldots) = (P_n T P_n)^{-1} P_n y \quad (n \geq n_0)$$

converges to a solution x of $Tx = y$. In other words the finite section method converges for T if and only if $T \in \Pi\{P_n\}$.

Recall (see Section 1) that $\mathcal{A}$ stands for the class of all complex-valued functions that are analytic on some annulus containing the unit circle.

Theorem 5.1 *For an invertible Toeplitz operator on ℓ_p, $1 \leq p < \infty$, with symbol from the class $\mathcal{A}$, the finite section method converges.*

Proof: Let $\alpha \in \mathcal{A}$ be the symbol of T. Since T is invertible, Theorem 2.4 shows that α does not vanish on the unit circle, and the winding number κ of α relative to zero is equal to zero. Thus, by Theorem 2.1, the function α admits a factorization $\alpha = \alpha_- \alpha_+$, where α_- and a_+ belong to $\mathcal{A}$ and do not vanish on $\mathbb{T}$, and conditions (i) and (ii) in the definition of a Wiener-Hopf factorization (see the paragraph after Theorem 2.4) are satisfied. From Theorem 2.2 it then also follows that

$$T = T_{\alpha_-} T_{\alpha_+}$$

with

$$(T_{\alpha_-})^{-1} = T_{\alpha_-^{-1}}, \quad (T_{\alpha_+})^{-1} = T_{\alpha_+^{-1}}.$$

Put $F = T_{\alpha_+} T_{\alpha_-}$. Then F is invertible, and by Theorem 2.5 the operator T and F differ by a compact operator. Therefore, by Theorem XIII.3.5, it suffices to prove that the finite section method converges for F.

From the analyticity properties of the functions $\alpha_+^{\pm 1}$ and $\alpha_-^{\pm 1}$ it follows that for each $n \in \mathbb{Z}$ we have

$$P_n T_{\alpha_+} = P_n T_{\alpha_+} P_n, \qquad T_{\alpha_-} P_n = P_n T_{\alpha_-} P_n, \tag{5.2}$$

$$P_n T_{\alpha_+}^{-1} = P_n T_{\alpha_+}^{-1} P_n, \qquad T_{\alpha_-}^{-1} P_n = P_n T_{\alpha_-}^{-1} P_n. \tag{5.3}$$

It is now straightforward to check that $P_n T_{\alpha_+} P_n$ and $P_n T_{\alpha_-} P_n$ are invertible as operators on Im P_n and

$$(P_n T_{\alpha_+} P_n)^{-1} P_n = P_n T_{\alpha_+}^{-1}, \qquad (P_n T_{\alpha_-} P_n)^{-1} P_n = T_{\alpha_-}^{-1} P_n \tag{5.4}$$

for each $n \in \mathbb{Z}$. Indeed, since $P_n^2 = P_n$, the first identities in (5.2) and (5.3) yield

$$(P_n T_{\alpha_+} P_n)(P_n T_{\alpha_+}^{-1} P_n) = P_n T_{\alpha_+} P_n T_{\alpha_+}^{-1} P_n = P_n T_{\alpha_+} T_{\alpha_+}^{-1} P_n = P_n,$$

and

$$(P_n T_{\alpha_+}^{-1} P_n)(P_n T_{\alpha_+} P_n) = P_n T_{\alpha_+}^{-1} P_n T_{\alpha_+} P_n = P_n T_{\alpha_+}^{-1} T_{\alpha_+} P_n = P_n.$$

Therefore $P_n T_{\alpha_+} P_n$ is invertible on Im P_n and

$$(P_n T_{\alpha_+} P_n)^{-1} P_n = P_n T_{\alpha_+}^{-1} P_n P_n = P_n T_{\alpha_+}^{-1},$$

which proves the first identity in (5.4). The invertibility of $P_n T_{\alpha_-} P_n$ on Im P_n and the second identity in (5.4) are proved in the same way.

Notice that

$$P_n F P_n = P_n T_{\alpha_+} T_{\alpha_-} P_n = (P_n T_{\alpha_+} P_n)(P_n T_{\alpha_-} P_n), \tag{5.5}$$

and hence $P_n F P_n$ is invertible on Im P_n for each n. Now take $y \in \ell_p$. Then, by (5.5) and (5.4),

$$(P_n F P_n)^{-1} P_n y = (P_n T_{\alpha_-} P_n)^{-1}(P_n T_{\alpha_+} P_n)^{-1} P_n y$$

$$= (P_n T_{\alpha_-} P_n)^{-1} P_n T_{\alpha_+}^{-1} y = T_{\alpha_-}^{-1} P_n T_{\alpha_+}^{-1} y.$$

But $P_n T_{\alpha_+}^{-1} y \to T_{\alpha_+}^{-1} y$ if $n \to \infty$. Thus

$$\lim_{n \to \infty} (P_n F P_n)^{-1} P_n y = \lim_{n \to \infty} T_{\alpha_-}^{-1} P_n T_{\alpha_+}^{-1} y = T_{\alpha_-}^{-1} T_{\alpha_+}^{-1} y = F^{-1} y,$$

and hence the finite section method for F converges. $\square$

We proceed with an analysis of the finite section method for Laurent operators and pair operators on $\ell_p(\mathbb{Z})$, where $1 \leq p < \infty$ is fixed. For each n let Q_n be the projection of $\ell_p(\mathbb{Z})$ defined by

$$Q_n((x_j)_{j \in \mathbb{Z}}) = (\ldots, 0, 0, x_{-n}, \ldots, x_n, 0, 0, \ldots). \tag{5.6}$$

The projections Q_n have all norm one, and $Q_n x \to x$ $(n \to \infty)$ for each $x \in \ell_p$ (because $1 \leq p < \infty$). We say that the *finite section method converges* for an invertible operator L on $\ell_p(\mathbb{Z})$ if for n sufficiently large, $n \geq n_0$ say, the operator $Q_n L Q_n$ on Im Q_n is invertible and for each $y \in \ell_p(\mathbb{Z})$ the vector $x(n) = (Q_n L Q_n)^{-1} Q_n y$, $n \geq n_0$, converges to a solution x of $Lx = y$. In other words, the finite section method converges for L if and only if $L \in \Pi\{Q_n\}$. We shall prove the following theorem.

Theorem 5.2 *Let L_α be an invertible Laurent operator on $\ell_p(\mathbb{Z})$, $1 \leq p < \infty$, with symbol $\alpha \in \mathcal{A}$ (i.e., α does not vanish on $\mathbb{T}$). In order that the finite section method converge for L_α it is necessary and sufficient that the winding number of α with respect to zero is equal to zero.*

Proof: The fact that α does not vanish on the circle follows from the invertibility of L_α (use Theorem 1.1). Let P_n be the projection of ℓ_p defined by (5.1), and define J_n to be the map from Im Q_n to Im P_{2n+1} given by

$$J_n((x_j)_{j \in \mathbb{Z}}) = (x_{-n}, x_{-n+1}, \ldots, x_n, 0, 0, \ldots). \tag{5.7}$$

From the definitions of Q_n and P_n it follows that J is a linear operator which maps Im Q_n in a one to one way onto Im P_{2n+1}. Moreover, $\|J_n x\| = \|x\|$ for each $x \in$ Im Q_n. A straightforward calculation shows that

$$J_n(Q_n L_\alpha Q_n) = (P_{2n+1} T_\alpha P_{2n+1}) J_n, \quad n = 0, 1, 2, \ldots, \tag{5.8}$$

where T_α is the Toeplitz operator with symbol α. Since J is one to one and onto, the operator $Q_n L_\alpha Q_n$ is invertible on Im Q_n if and only if $P_{2n+1} T_\alpha P_{2n+1}$ is invertible on Im P_{2n+1}, and in that case

$$\|(Q_n L_\alpha Q_n)^{-1}\| = \|(P_{2n+1} T_\alpha P_{2n+1})^{-1}\|, \tag{5.9}$$

because J_n is norm preserving.

We shall also need the map $J_n^\# :$ Im $Q_n \to$ Im P_{2n+1} defined by

$$J_n^\#((x_j)_{j \in \mathbb{Z}}) = (x_n, x_{n-1} \ldots x_{-n}, 0, 0, \ldots). \tag{5.10}$$

Notice that the right hand side of (5.10) may be obtained from the right hand side of (5.7) by reversing the order of the first $2n + 1$ entries. Again $J_n^\#$ is a linear operator which maps Im Q_n in a one to one way onto Im P_{2n+1}, and $J_n^\#$ is norm preserving. Moreover,

$$J_n^\#(Q_n L_\alpha Q_n) = (P_{2n+1} T_{\alpha^\#} P_{2n+1}) J_n^\#, \quad n - 0, 1, 2, \ldots, \tag{5.11}$$

where $T_{\alpha^\#}$ is the Toeplitz operator on ℓ_p with symbol $\alpha^\#$ given by

$$\alpha^\#(\lambda) = \alpha\left(\frac{1}{\lambda}\right), \quad \lambda \in \mathbb{T}. \tag{5.12}$$

As mentioned in the previous section (see the paragraph before the proof of Theorem 4.1), the function $\alpha^\# \in \mathcal{A}$. From (5.11) it follows that $Q_n L_\alpha Q_n$ is invertible on Im Q_n if and only if the same holds true for $P_{2n+1} T_{\alpha^\#} P_{2n+1}$ on Im P_{2n+1}, and in that case

$$\|(Q_n L_\alpha Q_n)^{-1}\| = \|(P_{2n+1} T_{\alpha^\#} P_{2n+1})^{-1}\|. \tag{5.13}$$

Now, assume that the finite section method converges for L_α. Then, by Theorem XII.7.1, the operator $Q_n L_\alpha Q_n$ is invertible on Im Q for n sufficiently large, $n \geq N$ say, and $\sup_{n \geq N} \|(Q_n L_\alpha Q_n)^{-1}\| < \infty$. Using (5.9) and (5.13) it follows that

$$\sup_{n \geq N} \|(P_{2n+1} T_\alpha P_{2n+1})^{-1}\| < \infty, \quad \sup_{n \geq N} \|(P_{2n+1} T_{\alpha^\#} P_{2n+1})^{-1}\| < \infty. \tag{5.14}$$

By the remark made at the end of Section II.17 (which carries over to a Banach space setting) it follows that both T_α and $T_{\alpha^\#}$ are one to one. Let κ be the winding number of α with respect to zero. Since T_α is one to one, Theorem 2.2 implies that $\kappa \geq 0$. Notice that $\alpha^\#$ does not vanish on $\mathbb{T}$ and its winding number relative to zero equals to $-\kappa$. Then Theorem 2.2 applied to $T_{\alpha^\#}$ yields $-\kappa \geq 0$. Hence $\kappa = 0$ as desired.

Finally, let us assume that the winding number of α relative to zero is equal to zero. Then T_α is invertible (by Theorem 2.2), and according to Theorem 5.1 we have $T_\alpha \in \Pi\{P_n\}$. But then we can apply Theorem XII.7.1 to show that the first inequality in (5.14) holds for some positive integer N, and hence (5.9) yields $\sup_{n \geq N} \| Q_n L_\alpha Q_n)^{-1} \| < \infty$. So, using Theorem XII.7.1 again, we can conclude that $\tilde{L}_\alpha \in \Pi\{Q_n\}$. $\qquad\qquad\square$

We conclude this section with an analysis of the convergence of the finite section method for the pair operator $M_{\alpha,\beta}$ introduced in the previous section. Here α and β belong to $\mathcal{A}$. Recall that for an invertible operator L on $\ell_p(\mathbb{Z})$ the convergence of the finite section method is defined in the paragraph preceding Theorem 5.2.

Theorem 5.3 *Assume the operator $M_{\alpha,\beta}$ on $\ell_p(\mathbb{Z})$, $1 \leq p < \infty$, is invertible (i.e., α and β do not vanish on the unit circle). In order that the finite section method converge for $M_{\alpha,\beta}$ it is necessary and sufficient that the winding numbers of α and β relative to zero are equal to zero.*

Proof: Since $M_{\alpha,\beta}$ is invertible, we know from Theorem 4.2 that α and β do not vanish on $\mathbb{T}$, and the winding number of α relative to zero is equal to the winding number of β relative to zero. We denote this winding number by κ. It remains to prove the statement about the convergence of the finite section method. We split the proof in two parts.

Part 1. In this part we assume that the finite section method converges for $M_{\alpha,\beta}$. We have to show that the number κ introduced in the previous paragraph is equal to zero. Assume $\kappa > 0$. Let $\tilde{P}$ be the projection of $\ell_p(\mathbb{Z})$ defined by

$$\tilde{P}((x_j)_{j\in\mathbb{Z}}) = (y_j)_{j\in\mathbb{Z}}, \qquad y_j = \begin{cases} x_j, & j \geq \kappa, \\ 0, & \text{otherwise.} \end{cases}$$

Put $\tilde{Q} = I - \tilde{P}$. We claim that the operator $\tilde{H}_{\alpha,\beta} = \tilde{Q}L_\alpha P + \tilde{P}L_\beta Q$ is compact. Here P and Q are the complementary projections appearing in (3) of the previous section. To see this, write

$$\tilde{H}_{\alpha,\beta} = QL_\alpha P + (P - \tilde{P})L_\alpha P + PL_\beta Q + (\tilde{P} - P)L_\beta Q.$$

The fact that α and β belong to $\mathcal{A}$ implies that $QL_\alpha P$ and $PL_\beta Q$ are compact (as we have seen in the proof of Theorem 2.5). Since $P - \tilde{P}$ is an operator of finite rank, the operators $(P - \tilde{P})L_\alpha P$ and $(\tilde{P} - P)L_\beta Q$ are also compact. Hence $\tilde{H}_{\alpha,\beta}$ is compact.

Next, we show that the operator $\tilde{M}_{\alpha,\beta} = M_{\alpha,\beta} - \tilde{H}_{\alpha,\beta}$ is invertible. Notice that $\tilde{M}_{\alpha,\beta} = \tilde{P}L_\alpha P + \tilde{Q}L_\beta Q$. Thus it suffices to show that $\tilde{P}L_\alpha P$ from Im P to Im $\tilde{P}$ and $\tilde{Q}L_\beta Q$ from Im Q to Im $\tilde{Q}$ are invertible. To see that $\tilde{P}L_\alpha P$ is invertible from Im P to Im $\tilde{P}$, let $J :$ Im $P \to \ell_p$ and $\tilde{J} :$ Im $\tilde{P} \to \ell_p$ be defined by

$$Jx = (x_0, x_1, x_2, \ldots), \qquad \tilde{J}x = (x_\kappa, x_{\kappa+1}, x_{\kappa+2}, \ldots)$$

for each $x = (x_j)_{j\in\mathbb{Z}}$ in Im P and Im $\tilde{P}$, respectively. Then

$$\tilde{J}(\tilde{P}L_\alpha P) = T_{\lambda^{-\kappa}\alpha}JP. \tag{5.15}$$

Notice that the function $\tilde{\alpha}(\lambda) = \lambda^{-\kappa}\alpha$ does not vanish on $\mathbb{T}$ and has winding number zero relative to zero. By Theorem 2.2 this implies that the Toeplitz operator $T_{\lambda^{-\kappa}\alpha}$ is invertible. Next, observe that both J and $\tilde{J}$ are invertible. Hence (5.15) shows that $\tilde{P}L_\alpha P$ is an invertible operator from Im P onto Im $\tilde{P}$. In a similar way one shows that up to invertible factors the operator $\tilde{Q}L_\beta Q$ from Im Q to Im $\tilde{Q}$ is equal to the Toeplitz operator $T_{\lambda^\kappa \beta^\#}$, where $\beta^\#(\lambda) = \beta(1/\lambda)$. We know that $\lambda^\kappa \beta^\#$ does not vanish on $\mathbb{T}$ and its winding number relative to the origin is equal to zero. So $T_{\lambda^\kappa \beta^\#}$ is invertible, and hence $\tilde{Q}L_\beta Q :$ Im $Q \to$ Im $\tilde{Q}$ is invertible.

Recall that the finite section converges for $M_{\alpha,\beta}$. By Theorem XIII.3.5, the compactness of $\tilde{H}_{\alpha,\beta}$ and the invertibility of $\tilde{M}_{\alpha,\beta} = M_{\alpha,\beta} - \tilde{H}_{\alpha,\beta}$ imply that the finite section method converges for $\tilde{M}_{\alpha,\beta}$. Thus for n sufficiently large, the operator $Q_n\tilde{M}_{\alpha,\beta}Q_n$ on Im Q_n is invertible. We shall show that this is impossible. Indeed, let $e_i = (\delta_{ij})_{j\in\mathbb{Z}}$, where δ_{ij} is the Kronecker delta, and take $n > \kappa$. Notice that the vectors $e_{-n}, e_{-n+1}, \ldots, e_{n-1}, e_n$ form a basis of Im Q, and the matrix of the operator $Q_n\tilde{M}_{\alpha,\beta}Q_n$ on Im Q_n relative to this basis has the form:

$$\begin{bmatrix} b_0 & \cdots & b_{-n+1} & 0 & \cdots & 0 \\ \vdots & & \vdots & \vdots & & \vdots \\ b_{n-1} & \cdots & b_0 & 0 & \cdots & 0 \\ b_n & \cdots & b_1 & 0 & \cdots & 0 \\ \vdots & & \vdots & \vdots & & \vdots \\ b_{n+\kappa-1} & \cdots & b_\kappa & 0 & \cdots & 0 \\ 0 & \cdots & 0 & a_\kappa & \cdots & a_{\kappa-n} \\ \vdots & & \vdots & \vdots & & \vdots \\ 0 & \cdots & 0 & a_n & \cdots & a_0 \end{bmatrix}. \tag{5.16}$$

Here a_ν and b_ν are the coefficients of λ^ν in the Laurent series expansions of α and β, respectively. The matrix (5.16) is a square matrix of order $2n + 1$, and it partitions as a 2×2 block matrix

$$\begin{bmatrix} B & 0 \\ 0 & A \end{bmatrix},$$

where A has size $(n - \kappa + 1) \times (n + 1)$ and B has size $(n + \kappa) \times n$. Thus

$$\text{rank } A + \text{rank } B \leq n - \kappa + 1 + n < 2n + 1.$$

Therefore the matrix (5.16) is not invertible, and hence $Q_n \tilde{M}_{\alpha,\beta} Q_n$ on Im Q_n is not invertible. We reached a contradiction. Hence $\kappa > 0$ is impossible. In a similar way one shows that $\kappa < 0$ is impossible. Thus $\kappa = 0$, as desired.

Part 2. In this part we assume that the winding number of α and β relative to zero is equal to zero. We want to show that the first section method converges for $M_{\alpha,\beta}$. Let P be the projection of $\ell_p(\mathbb{Z})$ defined by

$$P((x_j)_{j \in \mathbb{Z}}) = (y_j)_{j \in \mathbb{Z}}, \qquad y_j = \begin{cases} x_j, & j = 0, 1, 2, \ldots, \\ 0, & \text{otherwise.} \end{cases}$$

Put $Q = I - P$. Then

$$M_{\alpha,\beta} = PL_\alpha P + QL_\beta Q + QL_\alpha P + PL_\beta Q.$$

As we have seen in the previous section, the operator $QL_\alpha P + PL_\beta Q$ is compact and up to invertible factors the operators $PL_\alpha P$ on Im P and $QL_\beta Q$ on Im Q are equal to the Toeplitz operators T_α and $T_{\beta^\#}$, respectively. By our assumptions on α and β the Toeplitz operators T_α and $T_{\beta^\#}$ are invertible, and hence the operator $PL_\alpha P + QL_\beta Q$ is invertible too. By Theorem 5.1 the invertibility of T_α and $T_{\beta^\#}$ also implies that these operators belong to $\Pi\{P_n\}$, where P_n is the projection of ℓ_p defined by (5.1). Thus for n sufficiently large, $n \geq N$ say, the operators $P_n T_\alpha P_n$ and $P_n T_{\beta^\#} P_n$ on Im P_n are invertible and

$$\sup_{n \geq N} \|(P_n T_\alpha P_n)^{-1}\| < \infty, \qquad \sup_{n \geq N} \|(P_n T_{\beta^\#} P_n)^{-1}\| < \infty.$$

But this implies (use formulas (5.5) and (5.7) in the previous section) that for $n \geq N + 1$ the operator $Q_n(PL_\alpha P + QL_\beta Q)Q_n$ is invertible and

$$\sup_{n \geq N} \|\{Q_n(PL_\alpha P + QL_\beta Q)Q_n\}^{-1}\| < 0.$$

Since $PL_\alpha P + QL_\beta Q$ is invertible, we can use Theorem XII.7.1 to show that the finite section method converges for this operator. Our assumptions on α and β also imply that the operator $M_{\alpha,\beta}$ is invertible (by Theorem 4.2). Thus $M_{\alpha,\beta}$ is an invertible operator which is a compact perturbation of an invertible operator for which the finite section method converges. But then Theorem XIV.3.5 implies the the finite section method converges for $M_{\alpha,\beta}$.

16.6 Singular Integral Operators on the Unit Circle

In this section we study the inversion of the following singular integral equation on the unit circle $\mathbb{T}$:

$$w(\lambda) f(\lambda) + \frac{1}{\pi i} \int_\mathbb{T} \frac{k(\lambda, z)}{z - \lambda} f(z) \, dz = g(\lambda), \qquad \lambda \in \mathbb{T}. \tag{6.1}$$

Here w is a continuous function on $\mathbb{T}$, the kernel function k is continuous on the torus $\mathbb{T} \times \mathbb{T}$, and k is assumed to satisfy the following Lipschitz type continuity condition:

$$|k(\lambda, \mu) - k(\lambda, \lambda)| \leq c|\mu - \lambda|, \qquad \lambda, \mu \in \mathbb{T}, \qquad (6.2)$$

for some constant c. The right hand side g of (6.1) belongs to $L_2(\mathbb{T})$ which consists of all functions f such that $t \mapsto f(e^{it})$ is Lebesgue measurable and square integrable on $-\pi \leq t \leq \pi$, endowed with the inner product

$$\langle f_1, f_2 \rangle = \frac{1}{2\pi} \int_{-\pi}^{\pi} f_1(e^{it}) \overline{f_2(e^{it})} \, dt.$$

The integral in (6.1) has to be understood as a principal value integral. The precise meaning of the left hand side of (6.1) will be explained by the next lemma and its proof. ☐

Lemma 6.1 *Condition (6.2) implies that the operator B is given by*

$$(Bf)(\lambda) = \int_{\mathbb{T}} \frac{k(\lambda, z)}{z - \lambda} f(z) \, dz, \quad \lambda \in \mathbb{T}, \qquad (6.3)$$

defines a bounded linear operator on $L_2(\mathbb{T})$.

Proof: First let us remark that

$$\int_{\mathbb{T}} \frac{1}{z - 1} \, dz = \lim_{\varepsilon \downarrow 0} \left(\int_{-\pi}^{-\varepsilon} \frac{ie^{it}}{e^{it} - 1} \, dt + \int_{+\varepsilon}^{\pi} \frac{ie^{it}}{e^{it} - 1} \, dt \right) = \pi i. \qquad (6.4)$$

This follows directly from the identity

$$ie^{it}(e^{it} - 1)^{-1} = \frac{1}{2}i + \frac{1}{2}\sin t \,(1 - \cos t)^{-1},$$

and the fact that the second term in the right hand side of this identity is an even function. The integral in the left hand side of (6.4) does not change if we replace z by $\lambda^{-1}z$ where $\lambda \in \mathbb{T}$ is fixed. Thus

$$\int_{\mathbb{T}} \frac{1}{z - \lambda} \, dz = \pi i \qquad (\lambda \in \mathbb{T}). \qquad (6.5)$$

Next, notice that

$$\frac{1}{\pi i} \int_{\mathbb{T}} \frac{z^m}{z - \lambda} \, dz = \begin{cases} \lambda^m, & m > 0, \\ -\lambda^m, & m < 0. \end{cases} \qquad (6.6)$$

To see this we first take $m > 0$ and write

$$z^m = \lambda^m + (z^{m-1} + z^{m-2}\lambda + \cdots + z^0\lambda^{m-1})(z - \lambda). \qquad (6.7)$$

Since

$$\int_{\mathbb{T}} z^k \, dz = \int_{-\pi}^{\pi} i e^{ikt} e^{it} \, dt = \begin{cases} 2\pi i, & k = -1, \\ 0, & \text{otherwise,} \end{cases} \tag{6.8}$$

we see from (6.7) and (6.5) that (6.6) holds for $m \geq 0$. For $m < 0$ we get the desired equality by replacing (6.7) by

$$z^m = \lambda^m + (z^m \lambda^{-1} + z^{m+1}\lambda^{-2} + \cdots + z^{-1}\lambda^m)(\lambda - z). \tag{6.9}$$

The functions $\varepsilon_n(e^{it}) = e^{int}$, $n \in \mathbb{Z}$, form an orthonormal basis of $L_2(\mathbb{T})$. We let $\mathbb{P}$ be the orthogonal projection of $L_2(\mathbb{T})$ defined by

$$\mathbb{P}\varepsilon_n = \begin{cases} \varepsilon_n, & n = 0, 1, 2, \ldots, \\ 0, & \text{otherwise.} \end{cases} \tag{6.10}$$

Put $\mathbb{Q} = I - \mathbb{P}$. Formulas (6.5) and (6.6) show that for each trigonometric polynomial φ we have

$$\frac{1}{\pi i} \int_{\mathbb{T}} \frac{\varphi(z)}{z - \lambda} \, dz = ((\mathbb{P} - \mathbb{Q})\varphi)(\lambda), \quad \lambda \in \mathbb{T}. \tag{6.11}$$

In the sequel for any φ in $L_2(\mathbb{T})$ the function given by the left hand side of (6.11) will by definition be equal to $(\mathbb{P} - \mathbb{Q})\varphi$. Thus we define for each $\varphi \in L_2(\mathbb{T})$

$$\frac{1}{\pi i} \int_{\mathbb{T}} \frac{\varphi(z)}{z - \lambda} \, dz := ((\mathbb{P} - \mathbb{Q})\varphi)(\lambda), \quad \lambda \in \mathbb{T} \text{ a.e..} \tag{6.12}$$

Now, let us consider the integral in (6.3). Put

$$h(\lambda, z) = \frac{k(\lambda, z) - k(\lambda, \lambda)}{z - \lambda}.$$

The function h is continuous on $\{(\lambda, z) \in \mathbb{T} \times \mathbb{T} \mid \lambda \neq z\}$ and, according to condition (6.2), the function h is bounded on $\mathbb{T} \times \mathbb{T}$. Hence the function $\tilde{h}(t, s) = h(e^{it}, e^{is})$ is square integrable on $[-\pi, \pi] \times [-\pi, \pi]$. This shows that the operator C given by

$$(Cf)(\lambda) = \frac{1}{\pi i} \int_{\mathbb{T}} \frac{k(\lambda, z) - k(\lambda, \lambda)}{z - \lambda} f(z) \, dz, \quad \lambda \in \mathbb{T} \text{ a.e.,} \tag{6.13}$$

is a Hilbert-Schmidt integral operator on $L_2(\mathbb{T})$. In what follows we understand the function Bf in (6.3) to be defined by

$$(Bf)(\lambda) = \pi i(Cf)(\lambda) + k(\lambda, \lambda) \int_{\mathbb{T}} \frac{1}{z - \lambda} f(z) \, dz, \quad \lambda \in \mathbb{T} \text{ a.e..} \tag{6.14}$$

In this way B is a well-defined bounded linear operator on $L_2(\mathbb{T})$. $\qquad \square$

The operator in the left hand side of (6.12) is usually referred to as the *operator of singular integration*, and it is denoted by $\mathbb{S}_{\mathbb{T}}$. Notice that equation (6.1) can now be rewritten in the following equivalent form

$$w(\lambda)f(\lambda) + k(\lambda, \lambda)(\mathbb{S}_{\mathbb{T}}f)(\lambda) + (Cf)(\lambda) = g(\lambda), \quad \lambda \in \mathbb{T} \text{ a.e.,} \qquad (6.15)$$

where C is the Hilbert-Schmidt integral operator given by (6.12). Next, put

$$\alpha(\lambda) = w(\lambda) + k(\lambda, \lambda), \quad \beta(\lambda) = w(\lambda) - k(\lambda, \lambda). \qquad (6.16)$$

Then both α and β are continuous functions on $\mathbb{T}$, and using $\mathbb{S}_{\mathbb{T}} = \mathbb{P} - \mathbb{Q}$, we can rewrite (6.14) as

$$\alpha(\cdot)\mathbb{P}f + \beta(\cdot)\mathbb{Q} + Cf = g.$$

Now put

$$Af = \alpha(\cdot)\mathbb{P}f + \beta(\cdot)\mathbb{Q}f + Cf. \qquad (6.17)$$

In what follows we refer to A as the *operator* on $L_2(\mathbb{T})$ *associated with* (6.1).

The next theorem is the first main result of this section.

Theorem 6.2 *The operator A associated with the singular integral equation (6.1) is Fredholm if and only if the functions α and β defined by (6.16) do not vanish on the unit circle, and in that case* ind A *is the negative of the winding number of $\gamma = \alpha/\beta$ relative to zero.*

Proof: Since the operator C defined by (6.13) is compact, formula (6.17) shows that A is a compact perturbation of the operator $A_0 = \alpha(\cdot)\mathbb{P} + \beta(\cdot)\mathbb{Q}$. But then, by Theorem 4.1 in the previous chapter, it suffices to prove the theorem for A_0 in place of A.

Next, let U be the operator on $L_2(\mathbb{T})$ that assigns to each $f \in L_2(\mathbb{T})$ its sequence of Fourier coefficients $(c_n(f))_{n \in \mathbb{Z}}$ with respect to the orthonormal basis $\ldots, \varepsilon_{-1}, \varepsilon_0, \varepsilon_1, \ldots$, where $\varepsilon_n(e^{it}) = e^{int}$ for $n \in \mathbb{Z}$. Thus

$$(Uf)_n = c_n(f) = \frac{1}{2\pi}\int_{-\pi}^{\pi} f(e^{it})e^{int}\, dt = \langle f, \varepsilon_n \rangle, \; n \in \mathbb{Z}.$$

It follows that $Uf \in \ell_2(\mathbb{Z})$, and U is a unitary operator from $L_2(\mathbb{T})$ onto $\ell_2(\mathbb{Z})$. From the definition of $\mathbb{P}$ in (6.10) and $\mathbb{Q} = I - \mathbb{P}$ we see that $U\mathbb{P}U^{-1} = P$ and $U\mathbb{Q}U^{-1} = Q$, where P is the orthogonal projection of $\ell_2(\mathbb{Z})$ defined by

$$P((x_j)_{j \in \mathbb{Z}}) = (\ldots, 0, 0, x_0, x_1, x_2, \ldots), \qquad (6.18)$$

and $Q = I - P$. Furthermore, for each $f \in L_2(\mathbb{T})$ we have

$$U\alpha(\cdot)f = L_\alpha Uf, \qquad U\beta(\cdot)f = L_\beta Uf, \qquad (6.19)$$

where L_α and L_β are the Laurent operators on $\ell_2(\mathbb{Z})$ with symbols α and β (or in the terminology of Section III.1 the Laurent operators on $\ell_2(\mathbb{Z})$ defined by the functions $a(t) = \alpha(e^{it})$ and $b(t) = \beta(e^{it})$ on $-\pi \le t \le \pi$). We conclude that the operator $A_0 = \alpha(\cdot)\mathbb{P} + \beta(\cdot)\mathbb{Q}$ is unitarily equivalent to the pair operator $M_{\alpha,\beta}$ on $\ell_2(\mathbb{Z})$.

From the remark made after the proof of Theorem 3.1 we know that for α and β continuous on $\mathbb{T}$ the pair operator $M_{\alpha,\beta}$ is Fredholm on $\ell_2(\mathbb{Z})$ if and only if α and β do not vanish on $\mathbb{T}$, and in that case ind $M_{\alpha,\beta} = -\kappa$, where κ is the winding number of α/β relative to zero. Since A_0 and $M_{\alpha,\beta}$ are unitarily equivalent, the same holds true for A_0, which completes the proof. $\qquad\qquad\square$

The operator $A_0 = \alpha(\cdot)\mathbb{P} + \beta(\cdot)\mathbb{Q}$ appearing in the above proof is usually referred to as the *main part* of the operator associated with (6.1). Notice that the action of A_0 is also given by

$$A_0 f = w(\cdot)f + v(\cdot)\mathbb{S}_{\mathbb{T}} f, \qquad (6.20)$$

where $v(\lambda) = k(\lambda, \lambda)$, the function w is as in (6.1), and $\mathbb{S}_{\mathbb{T}}$ is the operator of singular integration. As we have seen, Theorem 6.2 remains true if A is replaced by A_0. The next two theorems concern invertibility of the operator A_0.

Theorem 6.3 *Let A_0 be the main part of the operator associated with the singular integral equation (6.1), and let α and β be given by (6.16). Then A_0 is invertible if and only if α and β do not vanish on the unit circle $\mathbb{T}$ and the winding number of $\gamma = \alpha/\beta$ relative to zero is equal to zero. Let these conditions be satisfied, and assume in addition that α and β are analytic in an annulus containing $\mathbb{T}$. Then γ admits a Wiener-Hopf factorization $\gamma = \gamma_-\gamma_+$ and the inverse of A is given by*

$$A_0^{-1} f = (\gamma_+(\cdot)^{-1}\mathbb{P} + \gamma_-(\cdot)\mathbb{Q})\gamma_-(\cdot)^{-1}\beta(\cdot)^{-1} f, \ f \in L_2(\mathbb{T}),$$

where $\mathbb{P}$ is the projection defined by (6.10) and $\mathbb{Q} = I - \mathbb{P}$.

From (6.16) we see that the additional assumption requiring α and β to be analytic in an annulus containing $\mathbb{T}$ (which appears in the above and in the next theorem) is satisfied if and only if the functions $w(\lambda)$ and $k(\lambda, \lambda)$ are analytic at each point of $\mathbb{T}$. The main reason for putting this additional analyticity requirement is the use of the Wiener-Hopf factorization. If the functions are merely continuous, the Wiener-Hopf factorization may not exist.

Theorem 6.3 appears as a special case of the next theorem.

Theorem 6.4 *Let A_0 be the main part of the operator associated with the singular integral equation (6.1), and let α and β be the functions defined by (6.16). Then A_0 is left or right invertible if and only if α and β do not vanish on the unit circle $\mathbb{T}$. Assume the latter condition holds, and let κ be the winding number of $\gamma = \alpha/\beta$ relative to zero. Then*

   (i) A_0 *is left invertible if and only if $\kappa \geq 0$, and in that case* codim Im $A_0 = \kappa$,

  (ii) A_0 *is right invertible if and only if $\kappa \leq 0$, and in that case* dim Ker $A_0 = -\kappa$.

If, in addition, α and β are analytic in an annulus containing the unit circle, then γ admits a Wiener-Hopf factorization

$$\gamma(\lambda) = \gamma_-(\lambda)\lambda^\kappa \gamma_+(\lambda), \qquad \lambda \in \mathbb{T},$$

and a left or right inverse of A_0 is given by

$$A_0^{-1}f = (\gamma_+(\cdot)^{-1}\mathbb{P} + \gamma_-(\cdot)\mathbb{Q})(\varepsilon(\cdot)^{-\kappa}\mathbb{P} + \mathbb{Q})\gamma_-(\cdot)^{-1}\beta(\cdot)^{-1}f, \quad f \in L_2(\mathbb{T}).$$

Here $\mathbb{P}$ is the orthogonal projection defined by (6.10), the operator $\mathbb{Q} = I - \mathbb{P}$, and $\varepsilon(\lambda) = \lambda$.

Proof: Let U be the unitary operator from $L_2(\mathbb{T})$ onto $\ell_2(\mathbb{Z})$ introduced in the second paragraph of the proof of Theorem 6.2. Then $UA_0U^{-1} = M_{\alpha,\beta}$, where $M_{\alpha,\beta} = L_\alpha P + L_\beta Q$. Here P is the projection defined by (6.18), the operator $Q = I - P$, and L_α and L_β are the Laurent operators on $\ell_2(\mathbb{Z})$ with symbols α and β, respectively. Since $UPU^{-1} = P$, $UQU^{-1} = Q$, and $U\omega(\cdot)U^{-1} = L_\omega$ for any continuous function on ω, Theorem 6.4 immediately follows from Theorem 4.2 when the functions α and β belong to the class $\mathcal{A}$ (that is, are analytic in an annulus containing the unit circle). To complete the proof for the general case, we use the same unitary equivalence and apply Theorem 2.2. $\qquad\qquad\square$

As was noted earlier, this chapter is a continuation of the material in Chapter III. Further developments can be found in the paper [Kre], and in the monographs [GF], [BS], [GGK2]. The requirement that p is finite is not essential in the first four sections. All results in these sections remain true for $p = \infty$ with some minor modifications in the proofs.

Exercises XVI

  1. Let T be the Toeplitz operator on ℓ_p $(1 \leq p < \infty)$ with symbol

$$\alpha(\lambda) = \frac{5a}{2}\lambda^{-1} - \left(a + \frac{5}{2}\right) + \lambda.$$

  Here a is a complex parameter.

   (a) Determine the values of the parameter a with $|a| \neq 1$ for which the equation

$$Tx = 0, \qquad x \in \ell_p,$$

  has a non-trivial solution. Compute all solutions of this homogeneous equation.

(b) For which values of the parameter a is the operator T invertible? Find the inverse of T if it exists.

(c) What is the spectrum of T? Does it depend on p?

2. Let T be the Toeplitz operator on ℓ_p $(1 \le p < \infty)$ with symbol

$$\alpha(\lambda) = \frac{a}{2}\lambda - \left(a + \frac{1}{2}\right) + \lambda^{-1}.$$

Again a is a complex parameter.

(a) For which values of a is the operator T invertible? When is T one-sided invertible?

(b) Solve the nonhomogeneous equation

$$Tx = e_0, \qquad x \in \ell_p,$$

where $e_0 = (1, 0, 0, \ldots)$.

(c) Solve the same problem as in (b) with $y = (1, q, q^2, \ldots)$, $|q| < 1$, in place of e_0.

3. Let T be the Toeplitz operator on ℓ_p $(1 \le p < \infty)$ with symbol $\alpha \in \mathcal{A}$. Assume α admits a Wiener-Hopf factorization

$$\alpha(\lambda) = \alpha_-(\lambda)\lambda^\kappa\alpha_+(\lambda), \qquad \lambda \in \mathbb{T}.$$

(a) When has the homogeneous equation

$$Tx = 0, \qquad x \in \ell_p,$$

a non-trivial solution? Find all solutions of this equations. Hint: use the Taylor expansion of $\alpha_+(\cdot)^{-1}$ at zero.

(b) Prove that Ker T has a basis of the form

$$(\gamma_0, \gamma_1, \ldots, \gamma_{r-1}, \gamma_r, \gamma_{r+1}, \ldots),$$

$$(0, \gamma_0, \ldots, \gamma_{r-2}, \gamma_{r-1}, \gamma_r, \ldots),$$

$$\vdots$$

$$(\underbrace{0, 0, \ldots, 0}_{r}, \gamma_0, \gamma_1, \ldots).$$

Determine the number r.

(c) Solve the nonhomogeneous equation

$$Tx = e_0, \qquad x \in \ell_p,$$

where $e_0 = (1, 0, 0, \ldots)$. Express the solution(s) in terms of the functions $\alpha_\pm(\cdot)^{-1}$.

(d) Solve the same problem as in (c) with $y = (1, q, q^2, \ldots)$, $|q| < 1$, in place of e_0.

4. Let T be the Toeplitz operator on ℓ_p $(1 \le p < \infty)$ with symbol $\alpha \in \mathcal{A}$, and let T' be the Toeplitz operator with symbol $\beta(\lambda) = \alpha(\lambda^{-1})$.

   (a) What is the relation between the matrices of T and T' relative to the standard basis of ℓ_p?

   (b) Assume α admits a Wiener-Hopf factorization, and let $e_0 = (1, 0, 0, \ldots)$. Show that T is invertible if and only if the equations

$$Tx = e_0, \qquad T'x' = e_0$$

   have solutions in ℓ_p.

   (c) Let α be as in (b), and assume the two equations in (b) are solvable in ℓ_p. Express the inverse of T in terms of the solutions x and x'.

5. Let T be a Toeplitz operator with symbol $\alpha \in \mathcal{A}$. Assume that the two equations in (b) of the previous exercise have solutions in ℓ_1. Show that $\alpha(\lambda) \ne 0$ for each $\lambda \in \mathbb{T}$ and that relative to zero the winding number of the oriented curve $t \to \alpha(e^{it})$ with t running from $-\pi$ to π is equal to zero.

6. Let U and U^+ be the operators on ℓ_p $(1 \le p < \infty)$ defined in the first exercise to Chapter XV, i.e.,

$$U(x_0, x_1, x_2 \ldots) = (\underbrace{0, 0, \ldots, 0}_{r}, x_k, x_{k+1}, \ldots),$$

$$U^+(x_0, x_1, x_2 \ldots) = (\underbrace{0, 0, \ldots, 0}_{k}, x_r, x_{r+1}, \ldots),$$

Consider the operator

$$A = \sum_{j=1}^{\infty} a_{-j}(U^+)^j + a_0 I + \sum_{j=1}^{\infty} a_j U^j,$$

where a_j $(j \in \mathbb{Z})$ are complex numbers such that $|a_n| \le c\rho^{|n|}$, $n \in \mathbb{Z}$, for some $c \ge 0$ and $0 \le \rho < 1$. Assume $k \ne r$.

(a) Show that A is Fredholm if and only if

$$\alpha(\lambda) = \sum_{j=-\infty}^{\infty} a_j \lambda^j \ne 0 \qquad (|\lambda| = 1).$$

(b) If the latter condition on α holds, show that ind $A = (k-r)m$, where m is the winding number relative to zero of the oriented curve $t \mapsto \alpha(e^{it})$ with t running from $-\pi$ to π.

7. Fix $0 \le t < \infty$. Let V be the operator on $L_p[0, \infty)$, $1 \le p < \infty$, defined by

$$(Vf)(x) = f(x+t), \qquad x \ge 0.$$

Also consider the operator V^+ on $L_p(0, \infty)$ given by

$$(V^+ f)(x) = \begin{cases} f(x-t) & \text{for} \quad x \ge t, \\ 0 & \text{for} \quad 0 \le x < t. \end{cases}$$

(a) Compute the operators V^+V and VV^+.

(b) Let a be a complex parameter, $|a| \ne 1$. Solve in $L_p[0, \infty)$ the homogeneous equation

$$\frac{5a}{2} V^+ f - \left(a + \frac{5}{2} \right) f + Vf = 0.$$

8. Let V and V^+ be as in the previous exercise, and consider the operator

$$A = \sum_{j=1}^{\infty} a_{-j} (V^+)^j + a_0 I + \sum_{j=1}^{\infty} a_j V^j,$$

where a_j $(j \in \mathbb{Z})$ are as in Exercise 6. Solve in $L_p[0, \infty)$, $1 \le p < \infty$, the equation $Af = 0$ under the assumption that $\sum_{j=-\infty}^{\infty} a_j \lambda^j \ne 0$ for each $\lambda \in \mathbb{T}$.

9. Solve the same problem as in the previous exercise with V and V^+ being replaced by the following operators:

$$(Vf)(x) = f(ax), \qquad x \ge 0,$$

$$(V^+ f)(x) = \begin{cases} f(x/a), & x > 1, \\ 0, & 0 \le x \le 1. \end{cases}$$

Here a is a fixed positive integer.

10. Let $M = L_\alpha P + L_\beta Q$ be the pair operator on $\ell_p(\mathbb{Z})$, $1 \le p < \infty$, with

$$\alpha(\lambda) = \lambda^{-1} - \eta, \quad \beta(\lambda) = \lambda - \frac{5}{2}.$$

Here η is a complex parameter.

(a) For which values of η is the operator M invertible? Find the inverse if it exists.

(b) Determine the spectrum of M.

(c) When is M left or right invertible?

(d) Let $|\eta| \neq 1$. Solve in $\ell_p(\mathbb{Z})$ the equation $Mx = 0$. Also, solve $Mx = e_0$, with $e_0 = (1, 0, 0, \ldots)$.

11. Do the previous exercise with

$$\alpha(\lambda) = \lambda^{-1} - \frac{2}{5}, \qquad \beta(\lambda) = \lambda - \eta.$$

12. Let $K = PL_\alpha + QL_\beta$ be the associate pair operator on $\ell_p(\mathbb{Z})$, $1 \le p < \infty$, with α and β as in Exercise 10.

   (a) For which values of η is the operator K invertible? Find the inverse if it exists.

   (b) Determine the spectrum of K.

   (c) When is K left or right invertible?

   (d) Let $|\eta| \neq 1$. Solve in $\ell_p(\mathbb{Z})$ the equation $Kx = 0$. Also, solve $Kx = e_0$, with $e_0 = (1, 0, 0, \ldots)$.

13. Do the previous exercise with α and β as in Exercise 11.

14. Let A be the operator on $L_2(\mathbb{T})$ associated with the singular integral equation

$$\eta f(\lambda) + \frac{1}{\pi i} \int_{\mathbb{T}} \frac{\lambda z}{z - \lambda} f(z) \, dz = g(\lambda), \qquad \lambda \in \mathbb{T}.$$

Here η is a complex number. For which values of η is the operator A Fredholm? For these values of η compute the index of A.

Chapter XVII
Non Linear Operators

Linear operators are the simplest operators. In many problems one has to consider more complicated nonlinear operators. As in the case of linear operators, again the main problem is to solve equations $Ax = y$ for a nonlinear A in a Hilbert or Banach space. Geometrically, this problem means that a certain map or operator B leaves fixed at least one vector x, i.e.,

$$x = Bx, \quad \text{where} \quad Bx = x + Ax - y,$$

and we have to find this vector. Theorems which establish the existence of such fixed vectors are called *fixed point* theorems. There are a number of very important fixed point theorems. In this chapter we present one of the simplest; the Contraction Mapping Theorem. This theorem is very powerful in that it allows one to prove the existence of solutions to nonlinear integral, differential and functional equations, and it gives a procedure for numerical approximations to the solution. Some of the applications are also included in this chapter.

17.1 Fixed Point Theorems

A function f which maps a set S into S is said to have a *fixed point* if there exists an $s \in S$ such that $f(s) = s$.

Contraction Mapping Theorem 1.1 *Let S be a closed subset of a Banach space and let T map S in to S. Suppose there exists a number $\alpha < 1$ such that for all x, y in S,*

$$\|Tx - Ty\| \le \alpha \|x - y\|. \tag{1.1}$$

Then T has a unique fixed point in S.

Proof: Given an arbitrary vector $x_0 \in S$, let

$$x_n = T^n x_0, \quad n = 1, 2, \dots.$$

We shall show that $\{x_n\}$ converges to a fixed point of T. For convenience, we write $d(x, y)$ instead of $\|x - y\|$.

By hypothesis and the definition of x_n,

$$d(x_n, x_{n-1}) = d(Tx_{n-1}, Tx_{n-2}) \le \alpha d(x_{n-1}, x_{n-2})$$
$$\le \alpha^2 d(x_{n-2}, x_{n-3}) \le \cdots \le \alpha^{n-1} d(x_1, x_0).$$

Hence for $n > m$,

$$d(x_n, x_m) \leq d(x_m, x_{m+1}) + d(x_{m+1}, x_{m+2}) + \cdots + d(x_{n-1}, x_n)$$

$$\leq d(x_1, x_0) \sum_{k=m}^{\infty} \alpha^k = d(x_1, x_0) \frac{\alpha^m}{1 - \alpha} \to 0$$

as $n, m \to \infty$. Since S is a closed subset of a Banach space, $\{x_n\}$ converges to some $x \in S$. Clearly, T is continuous. Hence we have $x_n \to x$ and $x_{n+1} = Tx_n \to Tx$, which implies that $Tx = x$.

If z is also a fixed point of T, then

$$d(z, x) = d(Tz, Tx) \leq \alpha d(z, x)$$

which can only be if $d(z, x) = 0$ or, equivalently, $z = x$. $\square$

The operator T in Theorem 1.1 is called a *contraction*.

17.2 Applications of the Contraction Mapping Theorem

In this section the contraction mapping theorem is used to prove the existence and uniqueness of solutions to certain non linear integral and differential equations. In addition, we give a proof of the implicit function theorem.

Theorem 2.1 *Let k be continuous on $[a, b] \times [a, b] \times \mathbb{C}$. Suppose there exists a number m such that*

$$|k(t, s, \xi) - k(t, s, \xi')| \leq m|\xi - \xi'|$$

for all $\xi, \xi' \in \mathbb{C}$. Then for $|\lambda| < \frac{1}{m(b-a)}$, the equation

$$f(t) - \lambda \int_a^b k(t, s, f(s)) \, ds = g(t) \tag{2.1}$$

has a unique solution in $C([a, b])$ for every $g \in C([a, b])$.

Proof: Define T on $C([a, b])$ by

$$(Tf)(t) = \lambda \int_a^b k(t, s, f(s)) \, ds + g(t).$$

It is easy to see that $\operatorname{Im} T \subset C([a, b])$. T is a contraction. Indeed, for all f, $h \in C([a, b])$,

$$|(Tf)(t) - (Th)(t)| \leq |\lambda| \int_a^b |k(t, s, f(s)) - k(t, s, h(s))| \, ds$$

$$\leq |\lambda| \int_a^b m|f(s) - h(s)| \, ds \leq |\lambda| m(b - a)\|f - h\|.$$

Thus

$$\|Tf - Th\| \leq |\lambda| m(b - a) \|f - h\|,$$

and $|\lambda| m(b - a) < 1$. Hence T has a unique fixed point $f_0 \in C([a, b])$ by Theorem 1.1. Obviously, f_0 is the solution to (2.1).

Equation (1.1) includes linear integral equations of the second kind. For given $k_0(t, s)$, let $k(t, s, \xi) = k_0(t, s)\xi$.

Then

$$\int_a^b k(t, s, f(s)) \, ds = \int_a^b k_0(t, s) f(s) \, ds.$$

$\square$

Theorem 2.2 *Let f be a complex-valued function which is continuous and bounded on some open subset 0 of the plane. Suppose there exists a number C such that*

$$|f(x, y) - f(x, z)| \leq C|y - z| \tag{2.2}$$

for all (x, y), (x, z) in 0. Then for any $(x_0, y_0) \in 0$, the differential equation

$$\frac{dy}{dx} = f(x, y(x)) \tag{2.3}$$

with intitial condition

$$y(x_0) = y_0 \tag{2.4}$$

has a unique solution $y = y(x)$ on some imterval containing x_0.

Proof: Equation (2.3) together with (2.4) is equivalent to the equation

$$y(x) = y_0 + \int_{x_0}^x f(t, y(t)) \, dt. \tag{2.5}$$

We now show that (2.5) has a unique solution on some interval containing x_0. To do this, we define a contraction map as follows. Let $M = \sup_{(x,y)\in 0} |f(x, y)| < \infty$. Choose $\rho > 0$ such that

$$C\rho < 1, \quad |x - x_0| \leq \rho \quad \text{and} \quad |y - y_0| \leq M\rho \quad \text{imply} \quad (x, y) \in 0.$$

Let S be the set of complex valued functions y which are continuous on the interval $J = \{x : |x - x_0| \leq \rho\}$ and have the property that

$$|y(x) - y_0| \leq M\rho.$$

It is clear that S is a closed subset of the Banach space $C(J)$. Define T on S by

$$(Tg)(x) = y_0 + \int_{x_0}^x f(s, g(s)) \, ds.$$

Now $TS \subset S$ since $TS \subset C(J)$ and for $g \in S$,

$$|(Tg)(x) - y_0| = \left| \int_{x_0}^x f(s, g(s)) \, ds \right| \leq M\rho, \quad x \in J.$$

Furthermore, T is a contraction. Indeed, given $g, h \in S$, (2.2) implies

$$|(Tg)(x) - (Th)(x)| = \left| \int_{x_0}^x f(s, g(s)) - f(s, h(s)) \right| \leq C\rho \|g - f\|,$$

whence

$$\|Tg - Th\| \leq C\rho \|g - h\|, \quad C\rho < 1.$$

Therefore, there exists a unique $y \in S$ such that $Ty = y$, i.e., (2.5) holds.

If y_1 is also a complex valued function on J which is a solution to (2.3) with initial condition (2.4), then y_1 is in $C(J)$ and is a solution to (2.5). Hence $|y_1(x) - y_0| \leq M\rho$ for all $x \in J$. Thus y_1 is in S and y_1 is a fixed point of T. Therefore, $y = y_1$.

$\square$

Implicit Function Theorem 2.3 *Let f be a real valued function which is continuous on some set Ω in the plane. Suppose $\frac{\partial f}{\partial y}$ exists at each point in Ω and is continuous at some point $(x_0, y_0) \in \Omega$. If*

$$\frac{\partial f}{\partial y}(x_0, y_0) \neq 0$$

and

$$f(x_0, y_0) = 0,$$

then there exists a rectangle $R : [x_0 - \delta, x_0 + \delta] \times [y_0 - \varepsilon, y_0 + \varepsilon]$ contained in Ω and a unique $y(x) \in C([x_0 - \delta, x_0 + \delta])$ such that $(x, y(x)) \in R$, $f(x, y(x)) = 0$ for all $x \in [x_0 - \delta, x_0 + \delta]$ and $y(x_0) = y_0$.

Proof: We shall show that for a suitable set of functions, the operator

$$(T\varphi)(x) = \varphi(x) - \frac{1}{m} f(x, \varphi(x)), \quad m = \frac{\partial f}{\partial y}(x_0, y_0) \tag{2.6}$$

has a fixed point.

It follows from the conditions on f that there exist $\varepsilon > 0$ and $\delta > 0$ such that the rectangle $R = \{(x, y) : |x - x_0| \leq \delta, |y - y_0| \leq \varepsilon\}$ is contained in Ω,

$$\left| \frac{1}{m} \frac{\partial f}{\partial y}(x, y) - 1 \right| < \frac{1}{2}, \quad (x, y) \in R, \tag{i}$$

and

$$\left| \frac{1}{m} f(x, y_0) \right| < \frac{1}{2}\varepsilon, \quad |x - x_0| \leq \delta. \tag{ii}$$

Let $J = \{x : |x - x_0| \le \delta\}$ and define

$$S = \{g \in C(J) : g(x_0) = y_0, |g(x) - y_0| \le \varepsilon, \quad x \in J\}.$$

Clearly, S is a closed subset of the Banach space $C(J)$. Define T on S by (2.6). We shall show that $TS \subset S$ and that T is a contraction. Now $TS \subset C(J)$ and for $(x, y) \in R$, we have from (i) that

$$\left| \frac{\partial f}{\partial y}\left(y - \frac{1}{m} f(x, y) \right) \right| = \left| 1 - \frac{1}{m} \frac{\partial f}{\partial y}(x, y) \right| < \frac{1}{2}.$$

Hence it follows from the Mean Value Theorem applied to the function $y - \frac{1}{m} f(x, y)$ that for g and h in S,

$$|(Tg)(x) - (Th)(x)| \le \frac{1}{2}|g(x) - h(x)|. \tag{iii}$$

Thus

$$\|Tg - Th\| \le \frac{1}{2}\|g - h\|.$$

To show that T has a fixed point in S, it remains to prove that $TS \subset S$. Given $g \in S$,

$$(Tg)(x_0) = g(x_0) - \frac{1}{m} f(x_0, g(x_0)) = y_0 - \frac{1}{m} f(x_0, y_0) = y_0.$$

Define $y_0(x) = y_0, x \in J$. From (iii), (ii) and the definition of S, we get

$$|(Tg)(x) - y_0| \le |(Tg)(x) - (Ty_0)(x)| + |(Ty_0)(x) - y_0|$$
$$\le \frac{1}{2}|g(x) - y_0| + \left| \frac{1}{m} f(x, y_0) \right| < \varepsilon.$$

Thus $Tg \in S$, which proves that $TS \subset S$. Hence T has a unique fixed point $y \in S$. Clearly, y satisfies the conclusions of the theorem. If y_1 also satisfies the conclusions, then y_1 is in S and $Ty_1 = y_1$. Therefore, $y_1 = y$.
For any $\varphi \in S$, the sequence $\{T^n \varphi\}$ converges uniformly on J to $y(x)$. $\square$

17.3 Generalizations

The function d defined on $S \times S$ by $d(x, y) = \|x - y\|$ has the following properties.

  (i) $d(x, y) > 0$ if $x \ne y$; $d(x, x) = 0$.
  (ii) $d(x, y) = d(y, x)$.
  (iii) $d(x, y) + d(y, z) \ge d(x, z)$.

An arbitrary set S, together with a function d defined on $S \times S$ with properties (i), (ii) and (iii), is called a *metric* space. The metric space (S, d) is called *complete* if $\lim_{m,n\to\infty} d(x_n, x_m) = 0$ implies the existence of an $x \in S$ such that $d(x_n, x) \to 0$.

The proof of the contraction mapping theorem shows that this theorem holds for complete metric spaces.

Theorem 3.1 *Suppose S is a complete metric space and $T : S \to S$ has the property that T^n is a contraction for some positive integer n. Then T has a unique fixed point.*

Proof: Since T^n is a contraction, it has a unique fixed point $x \in S$. Now

$$T^n T x = T T^n x = T x,$$

i.e., Tx is also a fixed point of T^n. But T^n has only one fixed point. Hence $Tx = x$. Any other fixed point of T is also a fixed point of T^n. Thus the fixed point of T is unique. $\qquad\square$

If we remove the requirement in Theorem 1.1 that T is a contraction and replace it by the weaker condition

$$\|Tx - Ty\| < \|x - y\|, \quad \text{for all} \quad x, y \in S, \quad x \neq y,$$

then T need not have a fixed point. However, if, in addition, S is compact, then T has a unique fixed point.

Definition: A subset S of a normed linear space is *compact* if every sequence in S has a subsequence which converges to a vector in S.

Theorem 3.2 *Let S be a compact subset of a normed linear space and let T map S into S. If*

$$\|Tx - Ty\| < \|x - y\| \quad \text{for all} \quad x, y \quad \text{in} \quad S, \quad x \neq y, \qquad (3.1)$$

then T has a unique fixed point in S.

Proof: Let $m = \inf_{x\in S} \|Tx - x\|$. There exists a sequence $\{x_n\}$ in S such that $\|Tx_n - x_n\| \to m$. Since S is compact, $\{x_n\}$ has a subsequence $\{x_{n'}\}$ which converges to some $x \in S$. Hence

$$\|x - Tx\| = \lim_{n'\to\infty} \|x_{n'} - Tx_{n'}\| = m.$$

Therefore $Tx = x$, otherwise

$$m \leq \|T^2 x - Tx\| < \|Tx - x\| = m.$$

It is obvious from (3.1) that T cannot have more than one fixed point. $\qquad\square$

The proof shows that the theorem can be extended to compact metric spaces.

If the set S in the Theorem 3.2 is also convex, then we can weaken the inequality (3.1) and obtain the following result.

Theorem 3.3 *Let S be a compact convex subset of a normed linear space and let T map S into S. If*

$$\|Tx - Ty\| \leq \|x - y\| \quad \text{for all} \quad x, y \quad \text{in} \quad S,$$

then T has a fixed point.

Proof: First we assume that $0 \in S$. For $n = 1, 2, \ldots$, define T_n on S by $T_n x = (1 - \frac{1}{n})x$. Since S is convex and $0 \in S$, $T_n x = (1 - \frac{1}{n})x + \frac{1}{n}.0 \in S$. Clearly, T_n is contraction. Hence by Theorem 1.1, there exists an $x_n \in S$ such that

$$\left(1 - \frac{1}{n}\right) T x_n = T_n x_n = x_n, \quad n = 1, 2, \ldots. \tag{3.2}$$

Since S in compact, $\{x_n\}$ has a subsequence $\{x_{n'}\}$ which converges to some $x \in S$. Thus $T x_{n'} \to Tx$ and (3.2) implies

$$Tx = \lim_{n' \to \infty} T x_{n'} = \lim_{n \to \infty} x_{n'} = x.$$

If $0 \notin S$, choose any $x_0 \in S$ and let $S_0 = -x_0 + S$. The theroem follows from the result we just proved applied to S_0 and the operator $T_0 : S_0 \to S_0$ defined by

$$T_0(-x_0 + x) = -x_0 + Tx, \quad x \in S. \qquad \square$$

Appendix 1
Countable Sets and Separable Hilbert Spaces

This appendix presents an alternative definition of separability of a Hilbert space. We start with some preliminary results about countable sets.

Definition: A set S is *countable* if it is either finite or can be put into one to one correspondence with the positive integers. Thus an infinite set is countable if there exists a sequence whose terms consist of all the members of S.

(i) *If S is a countable set, then the set consisting of all finite sequences of members of S is also countable.*

Indeed, since S can be put into one to one correspondence with a subset of positive integers it suffices to prove that the set Z which consists of all finite sequences of positive integers is countable.

For each positive integer n, let Z_n be the set of those $(m_1, \ldots, m_k) \in Z$ for which $\sum_{i=1}^{k} m_i = n$. Since Z_n is a finite set, the members of Z_n may be listed as we please. Thus we can list all the members of $Z = U_{n=1}^{\infty} Z_n$ by listing the members of Z_1, then Z_2, etc.

(ii) *The set of all finite sequences of rational complex numbers* (the real and imaginary parts of the number are rational) *is countable.* Indeed, by identifying the complex number $\frac{p_1}{q_1} + i\frac{p_2}{q_2}$ with (p_1, q_1, p_2, q_2), where p_j and q_j are integers, it follows from (i) that the set of finite sequences of rational complex numbers is a subset of a countable set and therefore is countable.

(iii) *The set of all polynomials with rational complex coefficients is countable* since we can identify the polynomial $a_n x^n + \cdots + a_1 x + a_0$ with $(a_n, \ldots, a_1, a_0)$.

(iv) *The set of real numbers is uncountable.* Indeed, it suffices to show that any sequence of real numbers in $[0, 1]$ does not contain some real number in $[0, 1]$.

Suppose $s_1, s_2, \ldots$ is a sequence of real numbers. Each s_i has the decimal expansion $s_i = a_1^{(i)} a_2^{(i)}, \ldots$, where infinitely many of the integers $a_1^{(i)}, a_2^{(i)}, \ldots$ are not 9. Let $s = \cdot a_1 a_2 \ldots$, where $a_k = 0$ if $a_k^{(k)} \neq 0$ and 1 if $a_k^{(k)} = 0$. Then s is in $[0, 1]$ but $s \neq s_i$ for every i.

Theorem: *A Hilbert space $\mathcal{H}$ is separable if and only if $\mathcal{H}$ contains a countable set which is dense in $\mathcal{H}$.*

Proof: If $\{w_1, w_2, \ldots\}$ is dense in $\mathcal{H}$, then obviously $\mathrm{sp}\{w_1, w_2, \ldots\}$ is dense in $\mathcal{H}$. Therefore, by definition, $\mathcal{H}$ is separable. On the other hand, if $\mathcal{H}$ is separable, then there exist vectors $v_1, v_2, \ldots$ such that $\mathcal{H} = \overline{\mathrm{sp}}\{v_1, v_2, \ldots\}$. Hence the set S of all linear combinations of the form $\sum_{i=1}^{n} \alpha_i v_i$, where each α_i is a complex rational number, is dense in $\mathcal{H}$. The countability of S now follows from (ii). $\square$

Corollary: *Every orthonormal set in a separable inner product space is countable.*

Proof: Let Ω be an orthonormal set in an inner product space E and let $S = \{x_1, x_2, \ldots\}$ be dense in E. We define an integer valued function f on Ω as follows:

For each $\varphi \in \Omega$, choose an integer j such that $\|\varphi - x_j\| < 1/2$ and define $f(\varphi) = j$. Now if ψ is in Ω and $\psi \neq \varphi$, then $\|\varphi - \psi\| = \sqrt{2}$, which implies that $f(\varphi) \neq f(\psi)$. Hence Ω is in one to one correspondence with a subset of the positive integers and therefore is countable. $\square$

Appendix 2
The Lebesgue Integral and L_p Spaces

In this section we give a very brief introduction to the Lebesgue integral which is necessary for the description and understanding of the spaces L_p.

A.2.1 Lebesgue Integral

A subset Z of the real line is said to have *Lebesgue measure zero* if for each $\varepsilon > 0$, there exists a countable set of intervals $I_1, I_2, \ldots$ such that $Z \subset U_j I_j$ and $\sum_j \mu(I_j) < \varepsilon$, where $\mu(I_j)$ is the length of I_j.

Every countable subset $\{x_1, x_2, \ldots\}$ of the line has Lebesgue measure zero. Indeed, given $\varepsilon > 0$, take $I_j = [x_j, x_j + \varepsilon/2^{j+1}]$.

A real valued function f which is defined on an interval J is called a *step function* if $f(x) = \sum_{k=1}^n \alpha_k C_{I_k}(x)$, where α_k is a real number, $I_1, \ldots, I_n$ are mutually disjoint subintervals, and $C_{I_k}(x) = 1$ if $x \in I_k$ and zero otherwise. The step function f is *Lebesgue integrable* if $\sum_{k=1}^n \alpha_k \mu(I_k) < \infty$, $(0 \cdot \infty = 0)$. In this case, we define the integral

$$\int_J f(x)\, dx = \sum_{k=1}^n \alpha_k \mu(I_k).$$

A non negative real valued function f defined on the interval J is *Lebesgue measurable* if there exists a nondecreasing sequence $\{f_n\}$ of step functions defined on J such that $f_n(x) \to f(x)$ for all $x \in J$ which lie outside some set of Lebesgue measure zero (possibly the empty set). We then say that $\{f_n\}$ converges to f *almost everywhere* (a.e.). If each step function f_n is Lebesgue integrable, then $\int_J f_1(x)\, dx \le \int_J f_2(x)\, dx \le \ldots$. The function f is *Lebesgue intergrable* if $\lim_{n \to \infty} \int_J f_n(x)\, dx < \infty$. The *Lebesgue integral* of f is defined by

$$\int_J f(x)\, dx = \lim_{n \to \infty} \int_J f_n(x)\, dx.$$

It turns out that the integral is independent of the choice of the increasing sequence $\{f_n\}$.

Given a real valued function f defined on J, define

$$f^+(x) = \max(f(x), 0), \quad f^-(x) = \max(-f(x), 0).$$

Then $f = f^+ - f^-$, $f^+ \geq 0$, $f^- \geq 0$. The function f is Lebesgue integrable if both f^+ and f^- are Lebesgue integrable. In this case we define

$$\int_J f(x) \, dx = \int_J f^+(x) \, dx - \int_J f^-(x) \, dx.$$

If f is a complex valued function on J, say $f = f_1 + if_2$, where f_1 and f_2 are real valued functions, then f is Lebesgue integrable if f_1 and f_2 are Lebesgue integrable, in which case we define

$$\int_J f(x) \, dx = \int_J f_1(x) \, dx + i \int_J f_2(x) \, dx.$$

A bounded Lebesgue measurable function is Lebesgue integrable. A real valued function which is Riemann integrable is Lebesgue integrable and the two integrals coincide. However, a Lebesgue integrable function need not be Riemann integrable. For example, if f is zero on the rationals and 1 on the irrationals in $[0, 1]$, then f is not Riemann integrable, but it is Lebesgue integrable with $\int_0^1 f(x) \, dx = 1$.

The following theorem is fundamental to the theory of Lebesgue integration.

Lebesgue Dominated Convergence Theorem. *Suppose $\{f_n\}$ is a sequence of Lebesgue integrable functions which converges a.e. to f on the interval J. If there exists a Lebesgue integrable function g such that*

$$|f_n(x)| \leq g(x) \quad a.e. \text{ on } J, \quad n = 1, 2, \ldots,$$

then f is Lebesgue integrable and

$$\int_J f_n(x) \, dx \mapsto \int_J f(x) \, dx.$$

The theory of Lebesgue integration of complex valued or real valued functions defined on a rectangle is very similar to the theory of Lebesgue integration of functions defined on an interval. One need only replace, in the discussion above, intervals by rectangles and lengths by areas. The integral of f over a rectangle R is written $\int_R f(x, y) \, dx \, dy$.

Fubini's theorem enables one to express the integral as an iterated integral.

Fubini's Theorem. *Let f be Lebesgue integrable on the rectangle R: $[a, b] \times [c, d]$. Then for almost every $x \in [a, b]$, the function $f(x, \ )$ is Lebesgue integrable on $[c, d]$ and $\int_c^d f(x, y) \, dy$ is integrable on $[a, b]$. Moreover,*

$$\int_a^b \left\{ \int_c^d f(x, y) \, dy \right\} dx = \int_R f(x, y) \, dx \, dy.$$

Similarly, for almost every $y \in [c, d]$, the function $f(\, , y)$ is Lebesgue integrable on $[a, b]$ and $\int_a^b f(x, y)\, dx$ is Lebesgue integrable on $[c, d]$. Moreover,

$$\int_c^d \left\{ \int_a^b f(x, y)\, dx \right\} dy = \int_R f(x, y)\, dx\, dy.$$

A.2.2 L_p Spaces

A complex valued Lebesgue measurable function defined on an interval J is said to be in $L_p(J)$, $1 \le p < \infty$, if $|f|^p$ is Lebesgue integrable. With the usual definition of addition and scalar multiplication of functions, $L_p(J)$ is a vector space. If we define

$$\|f\|_p = \left(\int_J |f(x)|^p\, dx \right)^{1/p},$$

the $\|\cdot\|_p$ has all the properties of a norm except that $\|f\|_p = 0$ only implies $f = 0$ a.e. To remedy this situation, functions are identified which are equal almost everywhere. To be specific, we decompose $L_p(J)$ into disjoint sets, called *equivalence classes*, as follows. For each $f \in L_p(J)$,

$$[f] = \{g \in L_p(J) : f = g \quad \text{a.e.}\}.$$

Then, either $[f] = [h]$ or $[f] \cap [h] = \emptyset$. The set of equivalence classes becomes a vector space under the operations

$$[f] + [h] = [f + h], \quad \alpha[f] = [\alpha f].$$

Moreover,

$$\|[f]\|_p = \left(\int_J |g(x)|^p\, dx \right)^{1/p},$$

where g is arbitrary in $[f]$, defines a norm on this vector space. For the sake of simplicity, we usually do not distinguish between functions and equivalence classes.

It is a very important fact that $L_p(J)$ is complete. For a detailed proof we refer the reader to [R].

A complex valued Lebesgue measurable f defined on J is called *essentially bounded* if there exists a number M such that $|f(x)| \le M$ a.e. The greatest lower bound of all such M is denoted by $\|f\|_\infty$. If functions which are equal almost everywhere are identified as above, then $\|\cdot\|_\infty$ is a norm on the vector space $L_\infty(J)$ of essentially bounded functions and $L_\infty(J)$ is a Banach space.

The following inequality is essential for our treatment of integral operators on L_p spaces.

Hölder's Inequality. If $f \in L_p(J)$ and $g \in L_q(J)$, where $1 \leq p \leq \infty$ and $\frac{1}{p} + \frac{1}{q} = 1(\frac{1}{\infty} = 0)$, then $f \cdot g$ is in $L_1(J)$ and

$$\|fg\|_1 \leq \|f\|_p \|g\|_q.$$

Equality holds if and only if there exists some non-zero α, β in $\mathbb{C}$ such that $\alpha|f|^p = \beta|g|^q$ a.e.

Suggested Reading

1. Akhiezer, N.I., and Glazman, I.M., *Theory of Linear Operators in Hilbert Space*, vol. I (1961) and vol. II (1963) Ungar, New York.

2. Böttcher, A., and Gudsky, S.M., *Toeplitz matrices, asymptotic linear algebra, and functional analysis*, Birkhäuser Verlag, Basel, 2000.

3. Böttcher, A., and Silbermann, B., *Analysis of Toeplitz operators*, Springer-Verlag, Berlin, 1990.

4. Douglas, R.G., *Banach Algebra Techniques in Operator Theory*, Academic Press, New York, 1972.

5. Dunford, N., and Schwartz, J.T., *Linear Operators, Part I: General Theory* (1958) *Part II: Spectral Theory* (1963), Interscience, New York.

6. Gelfand, I.M., Functional Analysis, in: *Mathematics – Its Content, Methods and Meaning*, 2nd ed., vol. 3, M.I.T. Press, Cambridge, 1969, 227–261.

7. Gohberg, I.G., and Krein, M.G., *Introduction to the Theory of Linear Non-Self Adjoint Operators in Hilbert Space*, Translations, Math. Monographs, vol. 18, Amer. Math. Soc., Providence, 1969.

8. Goldberg, S., *Unbounded Linear Operators*, McGraw-Hill, New York, 1966.

9. Halmos, P.R., *A Hilbert Space Problem Book*, Van Nostrand, Princeton, 1967.

10. Kato, T., *Perturbation Theory for Linear Operators*, 2nd ed., Springer Verlag, New York, 1976.

11. Lax, P.D., *Functional Analysis*, Wiley-Interscience, New York, 2002.

12. Riesz, F., and Sz.-Nagy, B., *Leçons d'Analyse Fonctionnelle*, Académie des Sciences de Honguè, 1955.

13. Rudin, W., *Functional Analysis*, McGraw-Hill, New York, 1973.

14. Schechter, M. *Principles of Functional Analysis*, Amer. Math. Soc., Providence, R.I., 2002.

15. Shilov, G.E., *An Introduction to the Theory of Linear Spaces*, Dover Pub., New York, 1974.

16. Sz.-Nagy, B., and Foias, C., *Harmonic Analysis of Operators on Hilbert Space*, North-Holland Publ. Co., Amsterdam-Budapest, 1970.

17. Taylor, A.E., and Lay, D.C., *Introduction to Functional Analysis, 2nd* ed., Wiley, New York, 1980.

18. Weidmann, J., *Linear Operators in Hilbert Spaces*, Springer Verlag, New York, 1980.

19. Zaanew, A.C., *Linear analysis*, North-Holland Publ. Co., Amsterdam, 1956.

References

[A]   S.S. Antman, The equations for large vibrations of strings, *Amer. Math. Monthly* **87** (1980), 359–370.

[Ah]   L.V. Ahlfors, *Complex Analysis*, McGraw-Hill, 1966.

[BS]   A. Böttcher and B. Silbermann, *Analysis of Toeplitz Operators*, Springer-Verlag, Berlin, 1990.

[CH]   R. Courant and D. Hilbert, *Methods of Mathematical Physics*, vol. I, Interscience, New York, 1953.

[DS1]   N. Dunford and J.T. Schwartz, *Linear Operators*, Part I: *General Theory*, Interscience, New York, 1958.

[DS2]   N. Dunford and J.T. Schwartz, *Linear Operators*, Part II: *Spectral Theory*, Interscience, New York, 1963.

[E]   P. Enflo, A counterexample to the approximation problem in Banach spaces, *Acta Math.* **130** (1973), 309–317.

[F]   I.A. Fel'dman, Some remarks on convergence of the iterative method, *Izvestia Akad. Nauk Mold. SSR* **4** (1966), 94–96.

[G]   S. Goldberg, *Unbounded Linear Operators*, McGraw-Hill, New York, 1966.

[GF]   I. Gohberg and I. Feldman, *Convolution Equations and Projection Methods for their Solution*, Transl. Math. Monograph, vol. 41, Amer. Math. Soc., Providence, R.I., 1974.

[GG]   I. Gohberg and S. Goldberg, *Basic Operator Theory*, Birkhäuser, Basel, 1981.

[GGKa1] I. Gohberg, S. Goldberg and M.A. Kaashoek, *Classes of Linear Operators, vol.* I, Birkhäuser, Basel, 1990.

[GGKa2] I. Gohberg, S. Goldberg and M.A. Kaashoek, *Classes of Linear Operators, vol.* II, Birkhäuser, Basel, 1993.

[GGKr]   I. Gohberg, S. Goldberg and N. Krupnik, *Traces and determinants of linear operators*, Birkhäuser, Basel, 2000.

[GKre]   I. Gohberg and M.G. Kreĭn, *Introduction to the theory of linear non-selfadjoint operators*, Transl. Math. Monographs, vol. 18, Amer. Math. Soc., Providence, R.I., 1969.

[GKre1] I. Gohberg and M.G. Krein, The basic propositions on defect numbers, root numbers and indices of linear operators, *Uspekhi Math. Nauk* 12, 2(74) (1957), 43–118 (Russian); English Transl., *Amer. Math. Soc. Transl.* (Series 2) **13** (1960), 185–265.

[H]   P.R. Halmos, *Finite-dimensional Vector Spaces,* 2nd ed., Van Nostrand, Princeton, 1958.

[K]   T. Kato, *Perturbation Theory for Linear Operators,* 2nd ed., Springer Verlag, New York, 1976.

[Kra]   M.A. Krasnosel'skii, Solving linear equations with selfadjoint operators by iterative method, *Uspehi – Matem. Nauk,* 15, 3 (1960), 161–165.

[Kre]   M.G. Krein, Integral equations on a half-line with kernel depending upon the difference of the arguments, *Uspekhi Math. Nauk* 13 (5) (1958), 3–120 (Russian); English Transl., *Amer. Math. Soc. Transl.* (Series 2) **22** (1962), 163–288.

[P]   H. Poincaré, Sur les déterminants d'ordre infini, *Bull. Soc. Math. France* **14** (1886), 77–90.

[PS]   C. Pearcy and A.L. Shields, A survey of the Lomonosov technique in the theory of invariant subspaces, *Topics in Operator Theory,* Mathematical Surveys, No. 13, American Mathematical Society, Providence, 1974.

[R]   H.L. Royden, *Real Analysis,* 2nd ed., Macmillan, New York, 1968.

[S]   G. Strang, *Linear Algebra and Its Applications,* 2nd ed., Academic Press, New York, 1980.

[Sc]   M. Schechter, *Principles of functional analysis,* Amer. Math. Soc., Providence, R.I., 2002.

[TL]   A.E. Taylor and D.C. Lay, *Introduction to Functional Analysis,* 2nd ed., Wiley, New York, 1980.

[W]   R. Whitley, Projecting m onto c_0, *Amer. Math. Monthly,* **73** (1966), 285–286.

List of Symbols

A^* adjoint of the operator A, 77
A' conjugate of the operator A, 307
$A|M$ restriction of the operator A to M
$A(\mathcal{H}_1 \to \mathcal{H}_2)$ operator A with domain a subspace of $\mathcal{H}_1$ and
    range a subspace of $\mathcal{H}_2$, 204, 279
$\mathbb{C}$ field of complex numbers
$\mathbb{C}^n$ complex Euclidean n-space, 1
$C([a, b])$ the space of complex valued continuous functions on $[a, b]$, 260
$C_0^\infty([a, b])$ the space of infinitely differentiable complex
    valued functions which vanish outside an open subinterval of $[a, b]$, 208
codim M codimension of a subspace M, 88
$d(A)$ codimension of the range of A, 347
∂S boundary of a set S
dim M dimension of a subspace M
$d(v, s)$ distance from a vector v to a set S, 8
δ_{jk} Kronecker delta, 33
$\mathcal{D}(A)$ domain of an operator A, 203, 279
$\det_{\mathcal{P}}(I - P)$ determinant of $I - P$, with P a Poincaré operator, 319
$\mathcal{F}(\ell_2)$ space of bounded linear operators of finite rank, 321
$g(y_1, y_2, \ldots, y_n)$ Gram determinant, 11
$G(A)$ graph of the operator A, 207
Im A range of an operator A, 63, 279
$\Im \alpha$ imaginary part of a complex number α
$\langle \cdot, \cdot \rangle_A$ inner product on the graph of A, 207
ind A index of the operator A, 347
κ winding number, 143
Ker A kernel of A, 65
ℓ_2, $\ell_2(Z)$ the Hilbert spaces of square summable sequences
    with entries in $\mathbb{C}$, 3, 5
$\ell_2(\omega)$ weighted ℓ_2 space, 40
$\ell_p, 1 \leq p \leq \infty$, 260
$L_p([a, b]), 1 \leq p \leq \infty$, 261, 413
$\mathcal{L}(\mathcal{H})$ space of bounded linear operators on $\mathcal{H}$, 52
$\mathcal{L}(\mathcal{H}_1, \mathcal{H}_2)$ space of bounded linear operators from $\mathcal{H}_1$ into $\mathcal{H}_2$, 52
$M_{\alpha,\beta}$ pair operator, 377
$n(A)$ dimension of the kernel of A, 347
$\Pi(P_n), \Pi(P_n, Q_n)$, 97, 289

$\mathcal{P}$ space of Poincaré operators, 317
$P^{(n)}$, 318
$\rho(A)$ resolvent set of the operator A, 110, 290
$\Re\alpha$ real part of a complex number α
$r(T)$ spectral radius of the operator T, 147
$S^{\perp}$ orthogonal complement of S
   and also the annihilator of S, 17, 308
$^{\perp}N$, 308
sp S span of the set S, 6
$\bar{S}$ closure of the set S, 16
$\sigma(A)$ spectrum of the operator A, 110, 290
$s_j(A)$ j-th singular value of the operator A, 248
$\mathbb{T}$ unit circle in the complex plane
tr A trace of the operator A, 320
X' conjugate space of the space X, 265
$\mathbb{Z}$ set of integers

Index

adjoint operator 77, 208
algebraic multiplicity 337
almost periodic 38
analytic operator valued
    function 295
associate pair operator 384

backward shift 56
Baire category theorem 281
ball 6, 309
Banach space 259
band matrix 94
band operator 94
basic harmonic motions 222
basic system of eigenvectors
    and eigenvalues 180
basis, orthonormal 28
   Schander 265
   standard orthonormal 28
bilateral shift 56
biorthogonal system 48, 271
bounded linear functionals,
    representation of 265–267
bounded linear operator 52, 277

Cauchy-Schwarz inequality 7
Cauchy sequence 7, 259
Cayley transform 255
characteristic frequencies 222
closed graph theorem 281
closed operator 203, 279
closed set 16
closure of a set 16
codimension 88
compact operator 91, 299
compact set 406
complement of a subspace 85, 271
complete 406
complete inner product space 7
complete normed linear space 259
complete orthonormal system 29
complex n-space 1

conjugate of an operator 307
conjugate space 265
continuity of an operator 57
contraction 402
contraction mapping theorem 401
convex set 16
countable set 409

defining function 136
diagonalization 178
Dini's theorem 197
direct sum 85, 271
distance from a point to a set 8

eigenvalue 172
eigenvector 172
equivalent norms 262, 283

finite section method 99, 159, 163, 385
fixed point 401
forward shift 56
Fourier coefficient 28, 31
Fourier series 31
Fourier transform 37
Fredholm alternative 306
Fredholm operator 347, 356
functional 60

Gram determinant 11
graph of an operator 207
graph norm 207, 280, 357
Green's function 205

Hadamard's inequality 13
Hahn-Banach theorem 267
Hilbert space 7
Hilbert-Schmidt operator 253
Hilbert-Schmidt theorem 193, 194

image of an operator 63
imaginary part of an operator 244
implicit function theorem 404
index of an operator 347, 356

injective map 65
inner product 6
inner product space 6
integral equation of the second kind 73
integral operator 56
interior point 281
invariant subspace 108
inverse operator 64, 204
invertible operator 64, 204
isometry 114, 246

kernel function 56
kernel of an operator 65
Kronecker delta 33

Laurent operator 135, 361
Lebesgue integral 411
left inverse 85
left invertible 85
Legendre polynomial 26
linear isometry 36
L_p space 261, 413

maximal element 267
Mercer's theorem 197
metric space 406
modified projection method 106
multiplicity of an eigenvalue 181

non negative operator 185
norm 6, 259
    of an operator 52
normal operator 244
normed linear space 259
nuclear operator 314
nth section 159, 318

one-one operator 65
operator of algebraic complements 334
operator of singular integration 393
orthogonal complement 17
orthogonal projection 81
orthogonal vectors 8
orthonormal basis 28
    stability of 34
orthonormal system 9

pair operator 377
parallelogram law 7
Parseval's equality 29
partially ordered set 267

Poincaré operator 317
    determinant 319
positive operator 185
product space 207
projection 84, 286
projection method 97, 289
    modified 106
Pythagorean theorem 9

quotient space 274

range of an operator 63
rank of an operator 63
real part of an operator 244
regular point 110, 290
resolvent 295
resolvent set 110, 290
Riesz representation
    theorem 61
right inverse 89
right invertible 89

Schauder basis 265
Schwarz inequality 7
self adjoint operator 80, 210
separable Hilbert space 35
    normed linear space 264
simple eigenvalue 213
simple harmonic oscillation 220
singular values 248
span of a set 6
spectral radius 147
spectral theorem 178
spectrum of an operator 110, 290
standard basis 19
    orthonormal basis 28
strictly positive operator 166
Sturm-Liouville system 211
    operator 211
subspace 6
    invariant 108
symbol 136, 142, 362, 364
symmetric operator 215
Szegö polynomials 19

Toeplitz matrix 20, 59
Toeplitz operator 141, 364
totally ordered set 267
trace class 253
trace of finite rank operator 321
    Poincaré operator 320

triangle inequality 2, 7
tridiagonal matrix 94

uniform boundedness principle 284
uniformly convex 274
unitarily equivalent 248
unitary operator 114, 246

vector space 5
Volterra integral operator 293, 294

Weierstrass approximation theorem 29
  second approximation theorem 30
Wiener-Hopf factorization 365
winding number 143, 152, 153
Wronskian 211

Zorn's lemma 267

Operator Theory - Advances and Applications

OT series

Edited by

Israel Gohberg, *School of Mathematical Sciences, Tel Aviv University, Ramat Aviv, Israel*

This series is devoted to the publication of current research in operator theory, with particular emphasis on applications to classical analysis and the theory of integral equations, as well as to numerical analysis, mathematical physics and mathematical methods in electrical engineering.

Kopachevsky, N.D. / Krein, S.
Operator Approach to Linear Problems of Hydrodynamics.
Volume 2: Nonself-adjoint Problems for Viscous Fluids (2003)
ISBN 3-7643-2190-3
Vol. 146

Albeverio, S. / Demuth, M. / Schrohe, E. / Schulze, B.-W. (Eds.)
Nonlinear Hyperbolic Equations, Spectral Theory, and Wavelet Transformations. A Volume of Advances in Partial Differential Equations (2003)
ISBN 3-7643-2168-7
Vol. 145

Belitskii, G. / Tkachenko, V.
One-dimensional Functional Equations (2003)
ISBN 3-7643-0084-1
Vol. 144

Alpay, D. (Ed.)
Reproducing Kernel Spaces and Applications (2003)
ISBN 3-7643-0068-X
Vol. 143

Böttcher, A. / Dos Santos, A.F. / Kaashoek, M.A. / Brites Lebre, A. / Speck, F.-O. (Eds.)
Singular Integral Operators, Factorization and Applications. International Workshop on Operator Theory and Applications, IWOTA 2000 - Portugal (2003) ISBN 3-7643-6947-7
Vol. 142

Dos Santos, A.F. / Gohberg, I., / Manojlovic, N. (Eds.)
Factorization and Integrable Systems. Proceedings of the Summer School, Faro, Portugal, 2000 (2003)
ISBN 3-7643-6938-8
Vol. 141

Ellis, R. / Gohberg, I.
Orthogonal Systems and Convolution Operators (2002)
ISBN 3-7643-6929-9
Vol. 140

Müller, V.
Spectral Theory of Linear Operators and Spectral Systems in Banach Algebras (2003)
ISBN 3-7643-6912-4
Vol. 139

Albeverio, S. / Demuth, M. / Schrohe, E. / Schulze, B.-W. (Eds.)
Parabolicity, Volterra Calculus, and Conical Singularities. A Volume of Advances in Partial Differential Equations (2002)
ISBN 3-7643-6906-X
Vol. 138

Dybin, V. / Grudsky, S.M.
Introduction to the Theory of Toeplitz Operators with Infinite Index (2002)
ISBN 3-7643-6728-8
Vol. 137

Wong, M. W.
Wavelet Transforms and Localization Operators (2002)
ISBN 3-7643-6789-X
Vol. 136

Böttcher, A. / Gohberg, I. / Junghanns, P. (Eds.)
Toeplitz Matrices, Convolution Operators, and Integral Equations. The Bernd Silbermann Anniversary Volume (2002)
ISBN 3-7643-6877-2
Vol. 135

Alpay, D. / Gohberg, I. / Vinnikov, V. (Eds.)
Interpolation Theory, Systems Theory and Related Topics. The Harry Dym Anniversary Volume (2002)
ISBN 3-7643-6762-8
Vol. 134

Krall, A. M.
Hilbert Space, Boundary Value Problems and Orthogonal Polynomials (2002)
ISBN 3-7643-6701-6
Vol. 133

Albeverio, S. / Elander, N. / Everitt, W. N. / Kurasov, P. (Eds.)
Operator Methods in Ordinary and Partial Differential Equations. S. Kovalevsky Symposium, University of Stockholm, June 2000 (2002)
ISBN 3-7643-6790-3
Vol. 132

Böttcher, A. / Karlovich, Y.I. / Spitkovsky, I.M.
Convolution Operators and Factorization of Almost Periodic Matrix Functions (2002)
ISBN 3-7643-6672-9
Vol. 131

For orders originating from all over the world except USA and Canada:

Birkhäuser Verlag AG
c/o Springer GmbH & Co
Haberstrasse 7, D-69126 Heidelberg
Fax: ++49 / 6221 / 345 42 29
e-mail: birkhauser@springer.de

For orders originating in the USA and Canada:

Birkhäuser
333 Meadowland Parkway
USA-Secaucus, NJ 07094-2491
Fax: ++1 201 348 4505
e-mail: orders@birkhauser.com

Birkhäuser

http://www.birkhauser.ch

Operator Theory - Advances and Applications

OT series

Edited by
Israel Gohberg, *School of Mathematical Sciences, Tel Aviv University, Ramat Aviv, Israel*

This series is devoted to the publication of current research in operator theory, with particular emphasis on applications to classical analysis and the theory of integral equations, as well as to numerical analysis, mathematical physics and mathematical methods in electrical engineering.

Gohberg, I., Tel Aviv University, Israel /
Goldberg, S., University of Maryland, USA /
Kaashoek, M.A., Vrije Universiteit, Amsterdam, The Netherlands

Classes of Linear Operators Vol. 1, 2

Vol. 1 (1993). 478 pages. Hardcover
ISBN 3-7643-2531-3

Vol. 2 (1993). 562 pages. Hardcover
ISBN 3-7643-2944-0

This book presents a panorama of operator theory. It treats a variety of classes of linear operators which illustrate the richness of the theory, both in its theoretical developments and its applications. For each of the classes various differential and integral operators motivate or illustrate the main results. The topics have been updated and enhanced by new developments, many of which appear here for the first time. Interconnections appear frequently and unexpectedly.
The first volume consists of four parts: general spectral theory, classes of compact operators, Fredholm and Wiener-Hopf operators, and classes of unbounded operators.
The second volume consists of five parts: triangular representations, classes of Toeplitz operators, contractive operators and characteristic operator functions, Banach algebras and algebras of operators, and extension and completion problems.
The exposition is self-contained and has been simplified and polished in an effort to make advanced topics accessible to a wide audience of students and researchers in mathematics, science and engineering.

"...Used as a graduate textbook, the book allows the instructor several good selections of topics to build a course. ...The authors took great care to polish and simplify the exposition; as a result, the book can serve also as an excellent basis for reading courses or for self-study. ... Besides being a textbook, the book is a valuable reference source for a wide audience of mathematicians, physicists and engineers. The specialists in functional analysis and operator theory will find most of the topics familiar, although the exposition is often novel or non-traditional, making the material more accessible. ... "
Zentralblatt für Mathematik und ihre Grenzgebiete

"This book ... shows for several times the fruitful application of complex analysis to problems in operator theory. ... Each part contains interesting exercises and comments on the literature of the topic."
Monatshefte für Mathematik\medskip

For orders originating from all over the world except USA and Canada:

Birkhäuser Verlag AG
c/o Springer GmbH & Co
Haberstrasse 7, D-69126 Heidelberg
Fax: ++49 / 6221 / 345 42 29
e-mail: birkhauser@springer.de

For orders originating in the USA and Canada:

Birkhäuser
333 Meadowland Parkway
USA-Secaucus, NJ 07094-2491
Fax: ++1 201 348 4505
e-mail: orders@birkhauser.com

Birkhäuser

http://www.birkhauser.ch